Possessing Nature

STUDIES ON THE HISTORY OF SOCIETY AND CULTURE
Victoria E. Bonnell and Lynn Hunt, Editors

Possessing Nature
Museums, Collecting, and Scientific Culture in Early Modern Italy

Paula Findlen

UNIVERSITY OF CALIFORNIA PRESS
Berkeley   Los Angeles   London

The publisher gratefully acknowledges
the contribution provided by the
General Endowment Fund of the
Associates of the University of California Press

University of California Press
Berkeley and Los Angeles, California

University of California Press
London, England

Library of Congress Cataloging-in-Publication Data

Findlen, Paula.
  Possessing nature : museums, collecting, and scientific culture in
early modern Italy / Paula Findlen.
    p.   cm. — (Studies on the history of society and culture ; 20)
  Includes bibliographical references and index.
  ISBN: 978-0-520-20508-6
  1. Science museums—Italy—History.   2. Science museums—Europe—
History.   3. Natural history museums—Italy—History.   4. Natural
history museums—Europe—History.   I. Title.   II. Series.
Q105.I8F56 1994
508'.074'45—dc20   93-14997
                                                                          CIP

Printed in the United States of America

09   08   07
12   11   10   9   8   7

The paper used in this publication is both acid-free and totally chlorine-free (TCF).
It meets the minimum requirements of ANSI/ NISO Z39.48-1992 (R 1997)
(*Permanence of Paper*). ⊚

*To my great-grandmother
Margaret Black Avery:
a museum in her own right*

The soul of man affects a kind of infinity in its objects. The affectations are always reaching after new pleasures, the desires carried forth after new possessions . . . the eye is never satisfied with seeing.

—JOHN SPENCER, *DISCOURSE CONCERNING PRODIGIES* (1665)

CONTENTS

ACKNOWLEDGMENTS

In a book about early modern scholarly rituals, it is tempting to view the writing of acknowledgments as an example of our own organic links with the culture that created this genre; it gives me great pleasure to think that we share something in common with Aldrovandi and Kircher, since most of their world is alien to our modern sensibilities. This project has metamorphosized from a graduate seminar paper at the University of Chicago into a dissertation at the University of California at Berkeley and (finally) into a book written at the University of California at Davis. Along the way, I have imposed on a number of people who deserve my thanks.

Randy Starn, Roger Hahn, and Barbara Shapiro survived the reading of a lengthy dissertation and offered much encouragement and acute criticism along the way. Lorraine Daston, Bill Eamon, Anthony Grafton, and Laurie Nussdorfer all reminded me that the leap from dissertation to book is neither quick nor easy. Elizabeth Knoll has been the best sort of editor; while encouraging me to get it done, she has always listened to the most recent transmutations of the project with patience and good humor. Other friends have offered comments, criticisms, and bibliographic suggestions along the way, in particular, Bill Ashworth, Martha Baldwin, Cristelle Baskins, Ann Blair, Mario Biagioli, Jim Bono, Ken Gouwens, Phyllis Jestice, Jeff Ravel, Nancy Siraisi, Pamela Smith, and Jay Tribby. Colleagues in the History Department and Program in the History and Philosophy of Science, particularly Bill Bowsky and Betty Jo Dobbs, have made Davis a particularly congenial environment in which to complete my revisions. Ryan Henson spent part of two summers excavating materials from the library. Jennifer Selwyn helped proof and index. Marta Cavazza and Giuliano Pancaldi kindly unearthed a plan of the *Studio Aldrovandi*. A special thanks to Cathy Kudlick and Katy Park for reading so many drafts, and to Michelle Nordon and Michelle Bonnice at U.C. Press and the copy editor, Kathy Walker, for seeing it into print. Their comments and queries have greatly improved the argument and clarity of this book.

This project could not have been conceptualized nor completed without the fine work done by many (mostly Italian) scholars on collecting, natural history, and the permutations of Cinquecento and Seicento culture. I would like to thank Lina Bolzoni, Gigliola Fragnito, Paolo Galluzzi, Adalgisa Lugli, Arthur MacGregor, Giuseppe Olmi, Krzysztof Pomian, and Lucia Tongiorgi Tomasi, among others, for showing me how such a project could be done well. In a different spirit, I also wish to acknowledge the work of two scholars with whom, unfortunately, I never had the pleasure of discussing this subject: Eric Cochrane and Charles Schmitt. No doubt they would have disagreed with my methodology and interpretation in many instances; yet I hope they would have appreciated my own fascination with the vagaries of sixteenth- and seventeenth-century Italian academic culture, which both explicated in such loving detail.

The research for this study in England and Italy was funded by the Fulbright Commission, the American Institute for the History of Pharmacy, and a Wellesley College Graduate Fellowship. Fellowships from the History Department and the University of California at Berkeley allowed me the time to conceptualize and complete the project as a dissertation; Faculty Research Grants and a Junior Faculty Fellowship from the University of California at Davis gave me the opportunity to rethink it, once again, and produce a publishable version.

Librarians at the Bancroft Library in Berkeley, the University of California at Davis, the Newberry Library, and the University of Chicago made books, both old and new, materialize when they were most needed. My research in England was greatly facilitated by the help of the staff at the British Museum Library and Manuscript Room, the Wellcome Institute for the History of Medicine and the Warburg Institute. In Italy, M. Cristina Bacchi, Irene Ventura Folli, Laura Miani, Patrizia Moscatelli, and Maria Cristina Tagliaferri, all of the Biblioteca Universitaria in Bologna, provided bibliographic and paleographic assistance and generously shared their knowledge of the Aldrovandi manuscripts with me during my stay in Bologna, as did the staff of the Biblioteca Comunale dell'Archiginnasio and the Archivio di Stato in Bologna. Father Monachino of the Pontificia Università Gregoriana allowed me to consult the Kircher correspondence, while the archivists at the Archivum Romanum Societatis Iesu kindly provided me with a copy of the unpublished catalogue to those manuscripts. The director of the Archivio Isolani in Bologna let me peer at the burned and crumbled fragments of the Paleotti papers, to make what sense I could of them. There are many others I would like to thank whom I omit only for the sake of brevity. As the list of abbreviations may indicate, this is a project fabricated out of many trips to different regional archives and libraries.

During different stays in Italy, the hospitality of Giuliano Pancaldi, Nicoletta Caramelli, and their family made my time in Bologna enjoyable in innumerable ways. Jeff Vance has put up with my absences, mental and geographical, and always managed to put my obsession in perspective. My own family increasingly has become aware that they will never be able to enter another museum without wondering where the sixteenth-century armadillo is—I hope that I have not ruined the experience of the museum too much for them. Finally, I would like to thank the Caramoor Foundation for the Arts in Katonah, New York, for employing me as a tour guide for two summers, some years ago. My time explaining this rather eclectic twentieth-century museum surely has influenced my interest in the history and practice of collecting. While there are many differences between modern museums and Renaissance "cabinets of curiosity," I am heartened to think that they are not always distinguishable.

ABBREVIATIONS

AI	Archivio Isolani, Bologna
Ambr.	Biblioteca Ambrosiana, Milano
ARSI	Archivum Romanum Societatis Iesu
ASB	Archivio di Stato, Bologna
ASMo.	Archivio di Stato, Modena
ASVer.	Archivio di Stato, Verona
BANL	Biblioteca dell' Accademia Nazionale dei Lincei
BAV	Biblioteca Apostolica Vaticana
BCAB	Biblioteca Comunale dell' Archiginnasio, Bologna
BEst.	Biblioteca Estense, Modena
BL	British Library, London
BMV	Biblioteca Marciana, Venezia
BNF	Biblioteca Nazionale, Firenze
BPP	Biblioteca Palatina, Parma
BUB	Biblioteca Universitaria, Bologna
Laur.	Biblioteca Laurenziana, Firenze
PUG	Pontificia Università Gregoriana, Roma
Ricc.	Biblioteca Riccardiana, Firenze
UBE	Universitätsbibliothek, Erlangen
Casciato	Mariastella Casciato, Maria Grazia Ianniello, and Maria Vitale, eds., *Enciclopedismo in Roma barocca: Athanasius Kircher e il Museo del Collegio Romano tra Wunderkammer e museo scientifico* (Venice, 1986).

CL   Giuseppe Gabrieli, "Il Carteggio Linceo della Vecchia Accademia di Federico Cesi (1603–30)," *Memorie della R. Accademia Nazionale dei Lincei. Classe di scienze morali, storiche e filologiche,* ser. 6, v. VII, fasc. 1-3 (Rome, 1938–1941).

De Sepi   Giorgio de Sepi, *Romani Collegii Societatis Iesu Musaeum Celeberrimum* (Amsterdam, 1678).

Discorso naturale   Sandra Tugnoli Pattaro, *Metodo e sistema delle scienze nel pensiero di Ulisse Aldrovandi* (Bologna, 1981).

Fantuzzi   Giovanni Fantuzzi, *Memorie della vita e delle opere di Ulisse Aldrovandi* (Bologna, 1774).

Mattirolo   Oreste Mattirolo, "Le lettere di Ulisse Aldrovandi a Francesco I e Ferdinanco I Granduchi di Toscana e a Francesco Maria II Duca di Urbino," *Memoria della Reale Accademia delle Scienze di Torino,* ser. II, 54 (1903–1904): 355–401.

MS   Athanasius Kircher, *Mundus subterraneus* (Amsterdam, 1664).

Origins of Museums   Oliver Impey and Arthur MacGregor, eds., *The Origins of Museums: Cabinets of Curiosities in Sixteenth- and Seventeenth-Century Europe* (Oxford, 1985).

Vita   Ludovico Frati, "La vita di Ulisse Aldrovandi scritta da lui medesimo," in *Intorno alla vita e alle opere di Ulisse Aldrovandi* (Imola, 1907), pp. 1–29.

PHOTO CREDITS

I would like to thank the following institutions for permission to publish the illustrations reproduced in this book: Archivio di Stato, Bologna (Museum Plan 5); Biblioteca Universitaria, Bologna (figs. 1, 2, 4, 7, 8, 10, 14–16, and 20); Biblioteca Ambrosiana, Milan (Museum Plan 2); Pinacoteca Ambrosiana, Milan (fig. 6); Museo Civico di Storia Naturale, Milan (fig. 26); Museo e Gallerie Nazionali di Capodimonte, Naples (fig. 23); Ugo Guanda Editore, Parma (fig. 21); Bayerische Staatsgemäldesammlungen, Alte Pinakotek, Munich (fig. 3); The British Library, London (figs. 9, 11, and 19); The Bancroft Library, University of California, Berkeley (figs. 17, 18, 24, and 28–30); Department of Special Collections, University of California Library, Davis (figs. 12 and 13); Department of Special Collections, The University of Chicago Library (fig. 5); Rare Books, University of Illinois, Urbana (fig. 27); and Department of Special Collections, Memorial Library, University of Wisconsin—Madison (figs. 22 and 25).

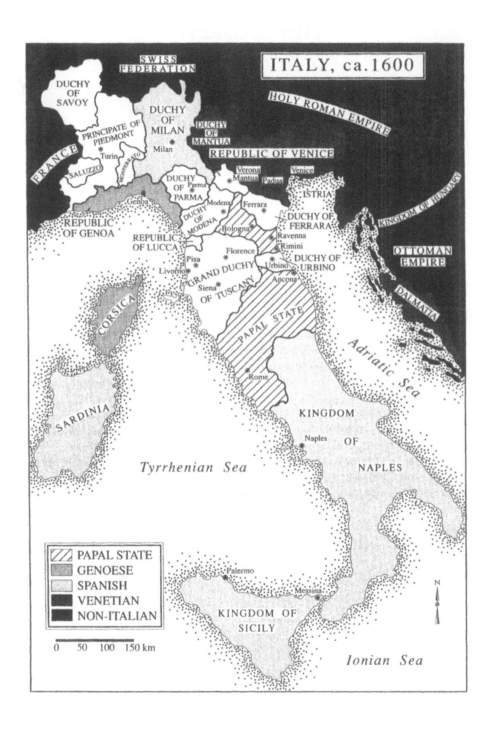

ITALY, ca. 1600

SWISS FEDERATION

DUCHY OF SAVOY

HOLY ROMAN EMPIRE

PRINCIPATE OF PIEDMONT

DUCHY OF MILAN

DUCHY OF MANTUA

Turin

MONFERRATO

Milan

REPUBLIC OF VENICE

FRANCE

SALUZZO

DUCHY OF PARMA

Parma

Verona
Mantua
Padua

Venice

Genoa

DUCHY OF MODENA

Modena

Ferrara

ISTRIA

KINGDOM OF HUNGARY

REPUBLIC OF GENOA

REPUBLIC OF LUCCA

Bologna

DUCHY OF FERRARA

Ravenna
Rimini

Florence

Pisa

Livorno

GRAND DUCHY OF TUSCANY

Urbino

Ancona

DUCHY OF URBINO

OTTOMAN EMPIRE

DALMATIA

Siena

CORSICA

PAPAL STATE

Adriatic Sea

Rome

SARDINIA

KINGDOM OF NAPLES

Naples

Tyrrhenian Sea

PAPAL STATE
GENOESE
SPANISH
VENETIAN
NON-ITALIAN

0 50 100 150 km

Palermo

Messina

KINGDOM OF SICILY

N

Ionian Sea

Introduction

This book recounts two overlapping histories. The first details the appearance of museums in early modern Europe, particularly collections whose purpose was to bring all of nature into one space. The second offers a reading of the development of natural history as a discipline. Both stories take Italy as their case study. There collecting first became a widespread practice, among an elite desirous to know the past, in all its forms, through the possession of its remnants. The collecting of antiquities and the passion for natural objects appeared in Italy before any other part of Western Europe; in both instances, a strong historicizing impulse on the part of Italian Renaissance humanists precipitated these activities. Simultaneously, at the universities of Pisa, Padua, and Bologna, and in the Italian courts, academies, and pharmacies, nature was subjected to an intensive inquiry in ways that she had not been since the time of Pliny and the great encyclopedic work of Albertus Magnus. These two activities—collecting and the interrogation of nature—met in the studies of naturalists such as Ulisse Aldrovandi (1522–1605) and Athanasius Kircher (1602–1680), resulting in new attitudes toward nature, as a collectible entity, and generating new techniques of investigation that subsequently transformed natural history.

Contemporaries were well aware of Italy's primacy in the renaissance of natural history. "And for that which regards natural history, the books brought to light by the Italians and the wonderful collections made in various times demonstrate how much they toiled before any other nation," wrote the eighteenth-century editor of Michele Mercati's *Metallotheca* (1717).[1] In Pisa, Padua, and Bologna, the first professors of natural history were appointed, the first botanical gardens founded, and museums were

---

[1] Biblioteca Angelica, Rome, ms. 1545, f. 219 (*Il museo di Michele Mercati compendiato, e riformato*).

1

made a regular part of the instructional materials available for the teaching of *materia medica*, the medicinal understanding of nature. These institutional signs of success built upon the philological and editorial work done by humanists and naturalists in such centers as Venice and Ferrara, where critical editions and translations of Galen and Pliny and criticism of Pliny poured forth from the presses. By 1669, Paolo Boccone could speak of "the esteem in which Museums of Natural History are held in Italy."[2]

Possessing nature was part of a more widespread delight in collecting objects of scientific worth. During the sixteenth and seventeenth centuries, the first science museums appeared—repositories of technology, ethnographic curiosities, and natural wonders.[3] They emerged at a time when all of Europe seemed to be collecting; museums, libraries, intricate gardens, grottos, and galleries of art filled the landscape of late Renaissance and Baroque Europe.[4] So enamored of collecting was the humanist secretary Antonio Giganti that he bemoaned the fact that time did not permit him the pleasure of producing a book worthy enough to grace the shelves of Ippolito

[2]In Corrado Dollo, *Filosofia e scienza in Sicilia* (Padua, 1979), p. 360 (Messina, 24 April 1669).

[3]Silvio Bedini's "The Evolution of Science Museums," *Technology and Culture* 6 (1965): 1–29, presents a positivistic but nonetheless informed overview of this subject. More recently, Carlo Maccagni's essay, "Le raccolte e i musei di storia naturale e gli orti botanici come istituzioni alternative e complementari rispetto alla cultura delle Università e delle Accademie," in *Università, accademie e società scientifiche in Italia e in Germania dal Cinquecento al Settecento*, ed. Laetitia Boehm and Ezio Raimondi (Bologna, 1981), pp. 283–310, presents a brief but suggestive view of the social and cultural significance of the early science museum as an "alternative institution." Giuseppe Olmi provides a synthetic look at natural history museums in the wider context of late Renaissance and Baroque collecting in his "Ordine e fama: il museo naturalistico in Italia nei secoli XVI e XVII," *Annali dell'Istituto storico italo-germanico in Trento* 8 (1982): 225–274. Readers wishing to consult an English version can look at his "Science-Honour-Metaphor: Italian Cabinets of the Sixteenth and Seventeenth Centuries," in *Origins of Museums*, pp. 5–16.

[4]The fundamental study of collecting is Julius Von Schlosser, *Die Kunst- und Wunderkammern der Spätrenaissance* (Leipzig, 1908). More recently, Luigi Salerno, "Arte e scienza nelle collezioni del Manierismo," in *Scritti di storia dell'arte in onore di Mario Salmi* (Rome, 1963), Vol. III, pp. 193–213, and Adalgisa Lugli's *Naturalia et mirabilia: il collezionismo enciclopedico nelle Wunderkammern d'Europa* (Milan, 1983) have refined Schlosser's thesis about the playful relationship between nature and art as seen in the iconography of collecting. For a more formal approach to the history of art collecting and to museums in general, see Paola Barocchi, "Storiografia e collezionismo dal Vasari al Lanzi," in *Storia dell'arte italiana* (Turin, 1981), Pt. I, Vol. II, pp. 5–81; also Simona Savini Branca, *Il collezionismo veneziano nel '600* (Padua, 1964), or a monograph such as Renato Martinoni, *Gian Vincenzo Imperiali. Politico, letterato e collezionista genovese del Seicento* (Padua, 1983). Giuseppe Olmi relates the institutionalization of collecting to the new political and cultural matrix of early modern Europe in his "Alle origini della politica culturale dello stato moderno: dal collezionismo privato al *Cabinet du Roy.*" *La Cultura* 16 (1978): 471–484; he also focuses on the relationship between collecting and encyclopedic traditions in his "Dal 'teatro del mondo' ai mondi inventariati. Aspetti e forme del collezionismo nell'eta moderna," in *Gli Uffizi: Quattro secoli di una galleria*, ed. Paola Barocchi and Giovanna Ragionieri (Florence, 1983), pp. 233–269. The influence of Krzysztof Pomian's work on the relationship between museums and the place of curiosity in early modern culture is evident throughout my work; see his *Collectors and Curiosities: Paris and Venice, 1500–1800* (London,

Agostini's museum in Siena.[5] The Tuscan noble Niccolò Gaddi, confidant of two Grand Dukes and friend of Aldrovandi, insisted that his "things . . . be shown by my heirs to all the Florentine and foreign gentlemen who wish to see them lovingly and courteously, upon request," so highly did he value this aspect of his patrimony.[6] Collecting, in short, had become an activity of choice among the social and educated elite. It filled their leisure hours and for some seemed to encompass every waking moment of their lives. Through the possession of objects, one physically acquired knowledge, and through their display, one symbolically acquired the honor and reputation that all men of learning cultivated.

Within the wider matrix of collecting, the possession of nature figured prominently. Along with art, antiquities, and exotica, nature was deemed a desirable object to own. Building upon Pliny's encyclopedic definition of nature as everything in the world worthy of memory and the narrower view of such writers as Dioscorides and Galen, who defined natural history as the study of objects useful in medicine, collectors brought ordinary and exotic nature into their museums. The alleged remains of legendary creatures—giants, unicorns, satyrs, basilisks—took their place next to real but puzzling phenomena such as fossils, loadstones, and zoophytes; previously unknown creatures such as the armadillo and the bird of paradise; and a plethora of ordinary artifacts that filled in the gaps between one paradox and the next. From the imaginary to the exotic to the ordinary, the museum was designed to represent nature as a continuum.

Why did so many Europeans envision collecting as the key to understanding their world? In a sense, the creation of the museum was an attempt to manage the empirical explosion of materials that wider dissemination of ancient texts, increased travel, voyages of discovery, and more systematic forms of communication and exchange had produced. While all of these factors contributed to the increased curiosity of the Europeans toward other cultures, and ultimately redefined the European world view as a relative rather than an absolute measure of "civilization," they also produced new attitudes toward nature and the discipline of natural history. "It is not to be

1990). The conference proceedings from the tercentennial of the Ashmolean Museum provide a survey and comprehensive bibliography of the most recent work in the history of early modern collecting; Oliver Impey and Arthur MacGregor, eds., *The Origins of Museums: The Cabinet of Curiosities in Sixteenth- and Seventeenth-Century Europe* (Oxford, 1985). For the perspective of a museum professional on this subject, see Eilean Hooper-Greenhill, *Museums and the Shaping of Knowledge* (London, 1992).

[5]BUB, *Aldovandi*, ms. 110. Under the entry for Agostino in his visitors' book, Ulisse Aldrovandi records the following poem by Giganti: "O utinam sic te liber oblectare legentem/ Possit, Musaeo vel decori esse tuo. Musaeo insigni omnigenis, quasi maxima mittet/India, vel Libya, vel novus orbis opes. Insigni et pictis tabulis, et marmore, et aere,/Art quod expolit prisca, recensque manus."

[6]Cristina Acidini Luchinat, "Niccolò Gaddi collezionista e dilettante del Cinquecento," *Paragone* 359–361 (1980): 167 (n. 10).

esteemed a small matter that by the voyages & travels of these later times, so
much more of nature has been discovered than was known at any former
period," wrote Francis Bacon, who made natural history the paradigm for a
new philosophy of nature in the seventeenth century:

> It would, indeed, be disgraceful to mankind, if, after such tracts of the mater-
> ial world have been laid open which were unknown in former times—so many
> seas traversed—so many countries explored—so many stars discovered—phi-
> losophy, or the intelligible world, should be circumscribed by the same bound-
> aries as before.[7]

Natural history, as Bacon observed, was a form of inquiry designed to record
the knowledge of the world for the use and betterment of mankind. As Europeans traveled farther with greater frequency, this knowledge expanded,
yielding new and unexpected results. Collecting was one way of maintaining some degree of control over the natural world and taking its measure.
If knowledge of the world could no longer be contained in a set of canonical texts, then perhaps it could be displayed in a museum. Thus, philosophers at the vanguard of the intellectual community, most significantly
Bacon, urged fellow investigators of nature to establish storehouses to mon-
itor the flow of objects and information. From these activities, a new philosophy of nature would emerge based on experience rather than erudition.
      Despite Bacon's admonitions, none of the collectors whom I discuss dis-
solved the boundaries of their world.[8] While the Italian naturalists would
have agreed with Bacon that experience was a necessary and often under-
valued part of knowledge, they would not have appreciated his dismissal of
ancient authority. Collectors such as Aldrovandi and Kircher understood
their activities to be the fulfillment of the work of Aristotle and those who
had followed him; for them, experience did not compete with authority but
rather complemented and enhanced it. The novelties they encountered did
not lead them to discard their philosophical framework but instead to mod-
ify it. As a paradigm of knowledge, collecting stretched the parameters of
the known to incorporate an expanding material culture. From the sixteenth-century naturalist's inclusion of New World plants in Old World categories, to the seventeenth-century Jesuit's Christian synthesis of all cultures
and their artifacts, the museum "saved appearances" rather than unsettling
ancient systems. By the mid-seventeenth century, it would become a symbol
of the "new" science, incorporated into scientific organizations such as the
Royal Society in England, the Paris Academy of Sciences, and later the Institute for Sciences in Bologna. In the preceding century, however, the museum emblematized the revitalization of Aristotelian natural philosophy

---

[7] In David Murray, *Museums: Their History and Their Use* (Glasgow, 1904), Vol. I, pp. 19–20.
[8] Michael T. Ryan's "Assimilating New Worlds in the Sixteenth and Seventeenth Centuries,"
Comparative Studies in Society and History 23 (1981): 519–538, elegantly elaborates on the way in
which Europeans read the "New World" into the "Old."

and Plinian natural history; it was about the reinvention of the old rather than the formation of the new.

While collecting did not immediately usher in a new philosophical definition of and structure for *scientia*, it certainly constituted a significant addition to the practice of philosophizing in early modern Europe. Organizing ideas around objects, naturalists increasingly saw philosophical inquiry as the product of a continuous engagement with material culture. The decision to display the fruits of collection led naturalists gradually to define knowledge as consensual, shaped in relation to the audience that entered the museum and therefore participated in the peculiar discursive practices that emerged within that context. Tactile as well as sociable, the philosophizing done in and around the museum enhanced the Aristotelian definition of knowledge as a product of sensory engagement with nature. Ultimately the value placed upon the experience of the senses would result in its uncoupling from this traditional philosophical framework. But at this point, naturalists perceived their museums to be a tangible sign of their commitment to the ancient study of nature. In the sixteenth century, this entailed little more than subsuming everything collected within a proper philosophical framework, as determined by the traditional classification of the sciences. By the seventeenth century, in the wake of the new experimental philosophies of Bacon, Descartes, and Galileo, naturalists still committed to the preservation of ancient views of nature now faced the challenge of responding to the hostile critics. Putting the techniques of the "new" philosophy in the service of the old, Aristotelian naturalists designated the museum as a site of critical synthesis. With hindsight, it is easy for us to predict their failure. At the time, they had the weight of more than 2000 years of authority on their side.

Telling this tale of failed encyclopedic dreams is a fairly complicated endeavor. In this study, I have used the rich font of materials available on this subject to sketch a broad portrait of naturalistic and collecting activities in the sixteenth and seventeenth centuries. Between Aldrovandi and Kircher lie a host of other patrons and collectors of nature who merit equal attention. Some—Pier Andrea Mattioli (1500–1577), Giovan Battista della Porta (1535–1615), Federico Cesi (1585–1630), Manfredo Settala (1600–1680), and Francesco Redi (1626–1697)—are well known, at least by name. Others have receded into obscurity over the course of time, as their publications have grown dusty and their museums have vanished. While many contemporaries knew of the famed apothecaries Francesco Calzolari (1521–1600) and Ferrante Imperato (1550–1625), only the latter merits a brief entry in the *Dictionary of Scientific Biography*. Similarly, the humanist broker Giovan Vincenzo Pinelli (1535–1601), the papal physician Michele Mercati (1541–1593), the grandducal botanist Paolo Boccone (1633–1704), and the Jesuit naturalist Filippo Bonanni (1638–1725) grace modern histories of science briefly, if at all. Their contributions have not, by our standards, merited a retrospective assessment. Perhaps it is worth asking ourselves why we remember the most prolific, measuring importance by the number of weighty tomes, when a different kind of documentation—one that takes into account the teaching and training of students and the collaborative process at work behind any publication—reveals that others, who published less, contributed as much if not more. In other instances, we might query why one form of copiousness—that of Galileo or Newton, to offer contemporaneous examples—merits attention, while another—that of Aldrovandi, Redi, or Kircher—goes relatively unremarked. In the process of writing the sort of history that speaks to our present image of "science," we often find it preferable to neglect those who do not fit comfortably into this category. One of the primary goals of this study is to bring to life individuals who, while marginal to our own view of science, were central to the early modern definition of scientific culture.

Naturalists such as Luca Ghini (ca. 1490–1556), holder of the first chair in natural history at both Bologna and Pisa and known as "Prince of the science of simples,"[9] were too busy training scholars and amassing the particulars of nature to have time to see their ideas into print.[10] The papal physician Mercati was undoubtedly too occupied tending his successive patrons's ailments, containing plague outbreaks, supervising cultural activities such as the raising of the obelisks, and escorting visitors through the Vatican mineralogical collection to ever finish his much anticipated museum catalogue, finally published in 1717.[11] And we can point to numerous other naturalists, mainly physicians, apothecaries, and custodians of botanical gardens, who had little inclination or ability to outline the philosophical framework for nature and yet contributed significantly to the discipline through their perseverance in the construction of its material base.

Others such as Giovan Vincenzo Pinelli found satisfaction in their ability to publicize the work of aspiring humanists, naturalists, and collectors, preferring their role as brokers to personal advancement.[12] Rather than publish themselves, they facilitated the work of other scholars. The profusion of activity that surrounded and followed them is a testimony to their success as promoters of the newly reformulated study of nature. Where it seems appropriate, I have tried to resuscitate the less famous, primarily to place them in their proper relationship to their more luminous counterparts, and to give the reader some understanding of the breadth of natural history as a

[9]Tommaso Garzoni, *Piazza universale di tutte le professioni del mondo* (1585; Venice, 1651 ed.), p. 155.

[10]Because so little is known about the particulars of Ghini's life, he usually appears in discussions of other naturalists, principally Aldrovandi and Mattioli.

[11]Bruno Accordi's "Michele Mercati (1541–1593) e la Metallotheca," *Geologica Romana* 19 (1980): 1–50, still is the only individual treatment of Mercati's work.

[12]On Pinelli, see Paolo Gualdo, *Vita Joannis Vincentii Pinelli* (Augsburg, 1607); and Marcella Grendler, "Book Collecting in Counter-Reformation Italy: The Library of Gian Vincenzo Pinelli (1535–1601)," *Journal of Library History* 16 (1981): 143–151.

discipline and a practice. Unfortunately, many will appear only briefly here since, predictably, they have left little or no documentation behind relative to the wealth of materials bequeathed to posterity by Aldrovandi, Kircher, and other equally self-conscious, self-promoting contemporaries.

Turning to the more well-known figures appearing in this study, we face a different set of problems. We have the case of Aldrovandi, who wrote reams and yet found himself mired in the difficulties of publishing the lavishly illustrated tomes that the new natural history demanded. For him, the burden of becoming the latest in a succession of new Aristotles was simply too much for one lifetime; on his death he left numerous manuscripts behind, as much a part of his museum as the objects described within them, but few publications. While Aldrovandi gathered and collated, his less prolix colleague Mattioli reaped the rewards by choosing to publish rather than collect—doing so, we might add, by making good use of the information painstakingly accumulated by his friends. "I have never had any use in keeping simples," wrote the unrepentant Mattioli in 1553, "always contenting myself with the garden of Nature and with those which I engraved in my book."[13] Unremarkably, we remember Mattioli's 1544 commentary on and translation of Dioscorides better than the work of contemporaries such as Aldrovandi, Calzolari, and Ghini who made it possible.[14] In contrast, naturalists such as Della Porta and Kircher, although successful in publishing, were, like Aldrovandi, unable to maintain an audience much beyond their lifetime; the changing tenor of the scientific community made their work outdated even as they wrote it. In this, ironically, they were joined by some of their sternest critics, since Paolo Boccone and Francesco Redi, court naturalists extraordinaires and tireless advocates for a more experimental natural history, are no better remembered than their sparring partners Kircher and Bonanni.[15] While Boccone and Redi were instrumental in articulating the problems with the Jesuit philosophy of nature, depicting it as a font of undifferentiated erudition, they nonetheless were identified with the sort of

[13]C. Raimondi, "Lettere di P. A. Mattioli ad Ulisse Aldrovandi," *Bullettino senese di storia patria* 13, fasc. 1–2 (1906): 16. "Simples" described the bits of nature, particularly herbs but potentially any natural object, that formed the basis of medicinal compounds.

[14]The most accessible discussion of Mattioli is Jerry Stannard, "P. A. Mattioli: Sixteenth Century Commentator on Dioscorides," *Bibliographical Contributions, University of Kansas Libraries* 1 (1969): 59–81; Richard Palmer's "Medical Botany in Northern Italy in the Renaissance," *Journal of the Royal Society of Medicine* 78 (1985): 149–157, gives a good overview of the dynamic between the different players and Mattioli's manipulation of the situation.

[15]Bruno Basile's *L'invenzione del vero. La letteratura scientifica da Galileo ad Algarotti* (Rome, 1987) represents some of the most recent work on Redi and his relationship to various opponents. For a general discussion of Redi, see Jay Tribby, "Cooking (with) Clio and Cleo: Eloquence and Experiment in Seventeenth-Century Florence," *Journal of the History of Ideas* 52 (1991): 417–439, and Paula Findlen, "Controlling the Experiment: Rhetoric, Court Patronage and the Experimental Method of Francesco Redi (1626–1697)," *History of Science* 31 (1993): 35–64. On Boccone and Bonanni, there is no good secondary literature, but see Bruno Accordi, "Paolo Boccone (1633–1704)—A Practically Unknown Excellent Geo-Paleontologist

Baroque scientific culture that the members of the Royal Society, spokesmen for the "new" philosophy, found so distasteful.

How do we bring all these naturalists, collectors, and patrons together? Drawing on sources such as museum inventories and catalogues, university and academy archives, correspondence between naturalists, travel journals, and the publications resulting from the collection of nature, this study offers an intellectual, cultural, and sociological portrait of the collectors of nature. To facilitate its presentation, I have divided the book into three parts. The first considers the linguistic, philosophical, and social matrices that gave the museum a precise intellectual and spatial configuration. It discusses the words, images, and locations that defined the museum as a place of learned activity. The second details the role of the museum as a laboratory. Beginning with the practices that brought nature into the museum and concluding with an analysis of the uses of nature in microcosm, for medicine as well as natural history, it explores the implications of "experiencing" nature in the museum. The final part elaborates on the sociology of collecting and the cultural logic in which naturalists invested. It considers the methods by which naturalists forged their identities as scholars, brokers, and patrons within the complex circumstances of their society and the language and protocols of behavior that guided them at every turn.

By focusing on the sixteenth and seventeenth centuries, a period in Italian history that has largely been ignored by Anglo-American historians of Italy,[16] though not by historians of science, I hope to demonstrate the complexity and vitality of a period commonly considered to be the nadir of Italy's political and cultural decline. Museums belonged to many different worlds; they were owned by princes, clerics, doctors, apothecaries, and virtuosi. Thus, they provided an axis through which all these different sectors of society intersected and in doing so reveal much about the social and cultural fabric of early modern Italy. The museum was not only a place in which objects were housed; it was also a setting in which relationships were formed. The human interactions that produced and maintained various collections remind us how much intellectual life was guided by patrician social conventions; patronage, civility, concern for prestige, and obsession with commemoration were all standard features of this world.

In presenting this work as an essay in scientific culture, I am particularly concerned with the circumstances, strategies, and contexts that made it pos-

---

of the Seventeenth Century," *Geologica Romana* 14 (1975): 353–359, and idem, "Illustrators of the Kircher Museum Naturalistic Collections," *Geologica Romana* 15 (1976): 113–122. The burgeoning field of Kircher scholarship will be cited throughout.

[16]A notable exception is the work of Eric Cochrane; see his *Tradition and Enlightenment in the Tuscan Academies 1690–1800* (Rome, 1961); *Florence in the Forgotten Centuries 1527–1800* (Chicago, 1973); and *Italy 1530–1630,* Julius Kirshner, ed. (London, 1988). Historians such as Judith Brown, Brendan Dooley, Elisabeth Gleason, John Marino, John Martin, Laurie Nussdorfer, and Anne Jacobson Schutte have also begun to fill out the portrait of early modern Italy.

sible for scientific activity to thrive in early modern Italy. We know something of the reasons why professional naturalists—university professors, physicians, and apothecaries—found new ideas about and techniques of investigating nature so attractive. Knowing nature was their livelihood. But what persuaded courtiers and the urban patriciate of the value of scientific knowledge?[17] Natural history, as Gaspare Gabrieli proclaimed in his inaugural lecture as the first professor of medicinal simples at the University of Ferrara in 1543, was appropriate for "not only humble and lowly men, but people coming from every social class conspicuous for political power, wealth, nobility and knowledge, such as kings, emperors, princes, heroes, poets, philosophers, and similar men."[18] He and other naturalists urged their students to see the nobility as well as the utility of their chosen discipline. The success of their efforts is borne out by Paolo Boccone's comment in his *Museum of Physic and Experiences* (1697): "I observed that in Florence, in imitation of their Ruler, all the Nobles are enamored of the studies of Physic and of experiences."[19]

In shaping this project, I have implicitly defined the formation of a *scientific culture* as the broadening of the activities that we subsume under the category of science into new arenas: the court, the academy, the pharmacy, the piazza, the marketplace, the museum. By doing this, I hoped to avoid the pitfall of describing what these naturalists did as "science," a category that had no relevance for them. Most naturalists aspired to be "natural philosophers," an appellate that allowed them to traverse the boundaries between medicine, natural history, and natural philosophy, in short, to be qualified to comment on all domains of scientific knowledge. They would have been puzzled by and indeed disparaging of our own attempts to reduce their expansive, encyclopedic activities by labeling them "scientists." Our own conception of science has discarded many of the practices that they saw as essential to the definition of *scientia*. While we perceive the museum of natural history to be alternately a research laboratory or a place of public education, they understood it to be a repository of the collective imagination of their society.

At the same time, the museum also publicized a scientific culture formerly confined to scholastic discourse. It was the centerpiece of the humanist vision of knowledge. Collecting provided an important mechanism to facilitate the transition of natural philosophy from a largely textual and bookish culture, difficult for all but the most learned to access, to a tactile, theatrical culture that spoke to a multiplicity of different audiences. It was

[17]Here I am paralleling some of the remarks made by Walter Houghton in his classic study "The English Virtuoso in the Seventeenth Century," *Journal of the History of Ideas* 3 (1942): 51–73, 190–219.

[18]Felice Gioelli, "Gaspare Gabrieli. Primo lettore dei semplici nello Studio di Ferrara (1543)," *Atti e memorie della Deputazione provinciale ferrarese di storia patria*, ser. 3, 10 (1970): 34.

[19]Paolo Boccone, *Museo di fisica e di esperienze* (Venice, 1697), p. 267.

circumscribed by a variety of newly reconstituted practices, among them experimenting, observing, and translating. The proliferation of words, images, and objects accompanying the formation of museums gives testimony to the fact that late Renaissance natural history was defined by its audience as well as by the books that outlined its shifting parameters. The new visibility of natural history was as much an act of cultural production as intellectual orchestration or institutional resolution; the centrality of collecting to the reformulation of this discipline had much to do with its ability to rearrange the boundaries of the scientific community.

The success of natural history as a profession for numerous doctors, apothecaries, and natural philosophers, in conjunction with its popularity as a learned pastime for patricians, points to a common problem in the history of science, where neat divisions between "scientists" and "amateurs" have often prevailed. Rather than perpetuating these cleavages, I wish to encourage the notion that old and new, sacred and secular, occult and scientific, professional and amateur systems of knowledge could and did coexist in the sixteenth and seventeenth centuries. While something we might approximately designate as the precursor to modern science was in the process of formation, it was assisted by, in fact derived from, a most "unscientific" world of philosophizing that did not privilege "science" because it had not yet identified it as a better, truer form of knowledge. The intellectual problem of understanding what Renaissance naturalists meant when they "did science" also has social consequences. The indistinguishability of natural history from other aspects of learned and courtly culture certainly calls into question the notion of a scientific community, as historians of science have commonly understood it. The museum was simultaneously a harbinger of new experimental attitudes and sociological formations, and the dominion of Aristotelian dicta and the magus who fabricated secrets in his laboratory. Standing at the crossroads between these two seemingly diverse scientific cultures, the museum provided a space common to all.

My purpose throughout has been to reconstruct the discipline of natural history as a whole, at least in the form that it took in Italy. I am interested in knowing not only *why* the study of nature was revived—the intellectual question most commonly asked in shaping such a study—but also *how* and *where*. What conditions allowed it to emerge from relative obscurity and neglect to become a practice lauded by scholars, promoted by academies and universities, and accepted at court? How did the formation of "theaters of nature" facilitate this transition? Natural history was a vibrant and multifaceted enterprise during this period. Despite the relative lack of attention that it has received in the historiography of the scientific revolution,[20] it attracted

---

[20] Here I indicate only a few of the most recent studies that attempt to redress this problem: William B. Ashworth, Jr., "Natural History and the Emblematic World View," in *Reappraisals of the Scientific Revolution,* David C. Lindberg and Robert S. Westman, eds. (Cambridge, 1990), pp. 303–332; Harold Cook, "Physick and Natural History in Seventeenth-Century England"

scholars as well as courtiers, those with a professional interest in nature as well as virtuosi who delighted in the intricacy, subtlety, and copiousness of her parts; in short, the learned and the curious. While naturalists came to the study of nature for a variety of different reasons, all found common grounds for interaction in the museum, a microcosm of elite society as well as nature herself. It is to this intersection of patrician values, scientific aspirations, and collecting practices that we now turn.

---

in *Revolution and Continuity: Essays in the History and Philosophy of Early Modern Science,* Peter Barker and Roger Ariew, eds. (Washington, DC, 1991); Karl H. Dannenfeldt, *Leonhardt Rauwolf: Sixteenth-Century Physician, Botanist and Traveller* (Cambridge, MA, 1968); F. David Hoeniger, *The Development of Natural History in Tudor England* (Charlotte, VA, 1969); idem, *The Growth of Natural History in Stuart England from Gerard to the Royal Society* (Charlotte, VA, 1969); Joseph M. Levine, "Natural History and the Scientific Revolution," *Clio* 13 (1983): 57–73; Karen Reeds, "Renaissance Humanism and Botany," *Annals of Science* 33 (1976): 519–542; Alice Stroup, *A Company of Scientists: Botany, Patronage, and Community at the Seventeenth-Century Parisian Royal Academy of Sciences* (Berkeley, 1990); Barbara Shapiro, "History and Natural History in Sixteenth- and Seventeenth-Century England," in Barbara Shapiro and Robert G. Frank, Jr., *English Scientific Virtuosi in the Sixteenth and Seventeenth Centuries* (Los Angeles, 1979), pp. 1–55. The only comprehensive treatment of premodern natural history is Scott Atran's *Cognitive Foundations of Natural History: Towards an Anthropology of Science* (Paris and Cambridge, U.K., 1989).

PART ONE

Locating the Museum

Part I locates the museum in time as well as space. It considers three main issues: the linguistic, epistemological, and social sites of collecting. Beginning with the layers of definition that helped to shape the Renaissance idea of collecting, the study explores the various intellectual structures that situated the museum firmly within the encyclopedic tradition and social structures that aligned it with patrician culture. Chapter 1 describes the emergence of the museum through an analysis of the first museum catalogues and earliest images of museums as well as through a reading of representative events that connected the activities of the collector to society. Chapter 2 introduces the reader to the vocabulary of collecting, its humanistic origins, and philosophical uses. It situates collecting within the encyclopedic projects of the period and the quest for total knowledge of nature. Chapter 3 identifies the social framework that gave meaning to language and shaped the ritual life of the museum. Whether through the lens of antiquity, curiosity, or civility, collectors defined the museum in relation to a scientific culture that was as humanistic as it was urbane. As I will suggest repeatedly throughout this book, the discursive culture of late Renaissance and Baroque Italy cannot be divorced from its social context. Collecting, like all humanistic practices, was a gentlemanly pastime. Indiscriminant in their choice of objects, collectors were highly selective in their choice of companions. As they demonstrated often in their words and actions, only individuals of privilege and learning had earned the right to collect and classify the world.

Nurtured in the patrician culture of sixteenth- and seventeenth-century Italy, collectors shared many common traits with other members of the social elite. Their concern for, indeed obsession with, memory reflected the broader interest in lineage. Humanists and aristocrats shared a similar interest in origins; searching the past gave meaning to their lives, defining them against an otherwise undifferentiated historical continuum. Likewise, curiosity, a sensibility formerly viewed with ambivalence as a sign of the prideful nature of humans, was elevated to a virtue. Guiding the programs of naturalists from Aldrovandi to Kircher, it also expressed an ethos peculiar to early modern society: the desire to make knowledge an expansive rather than a definitive enterprise. Who were the "curious" in this society? Individuals with the wit, learning, and experience to synthesize and analyze successfully the new data offered up to the Europeans, in other words, the humanists who revitalized the ancient paradigms of natural history as a framework for exploring their world. While curiosity allowed them to investigate the relative status of their identity, memory provided a bedrock from which to launch these explorations. Only those who *knew who they were*, as humanists reminded one another, could be open to the possibilities of reinvention.

The cultural framework that identified memory and curiosity as two central categories of analysis for naturalists also articulated a social vocabulary that defined relationships among collectors. Civility, the primary rubric under which all forms of gentlemanly behavior were subsumed, shaped the conventions by which collections were arranged and the conditions under which naturalists gained access to them. Museums, designed to facilitate social interaction, were laboratories of human behavior. A product of an eminently civil society, they remind us of the extent to which scientific culture was shaped by the early modern "civilizing process." Civility was yet another form of self-knowledge as social knowledge. Shared protocols of behavior, first formulated in the humanist republic of letters, ultimately defined the boundaries of the community of naturalists and collectors.

The combination of all three factors—memory, curiosity, and civility—defined one of the most important sites of knowledge in early modern Europe: the museum. Locating the museum is admittedly an elusive task, given the range of activities and settings defined by the term. All three chapters highlight key practices that made the definition of the museum social and culturally specific. In doing so, they reveal the dynamic nature of this peculiarly human institution.

ONE

"A World of Wonders in One
Closet Shut"[1]

ALDROVANDI'S MUSEUM

On 13 May 1572, the very day that Ugo Buoncompagni had chosen to return to his hometown to be invested as Gregory XIII (1572–1585), a fearsome dragon appeared in the countryside near Bologna, an omen of terrible times to come. Soon word of its presence spread, and a party was sent out to overtake it. The captured portent was duly carried inside the walls of the city for its citizens to inspect. Entrusted with its disposal, the senator Orazio Fontana consigned the large serpent to his brother-in-law Ulisse Aldrovandi, a collector of strange and wonderful things and an expert in draconology. As cousin of the newly elected Pope, Aldrovandi had an added claim to possess the prodigious serpent; in a strange way, his fortune too was bound to its discovery.

The naturalist promptly displayed his latest acquisition in his famous museum, where "an infinite number of gentlemen came to my house to see it."[2] The appearance of the dragon was an occasion for extemporaneous poems such as the one by Augustus Gottuvius, describing Aldrovandi as no less fortunate than the first Ulysses, encounterer of other quasi-mythical creatures,

---

[1] This quotation comes from the epitaph of the British collector John Tradescant (ca. 1577–1638):

As by their choice collection may appear,
Of what is rare, in land, in seas, in air:
Whilst they (as HOMER's Iliad in a nut)
A World of Wonders in one closet shut.

In Arthur MacGregor, "The Tradescants: Gardeners and Botanists," in *Tradescant's Rarities: Essays on the Foundation of the Ashmolean Museum 1683,* Arthur MacGregor, ed. (Oxford, 1983), p. 15.

[2] BUB, *Aldrovandi,* ms. 70, c. 24r.

for now his museum was complete.[3] Almost immediately Aldrovandi set about writing a treatise on the dragon, including examples of all other serpents he had ever seen or heard of.[4] "In that time, I wrote a Latin history of that dragon in less than two months, divided into seven books and entitled the *Dracologia*," he explained to the Grand Duke of Tuscany, Francesco I.[5] The naturalist also used this opportunity to reopen the debate about one of the many paradoxes that bedeviled the community of naturalists: whether or not serpents generated spontaneously from an alleged rooster's egg. In addition, he had one of the artists whom he employed full-time in his museum illustrate the dragon for posterity (fig. 1). Aldrovandi lost no time in maximizing the scholarly yield out of this well-publicized natural curiosity. Soon his authoritative descriptions were the talk of Italy.

While the appearance of the serpent was initially a local event, its connection with the election of a new pope ensured that it would be widely publicized beyond the confines of the city. Furthermore, the addition of the serpent to Aldrovandi's collection immediately elevated it from the realm of popular lore to the domain of science.[6] In a matter of a few weeks, naturalists, collectors, and the simply curious in all parts of Italy bombarded Aldrovandi with requests for information about the papal portent. From Imola, Filippo Sega wrote to inquire whether a sign— "almost an emblem or a hieroglyph"[7]—had been impressed in the dragon. Like many other correspondents, he desired a full description of the dragon's remarkable parts to read properly the significance of this unnatural phenomenon. Aldrovandi, in his capacity as the owner of the portent, was the one person authorized to have complete knowledge of this singular piece of nature.

By summer, the attributes of the papal portent had been duly magnified, as the story made its way through the Italian cities. "The other day cer-

---

[3]BUB, *Aldrovandi*, ms. 3, n.p.: "Lustris tercentum bis septum [*sic*] dum additur annus/Bis, Draco Felsineo sibilat ortus Agro. Sibilat eo quoque Coeli Cardine, eodem/Tempore, Romana Sceptriger Urbe DRAGO. Tertius ut Mundi Decus, estque Secundus Olympi. Sic Aldrovandi Primus Ulysses erit/Fortunate Draco Scriptoris Ulysses egebas: Qui celebrem toto redderet Orbe Gravis/Non minus indiguit te Fortunatus Ulysses, Omnia, Musaeum nunc Microcosmus habet."

[4]BUB, *Aldrovandi*, ms. 3; see also his *Serpentium et draconum historiae libri duo*, Bartolomeo Ambrosini, ed. (Bologna, 1639).

[5]BUB, *Aldrovandi*, ms. 82, cc. 372v–373r (Bologna, 6 September 1578). The naturalist sent the Grand Duke a precis of the book entitled *Del dragone da duoi piedi monstruoso*.

[6]Here I am paralleling an interpretation made by Ottavia Niccoli in her *Prophecy and People in Renaissance Italy*, trans. Lydia G. Cochrane (Princeton, 1990), pp. 189–196. However, I do not agree with Niccoli's argument about the complete disappearance of the prophetic overtones of prodigies after the Sack of Rome in 1527. Instead, I would describe the "secularization" and "scientization" of prodigious phenomena as a more gradual process that occurred in the course of the late sixteenth and seventeenth centuries.

[7]BUB, *Aldrovandi*, ms. 3, c. 21 (Imola, 1 June 1572).

Figure 1. The dragon of 1572. From BUB, *Aldrovandi, Tavole di animali*, IV, 130.

tain Monks from Certosa told me about the monstrous serpent with a bird's
feet . . . and fish's head found on the Bolognese," wrote the physician Alfonso Pancio from Ferrara on 6 July 1572, "and they said that Your Excellence has it and has had it illustrated. They begged me to write to you and
find out if such a thing were true."[8] Discussing the news with the good citizens of Ferrara, Pancio discovered that his nephew Francesco Anguilla had
already seen the serpent and could testify to its existence. Accordingly, Pancio asked his friend to send a picture to his patron Duke Alfonso II d'Este,
as a token of their mutual affection. Eager to please a powerful prince, the
Bolognese naturalist promptly discharged his obligation. "I received your
letter, together with the picture of the Dragon, which is most dear to me,"
reported Pancio on 25 November, "and I thank Your Excellence infinitely,
awaiting to see the history of it."[9] Once again, Aldrovandi's authority in the
true reporting of the phenomenon was acknowledged. His possession of the
object gave the illustrations that he sent to the physician Pancio and the
Duke of Ferrara a heightened credibility.

[8]BUB, *Aldrovandi*, ms. 38(2), II, c. 173 (Ferrara, 6 July 1572).
[9]BUB, *Aldrovandi*, ms. 38(2), II, c. 177 (Ferrara, 25 November 1572).

The Duke of Ferrara was not the only patron anxious to see the *Dracologia* in print. In Padua, naturalists such as Melchior Wieland (known to the Italians as Guillandino), prefect of the botanical garden, and the humanist Giovan Vincenzo Pinelli clamored for news of the latest marvel. "Now he and I greatly desire that Your Excellence favor us with your history of the two-footed Dragon," wrote Pinelli on 25 August 1572. In the winter of that same year, no doubt in response to Aldrovandi's protests that he had neither the time nor the scribes to disseminate adequately the manuscript version of the *Dracologia* to everyone who requested it, Pinelli urged him to publish it as soon as possible. "We will await the printed history of the dragon."[10] Despite these pleas, Aldrovandi's history of serpents was not published until 1639, more than thirty years after his death. By then it had lost its immediacy, not to mention its heightened moral significance. The spontaneity of the event was best captured in oral reports, occasionally committed to writing, that came from the mouths of people who had traveled to Aldrovandi's museum to see the serpent. Just as the Bolognese naturalist increased his authority by possessing the dragon, all who personally observed it enhanced their status among those who trafficked in natural curiosities.

While Aldrovandi could delay sending a copy of his manuscript treatise to his friend Pinelli, he needed to oblige the great and powerful immediately. Naturalists and patrician collectors would continue to be interested in the remarkable serpent; princes had a more limited attention span. Besides offering illustrations of the papal portent to friends, Aldrovandi also took advantage of this opportunity to strengthen his ties with patrons in Florence and Rome, all of whom consulted the collector for his opinion on the scientific and political implications of the dragon. As late as 1578, the Grand Duke of Tuscany, Francesco I, expressed interest in hearing a more detailed account of the peculiar serpent "that was killed by a peasant on the exact day that Gregory XIII was elected the highest Pontiff." Ten years later, when trying to secure Ferdinando I's patronage shortly after he had succeeded his brother as Grand Duke, Aldrovandi mentioned the *Dracologia* as one of his major scientific treatises.[11]

In the months following the appearance of the "monstrous dragon" (*Dragone mostroficato*), an uneasy court in Rome awaited Aldrovandi's judgment.[12] The naturalist strategically dedicated his *Dracologia* to the cardinal nephew Filippo Buoncompagni, Cardinal San Sisto, to insinuate himself further in the papal favor. Only five days after the appearance of the dragon, Aldrovandi rushed a brief description of it to the papal nephew; exactly one month later, he wrote again to assure the cardinal of the nobility of the unusual beast. Since Cardinal San Sisto had the ear of the pope, he was prob-

[10]G. B. De Toni, *Spigolature aldrovandiane XVIII*, p. 308 (Padua, 25 August 1572); p. 310 (Padua, 2 December 1572).

[11]BUB, *Aldrovandi*, ms. 82, cc. 372v–373r; Mattirolo, p. 382 (Bologna, n.d., 1588).

[12]BUB, *Aldrovandi*, ms. 38(2), IV, c. 349 (Rome, 13 May 1573); ms. 70, cc. 23v–24r.

ably the most important person to whom Aldrovandi communicated his findings. Undoubtedly San Sisto had been entrusted by Gregory XIII to inquire further into the matter; the pope himself, as a great patron and a man beyond reproach, would not deign to communicate personally any worry over the portent, not even to his cousin the collector. Yet surely, as someone nurtured in a world teeming with supernatural phenomena, Gregory XIII could not help but wonder at the meaning of it all. Popes had enough worries when they agreed to wear the tiara, particularly in the years following the splintering of Western Christendom into multiple christianities. While Gregory XIII could choose to maintain his distance from the discussions surrounding the portent, he could not ignore it. After all, the heraldic shield of the Buoncompagni had a "rising dragon" (*drago nascente*) emblazoned on it; people would not fail to note the familial significance of the portent. Inviting his cousin Aldrovandi to provide a complete philosophical analysis of the dragon deflected the attention away from its potentially ominous significance and toward its natural historical value.

Back in Bologna, Aldrovandi set himself the task of describing and explaining the portent's unusual anatomy, reporting periodically to the cardinal nephew. Puzzling over the inexplicable addition of two feet, he remarked that they could serve no visible purpose, "unless, in this way, the animal, through its imperfection, now proves to be of the highest perfection?"[13] Like many early modern philosophers, he delighted in paradoxes. The dragon, monstrous due to the excess of matter that had produced the unwanted feet, was not outside the realm of nature and therefore not beyond Aldrovandi's purview as a naturalist. It gave testament to the wonder of nature, which constantly yielded new and unexpected results. By the end of the year, he had developed this line of interpretation further. In a treatise composed for Giacomo Buoncompagni, son of Gregory XIII and castellan of Sant'Angelo, Aldrovandi underscored the value of this particular serpent to his collection. "Serpents naturally do not have feet," he explained,

> until the time and properly the day in which the most wonderful pontiff Gregory XIII was created Pope, when two serpents and dragons that had two feet were seen in the Bolognese countryside. I have copiously discussed the one which came into my hands and given my opinion [of it]. As soon as I can—for it is rather long—I will have the history transcribed and will send it to the most illustrious monsignor San Sisto, as promised, along with the picture of this dragon.[14]

In this fashion, the naturalist neatly turned a potentially disastrous occurrence into a providential act of patronage for his museum. By explaining away the serpent as an example of nature's fecundity rather than a diabolical catastrophe, he diffused its saturnine implications, scientifically securing the foundation of the new papacy for his patrons in Rome.

[13]BUB, *Aldrovandi*, ms. 3, c. 8r (Bologna, 13 June 1572).
[14]Aldrovandi, *Discorso naturale*, p. 186. Emphasis mine.

In his decision to interpret the dragon as a natural phenomenon, devoid of metaphysical implications but rich in anatomical meaning, Aldrovandi participated in a broad cultural trend to normalize the marvelous in the late sixteenth and seventeenth centuries. As Lorraine Daston observes, "Preternatural events always qualified as wonders, but only sometimes as signs."[15] She and many others have identified the museum, variously called a "cabinet of curiosity," "wonder room," and "theater of nature," as the principal site in which this process of demystification occurred. Aldrovandi, when asked to pass judgment on an animal of unusual significance, chose to subsume it within a largely naturalistic explanation, deeming it a "wonder" rather than a "sign." In response to Filippo Imola's question about the hieroglyphic significance of the dragon, Aldrovandi was probably very circumspect, since this sort of knowledge did not serve his purpose, at least in this instance. Concerned about the reaction that his interpretation would receive in Rome, he carefully framed his report in the proper philosophical language and used his humanist skills to make his observations eloquent and appropriate in the eyes of the pope.

While the discovery of the dragon was an inauspicious beginning for Gregory XIII, it might be described as somewhat of a "marvelous conjuncture"[16] for his kinsman Aldrovandi. Always in search of opportunities to gain new patrons and increase his visibility in the learned world, Aldrovandi briefly had the attention of all of Italy. The tense conferences held to determine the significance of the portent and to diffuse its ominous nature provided ample opportunity to showcase the Bolognese naturalist's formidable scientific erudition. Most likely Aldrovandi invited the archbishop, papal legate, senators, and his colleagues in the faculty of medicine at the University of Bologna to witness the dissection of the serpent in his museum at the hands of a protégé such as the surgeon Gaspare Tagliacozzi.[17] Turning his task into a civic spectacle, he encouraged many of the principal citizens of

[15]Lorraine Daston, "Marvelous Facts and Miraculous Evidence in Early Modern Europe," *Critical Inquiry* 18 (1991): 106; see also William B. Ashworth, Jr., "Remarkable Humans and Singular Beasts," in Joy Kenseth, ed., *The Age of the Marvelous* (Hanover, NH, 1991), pp. 113–144, esp. pp. 120ff; and Katharine Park and Lorraine Daston, "Unnatural Conceptions: The Study of Monsters in Sixteenth- and Seventeenth-Century France and England," *Past and Present* 92 (1981): 20–54. The literature on this subject is large, so I am only citing the most directly relevant work.

[16]This is the term that Galileo also used when describing the *mirabil congiuntura* of his discovery of the Medicean stars and the ascension of his patron Cosimo II as Grand Duke of Tuscany in 1609 and the similar intersection of his publication of the *Assayer* and the ascension of Urban VIII in 1624. See Pietro Redondi, *Galileo Heretic,* trans. Lydia G. Cochrane (Princeton, 1986), pp. 68–106; see also Mario Biagioli, "Galileo's System of Patronage," *History of Science* 28 (1990): 14–17.

[17]This is purely speculative on my part. In 1572, Tagliacozzi had recently been graduated from the University of Bologna (1570) and was embarking on a career of teaching there. However, in 1575 the anatomist was involved in the theriac controversies that arose between Aldrovandi and the College of Physicians, performing dissections and demonstrations of the true nature

the city to confirm what the hand of the dissector revealed: that the animal was indeed a natural occurrence. They collectively authorized his interpretation through their participation as observers. Thus it was in Aldrovandi's museum, and not in the backrooms of the Roman court at the hands of theologians, that whatever doubts occasioned by the portent were set to rest. Perhaps the number of dragons decorating the "Tower of Winds" (*Torre dei Venti*) that Gregory XIII added to the Vatican complex was not only a manifestation of the standard papal impulse to emblazon Rome with the family shield but also a sign of his defiance of the older tradition of portentious signs that he repudiated with the help of his cousin Aldrovandi.[18]

The arrival of the monstrous dragon neatly illustrates the significance of Aldrovandi's "theater of nature" (*teatro di natura*) to the political and civic life of sixteenth-century Italy. The pivotal role that Aldrovandi's museum played in the controversial events surrounding the election of the new pope was in fact the culmination of many instances in which the Bolognese naturalist was called upon to arbitrate important scientific problems and to demystify paradoxes like the noteworthy portent. In this respect, the function of his museum was not dissimilar to that of his patron, the Grand Duke of Tuscany Francesco I, who held court in his scientific laboratory. "He spends almost all of his time in a place they call the *Casino* (countryhouse) . . . ," reported the Venetian ambassador to Florence in 1576, "but nevertheless he intersperses . . . negotiations with secretaries regarding affairs of state, also expediting many requests for mercy as well as justice, in such a manner that he mixes pleasure with business, and business with pleasure."[19] Collecting was not just a recreational practice for sixteenth- and seventeenth-century virtuosi, but also a precise mechanism for transforming knowledge into power. While Aldrovandi allowed his museum to be shaped by the desires of his patrons, the Grand Duke Francesco created a scientific space in which affairs of state could be conducted. Each in different ways understood the setting that he had created to be essential to the formation of his identity as a natural philosopher.

What recognition did Aldrovandi earn for all his efforts? Contemporaries called Aldrovandi the "Bolognese Aristotle"; as late as the eighteenth century, he was hailed as "the Pliny of his time."[20] Labels do not reveal the en-

---

of the viper in the naturalist's *studio*, which lends some credence to this possibility; see chapter 6 of the present volume, and Maria Teach Gnudi and Jerome Pierce Webster, *The Life and Times of Gaspare Tagliacozzi Surgeon of Bologna 1545–1599* (New York, 1950), pp. 67–79.

[18]Cesare d'Onofrio reproduces a number of these dragons, both inside and outside the tower, in his *Roma val bene un'abiura: storie romane tra Cristina di Svezia, Piazza del Popolo e l'Accademia d'Arcadia* (Rome, 1976), pp. 26–27, 30.

[19]In Luciano Berti, *Il Principe dello studiolo: Francesco I dei Medici e la fine del Rinascimento fiorentino* (Florence, 1967), p. 58.

[20]Lorenzo Legati, *Museo Cospiano annesso a quello del famoso Ulisse Aldrovandi e donato alla sua patria dall'illustrissimo Signor Ferdinando Cospi* (Bologna, 1677), p. 8; Giovanni Cristofano Amaduzzi, Rome, 1 June 1782, in *Anecdota litteraria* (Rome, 1783), Vol. IV, p. 369. The

tire story, but they are indicative of the reputation Aldrovandi was able to cultivate, which certainly measured his success as a naturalist. Central to his fame was his museum of natural curiosities, housed in several rooms in his family palace and open to the learned and the curious of Europe. After his death in 1605, it was maintained as a civic museum by the Senate of Bologna. During Aldrovandi's own lifetime, contemporaries expressed their awe at his ability to amass natural objects. "After the return of Signor Contestabile, I did not have the opportunity to talk with him about you until last Friday . . . ," wrote a Milanese correspondent to Aldrovandi in 1598. "He told me that he had seen so many and various things in your *studio* that he remained stupefied. One can believe that there is no *studio* similar in all of Europe."[21] Learned naturalists such as Pier Andrea Mattioli proclaimed Aldrovandi's museum to be the most extensive microcosm of nature of its time. "Thus I remain always with heart aflutter and with baited breath until I see all the simples you have collected," confessed Mattioli in 1553, "and really I would like to come to Bologna only for this end, when I can."[22] While many nobles and scholars visited the museum simply to see its curiosities, others like Mattioli had more specific goals: they wished to examine specimens to complete the research that they too were doing in preparation for the writing of new and improved natural histories. Their presence in the museum only added to its luster. As a tribute to the fame and importance of his collection, Aldrovandi proudly described his museum as the eighth wonder of the world.[23]

In 1603, Aldrovandi formalized his agreement with the Senate of Bologna to have the contents of his *studio* transferred to the Palazzo Pubblico, the seat of government just off Piazza Maggiore in the center of town, after his death. "I will not say any more to you regarding the *studio* of Signor Aldrovandi," wrote Friar Gregorio da Reggio to the botanist Carolus Clusius in Leiden in 1602, "since I presume that you knew everything some time ago, that is, that he has left it to the Most Illustrious Senators of Bologna who have accepted it with great affection and with the firm intention of continuing the publication of his remaining works."[24] The network of communications that had spread the news of the papal portent in 1572, thirty years later publicized the fate of Aldrovandi's collection to scholars who perceived it to be a benchmark against which to measure their own encyclopedic schemes.

---

fundamental studies of Aldrovandi are Giuseppe Olmi, *Ulisse Aldrovandi: scienza e natura nel secondo Cinquecento* (Trent, 1976), and Sandra Tugnoli Pattaro, *Metodo e sistema delle scienze nel pensiero di Ulisse Aldrovandi* (Bologna, 1981).

[21]BUB, *Aldrovandi*, ms. 136, XXVII, c. 198r (Milan, 22 September 1598).

[22]Fantuzzi, p. 156 (Goritia, 12 July 1553).

[23]BAV, *Vat. lat.* 6192, II, f. 657r.

[24]G. B. de Toni, "Il carteggio degli italiani col botanico Carlo Clusio nella Biblioteca Leidense," *Memorie della R. Accademia di Scienze, Lettere ed Arti di Modena*, ser. III, Vol. X (1912): 146 (Bologna, 14 March 1602).

Frustrated by his inability to see his writings into print in his own lifetime, Aldrovandi stipulated, as a condition of his bequeathal to the Senate, that they perservere in publishing his work. By the end of the sixteenth century a new Aristotle could only shakily lay claim to this title if he did not create a body of knowledge made canonical by its appearance in print. For the next half century, the custodians of the *Studio Aldrovandi*, as it came to be called, meticulously edited, annotated and printed a few of the most important volumes that told "the History of the entire Museum."[25] Eight of his twelve published treatises were printed in this fashion.[26] Bartolomeo Ambrosini, custodian of the Aldrovandi museum from 1632 until 1657, supervised the publication of Aldrovandi's history of serpents and dragons in 1639. Five years earlier, Ambrosini had requested a sabbatical to finish preparing the manuscript for publication. His petition to the Senate reveals the continuing centrality of the Aldrovandi museum to the scientific and civic life of the city:

> Doctor Bartolomeo Ambrosini, having finished accommodating the public Museum in such a manner that it is visitable for any person whatsoever, [now wishes] to begin to compose the Histories of those Animals whose images are conserved in the above-mentioned Museum . . . that formerly belonged to the most excellent Signor Doctor Aldrovandi.[27]

Following the stipulations of Aldrovandi's donation, all of the research conducted by the custodian was done within the confines of the museum. "In the last [room] is the Most Excellent Doctor—who today is the Most Excellent Trionfetti—authorized by the *Reggimento* to have the said Works

---

[25]BUB, *Aldrovandi*, ms. 21, IV, c. 176 (Bologna, 23 June 1595), in Lodovico Frati, "Le edizione delle opere di Ulisse Aldrovandi," *Rivista delle biblioteche e degli archivi* 9 (1898): 163. The custodians of the *Studio Aldrovandi* were Johann Cornelius Uterwer (1610–1619), Bartolomeo Ambrosini (1632–1657), Ovidio Montalbani (1657–1671), Giovan Battista Capponi (1671–1675), Silvestro Bonfiglioli (1675–1696), Giovanni Domenico Guglielmini (1696–1698), Giovanni Lodovico Donelli (1698–1733), and Filippo Antonio Donelli (1734–1742). Others, such as Giacinto Ambrosini, Lorenzo Legati, and Lelio Trionfetti, assisted the custodian; see Christiana Scappini and Maria Pia Torricelli, *Lo Studio Aldrovandi in Palazzo Pubblico (1617–1742)*, Sandra Tugnoli Pattaro, ed. (Bologna, 1993).

[26]Aldrovandi published four volumes in his own lifetime: *Ornithologiae hoc est de avibus historiae libri XII* (Bologna, 1599–1603), 3 vols., and *De animalibus insectis libri septum* (Bologna, 1602). Prior to beginning his natural history, he published *Della statue romane antiche, che per tutta Roma, in diversi luoghi e case si veggono* (Rome, 1556). Aldrovandi's second wife, Francesca Fontana, saw into print his *De reliquis animalibus exanguibus libri quatuor* (Bologna, 1605). Uterwer edited Aldrovandi's *De piscibus libri V. et de cetis lib[rus] unus* (Bologna, 1612), *De quadrupedibus solidipedibus volumen integrum* (Bologna, 1616), and, with T. Dempster, *Quadrupedum omnium bisulcorum historia* (Bologna, 1621). Ambrosini edited *De quadrupedibus digitatis viviparis libri tres, et de quadrupedibus digitatis ovipars libri duo* (Bologna, 1637), *Serpentium et draconum historiae libri duo* (Bologna, 1639), *Monstrorum historia, cum parallipomenis historiae omnium animalium* (Bologna, 1642), and *Musaeum metallicum in libros IIII. distributum* (Bologna, 1648). Montalbani and Legati edited the last published work of Aldrovandi, *Dendrologiae naturalis scilicet arborum historiae libri duo* (Bologna, 1667).

[27]ASB, *Assunteria di Studio. Requisiti di Lettori*, Vol. I, n. 27 (1 April 1634).

printed and here he does his studies," observed the local chronicler Ghiselli in 1729.[28]

In 1672, Giovan Battista Capponi wrote to Francesco Redi about his recent promotion to the Aldrovandi chair, with all its related responsibilities, following the deaths of Ovidio Montalbani and Giacomo Zanoni, custodians of the museum and the botanical garden, respectively. "A few months ago the Chair in Natural History, held by the great Aldrovandi, together with the prefecture of the Garden of Simples, was conferred upon me by our Most Illustrious Senate," he informed Redi. "To this was added the custodianship of the Museum and the Library of the same Aldrovandi, and the charge of continuing his work, interrupted by the death of Signor Montalbani."[29] Unfortunately, we hear nothing further of Capponi's plans to continue the editing and publication of Aldrovandi's manuscripts. Perhaps he, like Montalbani, begrudged publishing what was essentially his work under another's name.[30] Invested with the responsibility of adding to the corpus of an early modern Aristotle, the custodians of the museum could not resist inserting their own erudition into the text, sometimes overwhelming Aldrovandi's late Renaissance style of natural history with their Baroque embellishments.

By the 1670s, the value of the publishing project had diminished, as Aldrovandi's observations and conclusions increasingly began to look outdated to a scientific community less enamored with the traditional framework of natural history. No doubt naturalists such as Bologna's famous anatomist Marcello Malpighi suggested to the Senate of Bologna that it was somewhat embarrassing to continue to publish Aldrovandi's treatises under the city's imprimatur. Surely the Royal Society, of which Malpighi was a celebrated foreign member, would think that the Italians had never heard of Francis Bacon if they continued to publish the unwieldy and endlessly erudite tomes of Aldrovandi? For these and perhaps financial reasons, the publications ceased. Despite the decline of editorial responsibilities, Capponi's description of his duties demonstrates the continued relationship between the teaching of natural history and the tending of the *Studio Aldrovandi* in Bologna until the eighteenth century.

In 1657, the museum of the Marchese Ferdinando Cospi was annexed to the *Studio Aldrovandi*. A senate decree of 1660 permanently allotted one room for the *gonfaloniere's* collection, and it was formally donated to the Sen-

---

[28]BUB, Cod. 559 (770), XXI, f. 406 (Antonio Francesco Ghiselli, *Memorie antiche manoscritti di Bologna*).

[29]Laur., *Redi* 211, f. 426 (Bologna, n.d.)

[30]Montalbani's edition of the *Dendrologiae* (Bologna, 1667), like many of the works published under Aldrovandi's name, relied very little on Aldrovandi's research, and it is apparent from Montalbani's correspondence with Kircher and Redi that he begrudged the fact this his own name would not appear first. In Capponi's case, he only held the position for four years (1671–1675), undoubtedly too short a tenure to complete any project.

ate in 1667.[31] No doubt the proximity of the *studio* to Cospi's chambers as the elected head of government—according to contemporary descriptions, it was located "at the door next to those of the *Gonfaloniere* to which one ascends by the stairs that go to the great tower"[32]—influenced the nobleman's decision to donate his collection to the city. Situated in the symbolic center of the city, it provided a way of linking private interests with public duties. Moreover, the association with Aldrovandi's museum lent Cospi's collection philosophical credibility by associating it with the activities of the Bolognese Aristotle. Contemporaries celebrated the union of the two collections as an accumulation of wonders upon wonders:

> If there was already one Aldrovandi Ulisse
> Master of Portents, indeed of Monsters,
> He is no longer alone, nor for our times
> The *non plus ultra* prescribed by his Museum.[33]

Or so claimed Giovanni Battista Ferroni in his dedication to Lorenzo Legati's 1667 catalogue of the museum. Ferroni, deliberately misunderstanding Aldrovandi's own intent, created an image of the Renaissance naturalist, as "master of portents," that the Baroque collector Cospi could challenge. While Aldrovandi had insisted that the custodian be a professor of natural history, Cospi instead chose a dwarf, a living marvel, to guide visitors through the museum. If sixteenth-century collections were noteworthy for their encyclopedism, resulting in the indiscriminate inclusion of every natural object, mid-seventeenth century collections were distinguished by their exoticism, which invested such categories as "wonder" and "marvel" with new meaning.

The reinvention of the museum in the 1660s as a Baroque theater of marvels was yet another reason why the publications of Aldrovandi's manuscripts ceased. Ovidio Montalbani's *Dendrologia* (1667), a comprehensive and delightfully whimsical study of trees published under Aldrovandi's name, was more in the spirit of the Cospi museum than the *Studio Aldrovandi*. Montalbani, whose assistant Lorenzo Legati wrote the catalogue

[31]BUB, Cod. 738 (1071), XXIII, n. 14. (*Decreto per la concessione di una sala al Marchese Ferdinando Cospi appresso lo Studio Aldrovandi*, 28 June 1660). A 1696 inventory of the museum places Aldrovandi's *studio* in the first room, a library in the second, a room for the Vicelegate of Bologna in the third, and Cospi's collection in the fourth; BUB, Cod. 384 (408), Busta VI, fasc. II (*Inventario dei mobili grossi, che si travano nello Studio Aldrovandi e Museo Cospiano*, 12 March 1696). For a brief summary of the donation of the museum, see Laura Laurencich-Minelli, "Museography and Ethnographic Collections in Bologna during the Sixteenth and Seventeenth Centuries," in *Origins of Museums*, p. 22; also see G. B. Comelli, "Ferdinando Cospi e le origini del Museo civico di Bologna," *Atti e memorie della R. Deputazione di Storia Patria per le Provincie di Romagna*, ser. 3, Vol. VII (1889): 96–127.

[32]BUB, Cod. 559 (770), XXI, f. 406.

[33]Lorenzo Legati, *Breve descrizione del museo dell'Illustrissimo Signor Cavaliere Commendatore dell'Ordine di San Stefano Ferdinando Cospi* (Bologna, 1667), p. 10.

to Cospi's museum, was a great admirer of the Jesuit Kircher, and they exchanged tales of marvelous images carved naturally in fruits and trees and images of Etruscan tablets as preparation for their publications. Montalbani provided a strong link between the Aldrovandi and Cospi collections, between Renaissance curiosity and Baroque wonder. Surely Malpighi and his group of followers must have been relieved when naturalists inclined to support new, more experimental paradigms of nature succeeded Montalbani to the custodianship of the museum and increasingly divorced themselves from its more outdated responsibilities.[34]

Throughout the sixteenth and seventeenth centuries, visitors flocked to the museum in Bologna. Many even highlighted the dragon, so lovingly described by Aldrovandi, as one of the more noteworthy items in the collection. In 1664, Philip Skippon described it as a "dragon or snake, with wings and legs, kill'd nigh this city." Athanasius Kircher included an illustration of it in his discussion of famous dragons in his *Subterranean World*, while Francesco Redi used it as a point of comparison for his discussion of two-headed dragons: "In the Bolognese Museum of the famous Aldrovandi an embalmed one is still preserved." As late as 1693, William Bromley felt that the "dragon captured in the Bolognese countryside in 1572 at the time of the Election of Gregory XIII" was one of the more interesting of the "712 Natural Rarities in little Glass-Bottles, with their names" displayed in the Aldrovandi museum.[35]

Scholars such as Malpighi plied the custodians with gifts of *naturalia* and *mirabilia*. While the anatomist may have disparaged the publications coming out of the museum, he nonetheless recognized its importance as a repository of natural objects and as one of the primary social centers of scientific culture in Bologna. "I beg you to favor me by procuring some aquatic curiosity, rarity or petrified thing to place in the public *Studio* and Museum of Aldrovandi," wrote Malpighi to Giovan Battista Capucci. As part of the ceremonial entrance of the new governor into the city, Bologna's papal legate made a point of visiting the museum to inspect its marvelous civic treasures. The visit ceremonially acknowledged the centrality of the Aldrovandi museum to the cultural identity of the city. Within weeks of this arrival in 1690, Cardinal Pamfili toured the museum. "He discussed some curious materials

[34] I have been speculating quite a bit on what Malpighi may have thought of these activities simply to give readers a focus for the shifting philosophical climate and the increased diversity in attitudes toward the study of nature. For a discussion of the impact of the "new" philosophy on Bologna, see Marta Cavazza, *Settecento Inquieto. Alle origini dell'Istituto delle Scienze di Bologna* (Bologna, 1990).

[35] Philip Skippon, *An Account of a Journey Made thro' Part of the Low-Countries, Germany, Italy and France*, in *A Collection of Voyages and Travels*, A. Churchill and S. Churchill, eds. (London, 1752 ed.), Vol. VI, p. 572; Kircher, MS, Vol. II, p. 93; Francesco Redi, *Osservazioni di Francesco Redi Academico della Crusca intorno agli animali viventi che si trovano negli animali viventi* (Florence, 1684), p. 2; William Bromley, *Remarks made in Travels through France and Italy* (London, 1693), p. 123.

with Signor Doctor Bonfiglioli on the occasion of seeing the *Studio Aldrovandi*," remarked Malpighi to the legate's nephew, Marcantonio Borghese, "and His Eminence displayed the judicious curiosity of a rather learned man."[36]

In describing the legate's conduct, Malpighi offered a normative view of what he considered to be the proper demeanor of a visitor to a museum. The ideal museum-goer was a man capable of understanding the experience of seeing a museum. "Judicious curiosity," rather than an unbridled appetite for wonder, defined him. Malpighi's assessment of the qualities of judgment that the papal legate had displayed evoked the sensibilities of the late seventeenth century, when the critiques of curiosity by natural philosophers such as Galileo and Descartes had begun to inform the study of nature. By the 1690s, the conduct of learned visitors to the museum had been "civilized," not according to the protocols of gentlemanly behavior, already in place in the sixteenth century, but according to the etiquette of the new experimental philosophy that castigated the inappropriate uses of wonder.[37]

Yet despite the efforts of the Malpighis, the emotional reaction to wonder could not fully be contained. At the end of the eighteenth century, when both the Aldrovandi and Cospi collections had been long subsumed within the museum of the Institute for Sciences (*Instituto delle Scienze*), virtuosi, those men of leisure who thrived on curiosities, still visited Bologna to see the marvels of nature that Aldrovandi had brought together. "That singular museum, left by him in his will to the senate of his homeland Bologna, is even now an ornament of that illustrious city and the principal sight for learned travelers."[38] The remnants of his enormous collection can be seen today in a special room of the University Library in Bologna that houses the books, manuscripts, illustrations, and woodcuts of this remarkable collector and the few remaining artifacts (in 1988 the Museo Civico Medievale opened up a permanent exhibition of the extant contents of

[36]Howard Adelmann, ed., *The Correspondence of Marcello Malpighi*, 5 vols. (Ithaca, 1975), Vol. II, p. 789; Vol. IV, p. 1655. Bonfiglioli was the custodian during this period.

[37]On the role of the "civilizing process" in early modern science, I refer readers to Mario Biagioli, "Scientific Revolution, Social Bricolage and Etiquette," in *The Scientific Revolution in National Context*, Roy Porter and Mikulas Teich, eds. (Cambridge, U.K., 1992), pp. 11–54; Daston, "Baconian Facts, Academic Civility and the Prehistory of Objectivity," *Annals of Scholarship* 8 (1991): 337–363; and Steven Shapin, " 'A Scholar and a Gentleman': The Problematic Identity of the Scientific Practitioner in Early Modern England," *History of Science* 29 (1991): 279–327. The point of departure in all instances in Norbert Elias, *The Civilizing Process, Vol. I: The History of Manners*, and *Vol. II: Power and Civility*, trans. Edmund Jephcott (New York: 1982).

[38]*Anecdota litteraria*, Vol. IV, p. 369. For further information about the reconstruction of Aldrovandi's museum in the University Library in the early twentieth century, see F. Rodriguez, "Il museo aldrovandiano della Biblioteca Universitaria di Bologna," *Memorie della accademia delle scienze dell'Istituto di Bologna. Classe di scienze fisiche* 8 (1958): 5–55; Carlo Gentili, "I musei Aldrovandi e Cospi e la loro sistemazione nell'Istituto," in *I materiali dell'Istituto delle Scienze* (Bologna, 1979), pp. 90–99.

Cospi's museum). Very few objects have survived in their three-dimensional form, save for the more durable ones, such as fossils, and random others, such as an armadillo. The majority "exist" in the form of illustrations and written descriptions, the fragments of a collection that reflected Aldrovandi's desire to preserve all parts of nature for his personal glory and philosophical fulfillment.

Aldrovandi was aware that the enormity of his task would stretch beyond his lifetime; just as Aristotle had passed his work on to his disciples, Aldrovandi hoped that his own pupils would continue the unfinished business of writing the definitive history of nature. At the end of his life, his literary production—mostly based on the objects of his museum—totaled more than 400 volumes, a number that he claimed would take more than a century to print. "And with all this I have kept three scribes in my house, excellent painters, designers, engravers, and have spent much on transportation [of artifacts] and on a library that can stand up to any other other particular library in Italy."[39] While it is hardly surprising that we should find Aldrovandi's attempts to bring all knowledge under one roof overwhelming, given our own assumption that there is always more to know, the number and size of his many unfinished projects amazed even contemporaries. On a trip to Italy in 1687, Maximilian Misson commented:

> In the [Palazzo Pubblico] we saw the Cabinet of Curiosities of *Aldro[v]andus*. That of the Marquis of *Cospi* is united to it, and the whole belongs to the city. . . . But there is nothing in both these Cabinets so rare and surprising as what I am going to relate to you. In a Chamber at the side of this we saw a 187 Volumes in *folio*, all written by *Aldro[v]andus'* own Hand, with more than 200 Bags full of loose Papers.[40]

Separate from the thirteen printed volumes, which covered a range of topics—birds, insects, fish, quadrupeds, serpents, monsters, and hard objects, that is, stones, metals, gems, and fossils—the manuscripts included Aldrovandi's voluminous correspondence, drafts of treatises, tables that classified and tabulated the natural world, and his own encyclopedic compendia that organized all his reading notes alphabetically, topically, and geographically. The garrulous Richard Lassels, visiting the museum several times in the mid-seventeenth century, observed, "Seeing these *Manuscripts* I asked whether the man had lived three hundred years, or no . . . but it was answered me, that he lived only fourscore and three: a short age for such a long work: but it sheweth us how farre a man may travel in sciences in his life time, if he rise but betimes, and spurr on all his life time with obstinate labour."[41]

---

[39] BUB, *Aldrovandi*, ms. 70, cc. 16v–17v.

[40] Maximilian Misson, *A New Voyage to Italy* (London, English trans., 1695), Vol. II, pp. 186–187.

[41] Richard Lassels, *The Voyage of Italy, or a Compleat Journey through Italy. In Two Parts* (London, 1670), Vol. I, p. 148.

We have come a long way from the papal portent with which I began this chapter. There were thousands of objects in Aldrovandi's collection, each with its own story. Stuffed into drawers and bottles, hung on the walls and the ceiling, they beckoned viewers to single them out, one by one, in contemplating how so many things had arrived all in one place. Yet it was the collective enormity of Aldrovandi's project that most struck his contemporaries. Themselves engaged in various encyclopedic enterprises, they nonetheless acknowledged that Aldrovandi, rivaled only by the Swiss naturalist Conrad Gesner, whose similarly ambitious undertaking was cut short by his untimely death in 1565, had used his museum to write a complete history of nature in accordance with ancient philosophical principles.

AFTER ALDROVANDI

Aldrovandi's museum was not the only noteworthy collection in sixteenth- and seventeenth-century Italy. It took its place alongside several other early science museums whose memories, though not their artifacts, have stood the test of time.[42] Such contemporaries as the apothecaries Francesco Calzolari in Verona and Ferrante Imperato in Naples and the papal physician Michele Mercati in Rome created museums of nature considered the wonders of the sixteenth century. "There is not a Signor or great person who comes to Naples from far and remote parts who does not wish to see [the museum] out of curiosity and who is not overcome with amazement, after having seen it," boasted Francesco Imperato of his father's collection.[43] A century later, a chance encounter with the manuscript edition of Mercati's catalogue to the Vatican mineralogical collection led Paolo Boccone to recommend, "this manuscript . . . merits being seen by every traveling virtuoso and lover of physic for the natural history and erudition that it contains."[44] Calzolari, Imperato, and Mercati, all acquainted with Aldrovandi, perhaps did not equal the Bolognese collector's unbounded ambition. Yet they created museums of natural objects that testified to the diffusion of collecting as a common practice for naturalists by the late sixteenth century.

In Naples, the collecting activities of Imperato and his son Francesco brought them into close contact with the members of the Accademia dei Lincei (Academy of the Lynx-Eyed, 1603–1630), founded by the Roman noble Federico Cesi in 1603 and counting the mathematician Galileo among

[42]Random objects from the Calzolari and Moscardo collections can be found in the Museum of Natural History in Verona. Likewise, many of the artifacts from the Settala collection are dispersed between the Museum of Natural History and the Biblioteca Ambrosiana in Milan. There has been some discussion about putting Kircher's collection back together again, as was recently done with the Cospi collection, but to my knowledge, nothing has yet come to pass.

[43]Francesco Imperato, *Discorsi intorno a diverse cose naturali* (Naples, 1628), sig. 3.

[44]Paolo Boccone, *Osservazioni naturali* (Venice, 1697), p. 295.

its members. Imperato, like many members of the Accademia dei Lincei, was in the process of rewriting the history of nature based on observation. Less reliant on ancient authority than Aldrovandi—as an apothecary he lacked the humanistic skills necessary to mine the classical sources—he used the materials in his collection to describe anew the contours of the natural world. Similarly engaged in the creation of a new encyclopedia of knowledge, many members of the Accademia dei Lincei visited Imperato's museum when they came to Naples, while most of the Neapolitan members depended upon Imperato's repository to pursue their own investigations. In turn, the apothecary expected the Linceans to persuade their most famous member, Galileo, to enter his museum. "I have heard that Signor Galileo may come to Naples, and I hope to discuss many things with him, in particular the stone that absorbs and retains light," wrote the apothecary to the naturalist Johann Faber in 1611.[45] If he could not receive a telescope, as Cesi and Barberini did, Imperato hoped at least to enjoy the pleasure of a philosophical discussion about luminescence.

Taken as a whole, the activities of the Linceans more closely resembled an extension of the encyclopedic projects of Aldrovandi than the reasoned mathematical calculations of the experimental philosophy promulgated by men such as Galileo. Like Aldrovandi, they also recognized the importance of museums to the study of nature, although they desired to collect nature to dismantle rather than enhance Aristotelian natural history. Cesi, Faber, Della Porta, and their patron Francesco Barberini all had museums, while Cassiano dal Pozzo was probably the foremost art collector in Baroque Rome.[46] Returning to Rome from a trip to Madrid in 1608, Jan Eck informed Francesco Stelluti, "With me there are about one hundred plants . . . . I am collecting a museum of natural things."[47] Collecting was not exclusively the domain of naturalists such as Aldrovandi, Calzolari, and Imperato, who belonged to traditional institutions—the university and the professional guilds of doctors and apothecaries. It also animated the intellectual life of the new scientific academies that sprang up throughout Europe in the seventeenth century. In a similar spirit as Bacon, who dismissed Aristotle by attempting to replace him, Cesi used his collection of natural objects to redraw the map of knowledge. Yet the novelty of the Lincean's activities lay less in their actions than in their rhetoric; loudly proclaiming to do something new, they utilized essentially the same techniques as Aldrovandi to achieve their results.[48]

[45]CL, Vol. II, p. 163 (Naples, 10 June 1611).

[46]For a recent overview of Cassiano dal Pozzo's activities, see Francesco Solinas, ed., *Cassiano dal Pozzo. Atti del Seminario di Studi* (Rome, 1989).

[47]CL, Vol. I, p. 110 (Madrid, 2 June 1608).

[48]The most recent study of the Accademia dei Lincei is Richard Lombardo, *"With the Eyes of a Lynx": Honor and Prestige in the Accademia dei Lincei* (M.A. thesis, University of Florida, Gainesville, 1990).

While the Linceans, aided by Imperato, shaped the face of natural history in the early seventeenth century, the disintegration of the academy upon Cesi's death in 1630 curtailed the expansion of their efforts in the subsequent decades. In the wake of Galileo's condemnation for his adherence to Copernican astronomy in 1633, Italian natural philosophers were more cautious about publicizing intellectual programs that proclaimed philosophical novelty to be their central feature. The age of the Linceans had ended, and the era of the Jesuits and the Accademia del Cimento (Academy of the Experiment, 1657–1667) had begun. Even as Francesco Barberini was drying the ink on his signature to the condemnation of his fellow academician Galileo, he was in the process of negotiating to bring Athanasius Kircher to Rome to translate certain Coptic manuscripts and inspect the Egyptian hieroglyphs that so fascinated him. Philosophers such as Aldrovandi, Galileo, and Kircher might choose to remain committed to a particular form of inquiry; for them it was an intellectual investment. Virtuosi such as Barberini perceived knowledge to be a more mutable, socially defined category that transformed itself according to the dictates of the learned and courtly world. Galileo had been forced to sever his association with the papal court, where Barberini, as the nephew of Urban VIII, remained.[49] The cardinal needed a new philosopher to amuse him at court. The Jesuit Kircher—humanist, philosopher, *and* good Catholic—proved to be an eminently neutral choice. Kircher would become the emblem of the Baroque scientific culture that contemporaries such as Malpighi found so reprehensible and members of the Royal Society such as Robert Boyle and the secretary Henry Oldenburg viewed with great suspicion.

In seventeenth-century Italy, virtuosi vied for the attention of Athanasius Kircher in Rome and Manfredo Settala in Milan. The Jesuit Kircher, professor of mathematics at the Roman College until his death in 1680, was "above all conspicious in the interpretation of the most difficult languages, characters, and signs."[50] Unlike Aldrovandi, intent on exercising his curiosity eventually to demystify nature, Kircher perceived nature to be a glorious, divinely inspired hieroglyph, whose signs he displayed in the Roman College museum. His studies of ancient and universal languages, archaeology, astronomy, magnetism, and Chinese and Egyptian culture were greatly advanced by his collection of scientific and ethnographic rarities, which remained in the Roman College as a showpiece and pedagogical tool well into the nineteenth century.[51] After Kircher's death, the museum disintegrated until the Jesuit Filippo Bonanni, the "Wise Aristotelian," became its curator

[49]For more on Galileo's predicament, see Mario Biagioli, *Galileo Courtier* (Chicago, 1993).

[50]PUG, *Kircher*, ms. 565 (XI), f. 84 (Rimini, 13 November 1672).

[51]The main studies of Kircher are P. Conor Reilly, S.J., *Athanasius Kircher S. J. Master of a Hundred Arts 1602–1680* (Wiesbaden, 1974); Valerio Rivosecchi, *Esotismo in Roma barocca: studi sul Padre Kircher* (Rome, 1982); Mariastella Casciato, Maria Grazia Ianniello, and Maria Vitale, *Enciclopedismo in Roma barocca: Athanasius Kircher e il Museo del Collegio Romano tra Wunder-*

in 1698.[52] "Father Kircher's Cabinet in the Roman College was formerly one of the most curious in Europe, but it has been very much mangl'd and dismember'd," observed Misson in 1687, "yet there remains still a considerable collection of natural Rarities, with several mechanical Engines."[53] Because he lacked the sort of civic ties that preserved and enhanced the Aldrovandi museum, Kircher's collection was in danger of disappearing altogether. Bonanni subsequently restored it, adding his specialized collection of shells and his numerous well-crafted microscopes to the original museum. Perhaps the last natural philosopher to consider himself a complete, epistemologically whole Aristotelian, Bonanni understood the preservation of the Kircher museum to be the last defense against the experimental philosophy that threatened to dissolve the scope and methods of natural history as it had been practiced since Aristotle and Pliny.

It was Bonanni who described the Settala Gallery in Milan in 1681 as "the most famous among all of these [museums] in Italy for the variety of works, both of Nature and of Art, that are conserved there."[54] Son of the well-known physician Lodovico Settala, the cleric Manfredo inherited his father's passion for scientific curiosities. A self-described pupil of the Jesuit rhetorician Emanuele Tesauro, Settala understood his collection to be a supremely witty text.[55] Every object was replete with hidden meanings, designed more to surprise than to inform. Settala created a museum filled with natural and ethnographic curios, that arrived from the far reaches of the formidable Catholic missionary networks, as well as mechanical and optical inventions of his own device. He had the good fortune to find a willing patron in Federico Borromeo, Archbishop of Milan in the early decades of the seventeenth century and a great art collector himself. After Manfredo's death in 1680, the museum stayed in the Settala family for several generations, ultimately passing in 1751 to the Biblioteca Ambrosiana, which Borromeo had initiated in 1603.[56]

---

kammere e museo scientifico (Venice, 1986); and Dino Pastine, *La nascita dell'idolatria: l'Oriente religioso di Athanasius Kircher* (Florence, 1978).

[52]"Elogio di P. Philippo Buonanni," *Giornale de' letterati d'Italia* 37 (1725): 370, 375.

[53]Misson, *A New Voyage to Italy*, Vol. II, p. 139.

[54]Filippo Bonanni, *Ricreatione dell'occhio e della mente nella osservatione delle chiocciole* (Rome, 1681), p. 129.

[55]Laur., *Redi*, 211, f. 410 (Settala to Redi, Milan, 12 March 1675). In this wonderful letter, Settala wrote to inform him of the deaths of Robert Boyle and Emanuele Tesauro, "two hardly ordinary virtuosi. Poor Tesauro, who was my Teacher (*Maestro*), died of an Apoplexy at the age of eighty plus."

[56]Silvia Rota Ghibaudi, *Ricerche su Lodovico Settala* (Florence, 1959), p. 47; Angelo Paredi, *Storia dell'Ambrosiana* (Milan, 1981), pp. 7, 12. On Settala's museum in general and its reconstruction in 1984 at the Museo Civico di Storia Naturale in Milan, see Vincenzo de Michele, Luigi Cagnolaro, Antonio Aimi, and Laura Laurencich, *Il museo di Manfredo Settala nella Milano del XVII secolo* (Milan, 1983); Antonio Aimi, Vincenzo de Michele, and Alessandro Morandotti, *Musaeum Septalianum: una collezione scientifica nella Milano del Seicento* (Milan, 1984). For a more scholarly treatment of Settala's collecting activities, see Carla Tavernari, "Manfredo Settala,

Despite the good fortune of the objects, which were supervised with greater care than those in Kircher's museum, Settala himself was frustrated in his attempts to use his museum as the basis for a local scientific culture, just as Aldrovandi had done in Bologna, Redi in Florence, and Kircher in Rome. Successful in drawing foreigners to his museum, he had few local scholars to depend upon for intellectual exchange. Writing to Henry Oldenburg in 1667, he bemoaned the dearth of scientific activity in Milan. "We lack patrons of scientific investigators and the powerful strive to amass treasure rather than to illuminate nature."[57] While Milan could boast a few academies and colleges, it lacked a prominent university, a drawing point for scholars to Bologna, a political center such as the Medici court in Florence, or the sort of overlapping institutions that one found in Rome, the heart of Catholic Christendom. Visitors and correspondents ensured Settala's full participation in the republic of letters. But they could not compensate for the absence of political and cultural institutions that defined the museum as a civic entity.

Surrounding collectors such as Kircher and Settala were cardinals, princes, and dilettantes—most notably Francesco Barberini, Flavio Chigi, Cassiano dal Pozzo, and Lodovico Moscardo[58]—whose interest in the collection of art and nature fueled the more scholarly and spectacular activities of these learned Catholics. Following their lead, Baroque collectors exhibited a stronger interest in exotica and scientific instruments, populating their museums with ethnographic curios, telescopes, microscopes, and a thousand other playful devices alongside the wonders of nature. For them, it was the dialectic between nature and art, made manifest in the confusing juxtaposition of objects and instruments, rather than the vast compendium of the natural world, that they wished to represent. Renaissance naturalists had established the encyclopedia of nature as the ultimate goal of their collecting. Enhancing this pursuit, Baroque naturalists pondered the form the encyclopedia should take and constantly tested its boundaries. Equally engaged in revitalizing the pursuits of a traditional culture, they nonetheless differed in the methods they employed.[59] Yet the lineage of naturalists stretching

collezionista e scienziato milanese del '600," *Annali dell'Istituto e Museo di Storia della Scienza di Firenze* 1 (1976): 43–61; idem, "Il Museo Settala. Presupposti e storia," *Museologia scientifica* 7 (1980): 12–46.

[57] *The Correspondence of Henry Oldenburg*, A. Rupert Hall and Marie Boas Hall, eds. and trans., 9 vols. (Madison, WI, 1965–1973), Vol. III, p. 456 (Milan, 1 August 1667).

[58] See, for example, G. Inchisa della Rocchetta, "Il museo di curiosità del Card. Flavio I Chigi," *Archivio della società romana di storia patria*, ser. 3, 20 (1966): 141–292.

[59] For an illuminating discussion of Baroque science, with particular attention to the Jesuits, see Clelia Pighetti, "Francesco Lana Terzi e la scienza barocca," *Commentari dell'Ateneo di Brescia per il 1985* (Brescia, 1986): 97–117. I have taken much of the material in this paragraph from her essay. See also the articles in Timothy Hampton, ed., "Baroque Topographies: Literature/History/Philosophy," *Yale French Studies* 80 (New Haven, 1991); and Pamela Smith, *The Business of Alchemy: Science and Culture in Baroque Europe* (Princeton, in press).

from Aldrovandi to Bonanni was unwavering in the belief that erudition combined with experience was the most credible form of knowledge.

When Francesco Redi proclaimed his rejection of ancient authority in a
series of treatises on insects and vipers, emanating from the naturalistic collections at the Medici court in the late seventeenth century, he resembled
Cesi whose novelty lay more in his rhetoric than in his practices. Despite
Redi's violent disagreements with Kircher and Bonanni about most philosophical issues, he nonetheless was guided by the words of the ancients at
every step of the way. More importantly, the experiments he cultivated in
Florence operated according to the same rules of display informing
Kircher's own demonstrations at the Roman College. However much Redi
might attempt to regularize nature, she continued to resist him, capricious
to the end. Reluctantly perhaps, but decisively, Redi aligned himself with
the other Baroque collectors of nature, leaving it to later naturalists to un-
couple the act of collecting from the pursuit of wonder.

MUSEUMS AND THEIR CATALOGUES

There are many other museums and collectors that could be singled out,
but I will refrain from writing an individual history of each one.[60] The preceding pages have sketched out the appearance of museums of natural history by looking at the intentions of various naturalists. This section turns instead to the catalogues they left behind. We cannot help but think of
catalogues when we think of museums. They are the most important object
produced from a collection. While inventories were in evidence in the Middle Ages, the catalogue was an early modern invention. Inventories record
the contents of a museum. They quantify its reality, listing the objects without attaching analytical meaning to them. Catalogues purport to interpret.
Their appearance in the late sixteenth century further suggests how novel
the practices of Renaissance collectors were.

More than an unadorned list, the catalogue provided a self-conscious presentation of a collection. Catalogues were repositories of multiple intersecting stories that textualized and contextualized each object.[61] Descriptions
generally served two basic functions. First, they recounted the circumstances
by which an object entered a museum, often heroic tales of great deeds—the
capture of the 1572 dragon or Kircher's descent into the fiery craters of Vesuvius and Etna in 1637–1638—distant conquests, and signal visits of impor-

---

[60]Such numbers have only heuristic value, but Mario Cermenati estimated that there were
more than 250 museums of natural history in early modern Italy; Cermenati, "Francesco Calzolari e le sue lettere all'Aldrovandi," *Annali di botanica* 7 (1908): 83.

[61]The work of Jay Tribby on the museum catalogue as a form of "civil conversation" is es-
sential to this point; see his "Body/Building: Living the Museum Life in Early Modern Europe,"
Rhetorica 10 (1992): 139–163. For an interesting theoretical overview of this subject, see Werner
Hüllen, "Reality, the Museum and the Catalogue," *Semiotica* 80 (1990): 265–275.

tant patrons. Second, they situated an object historically, philologically, and comparatively. Collectors always wished to know the etymology of a name and the circumstances of its production; in this fashion, an artifact was located within a literary as well as scientific canon, defined as much by Ovid and Horace as by Aristotle and Pliny. The addition of a new artifact predictably occasioned speculation on its ability to maintain or dismantle long-standing interpretations of its scientific and medicinal properties. Finally collectors could not resist comparing an object to others of its kind, preferably in museums of equivalent or greater stature. Putting their latest acquisition to the test, they asked, "Is it bigger, better, stronger, nobler, or—best of all—incomparable?" This is a sample of the different methods by which collectors interrogated each object that came into their possession.

The development of the catalogue, from simple list to scientific corpus to literary production, occurred gradually. The first published description of a museum of natural history was that of Johann Kentmann's *Ark of Fossil Objects*, appended to Conrad Gesner's *On Fossil Objects* (1565).[62] Many sixteenth-century museums— for example, Aldrovandi's and Antonio Giganti's[63]—were bereft of any published description. By the late sixteenth century, collectors increasingly chose to publicize the contents of their museums. The medium of print allowed them to reach an audience beyond the individuals who personally toured their museums. A published catalogue conveyed a new level of status for the collector. Written by the naturalist himself, it displayed his erudition. Written by another scholar, it conveyed the status of a collector who had earned the right to commission a description of his work. Both Francesco Calzolari and Ferrante Imperato, perhaps satisfied with more modest contributions to the realm of learning than the tireless Aldrovandi or indefatigable Della Porta, produced catalogues describing the contents of their museums in lieu of writing a multivolume *Natural History*. For these two apothecaries, the publication of a catalogue became their main intellectual endeavor.

In the case of Calzolari's and Imperato's collections, we can trace the evolution of the catalogue as a literary form since two catalogues were published for each museum. The sixteenth-century versions differ markedly from their seventeenth-century counterparts. The 1584 catalogue of Calzolari's museum by the physician Giovan Battista Olivi was a spare, unillustrated text dedicated to their mutual friend, the physician Girolamo Mercuriale, then teaching at Padua.[64] Olivi's primary goal was to advertise the breadth of the pharmacopeia that Calzolari had amassed; hence the presence of a

---

[62]Martin Rudwick, *The Meaning of Fossils: Episodes in the History of Paleontology* (Chicago, 1985, 1976), p. 12.

[63]Both, however, were inventoried; the Giganti inventory is reproduced in Gigliola Fragnito, *In museo e in villa: Saggi sul Rinascimento perduto* (Venice: Arsenale, 1988), pp. 175–201.

[64]Giovan Battista Olivi, *De reconditis et praecipuis collectaneis ab honestissimo, et solertissimo Francisco Calceolari Veronensi in Musaeo adservatis* (Verona, 1584).

remarkable object was an occasion for a considered analysis of its medicinal properties rather than an effusion of poetry and wit. Each artifact reaffirmed Calzolari's virtuosity as an apothecary; rare spices, unicorn's horns, and other exotic bits of nature all compounded to produce better medicines. The catalogue was an advertisement for his trade. In contrast, Ferrante Imperato's 1599 "catalogue" was no précis nor did it discuss his skill in pharmacy (although the frequent allusions to distillation techniques reminded readers of his profession). Instead, Imperato chose to write an abbreviated natural history, culled from his lengthy meditation upon the materials of nature, all displayed in his museum.[65] In many respects, it represented his contribution to the Lincean revision of natural history. While the Olivi catalogue was written in Latin, Imperato's natural history appeared in Italian and was well illustrated, including the first representation of a museum (fig. 2). Republished in 1672, it was a great success.

While Imperato divided his materials into various elemental properties—salts, fats, metals, earths, waters, airs, and fiery substances—and Olivi remained intoxicated by the smells of Calzolari's precious medicaments and admired his enormous collection of distillation equipment, their successors used a different strategy to assign worth to objects. In 1622, the second catalogue of the Calzolari museum appeared. Commissioned by his grandson, Francesco Junior, and written by two local physicians, Benedetto Ceruti and Andrea Chiocco (possibly a relative of the Nicolò Chiocco whose sonnet graced the preface of the Olivi catalogue?), the catalogue testified to the elevated status of the museum and its new owner. Olivi had barely filled fifty pages, yet his successors summoned up more than ten-fold that number to do justice to essentially the same artifacts. Gracefully embellished with illustrations, the new catalogue displayed the choice philosophical wit of Ceruti and Chiocco; as the title indicated, the parameters of the museum had been greatly expanded to include "many samples belonging to natural and moral philosophy, and not a few others pertinent to the art of medicine ... exhibited and described with erudition, not without a great abundance of exotic specimens." Coral, for example, was cited not simply for its medicinal uses but for the delightful complexity of its etymology: was it the origin of Ovid's tale about the Gorgon or rather the result of this well-known fable?[66] Did nature imitate poetry or poetry nature? While the 1584 catalogue was best described as a medical treatise, the 1622 version was a philosophical excursus occasioned by the objects in the museum. The credentials of the authors had not changed; all three were physicians. Yet their understanding of the purpose of writing a catalogue differed greatly.

[65]Ferrante Imperato, *Dell'historia naturale* (Naples, 1599).

[66]Benedetto Ceruti and Andrea Chiocco, *Musaeum Francesci Calceolari Iunioris Veronensis* (Verona, 1622), p. 5. The section that I have extracted from the title reads, "in quo multa ad naturalem moralemque Philosophiam spectantia, non pauca ad rem medicam pertinentia erudite proponuntur, et explicantur, non sine magna rerum exoticarum supellectile ...."

Figure 2. The museum of Ferrante Imperato. From Ferrante Imperato, *Dell'historia naturale* (Venice, 1672 ed.).

With an eye for a wider audience, composed not only of physicians but also of virtuosi, Ceruti and Chiocco expanded the medical discussion of the properties of different naturalia into a more elaborate discourse on their rarity and curiosity. No doubt they and Francesco Junior realized that the Duke of Mantua, to whom the second catalogue was dedicated, would be more appreciative of witty conceits than an éloge to the most perfect theriac, the "antidote of antidotes" that the elder Calzolari made so well. Lacking the single-minded purpose of the 1584 publication, the 1622 volume branched out into a thousand different directions, enfolding the objects in numerous philosophical discourses on the paradoxes of nature and art. With evident delight, Ceruti recounted a debate in the museum between Pancrazio Mazzangha Bargoeus, professor of natural history at Pisa (1617–1628) and the local academicians of Verona, in which Mazzangha attested that birds could generate spontaneously from trees, thereby confirming the alleged origins of certain objects in the collection.[67] The first catalogue had been a product of debates about proper ingredients in medicines, a topic of great interest in the late sixteenth century. The second catalogue displayed all the hallmarks of the humanist erudition cultivated in the academies of late Renaissance and Baroque Italy. While Olivi explored the *possible uses* of nature, Ceruti and Chiocco explored the *imaginative possibilities* of natural phenomena.

Just as Olivi had been a prime candidate to author the first catalogue—he personally could attest to the virtue of Calzolari's theriac, which had saved his son's life—Chiocco was equally suited to complete the second version. Author of a treatise on *imprese*,[68] the emblems that so delighted learned patricians during this period, Chiocco quickly established relationships among phenomena such as a stone bearing the natural mark of a cross and the hieroglyphs that also promised to reveal hidden truths. His ability to see the verisimilitude between natural and artificial symbols enhanced his skill as a cataloguer.[69] The different proclivities of the two physicians, Olivi and Chiocco, mark the transition from the more narrowly conceived audience of the first catalogue, addressed primarily to physicians and apothecaries, to the courtly and patrician audience cultivated in the second instance. When Ceruti and Chiocco proclaimed Calzolari to be an example for the noble princes of Italy to follow, they counted upon the prestige of collecting to obscure the fact that they had invited the socially prominent to imitate the practices of a lowly, albeit famous, apothecary.[70]

While not as copious as the second Calzolari catalogue, Francesco Imperato's 1628 renarration of his father's museum indicated a similar shift

[67]Ibid., p. 26.

[68]Andrea Chiocco, *Discorso delle imprese* (Verona, 1601).

[69]Ceruti and Chiocco, *Musaeum Francesci Calceolari Iunioris Veronensis*, pp. 392–394.

[70]Ibid., sig. *2r: "Quantae dignitationis CALCEOLARIUM nostrum esse arbitramur; a quo Primarii Italiae Principes imitatione pernobilii [*sic*] exemplum sibi forsam [*sic*] habere voluerunt."

from utility to virtuosity. Both Ferrante and Francesco Imperato occupied visible positions in Neapolitan political culture. Ferrante was "captain of the people" (*capitano del popolo*) in Piazza di Nido during the 1584 revolt against the Spanish viceroy and twice *governatore popolare dell'Annunziata*. His son Francesco continued the family tradition of political activity, authoring a political treatise on the restructuring of local government in 1604.[71] Both Ferrante and Francesco publicized the frequent presence of the Spanish viceroy in their museum during his visits to Naples, no doubt to reinforce the elevation of their position in the social world of the city.[72] In this context, the description of Ferrante's passion for natural history is quite revealing. Due to the nobility and utility of natural history, Giovan Bernardino di Giuliano recounted, Ferrante became a naturalist, "putting aside the military and equestrian exercises which his father . . . performed under the banner of the Most Serene Aragonese King and the Imperial Majesty of Charles V, and devoting himself entirely to the study of philosophy, in which he progressed so far that he emulated the glory of his ancestors."[73] The social message is evident: Ferrante's story is a triumph of good breeding. Only an apothecary of a good family could rise so far in the profession of natural history. While Ferrante had said little about his social origins in the 1599 catalogue, his son was anxious to detail the family genealogy. Collecting and the study of nature, like many other practices in early modern Italy, reflected the progressive aristocratization of society.

Di Giuliano's preface to the second catalogue further indicated the role of the museum in the social ascension of the Imperato family. Turning to a description of Francesco's career, he underscored the importance of natural history as a leisure exercise: "Indeed for the greater satisfaction of the curious, in those hours in which he was engaged in the study of law and the management of public affairs of this City . . . he wished to shape the present Discourses." The circumstances of Francesco, now the Marchese di Spineto and "become exceedingly a Politician," highlighted the family's progression from soldiers to apothecaries to men of politics. They and their museum participated in the attempts on the part of the "principal citizens" (*principali cittadini*) to assimilate themselves into the Neapolitan aristocracy.[74] Possession of a museum was one way to climb the social ladder. Curiosity certainly was not a social leveling agent, but it did mitigate

---

[71] Rosario Villari, *La rivolta antispagnola a Napoli. Le origini 1585–1647* (Rome, 1987 ed.; 1967), pp. 51, 58, 107–108, 113–117. See also Enrica Stendardo, "Ferrante Imperato. Il collezionismo naturalistico a Napoli tra '500 e '600, ed. alcuni documenti inediti," *Atti e memorie dell'Accademia Clementina,* nuova serie, 28–29 (1992): 43–79.

[72] Ferrante Imperato, *Dell'historia naturale,* sig. a.2v; Francesco Imperato, *Discorsi,* p. 18. Ferrante dedicated his work to Giovanni di Velasco, Grand Constable of Castile, Governor of Milan, *Capitano Generale* in Italy, and son-in-law of the Viceroy.

[73] Giovan Bernardino di Giuliano, "Al Lettore," in Francesco Imperato, *Discorsi naturali,* n.p.

[74] This trend is discussed in Villari, *La rivolta antispagnola a Napoli,* pp. 113–117.

differences of birth and lineage, providing a point of intersection among socially differentiated groups.

If we turn to the content of Francesco's *Discourses on Diverse Natural Things*, we find further proof of the altered circumstances of the family. While Ferrante Imperato had addressed his catalogue to both professional naturalists and the "curious," his son wrote only for the latter category. The 1599 catalogue, *Natural History*, was the fruition of a lifetime's research into the hidden properties of nature. It was informed by Imperato's intimate knowledge of the structure and chemistry of most medicinal substances and by years of collecting fossils and other particularly interesting bits of nature. Instead, Francesco's *Discourses* showcased only the most visible and puzzling artifacts in the museum and indicated at best a cursory reading of the standard reference works on natural history that must have graced the family library. Pygmies, crocodiles, mandrakes, tarantula, bezoar stones, papyri, "animals and other things converted to stone," and other marvels of nature filled its pages. Little wonder that Federico Cesi, to whom the book was addressed, did not find it sufficient reason to invite the younger Imperato to become a member of the Accademia dei Lincei, as Francesco had hoped.[75] As with the second Calzolari catalogue, Francesco Imperato addressed his patrons more than his colleagues. Like the younger Calzolari, he no longer saw himself as belonging to the close-knit community of naturalists but to a wider patrician culture interested in the curiosities of nature.

The transformation of Calzolari's and Imperato's museums, in catalogue form and to a lesser degree in their choice of objects, demonstrates well the movement of collection from a professional activity to a noble pastime. The structure and presentation of other collections reinforces this trajectory. The catalogues of seventeenth-century collections such as those of the Veronese noble Lodovico Moscardo, Ferdinando Cospi, and Manfredo Settala were all models of wit and persuasion; dwarfs, chameleons, birds of paradise, and "abnormal trees formed by nature" happily coexisted alongside mathematical instruments, "Turkish knives and swords," and portraits and medals of famous men.[76] In the case of the Settala museum, two catalogues—one in Latin and the other in Italian—appeared within two years of each other, the first by the physician Paolo Maria Terzago and the second by the physician Pietro Francesco Scarabelli.[77] As Scarabelli noted, while the "terse Latinity of the erudite pen of signor Doctor Terzago" met the needs

[75]This incident is discussed further in Paula Findlen, "The Economy of Scientific Exchange in Early Modern Italy," in Bruce Moran, ed., *Patronage and Institutions: Science, Technology and Medicine at the European Courts, 1500–1750* (Woodbridge, U.K.: 1991), pp. 12–15.

[76]I have taken this list from the index of the *Note overo memorie del museo del Conte Lodovico Moscardo Nobile Veronese* (Verona, 1672). This index, along with many others, is reproduced in Barbara Balsinger, *The 'Kunst- und Wunderkammern': A Catalogue Raisonné of Collecting in Germany, France and England*, 2 vols. (Ph.D. diss., University of Pittsburgh, 1970), Vol. I, pp. 331–342.

[77]Paolo Maria Terzago, *Musaeum Septalianum Manfredi Septalae Patritii Mediolanensis industrioso Labore constructum* (Tortona, 1664); idem, *Museo o galleria adunata dal sapere, e dallo*

of a scholarly audience, "the multiplication of instances of Knights and curious Women" desirous of seeing the museum had obliged him to create a vernacular catalogue to meet the needs of a different audience.[78] Lorenzo Legati and Lodovico Moscardo, describing the museums of aristocrats, proceeded immediately to the vernacular in acknowledgment of the courtly and patrician audience that they sought.[79] At the turn of the century, only the Jesuits, defenders of the Latinate republic of letters to the end, persisted in publishing their catalogues in Latin.

While Olivi had devoted the majority of his catalogue to the efficacy of Calzolari's antidotes, particularly his famous theriac, the description of this prized medicine appeared on less than a page of Ceruti and Chiocco's voluminous catalogue of 1622, and graced Moscardo's 1656 and 1672 catalogues not at all.[80] The marginalization of medical information, in favor of antique urns, Egyptian idols, and nature's paradoxes tells us something about the changing status of the collection of nature between the sixteenth and seventeenth centuries. While collecting initially had been an activity for naturalists such as Aldrovandi, Calzolari, and Imperato interested in the *uses of nature,* within a century it had become a leisurely pastime for aspiring nobles and courtiers. Rather than singing the praises of curiosity as a utilitarian practice, seventeenth-century collectors made curiosity a virtue unto itself.[81] Both groups exhibited a fascination with paradoxes and difficult phenomena.[82] However, while late Renaissance naturalists delved into the secrets of nature to produce better medicines and complete the Aristotelian project of classifying the natural world, their Baroque counterparts, avid readers of such treatises as Emanuele Tesauro's *Aristotelian Telescope,* understood the appreciation and creation of subtlety as their ultimate goal. Even Filippo Bonanni's decision to include an entire section on microscopes in his *Kircherian Museum* of 1709 represented an extension of the category of wonder to the microscopic world.[83] As Stephen Greenblatt observes, "wonder" was a category that described both the inherent properties of a material object and the response to

studio del signor Canonico Manfredo Settala nobile milanese, Pietro Scarabelli, trans. (Tortona,1666). Although Scarabelli is listed as the translator, it is apparent that he did more than simply translate Terzago's text, so I hereafter refer to him as the author.

[78]Scarabelli, *Museo o galleria adunata dal sapere,* sig. +3r.

[79]Legati, *Museo Cospiano;* Moscardo, *Note overo memorie del museo.*

[80]Moscardo's collection contained a number of the objects from the Calzolari museum, which had been dispersed by the mid-seventeenth century.

[81]Here I am following an argument made by Houghton, "The English Virtuoso in the Seventeenth Century," *Journal of the History of Ideas* 3 (1942), pp. 192–211.

[82]For more on this subject, see Findlen, "Jokes of Nature and Jokes of Knowledge: The Playfulness of Scientific Discourse in Early Modern Europe," *Renaissance Quarterly* 43 (1990): 292–331.

[83]I owe this point to William Ashworth's "Remarkable Humans and Singular Beasts," in Kenseth, *The Age of the Marvelous,* p. 140.

it.[84] The museum catalogue, as it developed during the first century and a half of its appearance, embodied both definitions, codifying the culture of curiosity that defined the experience of collecting.

BAROQUE MIRRORS

Almost a century after the sighting of the dragon near Bologna, another papal story intrudes. It is the tale, however, of an image rather than an object, an appropriately emblematic shift since the mid-seventeenth century might be characterized as an age obsessively preoccupied with the properties of representation. While not an image produced in Italy, the Flemish artist Jan van Kessel's *Europa* (1664) is nonetheless one of the most comprehensive images of Italy as the site of collection (fig. 3). One of four panels depicting the four continents, Europe is represented specifically by an image of Rome. In the center stands the collector (in contrast to the images of the other three continents, which are peopled only to fulfill their ethnographic intent). Displaying a frame of butterflies, he is surrounded by the objects he has assembled: objects of leisure such as the backgammon board, playing cards, and tennis racket in the foreground; objects of curiosity such as the mandrake and monsters revealed behind a partially drawn curtain in the left corner, the insects arranged close to the ceiling, and the shells spilling out onto the floor; and objects of power, signified by the quantity of armor, drums, the flag, and shields, removed from the context of battle, and the papal tiara and scepter, lain discreetly on the table at which Roma sits, holding a cornucopia upright. Only two books are prominently displayed: Pliny's *Natural History* and the Bible.

In the background, the shadowy images of classicized female figures, set within their niches, alert us that the collector and his entourage inhabit the temple of the Muses. More precisely, it is a museum somewhere in Rome: looking beyond the table across which the papal artifacts are strewn, Castel Sant'Angelo appears in the distance. This is not a disembodied Rome, outside of time and place; it is the Rome of Alexander VII (1655–1667), the pope renowned after Urban VIII for his patronage of the arts and sciences.[85] Both the papal bull dangling over the edge of the table and the portrait behind the collector situate this image of Europe the collector as a representation of the Chigi papacy.

In this image, Van Kessel has collected the collector; by surrounding him with artifacts, the ability to display artifacts has also been put "on display." The assemblage of objects delimits not only *what* the collector has accumu-

---

[84]Stephen Greenblatt, *Marvelous Possessions: The Wonder of the New World* (Chicago, 1991), p. 22.

[85]For a brief discussion of Fabio Chigi, see Richard Krautheimer, *The Rome of Alexander VII 1655–1667* (Princeton, 1985), pp. 8–14.

Figure 3. Jan van Kessel, *Europa* (1664). From Bayerische Staatsgemäldesammlungen, Alte Pinakotek, Munich.

lated but *how* and *where* the collection has been produced; thus Van Kessel's portrait of "Europe" exhibits the same properties that we have already attributed to museum catalogues. The open letterbox and scattered coins in the foreground allude to the social and financial resources necessary to initiate a museum—friends and wealth. The repeated references to the Chigi papacy suggest the intricate patronage networks and the principle of accessibility that lay behind any successful museum. Rather than representing the museum as a Petrarchan temple of solitude, Van Kessel's *Europa* captures the humanist culture of late Renaissance and Baroque Italy, a visual and verbal cacophony that greeted the collector when he entered the *studio.*

Alexander VII was at the height of his powers when the artist Van Kessel created his quadritych. That Van Kessel chose to focus upon Rome, rather than Amsterdam, a city more proximate to his position as a court painter in Antwerp, attests to the power of its image as *theatrum mundi* in the seventeenth century.[86] Only Rome could symbolize all of Europe in a glance, notwithstanding the confessional divisions of the age. At the heart of this "theater of the world" lay the Roman College museum, orchestrated by the versatile "master of one hundred arts," Athanasius Kircher. We might even speculate that, despite the secular dress of Van Kessel's collector, it was Kircher whom the artist had in mind when he created this image of Baroque Rome. Certainly Kircher and his museum lay at the heart of what made seventeenth-century Rome both *urbs* and *orbis*, a city and a world in which power and privilege, wisdom and curiosity, were mutually exercised. Patronized by Protestants and Catholics alike, the Jesuit and his collection at times seemed to transcend confessional boundaries.

Turning from painterly portrayals to textual exegesis, the 1678 catalogue of Kircher's museum by his assistant Giorgio de Sepi offers yet another connection between Alexander VII's Rome and its image as the site of wonder and collection. While Van Kessel discreetly placed the pope's portrait on the table at which Europa reclines, Kircher offered a more playful homage. Nestled within one of his famous catoptric devices, a machine of multiple mirrors, was a portrait of Alexander VII.[87] The mirrors combined to produce not one but an infinite number of images of the pope, the centerpiece of the Jesuit's *Catoptrical Theater.* Here numerous devices produced startling visual effects. Like the mirror perched high above the objects in the Cospi museum in Bologna, which united the spectator with the collection in a highly protean maneuver,[88] Kircher's theater of illusions blurred the boundaries between inside and outside, between the museum and the world

---

[86]On the strong ties of Flemish painters such as Jan van Kessel and Jan Breughel to Italy, see Stefania Bedoni, *Jan Breughel in Italia e il collezionismo del Seicento* (Florence and Milan, 1983); D. Bodard, *Rubens e la pittura fiamminga del Seicento nelle collezioni pubbliche fiorentine* (Florence, 1977); and *I fiamminghi e l'Italia. Pittori italiani e fiamminghi dal XV al XVIII secolo* (Venice, 1951).

[87]De Sepi, pp. 3, 37.

[88]Legati, *Museo Cospiano*, p. 213.

that created it. At the same time, it reinforced the distance between collector and collected. The museum, Kircher might tell us, extended as far as the papal imperium. One pope and many lands was best depicted as a hall of mirrors whose "echoes" were produced through optical mimesis. While the impetus behind many sixteenth-century collections can be summed up in Aldrovandi's textual manipulation of an oddly formed serpent, a century later the *literal manipulation of form*, through technological intervention, characterized the Baroque museum.[89] "The Baroque . . . ," as Gilles Deleuze. observes, "endlessly creates folds."[90] The catoptric machine, which literally enfolded the image of the pope producing an infinite series of refractions, was a quintessentially Baroque artifact.

In the wake of Kircher, collectors such as Chigi and Settala amassed numerous machines, automata, and optical devices for the delight and instruction of their visitors. Inventories of the Chigi museum in Rome reveal, among other precious objects, a magic lantern, placed strategically next to the door leading into the family apartments, and centerstage, a puppet holding two pendula, leaning again the tomb of an Egyptian mummy "that demonstrates the facility of motion."[91] Settala, proud of his skills as an instrument maker, had so many contraptions scattered throughout his collection that Pietro Scarabelli described it as exhibiting an "extravagant diversity" and filled with "a thousand playful deceits."[92] There prismlike devices made an image of Christ float in mid-air, just as Kircher's *fonticulus sphaerum* replicated in miniature the submersion of Atlantis.[93]

Spanning the divide separating the world of Aldrovandi from the world of Kircher and their two respective patrons, Gregory XIII and Alexander VII, is a history of the transformation of collecting from a modest, individualistic practice into a lavish, corporate activity. From the delight in nature's plenitude to the thrill brought on by mankind's own contributions to that *copia*, naturalists and collectors took satisfaction in their ability to contain and produce wonder.

[89]This subject is dealt with in greater detail in Zakiya Hanafi's fascinating discussion of "Monstrous Machines" in her *Matters of Monstrosity in the Seicento* (Ph.D. diss., Stanford University, 1991), ch. 3.

[90]Gilles Deleuze, "The Fold," *Yale French Studies* 80 (1991): 227.

[91]Inchisa della Rocchetta, "Il museo di curiosità del Card. Flavio I Chigi," pp. 184–85, 187.

[92]Scarabelli, *Museo o galleria adunata dal sapere*, pp. 5, 8.

[93]Ibid., p. 8; De Sepi, p. 2. Kircher also had a device that reproduced the Resurrection (p. 39). Undoubtedly, Settala and Kircher had shared this particular "secret," as they did with many others, although it is difficult to identify with whom it originated.

TWO

Searching for Paradigms

And what good is it to know a multitude of things? Suppose you have learned all the circuits of the heavens and the earth, and the spaces of the sea, the courses of the stars, the virtues of herbs and stones, the secrets of nature, and then be ignorant of yourself?

—PETRARCH

"*Museum,*" wrote the Jesuit Claude Clemens, "most accurately is the place where the Muses dwell."[1] Early modern collectors were obsessively preoccupied with their ability to define the activities in which they engaged and the location in which they pursued them. As humanists, they celebrated the power of words, spinning elaborate etymologies and genealogies to acknowledge their indebtedness to the past. Knowing the origins of a word did not narrow its scope. If anything, it highlighted its complexity. As a category that expressed a pattern of activity transcending the strict confines of the museum itself, the idea of *musaeum* was an apt metaphor for the encyclopedic tendencies of the period. Most compelling about the term *musaeum* was its ability to be inserted into a wide range of discursive practices. Aldrovandi's collection of natural rarities in Bologna was simultaneously called *museo, studio, teatro, microcosmo, archivio,* and a host of other related terms, all describing the different ends that his collection served and, more importantly, alluding to the analogies between each structure.[2] The peculiar expansiveness of *musaeum* allowed it to cross and confuse philosophical categories such as *bibliotheca, thesaurus,* and *pandechion* with visual constructs such as *studio, casino, cabinet, galleria,* and *theatrum,* creating a rich

[1]Claude Clemens, *Musei sive bibliothecae tam privatae quam publicae extructio, cura, usus* (Leiden, 1635), sig. *4v. For a more detailed discussion of the language of collecting, see Paula Findlen, "The Museum: Its Classical Etymology and Renaissance Genealogy," *Journal of the History of Collections* 1 (1989): 59–78. For a broader discussion of the power of words and the search for origins in the early modern period, see Frank L. Borchardt, "Etymology in Tradition and in the Northern Renaissance," *Journal of the History of Ideas* 29 (1968): 415–429; and Marian Rothstein, "Etymology, Genealogy and the Immutability of Origins," *Renaissance Quarterly* 43 (1990): 332–347.

[2]References to the interchangeability of terms are voluminous; see, for example, BUB, *Aldrovandi,* ms. 38(2), Vol. I, c. 229, c. 259; ms. 41, c. 2r; ms. 136, Vol. XXVI, cc. 38–39; ASB, *Assunteria di Studio, Diversorum,* tome X, no. 6.

and complex terminology that described significant aspects of the intellectual and cultural life of early modern Europe.[3] The museum was variously a "room of books," a "treasure," and a "hold-all"; it was a space to fill, *cornucopia,* as well as a place for looking, *gazophylacium.* Simultaneously a "study," "countryhouse," "cabinet," "gallery," and "theater," it reflected a diverse understanding of the space that contained precious objects. Mediating between private and public space, among the monastic notion of study as a contemplative activity, the humanist notion of collecting as a textual strategy, and the social demands of prestige and display that a collection fulfilled, *musaeum* was an epistemological structure encompassing a variety of ideas, images, and institutions that were central to late Renaissance and Baroque culture.

The intense dissection of the word *musaeum* by collectors underscored their appreciation of the imaginative possibilities of language. The trail of words leading from the Greek ideal of the home of the muses and the μουσεῖον, the famous library of Alexandria, marked the transformation of the museum from a poetic construct into a conceptual system through which collectors interpreted and explored their world.[4] New activities required new definitions. "Those places in which one venerated the Muses were called Museums," explained Teodoro Bondini in his preface to the 1677 catalogue of Ferdinando Cospi's museum. "Likewise I know you will have understood that although a great portion of the Ancients approved of the name Muse only for the guardianship of Song and Poetry, nonetheless many others wished to incorporate all knowledge under such a name."[5] Bondini's expansion of the meaning of *musaeum,* as a nexus of all disciplines, finds its parallel in contemporary appreciation of the activities of particularly encyclopedic collectors. "At last my little Museum merits such a name," wrote Giacomo Scafili to Kircher upon the receipt of the latter's book, "now rich and complete with the *Musurgia,* your great work and gift, Father; even if there were nothing else in it save for this lone book, it could rightfully be called room of the Muses because the book contains them all."[6] A century after the appearance of the first museums in Italy, the language of collecting had evolved to such an extent that it designated the projects of

[3]Regarding the appearance of these and numerous other terms considered analogous to *musaeum,* see Berti, *Il Principe dello studiolo* pp. 194–195; Murray, *Museums,* Vol. I, pp. 34–38; Salerno, "Arte e scienza nelle collezioni del Manierismo," pp. 193–214; Wolfgang Liebenwein, *Studiolo. Storia e tipologia di uno spazio culturale,* Claudia Cieri Via, ed. (Modena, 1988). Other words that should be considered are *arca, cimelarchio, scrittoio, pinacotheca, metallotheca, kunstkammer, wunderkammer,* and *kunstschrank.*

[4]For a more detailed discussion of Timon's "cage of the muses," see Luciano Canfora, *The Vanished Library: A Wonder of the Ancient World,* Martin Ryle, trans. (Berkeley, 1989), and Steve Fuller and David Gorman, "Burning Libraries: Cultural Creation and the Problem of Historical Consciousness," *Annals of Scholarship* 4 (1987): 105–119.

[5]"Protesta di D. Teodoro Bondini a chi legge," in Legati, *Museo Cospiano,* n.p.

[6]PUG, *Kircher,* ms. 568 (XIV), f. 143r, (Trapani, 15 June 1652).

collectors such as Kircher as the "museums" that gave meaning to the rooms in which scholars hoarded and displayed knowledge. Taking this process of cultural transferral even further, another correspondent of Kircher's imagined his memory as a "gallery in which the rare and most exquisite things of the world are conserved."[7] Thus the definition and redefinition of *musaeum* and its attendant vocabulary became a primary point of departure for the definition of the encyclopedist whose museum expressed his mastery of all knowledge.

By the end of the seventeenth century, collectors had amassed a vast and bewildering vocabulary whose complexity mirrored the range of activities it described. Rejecting the classification of the Roman College museum as a *gallery*, a term referring primarily to its physical organization and to collections "made solely for their magnificence," the Jesuit Filippo Bonanni preferred to label Kircher's collection a *musaeum*. He justified his choice not only through copious citation of classical sources but also on the basis of the philological work of the sixteenth-century French scholar Dominique du Cange who provided the most appealing encyclopedic image for Bonanni. Drawing on Du Cange's false etymological comparison between museum and mosaic, Bonanni defined the Roman College museum in the following terms: "Let us say with Du Cange that, since the word 'Museal Work (*Opus Musiuum*) connotes that which is mosaicked by small stones of various colors,' thus in the places designated for the meanderings of the erudite there may be various things which not only delight the eyes with the Mosaic, but enrich the mind."[8] The museum, as mosaic, brought together pieces of a cosmology that had all but fallen apart in the course of several centuries. Organizing all known ideas and artifacts under the rubric of *musaeum*, collectors imagined that they had indeed come to terms with the crisis of knowledge that the fabrication of the museum was designed to solve.

REASSEMBLING THE MOSAIC

What were the "fragments" that comprised the mosaic of the museum? Bonanni's suggestive analogy, begun as an exercise in words, reveals an important aspect of the conceptual framework underlying the definition of *musaeum:* its ability to create patterns. Through the choice and juxtaposition of different objects, early modern naturalists formed "mosaics" that reflected the interpretive process underlying all collecting. Surveying the vast field of knowledge, they selected items that aided them in developing a meaningful understanding of the world. Their intellectual presuppositions

[7]PUG, *Kircher*, ms. 565 (XI), f. 292r (Cocumella, 9 September 1672). I have no idea where Cocumella is, but the letter makes it obvious that it was one of the distant missionary outposts.

[8]ARSI, *Rom.* 138. *Historia* (1704–1729), XVI, f. 182r (Filippo Bonanni, *Notizie circa la Galleria del Collegio Romano*, 10 January 1716). Dominique du Cange was the author of the *Glossarium Mediae et Infimae Latinitatis*, still one of the standard dictionaries of medieval Latin.

guided them at every turn, determining which artifacts naturalists found most appealing and regulating the meaning they extracted from them. Ultimately, the material acquisition of knowledge transformed the ideas that led scholars to collect in the first place. To understand what compelled sixteenth- and seventeenth-century naturalists to collect, we must first appreciate the philosophical aspirations that shaped their comprehension of the world. What did they hope to gain by concentrating the objects of nature into one place? How did they use material culture to bring their encyclopedic schemes to fruition?

By the mid-sixteenth century, natural philosophers had a variety of different approaches to knowledge from which to choose. Most traditional and canonical was the Aristotelian view of nature that favored the collecting of particular data only when directly pertinent to the universal axioms they created and reinforced. Aristotle also offered clear procedures by which knowledge would be disseminated, emphasizing the importance of deductive logic as the cornerstone of good scientific method.[9] With the exception of the Linceans and possibly Redi, all the naturalists discussed here were avowed Aristotelians, to some degree. But the qualification is important. Aristotelian philosophy underwent numerous transformations in the thirteenth and fourteenth centuries at the hands of Albertus Magnus and his pupil Thomas Aquinas. This dynamic approach to the words of the Philosopher continued in the sixteenth and seventeenth centuries, creating an even greater profusion of what Charles Schmitt aptly has dubbed "Aristotelianisms."[10] Just as Aristotelian philosophy was modified to meet the needs of late medieval Christianity, it underwent a similar metamorphosis in the context of late Renaissance humanism and Catholic Reformation culture. While Aldrovandi opened up Aristotelianism to a world of heightened sensory experience, Kircher used it as a point of departure for his Baroque meditations on the hidden meaning of the universe. They reflected a trajectory that began with the reedition and retranslation of the Aristotelian corpus at the end of the fifteenth century and ended with works such as Emanuele Tesauro's *Aristotelian Telescope* (1654).

Neither Aldrovandi nor Kircher confined themselves to one philosophical framework. Instead they combined different approaches to nature, reflecting the syncretist tendencies of the age. While maintaining a healthy respect for ancient authority, they eagerly embraced new philosophies of nature, secure in the knowledge that their openness would only enhance the traditions they upheld. Like many of their contemporaries, their novelty lay less in the creation of something radically new than in the reinvention

[9]The standard work on Aristotle is G. E. R. Lloyd, *Aristotle: The Growth and Structure of His Thought* (Cambridge, U.K., 1968).

[10]Charles B. Schmitt, *Aristotle and the Renaissance* (Cambridge, MA, 1983); also Edward Grant, "Aristotelianism and the Longevity of the Medieval World View," *History of Science* 16 (1978): 93–106.

of older forms of knowledge. Reconstituting Aristotle, they also reinvented Pliny, altering the philosophy of the former and giving the work of the latter greater centrality to the study of nature. Their expansive attitude toward the ancient canon also allowed them to include a variety of other authors who had not previously merited canonical status as philosophers of nature—Aristotle's pupil Theophrastus, the Greek physician Dioscorides, the Roman writers Ovid and Pliny, the mythical Hermes, and so on. This revised and increasingly eclectic list of "authorities" accompanied the heightened reverence for traditional medical writers who also observed nature, including Avicenna, whose commentaries on Aristotle were the staple of medieval and Renaissance universities, and the Roman physician Galen.[11] Like Aristotle, both gained in popularity as their works went into print. New forms of intellectual communication and exchange—the printing press and the editorial and epistolary activities that surrounded it—revitalized the time-honored enterprise of learning with the ancients as one's guide. Eventually this newfound intimacy would lead scholars to question the transcendent status of ancient authorities, casting them down from the Olympian heights to mortal ground. The rejection of the ancients was a slow and painful process, resisted by many of the people who, ironically, made it possible. In different ways, Renaissance and Baroque naturalists chose to extend the ancient paradigm of natural history rather than to dismantle it. Yet their decision to allow new influences to impinge upon this structure made it a precarious edifice indeed.

Beyond the classical framework defining the scope of natural history, other philosophies of nature, more specific to the early modern period, beckoned collectors. The fifteenth-century discovery of the Hermetic corpus, a body of allegedly pre-Christian writings attributed to the Egyptian God Hermes Trismegistus, greatly enhanced the symbolic study of nature.[12] Translated into Latin by Marsilio Ficino, it soon became a standard text for the humanists, particularly adherents of neo-Platonic and occultist natural philosophies. The hermetic view of nature, which presented it as a divinely encoded structure, held little attraction for most Renaissance naturalists who reveled more in the openness than in the secrecy of the universe. A century later, Kircher made it a central feature of the Jesuit system of knowledge. While Aldrovandi rejected the Hermetic corpus and Lullian

[11]On the fortunes of Avicenna, see Nancy Siraisi, *Avicenna in Renaissance Italy: The Canon and Medical Teaching in Italian Universities after 1500* (Princeton, 1987); on Galen, see Owsei Temkin, *Galenism: The Rise and Decline of a Medical Philosophy* (Ithaca, NY, 1973).

[12]The classic study of hermeticism remains Frances Yates, *Giordano Bruno and the Hermetic Tradition* (Chicago, 1964). I have drawn most of my material from it, for example, her discussion of Kircher, pp. 416–423. For a more recent view, see Brian P. Copenhaver, "Natural Magic, Hermeticism and Occultism in Early Modern Science," in *Reappraisals of the Scientific Revolution*, Lindberg and Westman, eds., pp. 261–301.

mnemonics as philosophical guides to the construction of his museum and his view of nature, he nonetheless benefited from the expanded role of symbolic discourse. As William Ashworth notes, Aldrovandi participated fully in the culture of emblematics that reached its height in the late sixteenth century.[13] His desire for *complete knowledge of every natural creation* led Aldrovandi to incorporate adages, morals, emblems, proverbs, sympathies, and antipathies alongside discussions of anatomy, physiology, and the various uses of the object under inspection.[14] Thus, the Bolognese naturalist was not interested in symbolic discourse per se but rather he studied and deployed it as a necessary part of the humanist definition of knowledge. He evidenced a common tendency among Renaissance encyclopedists to define inquiry not according to any *one* set of principles but as the interweaving of numerous and diverse philosophical systems.

If we turn to Kircher's treatises, for instance, his *Loadstone or Three Books on the Magnetic Art* (1641), *Great Art of Light and Shadow* (1646), or *Subterranean World* (1664), we find none of the subdivisions that characterized Aldrovandi's classification schemes. Kircher perceived the world as an organic, self-replicating entity. The structures that he sought lay not in Renaissance humanist categories of knowledge but in universal philosophical principles. Hermeticism, the most ancient of ancient wisdoms—Kircher apparently was undaunted by Isaac Casaubon's announcement in 1614 that the Hermetic corpus was a late antique rather than pre-Christian invention—offered one such framework. As close to the origins of mankind as any text could be, allegedly predating the Bible, it offered readers divine knowledge. For a natural philosopher who found communion with God in the study of nature, this represented priceless wisdom. In contrast to Aldrovandi's natural history, organized around the formal categories of humanist inquiry, Kircher's encyclopedia represented the culmination of the more mystical and allegorical strands of humanistic culture. Concerned with divine order and harmony, he had more affinities with neo-Platonists such as Ficino and Kepler than with the Renaissance Aristotelian Aldrovandi.

Hermeticism offered a new philosophical framework for natural philosophy, one particularly attractive to naturalists intent on making nature a moral object lesson for their contemporaries. Natural magic allowed the same individuals the opportunity to define their own role in explicating the operations of nature. Aldrovandi certainly knew something of this tradition. He had read Della Porta's *Natural Magic* (1558) and corresponded with the Neapolitan philosopher, somewhat younger than himself. Girolamo

[13]Ashworth, "Natural History and the Emblematic World View," pp. 313–316.

[14]Ashworth gives a full list of the different headings under which Aldrovandi classified each item. Readers wishing to sample Aldrovandi's style should consult *Aldrovandi on Chickens: The Ornithology of Ulisse Aldrovandi (1600) Volume II, Book XIV*, L. R. Lind, trans. (Norman, OK, 1963).

Cardano briefly had been his colleague at the University of Bologna.[15] Like both philosophers, he believed that nature contained numerous instances of hidden meaning and strove to identify and understand them. He was interested in hieroglyphics, physiognomy—Della Porta's specialty—and portents such as the 1572 dragon. But Aldrovandi typified sixteenth-century naturalists in his ambivalence toward the transformation of nature. This entailed a different level of knowledge than that toward which he strove. Eager to dissect nature, in a manner akin to Vesalius's willingness to open up the human body, he was more reluctant to create artificial demonstrations of natural operations. For someone interested in reviving the empiricist programs of Aristotle, Galen, and Pliny, Aldrovandi's primary goal, manipulating nature served no evident purpose.

Cardano's and Della Porta's most active followers were collectors such as Settala and Kircher and the latter's disciples Gaspar Schott and Francesco Lana Terzi. For seventeenth-century naturalists, knowledge of nature increasingly signified *power over nature*. Constantly critiquing Renaissance natural magic for being overly speculative in its claims to perform extravagant and unheralded experiments, they nonetheless drew inspiration from these activities. Their own attempts to replicate the operations of the natural world—volcanoes erupting, Atlantis submerging, light refracting and reflecting, sound dispersing, not to mention the standard array of perpetual motion machines, air pumps, telescopes, and microscopes that filled their galleries—created the bizarre technologies that we associate with Baroque museums.[16] While hermeticism defined the collector as a magus, natural magic made the philosopher a true master of nature. As Brian Copenhaver observes, "the theory of natural magic [was] instantiated by reference to a set of material objects."[17] For Aldrovandi and most Renaissance naturalists, those objects were found solely *in* nature; for Kircher and his Baroque counterparts, they were additionally derived *from* it.

As the previous examples illustrate, naturalists followed many different trajectories during this period, from the traditional to the esoteric to the experimental. Despite the divergent paths they took, all shared the common trait of curiosity. Curiosity led naturalists out into the world. It led them to define knowledge in terms of wonder and experience.[18] "Wonder" encompassed the emotions that confrontation with the unexpected aroused; "ex-

[15]On Della Porta, see Luisa Muraro, *Giambattista della Porta mago e scienzato* (Milan, 1978); on Cardano, see Alfonso Ingegno, *Saggi sulla filosofia di Cardano* (Florence, 1980).

[16]For more on this subject, see Zakiya Hanafi, "Monstrous Machines," in her *Matters of Monstrosity in the Seicento*, ch. 3.

[17]Copenhaver, "A Tale of Two Fishes: Magical Objects in Natural History from Antiquity through the Scientific Revolution," *Journal of the History of Ideas* 52 (1991):373–398.

[18]For discussions of curiosity in early modern culture, see Jean Céard et al., *La curiosité a la Renaissance* (Paris, 1986); Daston, "Neugier und Naturwissenschaft in der frühen Neuzeit," in Andreas Gröte, ed., *Macrocosmos im Microcosmo: Die Welt in der Stube* (in press); and Pomian, "The Age of Curiosity," in his *Collectors and Curiosities*, pp. 45–64. I would argue that curiosity,

perience" defined the knowledge gained from the repetition of such encounters. Aldrovandi, a product of the first wave of travel and exploration,
linked curiosity to his encyclopedism, allowing both to define the quest for
total knowledge. Kircher, nurtured in the missionary culture of the Jesuit
order, expressed his curiosity through the desire to master the most ancient
and exotic forms of knowledge. Aldrovandi hoped to know something of
the Americas; Kircher made it his goal to become the leading expert on
China and ancient Egypt. Aldrovandi struggled with the standard humanist
languages: Latin, Greek, some Arabic, and Hebrew; Kircher added Chinese, Sanskrit, Aramaic, and Etruscan to the list and attempted to unlock
the mysteries of the hieroglyphs. Both allowed curiosity to guide them in
their endeavors. Museums were fabricated out of the emerging dialectic between authority and curiosity, reverence for the wisdom of the past and excitement about the possibilities of the present. In his *Advancement of Learning* (1605), Francis Bacon defined knowledge as "a couch, whereupon to
rest a searching and restless spirit."[19] The museum was the place in which
the majority of "searching and restless spirits" congregated from the late
sixteenth century onward. In the museum, naturalists could imagine nothing less than complete mastery of all the things of the world, and it was to
this end that they strove.

Sixteenth- and seventeenth-century naturalists shared several common
tendencies. They revered authority and subsumed their philosophical speculations within a highly Christianized framework. In contrast to their medieval predecessors, they perceived the encyclopedia of knowledge to be
infinitely permeable, open rather than closed to multiple influences, dis-
continuous rather than continuous.[20] They also perceived nature to be
a text. Reading the "book of nature" was one of the primary activities for
early modern naturalists.[21] Collecting was an activity that contributed to the

an ambivalent sensation, became a positive sensation by taking on the attributes of wonder,
which had no negative connotations. Thanks to Pamela Smith for helping clarify this for me.

[19] In Houghton, "The English Virtuoso in the Seventeenth Century," p. 56.

[20] On medieval encyclopedism, see Lynn Thorndike, "Encyclopedias of the Fourteenth Century," in his *A History of Magic and Experimental Science*, Vol. III, pp. 546–567, and Maurice de
Gandillac et al., *La pensée encyclopédique au Moyen Age* (Neuchatel, 1966). For contrasting views
of early modern encyclopedism, see Ann M. Blair, *Restaging Jean Bodin: The "Universae naturae
theatrum (1596) in Its Cultural Context* (Ph.D. diss., Princeton University, 1990); Anthony
Grafton, "The World of The Polyhistors: Humanism and Encyclopedism," *Central European History* 28 (1985): 31–47; idem, "Humanism, Magic and Science," in *The Impact of Humanism on
Western Europe*, Anthony Goodman and Angus Mackay, eds. (London, 1990), p. 107; Daniel Defert, "The Collection of the World: Accounts of Voyages from the Sixteenth to the Eighteenth
Centuries," *Dialectical Anthropology* 7 (1982): 11–20. The classic study is Michel Foucault, *The Order of Things* (English trans., New York, 1970), pp. 17–45.

[21] On this subject, see Hans Blumenberg, *La leggibilità del mondo. Il libro come metafora della
natura*, Bruno Argenton, trans. (Bologna, 1984) [German original: *Die Lesbarkeit der Welt*, Frankfurt, 1981]; James J. Bono, *The Word of God and the Languages of Man* (in press); Ernst Curtius,
"The Book as Symbol," in his *European Literature and the Latin Middle Ages*, pp. 302–347;

reactivation and redefinition of the metaphor of the book. Possessing nature was a process that paralleled the humanists' possession of the wisdom of the ancients. In the late fifteenth century, Nicholas of Cusa compared the possession of nature to Petrarch's possession of certain Greek codices. While ignorance of Greek prevented Petrarch from deciphering their meaning, his sympathy toward their content gave his ownership a certain value.[22] Similarly, naturalists hoped that their possession of nature eventually would precipitate an understanding of her contents.

By the late sixteenth century, the timidness of the first humanists toward the paired books of nature and learning had been replaced with a new confidence. Emboldened by their ability to reactivate the languages of ancient philosophy, solving Petrarch's philological problem, naturalists used their humanist training to decipher the language of nature. The Dominican botanist Agostino del Riccio proposed alphabets for every part of nature to facilitate her legibility.[23] While Galileo contrasted his mastery of the book of nature with his rival Fortunio Liceti's mastery of the Aristotelian corpus, underscoring their incompatibility,[24] most naturalists found such a division unthinkable. Unlike Paracelsus, who had consigned the writings of any authority other than God to the flames, the venerable words of the Authorities of Nature guided them every step of the way. Careful perusal of ancient texts led philosophers to reread nature, the original Text. This was a debt that naturalists working in the traditions of Aldrovandi and Kircher fully acknowledged. Bonanni, for example, characterized the process of collecting shells as a form of reading, invoking the Ciceronian proverb *Conchas legere*.[25] He simultaneously "read shells" through the lenses of Ovid and Cicero as well as through the lenses of the microscopes he skillfully designed and used to confute all attacks on the Aristotelian program of natural history. The reciprocity between the collecting and reading of shells further accentuated the desire to see nature as a text. In Bonanni's universe, reading canonical texts such as the works of Aristotle and Pliny prepared naturalists to read nature.

By the time Bonanni affirmed these principles, the paradigm of nature that he upheld was already in decline. Throughout the seventeenth century,

---

Paula Findlen, "Empty Signs? Reading the Book of Nature in Renaissance Science," *Studies in the History and Philosophy of Science* 21 (1990): 511–518; Eugenio Garin, "La nuova scienza e il simbolo del 'libro,' " in his *La cultura filosofica del Rinascimento italiano* (Florence, 1961), pp. 451–465.

[22]This wonderful passage is recounted by Hans Blumenberg in *La leggibilità del mondo*, p. 58.

[23]BNF, *Cod. Magl.* II, 1, 13, f. 16r (Agostino Del Riccio, *Arte della Memoria*, 1595).

[24]Blumenberg, *La leggibilità del mondo*, p. 73.

[25]Bonanni, *Ricreatione*, p. 2. The expression *conchas legere* appears in Ovid, *Ars* (3.124), and Cicero, *De oratore* (2.22); *Thesaurus linguae latinae*, Vol. VII, pt. 2.2, p. 1123. Aldrovandi also included a discussion of this phrase under *proverbia* in his *De mollibus, crustaceis, testaceis, et zoophytis* (Bologna, 1606), p. 250. "Conchas legere proverbio dicuntur, qui ad animum remittendum nonnunquam ad ludicra quaedam sese demittunt."

proponents of various experimental philosophies emphasized the reading of nature without the mediation of any other text. "For us to accomplish anything," proclaimed Federico Cesi to the members of the Accademia dei Lincei, "it is necessary to read this great, truly named universal book of the world."[26] Expanding the metaphor of nature as a book, and concommitantly the role of collecting in the formation of a new book of nature, such naturalists as Cesi made a traditional image a locus of innovation. In his *Museum of Physic and Experiences* (1697), Paolo Boccone still insisted that nature was one great cipher, yet he described the language as "so many Hieroglyphics, adapted to the speculations of lovers of experimental Philosophy."[27] Nature was no longer a text for obdurate Aristotelians such as Aldrovandi, Kircher, and Bonanni, but instead revealed herself to naturalists who had absorbed the lessons of Bacon and Galileo. The possession of nature no longer refined and amplified the Aristotelian program. It could only undermine and ultimately destroy it.

While differing greatly on the nature and meaning of the contents, sixteenth- and seventeenth-century naturalists imagined the "book of nature" to be a text whose significance did not exceed their grasp. Possessing nature materially grounded this metaphor, just as it made many other aspects of natural philosophy visibly apparent. Searching for ways to come to terms with a shifting and unstable field of knowledge, the study of nature, collectors made the museum a "paradigmatic" space in which to philosophize.

CLASSICAL MODELS AND RENAISSANCE QUERIES

Renaissance museums were not indebted to the past for their name alone. The association between ancient models and contemporary practices shaped the philosophical programs that underlay the process of collecting. While museums of natural history are usually associated with the reformulation of the history of nature in the eighteenth century, they originated in a predominantly Aristotelian and Plinian framework. The Authorities of Nature—Aristotle, Theophrastus, Dioscorides, and the synthesizer Pliny—guided naturalists every step of the way. Describing the method by which he arrived at a proper name for a previously unidentified phenomenon, Aldrovandi explained, "a Philosopher is allowed to invent names where there are none, by testimony of our Philosopher Aristotle."[28] Even in the late seventeenth century, naturalists such as Filippo Bonanni continued to perceive the classical tradition as a benchmark against which to measure their own accomplishments. "Who knew more of Natural things than Pliny

[26]Federico Cesi, *Del natural desiderio di sapere et institutione de' Lincei per adempimento di esso*, in *Scienziati del Seicento*, Maria Luisa Altieri Biagi and Bruno Basile, eds. (Milan, 1980), p. 44.

[27]Paolo Boccone, *Museo di fisica e di esperienze* (Venice, 1697), p. 117.

[28]BMV, *Archivio Morelliano*, ms. 103 (=*Marciana* 12609), f. 20.

and Aristotle," wrote Bonanni in his *Recreation of the Eye and the Mind in the Observation of Snails* (1683), "Masters for many centuries?"[29]

The museum of natural history contributed significantly to the reconstruction of the Aristotelian research program in the Renaissance universities. For many naturalists, it represented nothing less than the revival of the original *musaeum*, populated by Aristotle's early disciples, and the reconstitution of the Aristotelian research program in that setting. At Padua in the late sixteenth century, the anatomist Hieronymus Fabricius constructed two theaters: the famous anatomical theater in which he dissected under the watchful eye of such students as William Harvey and his *Theater of the Whole Animal Fabric*, the "theater" composed of all his publications on the individual organs and diverse physiological processes that defined the "parts" of the animal world.[30] In 1564, some forty years before Fabricius published his famed study of chick embryos, Aldrovandi observed daily the transformation of the fetus in a hen's egg; this was to become part of the work that he ambitiously labeled the *Acanthology, or Universal History of Everything*.[31] Fabricius, in other words, was part of a continuum that included the research of his colleague in Bologna.

While Aldrovandi lacked the single-minded devotion to Aristotle displayed by Fabricius, who saw "Aristotle as his *only* predecessor" in his research on the generation of animals,[32] both shared the assumption that they were extending, if not completing, the scientific work of Aristotle. Texts such as Aristotle's *History of Animals, Generation of Animals,* and *On the Parts of Animals* were their guides. In this they were joined by such contemporaries as Andrea Cesalpino, professor of medicine at Pisa and author of the celebrated *Peripatetic Questions* (1571) and *On Plants* (1583), and the Paduan philosopher Jacopo Zabarella, whose publications were instrumental in the revitalization of Aristotelian method during the mid-sixteenth century. Cesalpino attempted to revitalize Aristotelian method as a means of determining the essential characteristics of plants, the *differentiae* that Aristotle made central to his philosophy of living kinds.[33] Instead, Aldrovandi strove

[29]Bonanni, *Ricreatione*, p. 21.

[30]Fabricius' project is thoroughly reconstructed in Andrew Cunningham, "Fabricius and the 'Aristotle Project' in Anatomical Teaching and Research at Padua," in *The Medical Renaissance of the Sixteenth Century*, A. Wear, R. K. French, and I. M. Lonie, eds. (Cambridge, 1985), pp. 195–222, esp. pp. 199–200.

[31]Howard B. Adelmann, *The Embryological Treatises of Hieronymus Fabricius of Aquapendente*, 2 vols. (New York, 1942, repr. 1967); Tugnoli Pattaro, *Metodo e sistema delle scienze nel pensiero di Ulisse Aldrovandi*, p. 151; BUB, *Aldrovandi*, ms. 86. "Acanthology" literally means the study of spines, derived from the name for a spiny plant, acanthus. This is the sort of playful neologism that appealed to Aldrovandi. Not only was it a naturalist's joke about identification but also a humanist joke about the thorny nature of knowledge.

[32]Cunningham, "Fabricius and the 'Aristotle Project,' " p. 211.

[33]Andrea Cesalpino, *Questions péripatéticiennes*, Maurice Dorolle, trans. (Paris, 1929). The best general study of Renaissance Aristotelianism is Charles Schmitt's *Aristotle in the Renaissance*

to reproduce the entire Aristotelian corpus through his own publications. While Cesalpino resisted the copiousness of the natural world by reaffirming the essentialist program, Aldrovandi reveled in the particulars of nature.

Aldrovandi and Zabarella both enjoyed the tutelage of Bernardino Tomitano, professor of theoretical medicine at Padua, who trained them in logic.[34] From Tomitano and other mentors in Bologna, Aldrovandi was initiated into a form of Aristotelianism that did not accept the sayings of the Stagirite verbatim but used them as a framework for developing a properly critical view of the nature of living things. As Zabarella remarked in 1585, "I will never be satisfied with Aristotle's authority alone to establish something, but I will always rely upon reason."[35] Even skeptical contemporaries such as the Portugese philosopher Francisco Sanches, who called into question the entire Aristotelian program, would have approved of this approach.[36] Aldrovandi's own search for "reason" led him to privilege observation and experience in understanding nature; he, too, criticized Aristotle for not checking all of his facts personally. Thus, the museum of nature became a logical extension of the empirical program laid out in Aristotle's biological writings and in the natural histories of his followers. For late Renaissance naturalists, it was a means of making visible the wisdom of the ancients and advertising their own role as mediators and ultimately defenders of tradition.[37]

The desire to imitate Aristotle manifested itself in several ways. Like Aldrovandi and Fabricius, many naturalists created printed theaters of nature, publishing their work in segments that imitated the divisions and the progression of Aristotle's own investigations of nature, particularly in the biological realm. Some also arranged the objects in their museums in accordance with ancient hierarchies, privileging animate objects over inanimate objects, and creatures closest to humans, quadrupeds, over those most alien to human sensibilities, namely, insects. Others extended the classifying impulse that led them to subdivide nature to encompass the organization of books. Aldrovandi, for example, dissected the subject organization of libraries with the same passion that he catalogued nature and every other part of the human experience. Like Gesner, Aldrovandi perceived his encyclopedia of nature to be dependent on the encyclopedia of knowledge. Thus, bibliographies were hoarded as if the names of the books themselves symbolically conveyed the possession of their contents, and books were orga-

---

(Cambridge, 1983). For an insightful discussion of Cesalpino's work, see Atran, *Cognitive Foundations of Natural History*, pp. 138–158.

[34] Tugnoli Pattaro, *Metodo e sistema delle scienze nel pensiero di Ulisse Aldrovandi*, pp. 37–39.

[35] In Schmitt, *Aristotle in the Renaissance*, p. 11.

[36] See Francisco Sanches, *That Nothing is Known (Quod nihil scititur)*, Elaine Limbrick, ed., and Douglas F. S. Thomson, trans. (Cambridge, U.K., 1988).

[37] The image of humanists as "defenders" of tradition is developed in Anthony Grafton, *Defenders of the Text: The Traditions of Scholarship in an Age of Science, 1450–1800* (Cambridge, MA, 1991).

nized in accordance with the standard classification of the sciences.[38] More generally, collectors used material culture to shed light on the "particulars" of nature, the discrete data contributing to the formation of certain knowledge that Aristotle had advised them to covet. At the end of this enterprise lay the elusive promise that, by collecting the world, one ultimately would master its Universal Truths.

Like many natural philosophers, Aldrovandi borrowed freely from other philosophical and medical traditions. His interest in medicine led him to give the words of Galen particular weight, as when he declared the necessity of being both a good natural philosopher *and* a good physician to be able to know nature properly. Creating imaginary divisions for his library, Aldrovandi placed natural history between natural philosophy and medicine, so that scholars consulting his books would understand their strong affinities.[39] Galen, Dioscorides, and Avicenna provided many insights into the uses of nature and the medical necessity of that knowledge, but the Aristotelian framework of natural history elevated it to the status of philosophy. Ancient medical writers enhanced Aldrovandi's understanding of "experience," a quality emphasized repeatedly by Galen, which he subsequently applied to the study of nature. Yet Aristotelianism gave his pursuit philosophical legitimacy. In this context, we should recall the importance that Aldrovandi gave to his "synoptic tables" whose organization provided the key to the signification of the objects in his museum. Like the *tabulae* that "became a common feature of textbooks and a literary form in their own right during the late sixteenth century,"[40] Aldrovandi's tables summarized the relationships among the different parts of nature and of knowledge. Only through this properly Aristotelian exercise could visitors gain any understanding of the relationships among objects in the museum. As the Bolognese naturalist wrote at the end of 1572, "in our *Universal Method of the Different Genuses of All Animate and Inanimate Objects* we have defined and explicated [these things] so that every visible thing known may be reduced to the nearest genus, in order to properly define and describe it."[41] For Aldrovandi, description yielded definition, definition order, and order knowledge.

Consisting of diagrams that situated everything from earths to fossils, monsters, zoophytes, angels, and even language in their proper relationship to one another, Aldrovandi's tables (*methodi*) established a hierarchy of

[38]BUB, *Aldrovandi*, ms. 97, cc. 440–443r; see also his *Bibliologia* (1580–1581), ms. 83, and his *Bibliotheca secondum nomina authorum*, ms. 147; Conrad Gesner, *Bibliotheca universalis* (Tiguri, 1545).

[39]Aldrovandi, *Discorso naturale*, p. 195; BUB, *Aldrovandi*, ms. 97, is filled with examples of his *methodi*. See cc. 440–443r for his library organization "according to the order and general division of the sciences."

[40]Schmitt, *Aristotle in the Renaissance*, p. 59. Sandra Tugnoli Pattaro reproduces several of these *tabulae* in her book.

[41]Aldrovandi, *Discorso naturale*, p. 193.

knowledge as well as kinds. Like the museum itself, they "collected and reduced" nature into a comprehensible entity that confirmed a philosophically ordained pattern. They established the "universal syntax" of the world.[42] Similar tables filled the pages of Mercati's *Metallotheca* and organized Cesalpino's *On Metals* (1596). Both authors corresponded frequently with Aldrovandi. Mercati began by defining his object of study and collection as "hard things" (*res durae*). He then proceeded to distinguish proper from improper stones; the former were invariate and composed of the four elements earth, water, air, and fire, while the latter encompassed more protean and complex forms such as magnets and asbestos.[43] In establishing these distinctions, Mercati was following the tradition of sorting out "particular differences, so that the difference joined with the genus makes the species and . . . gives it its proper name, which no other can have."[44] This was an exercise that any Aristotelian would have recognized and applauded.

Instructing the Senate of Bologna on the organization of his museum in 1603, Aldrovandi underscored the importance of keeping his synoptic tables with the specimen cabinets so that visitors could read properly the context of the museum. Their pattern, he felt, unfolded universal syntax of the world.[45] The product of over forty years of research, they represented the cumulative knowledge of an entire generation of Renaissance Aristotelians. Less than two decades after Aldrovandi's death, avowed anti-Aristotelians such as Federico Cesi would design tables, fashioned from their own collections, to dismember the ancient hierarchies of nature. If Cesi's *Theater of Nature* acknowledged no explicit debt to the work of such men as Aldrovandi, its insistence on the link between collecting and classifying nature made the association fully apparent. Kircher continued to scatter occasional tables throughout works such as the *Subterranean World*, and Bonanni invoked their logical structure when he classified shells. But only Renaissance naturalists, unencumbered by the criticisms leveled by seventeenth-century skeptics against the philosophical positioning of knowledge, imagined this practice to be a truly authoritative exercise.

While Aristotle provided a formal structure and philosophical purpose for the collecting of nature, Pliny's *Natural History* inspired naturalists to extend their curiosity to the farthest reaches of the known world in order to catalogue its wonders. As Federico Borromeo wrote at the beginning of his *Musaeum* (1625), "To begin this work, I think first of Pliny, above all others, not for the desire that I have to emulate him, which would be excessively foolish and audacious, but, in spite of myself, for the excellence of his ex-

[42]BAV, *Vat. Lat.* 6192, Vol. II, f. 656r; BUB, *Aldrovandi*, ms. 70. c. 9v.

[43]Michele Mercati, *Metallotheca* (Rome, 1717), p. 144. Mercati's observations are worth comparing with Aldrovandi's *Methodus fossilium*, BUB, *Aldrovandi*, ms. 92.

[44]BUB, *Aldrovandi*, ms. 70, c. 7r.

[45]See BUB, *Aldrovandi*, ms. 70, c. 9v, for a brief discussion of Aldrovandi's *syntaxes universali*.

ample."[46] Reading Pliny reminded naturalists just how vast and expansive the rubric of natural history could be in its ability to encompass all the things in the universe. As Aldrovandi put it, "There is nothing under the sun that cannot be reduced to one of the three genus, that is, inanimate things and fossils, extracted from the bowels of the earth, plants, or animals. Even artificial things may be included in one of these three genus according to the materials [of their composition]."[47] Following Pliny, most collectors of nature included art, antiquities, and scientific instruments in their museums. Mercati even included descriptions of some of the famous statues in the Belvedere in his *Metallotheca* as examples of marble.

Pliny's appeal lay in his expansive rather than synoptic approach to knowledge; in its most essential sense, his philosophy of nature completely undermined the premise of Aristotelian philosophy. While Aristotle helped naturalists reduce nature to First Principles, creating a formal philosophy of nature, Pliny allowed them to revel in the particularity and infinity of the world. Initially, observers of New World flora and fauna remarked that their marvelous properties made it *more possible* to believe Pliny; he served as a guide to the novelties that increased travel led the Europeans to uncover.[48] While Aristotle described the universe as eternal, a cosmos populated with timeless truths, Pliny showed them its expanse, offering a framework that could better accommodate the increase in natural knowledge.

The format of *Natural History* reminded collectors that no detail of nature was so insignificant that it deserved neglect. It also suggested that synthesis, not succinctness, was the hallmark of an encyclopedist. As Pliny outlined in the preface to his monumental work:

> It is not books but store-houses(*thesauros*) that are needed; consequently by perusing about 2000 volumes, very few of which, owing to the abstruseness of their contents are ever handled by students, we have collected in 36 volumes 20,000 noteworthy facts obtained from one hundred authors that we have explored, with a great number of other facts in addition that were either ignored by our predecessors or have been discovered by subsequent experience.[49]

The acquisitive nature of collecting revealed a similar desire to catalogue all the "noteworthy facts," now increased well beyond Pliny's original estimate of 20,000. Renaissance naturalists read the preface of *Natural History* as a challenge to their own ingenuity and perspicacity. If Pliny could surpass the ancients, then besting the Roman encyclopedist was an admirable goal, worthy of humanist ambitions.

Like Pliny, Aldrovandi was obsessed with the size of his collection; not a week passed without his recounting the total number of "facts" he had ac-

[46] *Il Museo del Cardinale Federico Borromeo*, Luigi Grasselli, trans. (Milan, 1909), p. 44.
[47] BAV, *Vat. Lat.* 6192, Vol. II, f. 656v.
[48] Antonello Gerbi, *Nature in the New World*, (Pittsburgh, 1985), pp. 61–63.
[49] Pliny, *Natural History*, H. Rackham, trans. (Cambridge, MA, 1938), Vol. I, p. 13 (preface, 17–18).

cumulated. In 1577, he possessed about 13,000 things; in 1595, 18,000; at
the turn of the century, approximately 20,000. No doubt he took satisfaction in the fact that he had at least equaled, if not surpassed, the greatest
collector of nature in antiquity. "If I wanted to describe the variety of fish
observed, depicted and dried by me, that can be seen by everyone in our
microcosm, truly it would be necessary to consume many pages simply to
name them."[50] Like Pliny, Renaissance encyclopedists took pride in the
length and quantity of their literary productions; if the number of "facts"
seemed large, the number of words produced in response to those facts was
even greater. When Richard Lassels toured the *Studio Aldrovandi* in the mid-
seventeenth century, he remarked, " in this *Pallace* I saw the rare *Cabinet* and
Study of Aldrovandus, to whom *Pliny the Second* if he were now alive would but
be *Pliny the Sixt[h]*; for he hath printed six great volumes of the natures of
all things in nature, each volume being as big as all *Plinyes* workes."[51] No
doubt Aldrovandi would have derived great pleasure from the fact that visitors to his museum perceived his encyclopedism to be a direct challenge to
that of the ancients.

In further imitation of Pliny, collectors represented themselves as literally absorbed by their pursuit of knowledge. When Jacopino Bronzino described Aldrovandi as "consumed in the history of natural things,"[52] he aptly
summarized the encyclopedic passion for working *within* one's material.
Consumed by the desire to possess all the facts of nature, naturalists organized all their activities around collecting. "[I am] hoping to see something
beautiful in your care," wrote Aldrovandi to Alfonso Pancio, physician to the
d'Este family in Ferrara, "not ever being sated by the learning of new things.
Not a week passes—I will not say a day—in which I am not sent something
special. Nor is it to be wondered at, because this science of nature is as infinite as our knowledge."[53] Having committed themselves to a lifetime of collecting, naturalists struggled to organize the information and objects they
possessed. Pliny created his natural histories by extracting information from
numerous other books; this technique gained popularity among sixteenth-
century humanists. As Ann Blair observes, many Renaissance scholars, particularly those engaged in encyclopedic projects, used commonplace books
designed to organize and sort their reading matter for future use in their
own writing.[54] Aldrovandi was no exception to this rule.

[50]Aldrovandi, *Discorso naturale*, p. 184. Regarding the number of objects in his collection, see
BAV, *Vat. Lat.* 6192, Vol. II, c. 656r; BUB, *Aldrovandi*, ms. 70, c. 66r; ms. 80, cc. 460–481.
    [51]Richard Lassels, *The Voyage of Italy, or A Compleat Journey Through Italy. In Two Parts* (Lon-
don, 1670), Vol. I, pp. 147–148.
    [52]BUB, *Aldrovandi*, ms. 136, Vol. XXVIII, c. 126r (Viadanae, 29 June 1599).
    [53]ASMo, *Archivio per le materie. Storia naturale*, Busta I (Bologna, 16 December 1577).
    [54]Blair, *Restaging Jean Bodin*, p. 4ff, esp. p. 51. For a particularly relevant discussion of this
subject in the context of academic traditions, see Richard J. Durling, "Girolamo Mercuriale's
De modo studendi," *Osiris*, ser. 2, 6 (1990): 195.

Drawing upon Pliny's list of Greek titles, Aldrovandi named his largest project, under which all others were to be subsumed, the *Pandechion Epistemonicon*, which he defined as "a universal forest of knowledge, by means of which one will find whatever the poets, theologians, lawmakers, philosophers and historians . . . have written on any natural or artificial thing one wished to know about or compose."[55] Throughout the half-century in which Aldrovandi actively collected, he constantly strove to fill the space he had created. Words, images, and texts were all incorporated into the universal encyclopedia of knowledge that he visualized. The omnipresence of Aldrovandi's *pandechion* evidenced itself in his flexible use of the term. Like other encyclopedic terms, it was a semantic structure organized to include "not only the notion of abundance itself but also the place where abundance is to be found, or, more strictly, the place and its contents."[56] Aldrovandi generally described his collection of objects as a "cimilarchion and pandechion of the things generated in this inferior world." Thus, the encyclopedia was defined by the experiential data that constituted one part of his collection. Although he rarely used this term to refer to any but his own collection—probably because, in his estimation, no other was so expansive nor so central to humanist epistemology—the Grand Duke of Tuscany's collection also merited the label *pandechion* because it was "full of an infinite number of experimental secrets."[57] Not surprisingly, the principle of plenitude was operative in Aldrovandi's decision to designate it an encyclopedic structure equal to his own. In the same spirit, Olivi called Calzolari's museum a cornucopia.[58] If nature were the "cornucopian text" that held the interest of the naturalist, then the museum itself was the receptacle of that *copia*.

Discussing some of his rarer dried plants with the Flemish naturalist Matthias Lobel, "which I conserve pasted in fifteen volumes in my Pandechion of nature for the utility of posterity," Aldrovandi reiterated the textual nature of the artifacts, which became "books" organized according to his Aristotelian and Plinian taxonomy of nature. "For a full supply of facts (*copia rerum*) begets a full supply of words," counseled Cicero.[59] Most importantly,

---

[55]Mattirolo, p. 381 (1588); regarding the origins and use of the term *Pandechion*, see Pliny, *Natural History*, p. 15 (preface, 28) where he discusses the πανσέκται ("Hold-alls"); Lewis and Short, *Latin Dictionary*, p. 1296. *Pandere* means to spread out, extend, expand, or unfold; to lay open. The idea of a forest, a *selva universale*, was a common trope. It was used frequently, for example, by Tommaso Garzoni, as Paolo Cerchi notes in his *Enciclopedismo e politica della riscrittura: Tommaso Garzoni* (Pisa, 1980), pp. 32–33; equally, Zenobbio Bocchi's botanical garden and museum at the Gonzaga court in Mantova was described by contemporaries as a "forest of natural things"; Ceruti and Chiocco, *Musaeum Francesci Calceolari*, sig. *4v.

[56]Terence Cave, *The Cornucopian Text: Problems of Writing in the French Renaissance* (Oxford, 1979), p. 6. I have taken this passage from his discussion of the definition of *copia*, which defined not only plenitude but also functioned as *thesaurus*.

[57]BUB, *Aldrovandi*, ms. 91, c. 522r; BMV, *Archivio Morelliano 103* (= *Marciana 12609*), f. 9.

[58]Olivi, *De reconditis et praecipuis collectaneis*, sig. ++4v, p. 2.

[59]Cicero, *De oratore* III.xxi.125, in Cave, *The Cornucopian Text*, p. 6.

there was the *Pandechion* proper: eighty-three volumes containing scraps of paper—excerpted passages and bits of information from every book Aldrovandi had ever read—which the naturalist, his wife, and assistants meticulously cut up and alphabetically organized until 1589.[60] The compendium functioned as a lexicon on almost any known subject. Responding to Lorenzo Giacomini's questions on wine making in 1587, Aldrovandi quoted Pliny but could not remember the exact citation. "But where he [Pliny] teaches it, for now I can't recall, though I have seen it and glossed it from head to foot. And if you were able to run through my *Epistemonicon*, you would have found it and infinite other observations."[61] For Aldrovandi, the encyclopedia was located neither in the text nor in the object alone; rather it was the dialectic between *res* and *verba* that fully defined the universality of his project.

Like Pliny, Aldrovandi created a *pandechion* that did not distinguish among categories of facts.[62] The real and the imaginary, the ordinary and the extraordinary, all found their place in his encyclopedic paradigm, undifferentiated by any criteria of truth. In 1653, Nicolò Serpetro defined natural history as a "marketplace of natural marvels."[63] Serpetro's imagery aptly summarized the impulse behind Aldrovandi's own collecting habits. Like the "universal piazza" in which Tommaso Garzoni has collected all the professions of the world,[64] the scientific marketplace was at once indiscriminant in its ability to let everything enter, and hierarchical in its privileging of the extraordinary. The marvelous contents of Serpetro's natural history read like the standard list of wonders in any respectable museum: mandrakes, giant's bones, pygmies, petrified objects, fabulous zoophytes such as the Scythian lamb, and a variety of metamorphosizing entities—precious stones formed inside animals, birds generated from trees, and so forth. Perhaps what was most remarkable about Serpetro's catalogue of rarities, by 1653, was its ordinariness. A similar repetoire could be found in just about any museum of natural history of the day. After visiting Calzolari's museum (Museum Plan 1) in Verona in 1571, Aldrovandi made a list of the most singular items. Among other objects of interest, he particularly noted the presence of a chameleon, a unicorn's horn, a bird of paradise, and a piece of papyrus.[65] In his notebooks, novel objects such as the chameleon and bird of paradise coexisted with venerable reminders of the humanist past (a sheet of papyrus)

---

[60]BUB, *Aldrovandi*, ms. 105; Tugnoli Pattaro, *Metodo e sistema delle scienze nel pensiero di Ulisse Aldrovandi*, p. 15.

[61]Ricc. Cod. 2438, I, f. 1r (Bologna, 27 June 1587).

[62]As Ann Blair notes in her study of Jean Bodin, "The commonplace book, furthermore, does not differentiate between categories of facts. . . . In this way, 'credulity' can coexist with direct observation." Blair, *Restaging Jean Bodin*, p. 4.

[63]Nicolò Serpetro, *Il mercato delle maraviglie della natura* (Venice, 1653).

[64]Garzoni, *Piazza universale*.

[65]BUB, *Aldrovandi*, ms. 136, Vol. III, c. 180.

The museum was located on the second floor of the apothecary's home. The rooms are arranged in the order in which they appeared to a visitor, without regard to their size or the relationship between them.

Sources: Olivi, sig. +3^V; Tergolina-Gislanzoni-Brasco, p. 12.

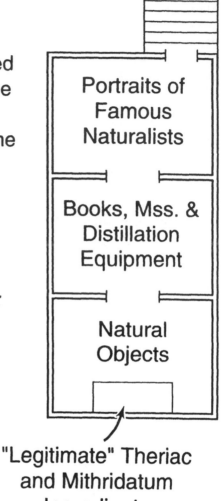

Museum Plan 1. The museum of Francesco Calzolari (1584).

and mythological artifacts (the unicorn's horn), making his *Pandechion Epistemonicon* a genuinely Plinian encyclopedia.

Added to the list of wonders, culled from Roman natural history, were peculiarly Christian objects. Not one to neglect the religious framework of his speculations, Aldrovandi, with the encouragement of the Tridentine reformer Paleotti, created a *Natural Biblical Theater* that glossed all references to nature found in Holy Writ; in a 1588 list of his major projects, he described it as an "elucidation of the Sacred Bible."[66] As part of this project, he drafted a treatise, *On the Cross;* an entire section was devoted to its natural history. At the bottom of a list of plants that he hoped to receive from Pietro Antonio Micheli, he wrote, "Before all other things, I would like to have those that are marked with the cross."[67] In the tradition of a good encyclopedist, Aldrovandi modified Pliny's category of *mirabilia* to include artifacts whose "wonder" addressed the concerns of his own audience. In the hands of Kircher, the impulse toward Christian natural history resulted in an entire treatise on "prodigious crosses."[68]

The novelty that such objects held in the sixteenth century had disappeared by the time that Serpetro wrote his *Marketplace of Natural Marvels.* Nonetheless, the list was still intact. This was not necessarily true for other objects. In a letter to Paleotti in 1594, Aldrovandi recalled the excitement that his planting of a "Peruvian chrysanthamum," the first sunflower seen in Italy, had occasioned in Bologna. "It was visited by all the gentlemen and gentlewomen for the size and beauty of its flower. But now it is so vulgar that no one cares about it."[69] Aldrovandi's remark to Paleotti reveals a certain nostalgia for the loss of novelty in the study of nature. For him, as for the popularizer Serpetro, the thrill of wonder was a sensation cultivated ad infinitum. When sunflowers were "in," Aldrovandi amazed visitors with the quantity and excellency of his samples. If his instincts were good, he timed their disappearance—and replacement with another object of worth—in anticipation of their consignment to the realm of the ordinary. Yet he could not help but long for the days when every item in nature was an object of wonder.

Certainly the books of illustrations that Aldrovandi preserved in the museum recaptured this Plinian sensibility.[70] The scope of the illustrations, even more than Serpetro's listing of natural wonders, indicates the different levels of curiosity at work. Sandwiched between sheets of perfectly ordinary phenomena—a parrot, a butterfly, a pear, or a sprig from a berry tree (fig. 4)—were extravagant creatures like the rainbow fish from which a solitary hand extruded. Many objects were illustrated from actual specimens, as was the case with the 1572 dragon (see fig. 1); still others were culled from popular prodigy books, natural histories or the latest broadsheets in circu-

[66]BUB, *Aldrovandi*, ms. 54; Mattirolo, p. 382; Paolo Prodi, *Il Cardinale Gabriele Paleotti* (Rome, 1959–1967), Vol. II, pp. 542–543; Giuseppe Olmi and Paolo Prodi, "Gabriele Paleotti, Ulisse Aldrovandi e la cultura a Bologna nel secondo Cinquecento," in *Nell'età di Coreggio e dei Caracci. Pittura in Emilia dei secoli XVI e XVII* (Bologna, 1986), pp. 215, 225.

[67]BUB, *Aldrovandi*, mss. 51 and 56, c. 446.

[68]Kircher, *Diatribe de prodigiosis crucibus* (Rome, 1661).

[69]AI, Fondo Paleotti. 59 (F 30) 29/11, f. 14 (Bologna, 30 July 1594).

[70]On Aldrovandi's illustrations, see Giuseppe Olmi, "Osservazioni della natura e raffigurazione in Ulisse Aldrovandi (1522–1605)," *Annali dell'Istituto Italo-Germanico in Trento* 3 (1977): 105–181.

Figure 4. Page of Aldrovandi's *naturalia*. From BUB, *Aldrovandi. Tavole di animali. U.V. 59.*

lation. In the Renaissance museum of nature, curiosity could accommodate both the commonplace and the unusual, the "real" and the probable. From this vantage point, Aldrovandi exhibited a greater indebtedness to Pliny than to Aristotle.

Aldrovandi and his associates represented the culmination of a tradition of intensive inspection of classical texts that generated emulation as well as criticism of ancient authority. Like the fifteenth-century cataloguers of Pliny's errors who attacked the Roman encyclopedist's credulity, inadequate observations, and poor knowledge of Greek, late Renaissance naturalists preferred to "save the proposition of Pliny" and other authorities rather than build from scratch.[71] Yet at the same time their refinements of textual and empirical methods altered their relationship to the past. Both Aldrovandi and Borromeo, for example, chastized the ancients for neglecting to illustrate their treatises, a defect they themselves remedied with on-going illustration projects. "Moreover how much light would we glean from interpreting the passages of writers, principally Pliny, if we had in sight those things which he told only with words," lamented Borromeo.[72] Even Aldrovandi, who clearly placed images in the lower realm of his philosophical hierarchy, conceded their utility when the original was lacking, as the existence of thousands of natural history illustrations in his collection attests.

> By the means of these pictures, together with the histories, scholars gain full knowledge of what [the plants and animals] were according to the ancients. And one cannot imagine anything more useful; if the ancients had drawn and painted all of the things which they described, one would not find so many doubts and endless errors among writers.[73]

Thus, while voices from the past continued to set the problems that Renaissance naturalists investigated—classical and biblical conundra such as whether a phoenix could regenerate itself from its own ashes, whether a bear licked its cubs into shape, or whether frogs generated spontaneously from menses and dust—they determined neither the conditions in which such questions were resolved nor the materials brought to bear in their resolution.

Upon returning from Cairo in 1561, Melchior Wieland wrote the following letter to Aldrovandi: "I left Europe with the desire to litigate not only with Pliny and Dioscorides, but all of antiquity, having proposed to deal with

---

[71]On the errors of Pliny, see Arturo Castiglione, "The School of Ferrara and the Controversy on Pliny," in E. A. Underwood, ed., *Science, Medicine and History* (London, 1953), pp. 593–610; Ricc. Cod. 2438, Pt. I, lett. 91 (Bologna, 27 June 1587).

[72]Arlene Quint, *Cardinal Federico Borromeo as a Patron and Critic of the Arts and His Musaeum of 1625* (New York, 1986), p. 233; I have modified her translation slightly to make it more readable.

[73]Aldrovandi, "Avvertimenti," in Barocchi, *Trattati*, p. 513; regarding the illustrations, see BUB, *Aldrovandi, Miscellanea di animali e piante depinte, Tavole di animali*, and *Tavole di piante*.

Pliny's subject, natural history." Yet after completing his trip, the Paduan botanist found himself bewildered by the "thousands of observations made during this voyage."[74] In attempting to correct the mistakes of the past, through greater travel and more precise observation, he had become more appreciative of the relative success and difficulty of ancient encyclopedic projects. No doubt pleased by the fact that friends found his names for plants and animals superior to Aristotle's, and his number of facts in excess of Pliny's, Aldrovandi nonetheless felt that he could not have envisioned nature in her entirety without the long tradition of natural history that preceded him.

THE LINCEAN ENCYCLOPEDIA

While Aldrovandi's encyclopedic schemes confined themselves to the territory that the Aristotelian and Plinian corpus had previously defined, the speculations of seventeenth-century natural philosophers claimed to move beyond this realm. In contrast to sixteenth-century encyclopedism, which attempted to fill the paradigms that the classical canons prescribed, the logic of seventeenth-century collecting precluded such an unmitigated acceptance of earlier categories. With the advantage of hindsight, seventeenth-century naturalists interpreted the failure of predecessors such as Aldrovandi to flesh out ancient collecting projects as an indication of the need for new methods and new questions. Despite Aldrovandi's own optimism about his intellectual work, he had left behind few philosophically satisfying results, at least in the eyes of his successors. While contributing greatly to the shape of natural history as it was practiced, in the end, he was still unable to subsume all the creatures in the world within his neo-Aristotelian categories nor had he reached the end of the fact-gathering process designed to complete Pliny's *Natural History*. The quantity of natural things had increased, but was the quality of understanding any better? Even as Aldrovandi organized the parts of nature in his new natural history, contemporaries such as the skeptic Francisco Sanches called into doubt the validity of his enterprise. "In my reasoning, I shall follow Nature alone," wrote the Portugese philosopher in his *That Nothing Is Known* (1581).[75] Dispensing with the need for any authority but that of the senses, he offered a radical epistemology that critiqued the traditional intellectual basis from which certainty was derived: texts and a historically contingent understanding of knowledge. While few philosophers adhered to Sanches's extreme position, it was a presage of things to come. By the early seventeenth century, the community of naturalists was in the process of redefining the scope and purpose of the encyclopedic enterprise.

[74]BUB, *Aldrovandi*, ms. 38(2), Vol. I, c. 76 (Padua, 19 September 1561).
[75]Sanches, *That Nothing Is Known*, p. 172.

The influx of artifacts into Europe from all parts of the globe paved the way for new models of knowledge as collectors found traditional explanations increasingly unable to accommodate the information and objects that they gathered. "A new world has been discovered—new realities—in New Spain or in the West and East Indies," wrote Sanches.[76] While Renaissance naturalists struggled to come to terms with these "new realities," often by denying their very novelty, the succeeding generation was increasingly less troubled by their presence and more reflective about their significance. In the wake of the political and religious upheavals of the Reformation era that called into question other forms of authority, the "old" no longer had the security of meaning it had formerly enjoyed. The fragmentation of Christendom and the erosion of traditional social rights and political privileges created a world *within* Europe that seemed potentially as strange and unpredictable as the New World itself. The seventeenth-century natural philosopher, creator of the new encyclopedia, searched for new models to explain a perplexing, increasingly expansive, and pluralistic universe. Unable to assume the superiority of the past in the face of the present, he had either to prove why that could be so, best exemplified in the work of the Jesuits, or argue for the creation of an alternative framework for the inspection of nature.

No one better exemplified the latter tendency than Federico Cesi, the Roman prince who founded the Accademia dei Lincei in 1603. In the early seventeenth century, Cesi organized a group of individuals—naturalists, mathematicians, and virtuosi—whose common purpose was the reformation of natural philosophy. As a requisite for membership in his academy, Cesi insisted that applicants be "slaves neither of Aristotle nor any other philosopher, but of noble and free intellect in regard to physical things." He attacked traditional philosophers, calling them "philodox" (*filodossi*), that is, sectarians who blindly followed an accepted path, rather than "philosophers" (*filosofi*), true lovers of wisdom.[77] Two thousand years of tradition met the optimism and criticism of several young Italian patricians who thought they could improve on the past. In direct opposition to text-bound scholars, the Linceans ostensibly "read" and collected the book of nature without the mediation of any authority, save for their own good instincts. By the 1610s, the academy numbered Galileo and Della Porta among its members; they corresponded and exchanged information with other Linceans such as the Neapolitan naturalist Fabio Colonna, the physician Johann Faber, who supervised the papal botanical gardens, and the mathematician Francesco Stelluti. In 1623, Francesco Barberini and

[76]Ibid., p. 222. On natural history during the age of contact, see Gerbi, *Nature in the New World*; Margaret T. Hodgen, *Early Anthropology in the Sixteenth and Seventeenth Centuries* (Philadelphia, 1964); and Wilma George, "Sources and Backgrounds to Discoveries of New Animals in the Sixteenth and Seventeenth Centuries," *History of Science* 18 (1980): 79–104.

[77]CL., Vol. II, p. 210 (Tivoli, 21 October 1611); Cesi, *Del natural desiderio di sapere*, p. 47.

Cassiano dal Pozzo, both influential figures at the court of Urban VIII, joined its ranks.[78] For two decades, the Accademia dei Lincei represented the vanguard of the "new" philosophy. They collected books, manuscripts, and botanical specimens and peered into the new instruments—the telescope and microscope—invented by Galileo. They corresponded with scholars throughout Europe and planned a network of Lincean "colonies" throughout Italy—only the one in Naples ever saw the light of day—where scholars would meet in museums, pharmacies, and botanical gardens to redraw the map of knowledge.

At the center of this flurry of activity lay Cesi's pet project: the *Theater of Nature*. This was the new encyclopedia of knowledge that Cesi worked to create. As Carlo Dati wrote in his eulogy of Cassiano dal Pozzo, the goal of the Accademia dei Lincei was "to compile natural history."[79] For Cesi, it was valuable enough to risk the disgrace of disinheritance and endure the wrath of his family, who saw no point in the young prince's speculative pastimes and was alarmed at his increasingly polemical activities.[80] "How truly enormous is the field of knowledge," exclaimed Cesi in his *On the Natural Desire for Knowledge and the Institution of the Linceans for the Fulfillment of This [Goal]*, "as large in the copiousness of speculations as in the copiousness of readings."[81] Cesi discovered in his youth what Aldrovandi learned in his maturity: collecting and the acquisition of knowledge were tasks without end. In response, he chose to create a new receptacle through which to funnel the wisdom of the age. While Renaissance humanists defined *encyclopedia* as a term broadly signifying their emulation of past philosophers, seventeenth-century naturalists designated it a term of innovation.[82]

What defined the Lincean encyclopedia? Like the museum that Cesi cultivated in his family palace, it contained "books, writings, secrets, natural things," in short, all the materials already evident in the work of Aldrovandi.[83] An early outline for the encyclopedia included prodigies and

[78]The story of the Accademia dei Lincei has been told often, mainly because of Galileo's association with it. There is still no definitive study of this academy, although readers should consult the numerous articles and edited manuscripts published by Giuseppe Gabrieli. More recent studies include Jean-Michel Gadair, "I Lincei: i soggetti, i luoghi, le attività," *Quaderni storici* 16 (1981): 763–787; Lombardo, *With the Eyes of a Lyn*; Giuseppe Olmi, " 'In esercitio universale di contemplatione, e prattica': Federico Cesi e i Lincei," in *Università, accademie e società scientifiche in Italia e Germania dal Cinquecento al Settecento*, Laetitia Boehm and Ezio Raimondi, eds. (Bologna, 1981), pp. 169–235; idem, "La colonia lincea di Napoli," in *Galileo e Napoli*, Fabrizio Lomonaco and Maurizio Torrini, eds. (Naples, 1987), pp. 23–58; Redondi, *Galileo Heretic*, pp. 80–97.

[79]Carlo Dati, *Delle lodi del commendatore Cassiano dal Pozzo* (Florence, 1664), n.p.

[80]This section is indebted to Pietro Redondi's discussion of Cesi's encyclopedia in his *Galileo Heretic*, pp. 86–88. I agree in many respects with Redondi's suggestive portrait of the Accademia dei Lincei, save for one important difference: I would distinguish more sharply the rhetoric of the academicians from their actual practice, something I detail throughout this section.

[81]Cesi, *Del natural desiderio di sapere*, p. 48.

[82]Redondi, *Galileo Heretic*, p. 83; Cesare Vasoli, *Enciclopedismo del Seicento* (Naples, 1978).

[83]CL, Vol. I, p. 89.

nature's ambiguities; Cesi was particularly fascinated with mixed categories and planned an entire study of "things in transition."[84] By the end of the 1620s, he could count books such as Fabio Colonna's *Ekphrasis* (1616), a study of plants, the Lincean edition of the Spanish physician Francisco Hernandez's *Treasure of Medical Things from New Spain* (1628), and his own study of bees, the *Apiarium* (1625), as contributing to the collective project of the encyclopedia. A multitude of other treatises were in progress. In the meantime, Cesi continued to refine his *Phytosophical Tables*, which summarized a new philosophy of botany. First published separately, they were eventually appended to the second edition of Hernandez's "Mexican Treasure," as were many of the unfinished Lincean projects.

The *Phytosophical Tables*, completed in 1628, fit within a recognizable tradition of botanical taxonomy then in the process of formation. When Cesi first founded the academy, he had Jan Eck write to Caspar Bauhin, Matthias Lobel, and Carolus Clusius, three of the leading Northern European botanists of the early seventeenth century.[85] Although none of them responded to the Linceans' letters, their work continued to inspire Cesi's own investigations into nature. In 1621, while enjoying various naturalistic excursions from his country residence in Acquasparta, Cesi wrote to Johann Faber to inquire about the state of Bauhin's classification projects. Excited about his own discovery of a peculiar form of fossilized wood, a "metallophyte," Cesi wondered if any other naturalist had included anything similar in his taxonomy. "I also wish to know if there is anyone who has distinguished and enumerated fossils in his classes in an orderly fashion, particularly metals and the said half-minerals. Similarly has anyone reduced synoptically the sciences into trees and tables?"[86]

Undoubtedly, Cesi had Bauhin in mind when he asked these questions. The Swiss botanist already had published his *Forerunner to the Theater of Botany* (1620), a descriptive treatise that dismissed the work of such Renaissance naturalists as Aldrovandi who included uses and virtues as important descriptors. Three years later, Bauhin's *Index to the Theater of Botany* (1623) appeared. Classifying 6000 plants and offering a radically decontextualized view of their essential characteristics, it became the standard botanical work of the seventeenth century.[87] Like Bauhin, Cesi recognized the importance of plants in advancing the science of taxonomy. Unlike Bauhin, he perceived the study of plants not as an end unto itself but as part of the larger

---

[84]Giuseppe Gabrieli, "L'orizzonte intellettuale e morale di Federico Cesi illustrato da un suo zibaldone inedito," *Rendiconti della R. Accademia Nazionale dei Lincei. Classe di scienze morali, storiche e filologiche,* ser. 6, 14, fasc. 7–12 (1938–1939): 676–678; Clara Sue Kidwell, *The Accademia dei Lincei and the "Apiarium": A Case Study of the Activities of a Seventeenth Century Scientific Society* (Ph.D. diss., University of Oklahoma, 1970), p. 307.

[85]CL, Vol. I, pp. 13–14.

[86]CL, Vol. II, pp. 732–733 (Acquasparta, 20 January 1621).

[87]M. M. Slaughter, *Universal Languages and Scientific Taxonomy,* p. 53; Atran, *Cognitive Foundations of Natural History,* pp. 135–137.

study of nature. While Cesi undoubtedly was disappointed that the Linceans would not enjoy the glory of publishing the definitive work on the "new" botany, he consoled himself with the fact that Bauhin had nothing to say about fossils and probably had never seen a metallophyte. Most importantly, the completion of the *Phytosophical Tables* reaffirmed his own ability to discern order in the chaos of the universe. "I find myself in that great Chaos of the methodical distribution of plants, and I seem to have totally overcome it," he wrote to Faber in 1622. "It will be a good part of my *Mirror of Reason* and *Theater of Nature.*" Two years later, he reported to Cassiano dal Pozzo, with evident pride, that he had "reduced the mineral wood into a table."[88] The academicians fostered the prince's sense of self-worth by singing the praises of his tables. "Certainly if Theophrastus came back from the dead, he would be amazed at them," wrote Colonna to Stelluti.[89] In the eyes of the Linceans, Cesi had surpassed the ancients.

Had Bauhin had the opportunity to peruse Cesi's *Phytosophical Tables*, he surely would have remained unimpressed. And at least Bauhin acknowledged Theophrastus as his master. Despite his interest in the Swiss naturalist's work, Cesi included an entire section on uses of plants in food, medicine, and the mechanical and liberal arts. Aldrovandi's program of natural history may have been disparaged, but it was not exactly excised from the Lincean encyclopedia. The category "liberal arts" included poetry, oratory, geometry, music, painting, and architecture as seen in public processions, funerals, and triumphs and as represented in enigmas, hieroglyphs, and images.[90] Despite their self-proclaimed novelty, the *Phytosophical Tables* contained many of the features that made Aldrovandi's natural history "emblematic," albeit in marginalized form. Where Aldrovandi lovingly had detailed each category of symbolic knowledge, Cesi now collapsed them all under the heading of the liberal arts.

Had Bauhin examined the second part of the synopsis, he would have been even more horrified. Here Cesi summarized all the sciences and arts pertinent to botany, and listed all the authors whose contributions to the field merited praise. Recall Cesi's condemnation of traditional natural philosophers: "This passionate friendship for authors, expressly prohibited by Aristotle, now exquisitely followed by the Aristotelians, impedes . . . the necessary reading of the book of the universe."[91] Fabio Colonna appeared more frequently than any other author. Naturalists such as Bauhin, Gesner, Clusius, Mattioli, and Nardo Antonio Recchi, the Neapolitan physican who had brought Hernandez's manuscripts to Italy, merited multiple entries. Others such as Joachim Camerarius, Andrea Cesalpino, Giovanni Pona, Cal-

[88]CL, Vol. I, p. 778 (Acquasparta, 19 November 1622); Vol. II, p. 947 (Acquasparta, 29 September 1624).

[89]CL., Vol. III, p. 1201 (Naples, 26 October 1629).

[90]Cesi, *Phytosophicae Tabulae*, in Altieri Biagi and Basili, *Scienziati del Seicento*, p. 72.

[91]Cesi, *Del natural desiderio di sapere*, p. 47.

zolari, and the redoubtable Della Porta appeared once, as did Galen, Dis-
corides, Pliny, and several other ancient writers and their commentators.
Undoubtedly, Cesi's association with Della Porta led him to give natural
magic a prominent place in his map of knowledge, a form of inquiry that
hardly fit Bauhin's criteria of "essential" knowledge.

Aristotle, Theophrastus, and Aldrovandi were the most noteworthy absences on the list of botanical authorities. While Cesi liberally included such
fields as plant harmony, plant physiognomy—one of Della Porta's most well-
known books was his *Phytognomonica* (1588)—botanical pharmacy, and flori-
legium, he refused to acknowledge one of the most well-known masters of
these genres, Aldrovandi. His table emphasized the importance of ethics,
politics, logic, and metaphysics to the study of plants, yet did not mention
Aristotle. Pliny appeared under the category of medicine, yet not under the
all-important category *Encyclopaedia* (which Cesi placed above metaphysics).
There he referred readers to Gesner and Bauhin. The more we inspect the
tables, the more puzzling they become and the less novel they appear. Certainly they reinforce the viewpoint that taxonomy was fundamentally an
Aristotelian exercise, even if its practitioners refused to acknowledge the
connection.[92] Cesi had liberated himself from the past with the tools of ancient philosophy and the techniques of Renaissance natural history.

The behavior of the Linceans only reinforces their conflicted relationship with authority. In 1602, Jan Eck, newly made Lincean, wrote twice to
Aldrovandi, calling himself a "disciple of the natural sciences."[93] In describing himself as *discipulus,* Eck acknowledged Aldrovandi as his "master"
in natural history. Eck was not the only Lincean to correspond with
Aldrovandi; Della Porta wrote several letters and Colonna at least one letter
to the Bolognese naturalist in the 1590s. Aldrovandi expressed great interest in the Hernandez manuscripts, then in the possession of Nardo Antonio
Recchi, and Della Porta attempted to procure a "list of Peruvian simples" for
him. In the interim, he sent a portrait of a remora, a fish fabled for its ability to stop ships dead in the water. More importantly, Della Porta asked
Aldrovandi if he had read his *Phytognomonica.* The Neapolitan naturalist
planned to reissue it and hoped to append the opinion of "some clever
man" to the new edition. If Aldrovandi did not own a copy, he would be
happy to send it to him.[94] Perhaps Aldrovandi's failing eyesight or his friendship with Imperato, Della Porta's rival in Naples, prevented the plan from
coming to fruition. Yet Della Porta's request, like Eck's homage, reflected
the status of Aldrovandi in the community of naturalists at the end of the

[92]Slaughter, *Universal Languages and Scientific Taxonomy*, p. 3.

[93]BANL, *Archivio Linceo*, ms. 18, cc. 12–15, 23–25r, esp. 12v (Jan Eck, *Epistolarum medicinalium*).

[94]BUB, *Aldrovandi*, ms. 136, XIII, c. 294r; XIV, c. 165; XIX, c. 156v; XXV, c. 83 (Naples, 7 June, 28 July, and August 1590, and 30 September 1595). For more on the remora, see Copen-haver, "A Tale of Two Fishes."

sixteenth century. By the 1620s, this had changed noticeably. Some twenty-five years after Eck's letters, Aldrovandi intrudes again in the Lincean correspondence, in a letter directed to Cesi by Fabio Colonna. "I saw the [book of] Aldrovandi," reported Colonna, who severely criticized its content and manner of production.[95] The proximity that Eck and Della Porta felt toward Aldrovandi contrasted sharply with Colonna's own distance. Aldrovandi earned the respect and admiration of the early Linceans; from their successors, only scorn.

In his discussion of the Accademia dei Lincei, Pietro Redondi described Cesi's project as an encyclopedia attempting to encompass "the entire Hermetic, magical, alchemist library without being contaminated by the traditional metaphysical speculations (sympathies and analogies) or by the purely qualitative principles of Renaissance naturalism and the Hermetic tradition."[96] From this vantage point, Della Porta was as essential to the project as Colonna, since both represented different philosophical trajectories that opened up new directions. In 1610, when Della Porta joined the academy, Cesi must have been quite excited about his addition to the ranks of the members. Della Porta was a figure of international renown, widely published and well known throughout the learned world. He would bring fame as well as intellectual legitimacy to the young academy. By the time of Della Porta's death in 1615, Cesi was less enamored of the Neapolitan magus. Galileo's star was on the rise, Della Porta's on the wane. While Cesi continued to be intrigued by natural magic, he looked less favorably upon applications for membership from alchemists and other possessors of "secrets" during the 1620s. They represented an older paradigm with which he no longer wished to be associated. Mathematics, experimental philosophy, and the world of sense-enhancing instruments were the wave of the future. He refused to be left behind. The *Theater of Nature* would simply have to accommodate all of these transformations.

In 1625, Cesi published the *Apiarium*, a definitive study of bees. While the *Phytosophical Tables* shed light on the organization of the Lincean encyclopedia, the *Apiarium* illuminates the content. Cesi began with a series of tables that classified the bee according to its principal divisions: solitary and civil, urban and rural, armed and unarmed, free and enslaved, singing and buzzing, and so forth.[97] To the extent that Cesi offered any biological or morphological description of bees, it was drawn primarily from Aristotle and Pliny, supplemented with a few random observations made by himself and Colonna.[98] While presenting the book to Maffeo Barberini, his patron,

[95]CL, Vol. III, pp. 1190–1191 (Naples, 16 December 1628). Colonna probably referred to Aldrovandi's *De mollibus, crustaceis, testaceis, et zoophytis* (Bologna, 1606).

[96]Redondi, *Galileo Heretic*, p. 87.

[97]These tables are reproduced in Kidwell, *The Accademia dei Lincei and the Apiarium*, pp. 134–139.

[98]Stelluti would publish more detailed observations in his *Persio tradotto* (Rome, 1630) as a footnote to his translation of Persius's *Satires*.

Cesi dedicated the *Apiarium* to Pliny, "master of the pleasing natural science," implicitly making himself the Roman encyclopedist's disciple. Pliny counseled that the secrets of nature lay in the smallest things, and Cesi now responded to this call:

> Wondering at and esteeming the structure of the little body you, Pliny, have agreed that the nature of things is never greater than when it is complete in its smallest possible form. If only you could have used the microscope, if you could have used the telescope, what could you have said earlier about the lion-maned, multi-tongued, hairy-eyed bee?[99]

Toward the end of this encomiastic natural history of the bee, Cesi thanked the "greatest philosophers, the greatest poets, and the outstanding men among the ancients and contemporaries, physicians and moralists" who contributed to his text. A mosaic of citations, culled mostly from Aristotle and Pliny, and even slightly from the unpraiseworthy Aldrovandi, it was a model of humanist erudition.

In March 1625, Johann Faber wrote to Cesi, "I am sending Your Excellency my Pliny."[100] Should we find it strange that Cesi, castigator of the "philodox," wished to read the *Natural History*? It was precisely because Cesi had freed himself from the shackles of ancient authority that he could read Pliny with alacrity. His new philosophy of nature, and the new techniques of observation that it generated, made him a profoundly uncanonical reader of a canonical text. "How far have I worn down my reed pen, more than in any part of [Pliny's] *Natural History*," he proclaimed at the end of the *Apiarium*.[101] Secure in the knowledge that he added more than mere words to the description of the bee—the task of an Aldrovandi, as Cesi would have explained—the Prince of the Linceans never even noticed that he had added nothing to its anatomical observation. His microscope, a gift of Galileo, had made it bigger, stranger, somehow different. Surely this was enough to prove his point?

Dismissive of Aristotle and suspicious of Aldrovandi, Cesi continued to work on his encyclopedia, right up until the end. Time was not as kind to Cesi as it had been to Aldrovandi. Aldrovandi almost lived to see his eighty-fifth birthday; Cesi did not even make it to forty-five. If most of a century had been inadequate to complete the Aristotelian program, less than a half-century surely did not allow enough time to dismantle it *and* create something new. In the end, Cesi left behind even fewer manuscripts than Aldrovandi and fewer directions for his disciples regarding the continuation of his encyclopedia. Hopes, dreams, and aspirations were not enough to keep the Linceans going, particularly in a scientific culture on the verge of condemning Galileo. Unwilling to acknowledge how much his encyclopedia owed to the past, Cesi died with his vision of a new scientific culture in-

[99]Kidwell, *Apiarium*, p. 261.
[100]CL, III, p. 1032 (Rome, 22 March 1625).
[101]Kidwell, *Apiarium*, p. 285.

tact. Of the members of the academy, only Stelluti, Colonna, Barberini, and dal Pozzo survived to mid-century. Stelluti, loyal to the end, completed the publication of Cesi's *Treatise on the Newly Discovered Wood-Fossil-Mineral* (1637). Colonna refined his ideas on the organic origins of fossils. The cardinal and his faithful servant survived the vicissitudes of a politically turbulent era as best they could and continued to expand the scope of their museums. As the years passed, they spoke less and less of the Accademia dei Lincei and its reformation of knowledge.

THE JESUITS PUT THEIR WORLD IN ORDER

Even as Cesi and his academicians published the latest installments of their new encyclopedia of nature, Jesuit philosophers such as Kircher reaffirmed the importance of ancient encyclopedic strategies by reorienting them to fit the demands of their culture. Only three years after the demise of the Accademia dei Lincei, Kircher arrived in Rome, recommended by the French *savant* Peiresc to Francesco Barberini. In September 1633, Peiresc praised Kircher to Barberini and Dal Pozzo for the "most curious inventions and most rare experiments that he practices and multiplies daily." The young Jesuit, he assured them, betrayed "the most evident signs of his great piety and innocence. No less, his magnanimity and sharp intelligence allow him to penetrate, well ahead of his time, the discoveries of many secrets, of nature, antiquity and the principal languages of Christianity."[102] By 1634, news had reached Galileo, then under house arrest after his unmasking as a Copernican, of the young polymath: "Again there is a Jesuit in Rome [who] stayed a long time in the Orient. Besides possessing twelve languages, [being] a good geometer, etc., he has a great many beautiful things with him," wrote Raffaelo Maggioti.[103] At the time of Galileo's death in 1642, Kircher already was hailed as the apostle of the new scientific order in the papal city. Cesi and Galileo represented the confidence of the new philosophy in the early seventeenth century; Kircher embodied the renewed strength of the "Monarchy of the Church" at mid-century, then in the process of extending the papal *imperium* to the far corners of the earth.[104] Central to this process was the redefinition of scientific orthodoxy.

Even as the activity surrounding the Lincean encyclopedia subsided, a new project was on the horizon. Scholars throughout Europe had begun to

---

[102]Nicolas-Claude Fabri de Peiresc, *Lettres à Cassiano dal Pozzo (1626–1637)*, Jean-François Lhote and Danielle Joyal, eds. (Clermont-Ferrand, 1989), p. 112 (10 September 1633); Peiresc to Barberini, 10 September 1633, in Cecilia Rizza, *Peiresc e l'Italia* (Turin, 1965), pp. 89–90.

[103]In Rivosecchi, *Esotismo in Roma barocca*, p. 49 (Rome, 18 March 1634).

[104]This phrase comes from Daniello Bartoli, *La Cina*, Bice Garavelli Mortara, ed. (Milan, 1975), p. 27. For a general overview of the role of the Jesuits in early modern scientific culture, see Steven J. Harris, "Transposing the Merton Thesis: Apostolic Spirituality and the Establishment of the Jesuit Scientific Tradition," *Science in Context* 3 (1989): 29–65.

focus their attention on the problem of language as the key to unlocking the hidden wisdom of the ages. Hieroglyphs particularly captured the Renaissance imagination; from the neo-Platonic circle that surrounded Ficino to the writers of emblem books and the humanists who fabricated new images of the papacy, everyone attempted to decode these *arcana arcanissima*.[105] Stumbling upon a book on the obelisks of Rome, perhaps the one written by the papal physician Mercati in 1589, Kircher recognized the value of the mysterious emblems for his studies of language, nature, and religion. "Immediately my curiosity was aroused and I began to speculate on the meaning of these hieroglyphs," he wrote in his autobiography.

> At first I took them for mere decoration, designs contrived by the imagination of the engraver, but then, on reading the text of the book I learned that these were the actual figures carved on ancient Egyptian monuments. From time immemorial these obelisks and their inscriptions have been at Rome and so far no one has been able to decipher them.[106]

Mercati had kept his research on the obelisks separate from his studies of nature; Kircher immediately saw the potential to integrate the two. His acquaintance with Peiresc, during a stay at the Jesuit college in Avignon, only fueled this interest.

By the 1630s, the humanist inquiry into the hieroglyphs, the most impenetrable of all symbols, was on the verge of yielding something. Reading the letters of Peiresc to Cassiano dal Pozzo about Kircher we can literally feel the excitement. "But if he can break the ice and penetrate some little thing, perhaps with time he can overcome other difficulties and, little by little, come to know . . . at least something if not everything [of the hieroglyphs]."[107] Two years later, Kircher published his *Coptic Forerunner* (1636), followed by his *Egyptian Language Restored* (1643). By the end of his life, he presented himself as the Christian Hermes who had restored order to the world.[108] The deciphering of the hieroglyphs was to be the basis upon which he claimed his authority for unlocking the secrets of nature.

With the help of disciples such as Gaspar Schott, Francesco Lana Terzi, Giorgio de Sepi, Gioseffo Petrucci, and his fellow collector Settala, Kircher deepened the roots of the ancient encyclopedia of knowledge and expanded its material base. Whereas Aldrovandi perceived Aristotle to be the font of knowledge, Kircher extended his chronological reach, moving ever

[105]Eric Iverson, *The Myth of Egypt and Its Hieroglyphs* (Copenhagen, 1961).

[106]In Reilly, *Athanasius Kircher*, p. 38. Kircher's publications on Egypt, Coptic, and the hieroglyphs are as follows: *Prodromus coptus sive Aegyptiacus* (Rome, 1636); *Lingua Aegyptiaca restituta* (Rome, 1643); *Rituale ecclesiae Aegyptiae sive cophtitarum* (n.p., 1647); *Obeliscus Pamphilius* (Rome, 1650); *Oedipus Aegyptiacus* (Rome, 1652–1654); *Obeliscus aegyptiacus* (Rome, 1666); and *Sphinx mystagoga* (Amsterdam, 1676).

[107]Peiresc, *Lettres à Cassiano dal Pozzo*, p. 134 (4 May 1634).

[108]De Sepi, p. 10. The importance of the hieroglyphs to Kircher's encyclopedia of knowledge is discussed in Rivosecchi, *Esotismo in Roma barocca*, p. 50ff.

backward to the original source. For him, Egyptian wisdom was the source of Western learning. Hermes *and* Aristotle would be his guides. In some thirty-eight books, Kircher mapped out the correspondences among different parts of the natural world and diverse forms of knowledge and addressed the most pressing intellectual problems of the age. Magnetism, hieroglyphics, mummies, universal and artificial languages, astronomy, mathematics, optics, acoustics, musical harmonies, plague, and natural history were all subjected to his scrutiny. Unable to travel to China, Kircher made himself the leading expert in print by collecting the reports of others. He published "archaeologies" of the Biblical world, reconstructing Noah's Ark and the Tower of Babel, with the same authority that he lent to the study of ancient Latium, the region surrounding Rome. Drawing upon his linguistic skills, Kircher attempted to erode the artificial divisions of language that made total knowledge an inaccessible and elusive goal. Where Aldrovandi hoped to recapture the Aristotelian and Plinian program of natural history, and Cesi envisioned a new encyclopedia of nature, Kircher set himself an even grander task: identifying the archetypal structures of the world. As Gioseffo Petrucci commented in his *Apologetic Forerunner to Kircherian Studies* (1677), Kircher was the centerpiece of the "great Theater of Knowledge," the Baroque encyclopedia.[109]

Kircher was by no means a typical Catholic nor even a typical Jesuit. The exotic and speculative propensities of his age were more highly cultivated in him than in most of his contemporaries, and he was quick to articulate his vision. Yet the success of Kircher's museum, as the center of a global missionary network and a scientific culture that crossed confessional as well as regional boundaries, made him the heart and soul of Baroque Europe. Whether earning the respect or disbelief of his contemporaries, Kircher commanded their attention; even creators of new experimental philosophies such as the members of the Accademia del Cimento or the Royal Society, who found his philosophy hopelessly outdated, could not afford to ignore him. Determined to absorb every old and new philosophy that came his way in order to subsume them within his great synthesis of knowledge, Kircher embodied the propensity of Baroque philosophers to bring the possibilities of Renaissance syncretism to their logical fruition.[110] The result was a delightful, alarmingly heterodox intellectual product, a cornucopia of ideas and information that threatened to overflow at any moment. Subsumed within a religious framework that gave his philosophy conceptual clarity, "to the greater glory of God" (*ad maiorem gloriam Dei*), Kircher's investigations of the natural world provided tangible demonstrations of the hidden affinities between objects and their properties. Only in the micro-

[109]Gioseffo Petrucci, *Prodromo apologetico alli studi chircheriani* (Amsterdam, 1677), preface.

[110]See Giuliana Mocchi, *Idea, mente, specie: Platonismo e scienza in Johannes Marcus Marci (1595–1667)* (Soverzia Manelli, 1990); Pighetti, "Francesco Lana Terzi e la scienza barocca."

cosm of the Roman College museum, and in the publications resulting from it, could the artifacts of nature—dispersed throughout the vast expanse of the world—be effectively catalogued and compared.

As R. J. W. Evans describes in his study of Habsburg intellectual life, the philosophical trajectories of Catholic Reformation culture lent an exoticism to intellectual discourse that was not evident in the scholarship of the previous century.[111] By the mid-seventeenth century, the parameters of knowledge had expanded considerably. The "new realities" of the Americas were no longer the novelties they had been for Aldrovandi, nor were they the definitive objects of experience that excited the Linceans who edited Hernandez's natural history of Mexico. Another century of travel, exploration, and conversion brought the boundaries of Western Europe more sharply into focus. Kircher's imagined geography included not only the New World but also Asia and Egypt, two settings that particularly fascinated the Jesuits. In one catoptric machine, Kircher placed an image of an elephant "that seemed collected from all Asia and Africa."[112] Multiplied an infinite number of times by the mirrors, the elephant emblematized the presence and power of the Catholic Church, whose most faithful members could gather a herd of elephants in their museums. While sixteenth-century naturalists faced the dilemma of incorporating the artifacts of the Americas into their cosmos, the seventeenth-century Jesuits attempted to develop a moral, religious, and philosophical framework that connected *all* the different regions of the world. Out of this exercise, they assured themselves, a new synthesis would emerge.

Strategies for collecting were not only designed to fulfill the humanistic desire for *prisca scientia;* museums also conveyed political and religious messages. Claude Clemens, librarian to Philip III of Spain, described the Escorial as "this Museum of Christendom"; attuned to the rhetoric of the Catholic Reformation, he proposed the creation of a structure that collected and ordered knowledge to control it.[113] Museums were not only necessary for their public utility for a growing community of scholars; they also protected the Catholic world from false erudition. Kircher's persistent attacks on demonic magic, impious arts such as alchemy and Paracelsian medicine, heliocentrism, and uncanonical physical tenets such as perpetual motion or the idea of a vacuum in nature, all demonstrated with the objects in his museum, illustrate the close connections between his scientific and reli-

[111]Evans, *The Making of the Habsburg Monarchy*, pp. 311–345, 419–450.

[112]De Sepi, p. 38.

[113]Clemens, *Musei sive bibliothecae* (Leiden, 1635), pp. 2–4, 523; as Pietro Redondi observes in his *Galileo Heretic*, Raymond Rosenthal, trans. (Princeton, 1987), pp. 80–81: "As instruments of intellectual monopoly, the great libraries created at the beginning of the century expressed the strength and prestige of traditional humanist and theological culture, which was forging new instruments of erudition and exegesis: the most modern weapons for sustaining, on all intellectual fronts, the effort of Catholic reform and religious struggle."

gious goals. Collecting provided the basis of a new intellectual synthesis and the foundation of a new Catholic order.[114] As Kircher affirmed in his *Egyptian Oedipus* (1652–1654), "Unity is the Essence of God."[115]

The encyclopedic impulse was not confined to the Catholic world alone, although it was more pervasive in an atmosphere in which the retention of ancient models of knowledge was linked to the persistence of orthodoxy and tradition. Kircher, linked to papal Rome and the Imperial court in Vienna, developed his ideas within the context of the two leading orthodoxies of the day, one political, the other religious. A product of a world disrupted by the turmoil of the Thirty Years' War (1618–1648) yet unified by the Catholic missionary networks, he was obsessed with the creation and maintenance of order. While Cesi turned botanical "chaos" into synoptic tables, drawing upon the collective knowledge of his academy to bring order to the natural world, Kircher relied on the resources of the Society of Jesus to gather the materials necessary for his own reinvention of the universe. Despite his strong ties to Catholic Reformation political and religious culture, Kircher embodied yet another trend of the mid-seventeenth century: the quest for *pansophia* that engaged scholars of varied religious backgrounds. Certainly the Jesuit's vow of obedience to the pope reminded him that knowledge was a powerful tool of conversion. Yet this did not restrain him from corresponding with scholars and collectors or from accepting gifts from patrons who remained Protestant; divided by religion, they shared common intellectual interests. Surely, Kircher reassured himself, openness to all forms of learning and an ability to communicate with all peoples, Catholic, Protestant, and pagan, was the only way to restore the monarchy of the church.

Situated in Rome, clearing house for the Jesuit missionary activities, Kircher sated his thirst for knowledge of all civilizations and all of nature; books, artifacts, and reports flowed incessantly into his museum. From this quantity of information, Kircher mapped out the *ars analogica,* the "analogical art" that revealed the pattern of the world as written by God. As Cesare Vasoli comments in his study of seventeenth-century encyclopedism, "The task of the Catholic scholar thus seemed to consist of making one's way through the encyclopedic 'forest' of mysteries, secrets, and 'sympathetic' virtues of the world, in search of a sort of archetypal language that collected, at its font, an unmoving, unchanging truth, beyond the flow of cultures, doctrines and civilizations."[116] The Jesuit collector, the embodiment of these encyclopedic proclivities, gathered and displayed all the secrets of the world.

[114]The most useful overview of this subject is William Ashworth, "Catholicism and Early Modern Science" in *God and Nature: Historical Essays on the Encounter Between Christianity and Science,* David Lindberg and Ronald Numbers, eds. (Berkeley and Los Angeles, 1986), pp. 136–166; see also Martha Baldwin, "Magnetism and the Anti-Copernican Polemic," *Journal for the History of Astronomy* 16 (1985): 155–174.

[115]Kircher, *Oedipus Aegyptiacus* (Rome, 1652–1654), Vol. III, p. 6.

[116]Cesare Vasoli, *Enciclopedismo nel Seicento,* (Naples, 1978), p. 45.

To invoke a favorite metaphor of Kircher and his contemporaries, the museum was the "Ariadne's thread" that led the faithful out of the labyrinth.[117]

Baroque Aristotelianism, of which Kircher was a sublime example, enhanced the syncretic tendencies of Renaissance philosophy.[118] It measured the teachings of Aristotle against a vaster field of ancient, pristine truths held up to the mirror of experience. As Charles Schmitt notes, by the end of the Renaissance, Aristotelianism had become a porous sponge that absorbed a variety of different philosophies. Not least of these were hermeticism and natural magic.[119] They appeared in the pedagogical programs instituted in the Jesuit classrooms and in the research of scholars such as Kircher. Naturalists who had begun with the quest for the authentic Aristotle, as a means of reviving his philosophy of nature, now found themselves reinventing Aristotle in light of the alternative natural philosophies that the wholesale excavation of antiquity had brought to light. By the time the *Ratio studiorum*, the Jesuit manual of pedagogy, appeared in 1599, it reflected these modifications. Advising the professor of philosophy, the Society of Jesus counseled, "In matters of any importance let him not depart from Aristotle unless something occurs which is foreign to doctrine or which academies everywhere approve of; much more so if it is opposed to orthodox faith."[120] Thus, the Jesuits advocated a philosophical "orthodoxy" that allowed ample room for other forms of knowledge to enter, either for reasons of faith or due to the consensus of the community of Catholic natural philosophers.

The ambivalent authority of Aristotle led Kircher to diverge significantly from the Greek philosopher's definition of knowledge. When Peiresc suggested to Dal Pozzo that he feared Kircher might "do violence to the authority of the ancients," he reflected a certain conservatism toward the Jesuit's intellectual program that many earlier philosophers would have shared.[121] Had Peiresc lived to see the publication of the *Egyptian Oedipus*, his worst fears would have been confirmed. Kircher advertised the three-volume tome, the summa of his knowledge about Egyptian philosophy and its permutations, as containing "Egyptian wisdom, Phoenician theology,

[117]Two examples: Kircher, *Ars magna sciendi* (Amsterdam, 1669), sig. ***2v; idem, *Arithmologia* (Rome, 1665), p. 73.

[118]The place of Aristotle in seventeenth-century culture is a sorely understudied subject; the standard study remains Guido Morpurgo Tagliabue, "Aristotelismo e Barocco," in Enrico Castelli, ed., *Retorica e barocca*, Atti del III Congresso Internazionale di Studi Umanistici, Vol. 3, 1954 (Rome, 1955), pp. 119–195. However, there is a large specialist literature on Emanuele Tesauro's *Cannocchiale aristotelico* that addresses this issue.

[119]Schmitt, *Aristotle in the Renaissance*, p. 97; Gabriele Baroncini, "L'insegnamento della filosofia naturale nei Collegi Italiani dei Gesuiti (1610–1670): Un esempio di nuovo aristotelismo," in Gian Paolo Brizzi, ed., *La "Ratio studiorum." Modelli culturali e pratiche educative dei Gesuiti in Italia tra Cinque e Seicento* (Rome, 1981), pp. 185–192, 213–215.

[120]*Ratio studiorum*, in Edward Fitzpatrick, ed., *Saint Ignatius and the Ratio Studiorum* (New York, 1933), p. 168. For a more detailed discussion of the theology of the Jesuit position toward knowledge, see Rivka Feldhay, "Knowledge and Salvation in Jesuit Culture," *Science in Context* 1 (1987): 195–213.

[121]Peiresc, *Lettres à Cassiano dal Pozzo*, p. 161 (29 December 1634).

Chaldean astrology, Hebrew cabbala, Persian magic, Pythagorean mathematics, Greek theosophy, Mythology, Arabic alchemy and Latin philology."[122] Like the magnetic virtues that linked together the diverse parts of the natural world, the wisdom of the ancients was also bound together by mysterious and unseen forces, accessible only to a Christian magus.

While agreeing with Aristotle that wisdom consisted of universal truths, made apparent in diverse forms of experience, Kircher differed sharply in his criteria for establishing knowledge and understanding its purpose. The primary function of Kircher's encyclopedia lay in the identification of *signs*. A symbol, he posited, "leads our mind through a kind of similitude to an understanding of something very different from the things which offer themselves to our external senses; whose property is to be hidden under a veil of obscurity."[123] The system of correspondences that linked these signs formed, as Kircher's colleague Daniello Bartoli put it, "the philosophy of nature, as if nature had written, almost in a cipher, her precepts everywhere."[124] Not unlike Paracelsus's doctrine of signatures, which Kircher roundly condemned as a good Catholic, or Ficino's astral magic, his *ars signata* presented nature as a divinely encoded structure that only a Catholic natural philosopher could read properly.[125]

Kircher's definition of the symbol paralleled his views on *contemplative magic*, defined as the study of "things inwardly concealed in the arcane majesty of Nature." Hieroglyphs, the monad, the philosopher's stone, and other metaphysical truths fell into this category. Publications such as his *Arithmology or the Hidden Mysteries of Numbers* (1665) revealed the "chain of natural things expressed by numbers," truths as intangible and as crucial to his philosophy of nature as the hieroglyphs.[126] In contrast, the loadstone, the cobra's stone (renowned for its remarkable curing powers), and other naturally occurring phenomena that visibly demonstrated the operative powers of nature fulfilled Kircher's definition of *effective magic*. Through it, "by little known means whatever is admirable in the various disciplines and arts is brought to life," observed the Jesuit in his *Great Art of Light and Shadow*.[127] While the former derived from his interest in alchemy and neo-Platonism,

---

[122]This is discussed briefly in Joscelyn Godwin, "Athanasius Kircher and the Occult," in John Fletcher, ed., *Athanasius Kircher und seine Beziehungen zum gelehrten Europa seiner Zeit*, Wolfenbüttler Arbeit zur Barockforschung, 17 (Wiesbaden, 1988), p. 17.

[123]Kircher, *Oedipus Aegyptiacus* (Rome, 1652–1654), ii, I, classis I, p. 6, in Evans, *The Making of the Habsburg Monarchy*, p. 437.

[124]Bartoli, *De' simboli trasportati al morale*, in Mario Praz, *Studies in Seventeenth-Century Imagery* (Rome, 1964), p. 19.

[125]For a broader discussion of the *ars signata*, see Massimo Luigi Bianchi, *Signatura rerum. Segni, magia e conoscenza da Paracelso a Leibniz* (Rome, 1987).

[126]Kircher, *Arithmologia*, p. 280.

[127]Kircher, *Ars magna lucis et umbrae* (1646 ed.), p. 769. This material and the translations are drawn from Godwin, "Athanasius Kircher and the Occult," in Fletcher, *Athanasius Kircher*, p. 23.

the latter was based upon a reading of Cardano and Della Porta. Here and elsewhere, Kircher compounded the wisdom of the ancients with the knowledge of more recent natural philosophers.

The objects in Kircher's museum reflected the emphasis he placed on symbolic knowledge and universal correspondences. Besides hieroglyphs, magnetism was the other subject that greatly preoccupied Kircher. His first publication, written before his arrival in Rome, was on the *Magnetic Art* (1631). This was followed by the *Loadstone or the Magnetic Art* (1641) and the *Magnetic Kingdom of Nature* (1667).[128] Many of his other treatises included a discussion of this phenomenon. Like the hieroglyph, magnetism expressed a tangible reality as well as being a metaphor for all natural operations. It emblematized effective magic, just as hieroglyphs embodied sympathetic magic, and therefore was part of *physica operativa*.[129] "The world is bound in secret knots," proclaimed the frontispiece of his *Magnetic Kingdom of Nature*. Magnetism was the golden chain, to invoke a favorite metaphor of Della Porta and Kircher, that manifestly linked together all segments of the universe. The objects in the Roman College museum served to make these connections apparent.

Examples of magnetic virtue were scattered throughout the museum. "Species of sympathetic matter," probably the heliotropic plants out of which Kircher created vegetable clocks, decorated the windowsills. The cobra's stone, experimentally proven by Kircher to draw forth poison sympathetically from a wound, was proudly displayed. "The virtues of all natural things imitate the power of the loadstone," wrote Kircher.[130] At the heart of the museum lay the loadstone itself, which De Sepi identified as "the center of the Kircherian Museum."[131] A tiny object, proportionally overwhelmed by all the machines and other large artifacts that more immediately caught the attention of visitors, it nonetheless was the "key" to the museum. Imbedded within various mechanical devices that demonstrated the occult powers of the loadstone, it literally put the collection in motion. Magnetism, as Kircher wrote, was "the path to the treasure of the entire world, the only guide and key to all motion whatsoever."[132] This was one of many instances in which Kircher extended Della Porta's definition of natural magic, the "practical part of Natural Philosophy."[133]

---

[128]For a more detailed discussion of Kircher's views on magnetism, see Baldwin, "Magnetism and the Anti-Copernican Polemic."

[129]Vasoli, "Considerazioni sull' 'Ars magna sciendi,' " in Casciato, p. 73.

[130]Kircher, *Magneticum naturae regnum* (Rome, 1667), p. 3.

[131]This, and the example of the heliotropic plants, are drawn from De Sepi, pp. 1, 18.

[132]Kircher, *Magnes sive de arte magnetica* (Cologne, 1643 ed.), in Ulf Scharlau, *Athanasius Kircher (1601–1680) als Musikschriftsteller: Ein Beitrag zur Musikanschauung des Barock* (Marburg, 1969), p. 6.

[133]Giambattista Della Porta, *Natural Magick*, Derek J. Price, ed. (New York, 1957; reproduction of 1658 ed.), p. 3.

Turning from the smallest objects in the collection to the largest, as the
frontispiece of Giorgio de Sepi's *Most Famous Museum of the Roman College of
the Society of Jesus* (1678) suggests, the five obelisks positioned strategically in
the corridors were the most monumental artifacts in the collection (fig. 5).
Representing in miniature the obelisks that Kircher had helped to restore in
Rome for various popes and cardinals, they embodied his "success" in de-
coding the Egyptian hieroglyphs. Hieroglyphs were more than man-made artifacts; as divine signs, they were impressed upon all natural and human creations. One of Kircher's favorite inventions was a talking statue of the Devil,
whose operation he described in the following terms: "By the statue opening
its mouth widely, our Hieroglyph is controlled mechanically."[134] Carved into
the obelisks, evident in the bits of nature on which God had left his divine
imprint and made manifest in the technological phantasms of the museum,
Kircher's "hieroglyph" represented the archetype of all knowledge.

Kircher's fascination with the hieroglyphs indicated a broader interest in
the problems of communication. Numerous objects in the Roman college
museum attested to its importance as a laboratory of communication. Chinese scrolls, Egyptian hieroglyphs, Etruscan tablets, and fragments of other
ancient and modern scripts were materials from which to reconstruct the
original language of humans. While late Renaissance philosophers created
enormous polyglot dictionaries, such as Claude Duret's *Treasure of the History of the Languages of This Universe* (1607), Kircher went several steps further. In addition to developing numerous proposals for artificial languages,
publicized in his *New and Universal Polygraphy* (1663) and *Great Art of Knowledge* (1669), Kircher, with his penchant for machines, designed an *Arca Glot-
totactica* ("linguistic chest"), a communicating device probably displayed in
his museum alongside his other marvelous inventions.[135] By moving the different levers, one could produce words in various languages, just as Pascal's
calculator manipulated numbers to produce more numbers.

While Kircher and many of his contemporaries felt that it would be impossible to restore the original language of humans, they nonetheless believed that knowledge of it was necessary to replace the corrupt offshoots of
Adamic language with a new universal language.[136] The urgency of his mis-

[134]Kircher, *Ars magna lucis et umbrae*, p. 772.

[135]On Kircher's interest in universal languages, see George E. McCracken, "Athanasius
Kircher's Universal Polygraphy," *Isis* 39 (1948): 215–228. Some of his most relevant writings include *Oedipus Aegyptiacus*, 4 vols. (Rome, 1652–1654); *Polygraphia nova et universalis* (Rome,
1663); and *Ars magna sciendi* (Amsterdam, 1669).

[136]From diverse perspectives, this general point is discussed by David E. Mungello, *Curious
Land: Jesuit Accommodation and the Origins of Sinology*, Studia Leibnitiana Supplementa, 26
(Stuttgart, 1985): 134–187; Céard, "De Babel à la Pentecôte: La transformation du mythe
de la confusion des langues au XVIe siècle," *Bibliothèque d'Humanisme et Renaissance* 42 (1980):
588–592; Thomas C. Singer, "Hieroglyphics, Real Characters, and the Idea of Natural Language in English Seventeenth-Century Thought," *Journal of the History of Ideas* 50 (1989): 50–51.

Figure 5. The Roman College museum. From Giorgio de Sepi, *Romani Collegii Societatis Iesu Musaeum Celeberrimum* (Amsterdam, 1678).

sion was reflected in letters written to him by such contemporaries as Johannes Vondelius, who pressed Kircher to complete his long anticipated work on the hieroglyphs lest the consequences of Babel become complete: "Already time and blight have almost consumed the Hieroglyphs, and practically destroyed the emblems," he wrote; "the lack of foreign languages will little by little obfuscate our intellects and deprive us of knowledge."[137] In the laudatory remarks of Vondelius and many others, Kircher emerged as the philosopher best equipped to lead society out of chaos. Through his museum and his publications, the Jesuit attempted to resolve the legacy of Babel. As the motto emblazoned on the frontispiece of his *New and Universal Polygraphy* read, "All in one."

No attempts were made to reconstruct the Tower of Babel inside the museum, although perhaps we can view the juxtaposition of different forms of writing and the fascination with ivory towers, such as those fashioned by Settala, as representations of Babel in miniature (fig. 6).[138] Instead, hypothetical experiments located it scientifically. Kircher, drawing upon the best research in physics and astronomy, calculated the impossibility of the Tower's reaching the moon, predicting that its weight alone would have catapulted the earth out of its orbit and into fiery ruin.[139] The diminished size of the Tower by no means negated its cultural and linguistic consequences, however. Describing it as a "theater of ambition," ostensibly the antithesis of his own monument to knowledge, Kircher outlined the history of communication from the unity of the *lingua sancta* to the appearance of 275 different languages after Babel, multiplied further over the course of time.[140] Fittingly, given its subject, the *Tower of Babel* (1679) was his last publication. The image of Babel, an unwieldy edifice rising majestically above Nimrod's city was a potent reminder of the dangers of knowledge, unchecked by the wisdom of God. It was an emblem of the perceived gulf that separated the encyclopedic speculations of Kircher from the religiously suspect activities of such philosophers as Paracelsus, Cardano, and Della Porta, whom he condemned.

Preceding his study of the Tower of Babel, Kircher investigated Noah's Ark. For many naturalists, the Ark was the greatest edifice to pure knowledge ever built, greater even than the Temple of Solomon and more successful than the infamous Tower of Babel (fig. 7).[141] Both Kircher and

[137]PUG, *Kircher*, ms. 563 (IX), f. 311r.

[138]Adalgisa Lugli reproduces some of these towers in *Naturalia et mirabilia*, figs. 144–147.

[139]Kircher, *Turris Babel* (Amsterdam, 1679), p. 38. Joscelyn Godwin summarizes and illustrates the contents of this book in his *Athanasius Kircher*, pp. 34–43.

[140]Kircher, *Turris Babel*, sig. 3v, pp. 10, 26. Kircher's contemporary Mersenne also was fascinated with these questions; see Peter Dear, *Mersenne and the Learning of the Schools* (Ithaca, NY, 1988), pp. 171–191ff.

[141]On this subject, the classic study remains Don Cameron Allen, *The Legend of Noah: Renaissance Rationalism in Art, Science and Letters* (Urbana, IL, 1963).

Figure 6. Daniello Crespi, *Portrait of Manfredo Settala* (seventeenth century). From Pinacoteca Ambrosiana, Milan.

Figure 7. Noah's Ark. From Athanasius Kircher, *Arca Noë* (Amsterdam, 1675).

Neickelius described it as the first museum of natural history, while collectors such as Johann Kentmann and John Tradescant referred to their own museums as "arks."[142] While Eden, and by extension the universe, was God's museum, the Ark represented mankind's first attempt to collect nature at the behest of his Maker. When the new papal legate made his ceremonial visit to the *Studio Aldrovandi* in 1635, he described it as "the work of another Noah."[143]

The dimension and contents of the Ark were a matter of high speculation by the seventeenth century. As early as 1500, Amerigo Vespucci wondered how the Ark had contained all the new animals he had seen in the course of his travels.[144] All over Europe, theologians, philosophers, and collectors debated the numerous—and infinitely delightful—paradoxes that the circumstances of the Ark engendered. Patricians with a dilettantish interest in nature, such as Galileo's friend Giovanfrancesco Sagredo, established menageries that playfully imitated the first Ark.[145] In Rome, the architect Borromini proposed the creation of a "place for animals in the form of Noah's Ark, with an apparatus that divides it into three orders, as was the Ark, so large that one could walk down the middle of each order on it" for the garden of Camillo Pamphili.[146] More purposefully, scholars such as Athanasius Kircher debated the scientific feasibility of the literal Ark. For Kircher, establishing the "reality" of the original museum of nature strengthened its association with his own collection. Describing and measuring the space into which all animals could fit increased the probability of success for his own replication of the Noachian project. Admirers such as the German naturalist Georg Caspar Kirchmayer, author of several treatises on the antediluvian and postdiluvian world, recounted a series of reconstructive experiments performed by Kircher at the Roman College to establish the Ark's measurements.[147] Babel might not be contained in the museum, but the Ark potentially could be replicated in this canonical setting.

One of the questions that plagued naturalists throughout the centuries was the ability of the Ark to house all creatures. Built near Eden, as

---

[142]Caspar Neickelius, *Museographia* (Leipzig, 1727), p. 9; Rudwick, *The Meaning of Fossils*, p. 12; Arthur MacGregor, "Collectors and Collections of Rarities in the Sixteenth and Seventeenth Centuries," in MacGregor, ed., *Tradescant's Rarities: Essays on the Foundation of the Ashmolean* (Oxford, 1983), p. 91; Salerno, "Arte, scienza e collezioni nel Manierismo," p. 193.

[143]BUB, *Aldrovandi*, ms. 41 (22 October 1635).

[144]Gerbi, *Nature in the New World*, p. 37.

[145]Antonio Favaro, ed., *Le Opere di Galileo Galilei*, 2d ed. (Florence, 1934), XII, pp. 246, 258. As Sagredo wrote to Galileo on 11 March 1616, "Il mio casino è fatto l'arca di Noè et è ben monito d'ogni sort di bestie, nè mi manca che questa sola" (p. 246).

[146]BAV, *Vat. lat.* 11258, in Marcello Fagiolo, "Il giardino come teatro del mondo e della memoria," in Fagiolo, ed., *La città effimera e l'universo artificiale del giardino*, p. 133.

[147]Georg Caspar Kirchmayer, *Dissertationes de paradiso, ave paradisi manucodiata, imperio antediluviano, & arca Noae, cum descriptione Diluvii*. In Thomas Crenius, *Fascis IV. exercitationum philologico-historicarum* (Leyden, 1700), Vol. II, p. 130.

Kircher's portrayal of Biblical topography indicates, the Ark was outside the moral universe of Paradise, yet contained its remnants. In his *Noah's Ark* (1675), Kircher attempted to reconcile various practical problems about the nature of the Ark and its contents. To counteract critics who implied that no man-made structure could possibly contain two of every kind, Kircher argued that only "pure species" had entered the Ark. This excluded real hybrids such as the mule and puzzling animals from recently discovered lands, such as the armadillo displayed in his museum, while leaving room for imaginary beasts whose reality Kircher reaffirmed, such as the unicorn and the gryphon.[148] The phoenix was excluded on the grounds that it violated the command to enter in pairs, while a host of other creatures—frogs, mice, scorpions, and all "insects" save for the serpent—were left adrift since, Kircher reasoned in good Aristotelian fashion, anything known to generate spontaneously had no need of a berth.[149] Sirens, however, received a special exemption: Kircher had one in his museum, and thus did not hesitate to include it among the survivors (unperturbed by the fact that it might possibly qualify as a mixed species). "There can be no doubt that such a creature exists, for in our museum we have its bones and tail," he affirmed.[150] Serpents, however, entered the Ark to remind humans of the reason for the Fall and to cure us of various illnesses during and after the voyage, due to their remarkable medicinal properties.[151] Thus, the parable of the Ark provided a means of classifying the animals that entered Kircher's museum, based on their relationship to the Universal Deluge. Like his restitution of languages, Kircher's ability to bring ante- and postdiluvian creatures into one space gave testimony to his success as a collector capable of reconciling nature at all her different stages of existence.

From the restoration of Eden in the botanical garden to the reconstruction of the Ark and the resolution of Babel in the museum, naturalists framed their collecting of nature within messages of redemption and salvation. When all else failed, nature, as the text that God alone had written and humans had failed to fully corrupt, provided the key to the great book of the universe. The museum was an investment in the new confidence of the late sixteenth- and seventeenth-century naturalists, who, at first timidly and then loudly, proclaimed that their gardens surpassed Eden, their Arks exceeded Noah's, and their reconfiguration of human language had reduced the shame of Babel into a parable of success. In the museum, naturalists effectively recreated the first Creation in miniature.[152] Kircher was the pinna-

---

[148]Both Don Cameron Allen and Joscelyn Godwin summarize this information well; Allen, *Legend of Noah*, pp. 182–191; Godwin, *Athanasius Kircher*, pp. 25–33. On the armadillo in Kircher's museum, see De Sepi, p. 27.

[149]As summarized in Allen, *Legend of Noah*, pp. 80–81, and Godwin, *Athanasius Kircher*, p. 26.

[150]Kircher, *Arca Noë*, in Reilly, *Athanasius Kircher*, p. 169.

[151]This interesting subject is dealt with in greater detail in the discussion of theriac debates in chapter 6.

cle of the renewed optimism in encyclopedism that characterized late Renaissance and Baroque collectors.

While upholding the Aristotelian and Biblical frameworks of natural history, two paradigms institutionally mandated by the Society of Jesus, Kircher used them to explore the possibilities of occultism, which R. J. W. Evans defines as "the crucial problem of this Baroque world view."[153] Perpetuating heterodoxies under the banner of orthodoxy, the Jesuit collector attempted to resolve the problems of all the encyclopedic schemes that had preceded his own, dissolving the encyclopedia itself in the process. As Adalgisa Lugli notes, Kircher's collection could not be contained in any *one* book nor even fully in his museum.[154] Returning to the frontispiece of the 1678 catalogue, what is striking about this image is its sense of openness and distance (see fig. 5). The obelisks draw our eyes upward to the vaulted ceiling, whose cosmological motifs bring us out of the space of the museum and into the heavens where the banner for the collection rests. Kircher and his visitors appear insignificant in comparison to the objects and the space of the museum; they are as small as the loadstone at its center. While Kircher, more than Aldrovandi or Cesi, proclaimed the success of his encyclopedia of knowledge—after all, he was able to see it into print—the message of his museum frontispiece contradicts his intent. Bringing all of nature into the museum, mankind has ultimately been dwarfed in the process. Kircher does not control the objects in his collection; they control him.

The particulars of nature that fueled Aldrovandi's indiscriminant and Cesi's more tempered curiosity appear almost incidental to Kircher's natural philosophy. Objects were not an end in themselves. They established a point of departure for Kircher's meditations upon the different ways in which the universe reflected God's wisdom and intervention. Like Aldrovandi, Kircher collected dragons; like Cesi, he displayed a sample of the "wood-fossil-mineral" which the Prince of the Linceans had discovered.[155] But he did not write treatises highlighting the remarkable characteristics of these individual phenomena. Instead the singularities of nature gave way to "the physical manifestations that scientific inquiry has miraculously succeeded in capturing."[156] In his own fashion, Kircher, an admirer of the Accademia dei Lincei and an enthusiastic experimenter, was a proponent of the new philosophy. Yet he diverged from it, as practiced by Galileo, Descartes, and many members of the Royal Society, in one important respect: his insistence that wonder was a category of analysis rather than simply a tool to lead men to the contemplation of higher truths.

[152]Lugli, *Naturalia et mirabilia*, p. 73.

[153]Evans, *The Making of the Habsburg Monarchy*, p. 340.

[154]Lugli, "Inquiry as Collection," *RES* 12 (1986): 120.

[155]I have already mentioned the dragon in chapter 1. On the *legno fossile minerale*, see Kircher, MS, Vol. II, pp. 65–66.

[156]Lugli, "Inquiry as Collection," p. 120.

PARADIGMS WITHOUT END?

By the end of the seventeenth century, the structures through which collectors viewed their world turned in on themselves, dissolving the patterns that they created. Galileo's condemnation of the disorderliness of the culture of the marvelous or Descartes' critique of curiosity as an ostensible method for organizing knowledge both point to their respective frustration with an image of knowledge without end.[157] When Krzysztof Pomian describes Moscardo's museum in Verona as "a universe peopled with strange beings and objects, where anything could happen, and where, consequently, every question could legitimately be posed," he summarized the tendency of Baroque collectors to allow their certainty in the structure of the museum to release them from any responsibility about the "truth" of its contents.[158] If anything, Baroque collectors had amplified their humanist predecessors' delight in the marvelous. Rather than judging their musings as "unscientific," as many have done,[159] we need to consider the extent to which they were a reflection of early modern scientific culture, which had not yet established any clear criteria for "truth" and "certainty."

Two entries from Legati's 1677 exposition of the Cospi Museum exemplify the ambiguity of these categories for seventeenth-century naturalists. The first, a "cat with two bodies and only one head, united until the navel" was deemed a possession of exquisite anatomical worth.[160] Brought into the museum by Legati's mentor, Ovidio Montalbani, after its birth in 1660, the cat provided a point of departure for numerous philosophical reflections on the indistinctness of generative matter, particularly in animals such as the cat that gave birth to several offspring simultaneously. "The probable cause of this monstrosity is the ability to confuse and unite in one Individual the generative material of two distinct bodies," the academicians who debated the reason for the monstrous cat in the museum gleefully affirmed. Like the headless dog accompanying it in the section on "Digited Quadrupeds," providing yet another opportunity for debate about the Aristotelian primacy of the heart over the brain in the generation of animals,

---

[157]The Galileian–Cartesian critique of curiosity is outlined in Pomian, *Collectors and Curiosities*, pp. 46–78.

[158]Ibid., p. 77.

[159]A few examples will suffice: Steven Mullaney contrasts wonder with inquiry, depicting curiosity as the antithesis of epistemology in his "Strange Things, Gross Things, Curious Customs: The Rehearsal of Cultures in the Late Renaissance," *Representations* 3 (1983): 42. In this, he is following the lead of Jean Céard, who sees indiscriminant curiosity as the antithesis of "science" in his *La nature et les prodiges*, p. 49. Even Krzysztof Pomian, in his fine study of the culture of curiosity, argues that curiosity, theology, and science are all distinct categories; thus, the late seventeenth-century physician Pierre Borel "offers a glimpse of nature prior to the scientific revolution"; Pomian, *Collectors and Curiosities*, pp. 47, 62–64, 77.

[160]All passages regarding the monstrous cat are from Legati, *Museo Cospiano*, pp. 26–30. For more detailed analysis of entries in the same catalogue, see Tribby, "Body/Building," and Hanafi, *Matters of Monstrosity in the Seicento*, ch. 2.

Montalbani's cat was proof of nature's ability to produce paradoxes. Both merited inclusion and mediation "for their extravagance." While Kircher subsumed his own reflections on wonder within a strictly hierarchical framework, in keeping with the Thomist teachings of the Jesuit order, secular collectors expanded the uses of scientific curiosity. Nature, just as Tesauro counseled in his *Aristotelian Telescope*, became a point of departure for Baroque rhetoric.

Also highlighted in Legati's catalogue was "the root of a tree with a human figure," proof positive of the elision of nature and art.[161] "The Torso, with its naturally proportioned back," observed Legati, "is so perfect that one could swear it was a work of Art rather than a simple joke of nature." Taking his comparison beyond simple metaphor, Legati proceeded to evaluate the quality of the composition, speaking approvingly of the branches that formed the legs and arm with a "closed hand" and more critically of the "arid branch, without a hand." The presence of such challenging phenomena in the Cospi museum reminded Legati of other witty conceits that he had encountered in the local naturalistic culture of Bologna. Among them were a juniper root resembling an owl in the botanist Zanoni's possession and "a bend in a maple sprig" representing a pigeon *al naturale*. Beneath their playful exterior, however, all of these phenomena reiterated the traditional framework of natural history that was being dismantled as Legati made these observations. As his catalogue demonstrated over and over again, the objects worthy of possession served to delight an audience that perceived curiosity as an end unto itself. Yet the virtuosity of every object in the Baroque museum did not prevent it from shoring up the defenses of the Aristotelian–Plinian program, revived in tangible form by Aldrovandi and continued throughout the seventeenth century. At a time when other natural philosophers were decrying the privileging of the exotic at the expense of the "nearest, most familiar things" as one of the "gravest errors that usually clutters the minds of most men,"[162] collectors of nature such as Kircher, Settala, Legati, and Montalbani defended the parameters of the marketplace of marvels that Aldrovandi had established and they were determined to maintain.

Discovering the limitlessness of nature, collectors responded by releasing the bounds of knowledge. The paradox of the museum lay in its attempt to confine knowledge while expanding its parameters. While the ostensibly rigid yoke of authority stabilized the process of collecting nature, the ever-flexible lens of curiosity adjusted its meaning by constantly finding gaps— undeveloped details and realms of speculation—that collectors could fill.

---

[161]Legati, *Museo Cospiano*, pp. 145–146.

[162]Giuseppe dal Papa, *Della natura dell'umido e del secco* (Florence, 1681), p. 7, in Brendan Dooley, "Revisiting the Forgotten Centuries: Recent Work on Early Modern Tuscany," *European History Quarterly* 20 (1990): 546 (n. 53).

The perplexity of contemporaries such as Galileo, who did not share this vision of the world, serves to illuminate the central features of encyclopedic discourse, which looked at the present primarily as a means of enlightening the past and embraced experimental culture as simply another form of proving what was already known. As Henry Oldenberg confessed to Robert Boyle after reading Kircher's *Subterranean World*, he was not certain where this all would lead. "In the meanwhile," he concluded, "I do much fear he gives us rather collections, as his custom is, of what is already extant and known, than any considerable new discoveries."[163] The end of certainty, as a philosophical premise, would bring about the dismantling of the Renaissance and Baroque encyclopedic paradigm.[164] But when Kircher died in 1680, he ascended to heaven with the security of knowing that his work had, at last, unfolded the divine plan of the cosmos.

---

[163] *Works of the Honourable Robert Boyle*, Vol. VI, pp. 195–196, in Clelia Pighetti, *L'influsso scientifico di Robert Boyle nel tardo, '600 italiano* (Milan, 1988), p. 95.

[164] Vasoli, *L'enciclopedismo del Seicento*, pp. 88–89.

THREE

Sites of Knowledge

Men take each other into the study.
—ANTONIO ROCCO

In a classic study of the relationship between humanism and space, Alfonse Dupront queried, "If humanism is most characteristic of [Renaissance values], where then does one find the humanists?"[1] Collectors, the embodiment of humanist culture in the sixteenth and seventeenth centuries, not only shaped the intellectual paradigms that governed their activities, they also defined the space they inhabited. Humanists, natural philosophers, and collectors were not just found *anywhere* in society. They inspected nature in a precisely demarcated setting, the museum, that took its place alongside the courts and academies of late Renaissance and Baroque Italy as a space in which learned and elite culture converged. Initially, museums appeared in the homes of urban elites and the courts of Renaissance princes. A 1492 inventory of the Medici palace in Florence after the death of Lorenzo the Magnificent clearly indicated the study as the place in which Lorenzo kept his most prized possessions. This tradition culminated with the *studiolo* of a later Medici, the second Grand Duke of Tuscany Francesco I.[2] Certainly Lorenzo was no innovator. Both patrician and humanist, he represented one stage in the process that linked the place of study with the activities of humanists and centers of political power. But the persistent interest of the Medici and other leading families in defining the museum suggests how much collectors, and the knowledge they produced, reflected the social and political transformations of this period. Returning to Dupront's question, "Where then does one find the humanists?," we might respond,

[1]Alfonse Dupront, "Espace et humanisme," *Bibliothèque d'Humanisme et Renaissance* 8 (1946): 8.

[2]Richard Goldthwaite, "The Empire of Things: Consumer Demand in Renaissance Italy," in *Patronage, Art and Society in Renaissance Italy,* F. W. Kent and Patricia Simons, eds. (Oxford, 1987), p. 171.

"In civil society." Civility, in all its forms, shaped the social paradigm of collecting. As Richard Goldthwaite argues, "By the end of the sixteenth century, Italians had created a style of living with its own world of goods that clearly set them off from other Europeans."[3] Terms such as *civiltà, sprezzatura,* and *buon creanza*—"civility," "nonchalance," and "good manners"—embodied the self-conscious connections between possession and power, display and self-display, that so preoccupied Italian elites. They also demarcated the social boundaries of knowledge.

By the end of the sixteenth century, museums defined a multitude of civil spaces. Botanical gardens such as those in Pisa and Padua integrated museums of natural history into their structure; learned organizations such as the Accademia dei Lincei made them a focal point of their activities. In Bologna, as we have already seen, the senators and citizens celebrated the *Studio Aldrovandi* as a centerpiece of the city, allowing it to shape the image of their local government through its placement in the Palazzo Pubblico. In nearby Florence, the Grand Duke of Tuscany's *studiolo* and later the Galleria degli Uffizi served a similar purpose. While the Aldrovandi and Cospi collections reflected the "mixed" governance of Bologna, simultaneously run by the papal governor and the local senate, the Medici collections reinforced the hereditary and personal rule of the Florentine state.[4] At the same time, both testified to the importance of collecting for the urban elite. The number of museums throughout Italy, in small towns as well as large cities, suggests that the social basis of collecting was more important than its political specificity, although such variations are worth noting. Whether living in the Republic of Venice, the Grand Duchy of Tuscany, the papal state, the Spanish-occupied Duchy of Milan and Kingdom of Naples, or the numerous other territories of the Italian peninsula, naturalists and collectors shared a common framework of reference: they were largely patrician men who inhabited the cities and celebrated the virtues of scholarship as a necessary part of civil life.

The presence of key cultural institutions in almost every major Italian city fostered this sense of shared identity. The well-known universities in Bologna, Padua, Pisa, Ferrara, and Rome and less celebrated ones in such cities as Florence, Siena, Parma, Naples, Messina, and Palermo forged a network of intellectuals among different urban centers, expanding beyond Italy to include the numerous foreign scholars who studied there because of the reputation of the faculties of law and medicine.[5] Likewise, the liter-

[3]Ibid., p. 173.

[4]Paolo Colliva, "Bologna dal XIV al XVIII secolo: 'governo misto' o signoria senatoria," in *Storia dell'Emilia Romagna,* Aldo Berselli, ed. (Imola, 1977), pp. 13–34; Eric Cochrane, *Florence in the Forgotten Centuries, 1527–1800* (Chicago, 1973).

[5]Paul Grendler is in the process of completing a study of the Italian universities to 1600. For a brief survey, see Richard Kagan, "Universities in Italy, 1500–1700," in *Les universités européennes du XVIe au XVIIIe siècles. Histoire sociale des populations étudiantes,* Dominique Julia, Jacques Revel, and Roger Chartier, eds. (Paris, 1986), Vol. 1, pp. 153–186.

ary, artistic, and scientific academies that appeared in great numbers established connections among men of learning that potentially transcended local interests; the Accademia dei Lincei, with members in Rome, Naples, and Florence, exemplified this trend.[6] On a more local level, pharmacies served as places for the civil exchange of ideas.

Even more crucial was the Jesuit college system that, as Gian Paolo Brizzi describes, educated young patricians to take their place in society.[7] The "noble seminaries" (*seminaria nobilium*) created a network of ruling elite who understood that their shared sense of identity derived from the urbane and civil education they had all enjoyed. Graduating from the Jesuit colleges, they attended universities and joined academies, further cementing their place in patrician male culture. By the seventeenth century, most Jesuit colleges had museums, less famous perhaps than the one at the Roman College, the showpiece of the Jesuit educational system, but nonetheless evident. Observing and participating in the scientific culture that grew up around these museums, young patricians were primed to create galleries in their own palaces, in imitation of the practices they had learned to value at school and seen in the homes of older men who had been their mentors. The formation of the Jesuit colleges, scientific societies, and museums reinvigorated the republic of letters, giving new meaning to the associations of honorable men.

Shared cultural practices, less tangible than educational institutions but equally influential, further articulated the boundaries of elite society and, by extension, society as a whole. "The full evolution of a thoroughly urbanized way of life eventually conferred on the Italian upper classes . . . a completely new concept of culture," writes Goldthwaite.[8] Collecting was one manifestation of this shift. Cultivated as part of the "civilizing process" that shaped the Italian patriciate of the sixteenth and seventeenth centuries, it provided yet another arena in which to articulate social difference through material possessions. For naturalists participating in the formation of a new scientific culture, the possession of nature and the objects necessary to its proper study—books, manuscripts, illustrations, and instruments—was a sign of their entry into civil society. In a world that measured social worth in terms of material objects and their display, collectors soon rose to preeminence. In a society that understood nature to be the largest object one could

[6]Note, however, that the only natural philosophers in Tuscany comprised the Accademia del Cimento. There is a large literature on Italian academies, so I will indicate only several of the most basic works; Gino Benzoni, "Per non smarrire l'identità: l'accademia," in his *Gli affani della cultura. Intellettuali e potere nell'Italia della Controriforma e barocca* (Milan, 1978), pp. 144–199; Eric Cochrane, "The Renaissance Academies in Their Italian and European Setting," in *The Fairest Flower* (Florence, 1985), pp. 21–39; idem, *Tradition and Enlightenment in the Tuscan Academies*, 1690–1800; and Michel Maylender, *Storia delle academie d'Italia*, 5 vols. (Bologna, 1926–1930).

[7]Gian Paolo Brizzi, *La formazione della classe dirigente nel Sei-Settecento. I seminaria nobilium nell'Italia centro-settentrionale* (Bologna, 1976).

[8]Goldthwaite, "The Empire of Things," pp. 173–174.

possess, following the Plinian definition that made it virtually synonymous with knowledge, naturalists enjoyed a high level of visibility. When Carlo Dati praised the "noble circumstances" that led Cassiano dal Pozzo to study nature, he defined the social world that made collecting and natural history accepted pastimes for the Italian upper classes.[9]

While the curiosity, rarity, and completeness of the objects in a museum determined their relative worth, challenging other collections to exceed their totality, the discursive practices surrounding artifacts further shaped the museum as a social paradigm. In the process of defining encyclopedism and the scientific culture that accompanied it, the collectors of nature also refined the norms that governed the behavior of scholars. Curiosity was not simply a philosophical lens through which to view the world. It was a *specific social attribute* that only naturalists who belonged to and associated with the patriciate could hope to attain. From this vantage point, the museum was not simply a laboratory of communication because collectors such as Kircher used it to work out their universal language schemes. As a *conversable space* it provided Italian patricians with a setting in which to practice the lessons they had learned in books and classrooms.[10] Writing to Galileo in 1614, Giovanni Bardi described the experience of visiting Cesi's museum. Bardi arrived at the Cesi palace with a copy of his book, which he had dedicated to Cesi:

> I stayed there almost two hours talking, and he showed me many of his curious things. . . . Yet another time when I went there to bring him a manuscript, I again stayed awhile to talk with great enthusiasm. Certainly I want to have the opportunity to go there often, because, other than what I learn by conversing with a person as knowledgeable as he is, I always leave with a miraculous desire to study these sciences in particular.[11]

Natural history, like all forms of philosophy at this time, was a communicative enterprise. Displaying nature was a prelude to conversing about natural history. The elaborate rituals in which collectors, their patrons, and their visitors engaged, all in the pursuit of knowledge, amply reflected this tendency.

Like the "princes" who regulated the academies, where virtuosi convened to exchange ideas and display their skills, collectors also established rules of conduct for the visitors who entered their museums. They controlled the conditions of access and shaped the meaning of the experience for those whose gender, social standing, and humanist credentials made them eligible to cross the threshold. In this respect, collectors exemplified the tendency of early modern society to articulate distinctions.[12] They were

[9]Dati, *Delle lodi del commendatore Cassiano dal Pozzo*, n.p.

[10]Here I am simply echoing a point made in greater detail by Jay Tribby.

[11]CL, Vol. II, p. 441 (Rome, 1 July 1614).

[12]On this subject, see Pierre Bourdieu, *Distinction: A Social Critique of the Judgement of Taste*, Richard Nice, trans. (Cambridge, MA, 1984).

among the preeminent cultural brokers of their society. Along with the authors of the new etiquette books and treatises on political and familial conduct, they governed the behavior of an elite in search of new models to fit a new lifestyle. Not unlike the artists, writers, musicians, mathematicians, architects, and other "producers of culture," they took advantage of their new circumstances to enhance their own standing in society. Through such maneuvers, the museum became a privileged site of knowledge.

CIVIL SOCIETY, CIVIL SPACE

The museum was located between silence and sound.[13] The quietude of the monastic *studium* and the eremetic retreat from society gave way in the late Renaissance to the visual and verbal cacophony of the museum, marking the transition from study to collecting. Humanists from Petrarch to Machiavelli had valued the dialectic between silence and eloquence.[14] For them, the study was a space of contemplation. Situated between the bedroom and the private chapel, it belonged to the inner recesses of the domicile. Dark, often windowless, the visual monotony relieved only by a table, a desk, a chair, and a niche for books or a chest to contain them, the earliest museums (in the original sense of the word) were spaces bereft of the signs of sociability that we have come to associate with the museum.[15] They lacked the tangible reminders of the commerce of scholarship, as seen in the gifts, visitor's books, and other artifacts connecting the museum to society, that were a staple of most early modern museums.

By the mid-sixteenth century, the space of learning reflected two conflicting models, one sociable, the other intimate. Not surprisingly, the latter persisted longer outside of Italy, in countries such as France and England where networks of academies and museums did not develop to quite the same extent or as early as they did in Italy. For Aldrovandi's well-known contemporary, the French humanist Michel de Montaigne, the library was not a "museum" in its social sense, but a *solitarium*. Montaigne, evidencing an awareness of the practices that challenged his own path to knowledge, wondered at scholars who could study in "pandemonium."[16] The English natural philosopher Robert Boyle sought every opportunity to clear his laboratory of visitors by emphasizing his duty as a "priest of nature." As Steven Shapin recently has noted, "the most far-reaching methodological insights of the

[13] Walter Ong, "System, Space and Intellect," p. 68. Jay Tribby ties Ong's image of the transition from a discursive to a visual culture firmly to the museum in his *Eloquence and Experiment*, chap. 1.

[14] Claudia Cieri Via, "Il luogo della mente e della memoria," in Wolfgang Liebenwein, *Studio. Storia e tipologia di uno spazio culturale* (Modena, 1988), p. xiv.

[15] Sebastian de Grazia reproduces a photo of Machiavelli's *scrittoio* in Sant'Andrea in Percussina, in his *Machiavelli in Hell* (Princeton, 1989), p. 26.

[16] Dupont, "Espace et humanisme," p. 8; Adi Ophir, "A Place of Knowledge Recreated: The Library of Michel de Montaigne," *Science in Context* 4 (1991): 163–189.

Scientific Revolution were also said to have been secured in solitude."[17] Undoubtedly, this is a sign of our preference for the Boyles, Descartes, and Newtons of this period, exemplars of the developments in mathematics and physics in which collaboration took on a different meaning. Certainly naturalists from Aldrovandi onward imagined natural history to be an implicitly cooperative enterprise, whose success depended upon the humanist republic of letters. However, this assumption was not in place at the beginning of the sixteenth century. We need to inspect the forms of humanist sociability that made it possible. What, in the course of two centuries, brought society into the private study and, conversely, led the scholars who defined this space to publicize it?

Despite the imprint of the Alexandrian museum as a paradigm of collective intellectual activity, as evidenced in the formation of humanist circles, the early sixteenth-century ideal of study was predominantly exclusionary.[18] This image of scholarship continued to be in evidence throughout the early modern period. Most famously within the Italian context, Machiavelli envisioned the activities in his *scrittoio* to be a means of reentering public life *in absentia* through the medium of literature. Yet the Florentine humanist did not conceive of scholarship per se as a socially grounded enterprise. The space of solitude that attracted scholars such as Machiavelli, Montaigne, and Boyle was replicated in images of museums. "*Museum* is a place where the Scholar sits alone, apart from other men, addicted to his Studies, while reading books," explained the seventeenth-century educator Comenius.[19] Scholarship was a process that absorbed its participants (*studiis deditus*), and the locus of study, the museum, created a seemingly impermeable physical barrier between the scholar and the outside world. Even as late as the eighteenth century, an age in which the museum had truly become a public spectacle, illustrations of museums reinforced their image as secretive and engrossing environments. The frontispiece of Caspar Neickelius's *Museographia* (1727), for example, situated the scholar within a gallery rather than the outdated *studio,* yet nonetheless imagined the museum as a location for solitary activity. While humanists increasingly advocated the social and utilitarian functions of a museum, their participation in a Christian culture that traditionally defined study as a form of spiritual communion continued to validate its eremetic image.

[17]Steven Shapin, "The House of Experiment in Seventeenth-Century England," *Isis* 79 (1988): 384–388; idem, " 'The Mind Is Its Own Place': Science and Solitude in Seventeenth-Century England," *Science in Context* 4 (1991): 194. The norm that Shapin has in mind does not apply well to the philosophical traditions developed in Italy. Instead it works quite well for France and England, applied to natural philosophers such as Descartes, Pascal, Boyle, and Newton. This contrast is yet another argument for the cultural specificity of scientific discourse.

[18]Châtelet-Lange, "Le Museo di Vanres," *Zeitschrift für Kunstgeschichte* 38 (1975): 279–280; Franzoni, "Rimembranze d'infinite cose," in *Memoria dell'antico nell'arte italiana,* Salvatore Settis, ed. (Turin, 1984), Vol. 1, p. 333.

[19]John Amos Comenius, *Orbis Sensualium Pictus* (London, 1659), p. 200.

The museums of the late Renaissance mediated between public and private space, straddling the humanistic world of collecting and the more religiously motivated images of the study as a site of contemplation. Both were valid ways to construct knowledge, yet provided different models for its constitution. The tension between retreat and sociability was already apparent in Paolo Cortesi's advice to cardinals on the formation of their households in 1510. Discussing the placement of a cardinal's Roman residence, Cortesi debated whether it should be "in the heart of the city" (*in oculis orbis*) or "far away from the crowds."[20] In his opinion, the former facilitated social life, the latter study. Yet by the end of the sixteenth century, the two activities seemed increasingly compatible. The relaxing of divisions between *negotium* and *otium* ("commerce" and "leisure") marked the expansion of the civil realm, which eventually encompassed significant aspects of the scholarly world. The evolution of the museum as a civil space reflected this transition.

Writing to Vincenzo Parpaglia from Ragusa, where he had recently been appointed archbishop, the prelate Ludovico Beccadelli observed, "Solitude makes men curious."[21] As Gigliola Fragnito amply makes apparent in her study of Beccadelli and his *sodalitas*, many sixteenth-century humanists emphasized the primacy of civil over contemplative life, revitalizing the time-old debate about the merits of the *vita activa* and the *vita contemplativa*. Beccadelli's patron, the Venetian Gasparo Contarini, affirmed the impossibility of living "removed from civil conversation."[22] It was not silence he pursued in his leisure, but sound. Such ideas found their normative expression in the treatises on etiquette that made their appearance in the sixteenth century. One particularly important member of Beccadelli's circle was the Florentine humanist Giovanni della Casa, best known for his book of manners, the *Galateo* (1558). As Della Casa observed, scholars could no more remove themselves from society than could anyone else. "Because of this no one will deny that knowing how to be gracious and pleasant in one's habits and manner is a very useful thing to whomever decides to live in cities and among men, rather than in desert wastes or hermit's cells."[23] In this image— avidly devoured by the urban elite who constituted the majority of the reading public—civility, urbanity, and the space of scholarship converged. Removed from the bustle of the Roman court, Beccadelli's enforced exile on the Dalmatian coast in the 1550s led him to collect as a means of communicating with the world. In solitude he found curiosity rather than tranquility. Collecting reaffirmed his connection to the particular community he

[20]Paolo Cortesi, *The Renaissance Cardinal's Ideal Palace: A Chapter from Cortesi's De Cardinalatu*, Kathleen Weil-Garris and John F. d'Amico, eds. and trans. (Rome, 1980), p. 71.

[21]BPP, *Ms. Pal.* 1010, c. 208r, in Fragnito, *In museo e in villa*, p. 161.

[22]Ibid., p. 13.

[23]Giovanni della Casa, *Galateo*, Konrad Eisenbichler and Kenneth R. Bartlett, trans. (Toronton, 1990), p. 4.

had left behind. Curiosity about the world provided a bridge between his current position, far from the center of patrician life, and his past pursuits that demanded the company of like-minded humanists.

By the sixteenth century, Della Casa's cursory suggestion became the basis of an entire civil ethos, a way of life articulated in courtesy books such as Stefano Guazzo's *Civil Conversation* (1574) and practiced in museums and academies throughout Italy. Writing to Aldrovandi, Mattioli recommended that naturalists cultivate "modesty, humanity, courtesy and politeness."[24] Castiglione, author of the well-known *Book of the Courtier* (1528), could not have put it better himself. While Della Casa concerned himself with bodily comportment and language, two central aspects to the fashioning of a successful gentlemen, Guazzo made "civil conversation" the essential criterion of success. Conversation, as Guazzo wrote, was the "beginning and end of knowledge."[25] Guazzo chastized earlier humanists such as Petrarch, who praised the virtues of solitude yet enjoyed the sociability of the papal court. Conversation, he concluded, progressed naturally from solitude because it extended the search for one's self that lay at the center of all learned inquiry. "Wherefore I conclude, that if the learned and students love solitarinesse for lacke of their like, yet they naturally love the companie of those which are their like," he wrote. "In so much that many of them have travailed farre with great labour to speak with other learned men, whose bookes they had at home in their houses."[26] Certainly he had in mind the pilgrimages of natural philosophers and virtuosi to each other's museums when he made this observation.

The physical organization of the museum also revealed its function as a conversable space. In Beccadelli's case, he allowed conversation to enter his study by decorating it with images of his humanist friends and famous scholars, so he could enjoy their "conversation" *in absentia*. His iconographic program drew inspiration from the impulse that led Petrarch to write letters to Cicero, precipitating numerous other humanist conversations with their particular muses. Unlike Beccadelli's gallery of famous men in his palace on the island of Šipan, most museums were located in urban space and could thus be connected more readily to the conversations of humanists that took place in the cities. Cortesi advised Roman cardinals, a category that included such individuals as Beccadelli's friend Contarini, to place "listening devices in the night study through which disputants in the auditorium can be heard."[27] More obliquely, Beccadelli's secretary Giganti used the placement of his *studio* in the Beccadelli family palace in Bologna, overlooking

[24]Raimondi, "Le lettere di P. A. Mattioli," p. 53 (Prague, 29 November 1560).
[25]*The Civil Conversation of M. Steven Guazzo*, George Petty and Bartholomew Young, trans. (London, 1581–1584; reproduction, New York, 1925), Vol. I, p. 39. For the most recent analyses of this text, see Giorgio Patrizi, ed., *Stefano Guazzo e la civil conversazione* (Rome, 1990).
[26]Guazzo, *Civil Conversation*, Vol. I, p. 31.
[27]Cortesi, *The Renaissance Cardinal's Ideal Palace*, p. 85.

Santo Stefano, to establish a continuous dialogue with his more famous neighbor Aldrovandi in the form of objects, queries, and learned treatises passed across the piazza. Physical proximity heightened the sodality that Aldrovandi and Giganti enjoyed. The case of Giganti is just one of many such instances in which the presence of a museum facilitated conversation with other learned men. Reading through the inventories, letters, and gifts that tell of multiple conversations between collectors, we can almost hear the sound that penetrated the museum. When naturalists such as Aldrovandi and Cesi signed their letters *ex Musaeo nostro* or "written from the Cesi museum," they reinforced the connection between the space of the museum and the production of humanist knowledge, mediated by the constant exchange of words that forged the bonds of civil society.[28]

Elaborating on the Mantuan apothecary Filippo Costa's *studiolino*, Giovan Battista Cavallara described it as "indeed a *genteel* theater of the rarest simples that our age has discovered."[29] Cavallara's choice of words was not unreflected. In ascribing gentility to Costa's museum, he signaled to his reader the precise social and moral ground in which it should be located. Collecting was an activity fit only for men of a certain status, and the close relationship between the two influenced its location and conditioned its practices. Costa, a contemporary of Aldrovandi who occasionally corresponded with the Bolognese naturalist, was a typical example of the "curious men of little significance" (*ometti curiosi*) who thrived in the culture of collecting.[30] For him, the increased interest in the study of nature made his pharmacy and museum one of the many centers of scientific activity in the Gonzaga state. It brought him to the attention of the Duke of Mantua and gave him the credentials to participate in the community of naturalists. Like the secretary Giganti, Costa used his museum to enhance his position; it probably determined his entry into civil society. He understood it to be as crucial to his success as the table manners and polite language that Della Casa and Guazzo outlined in their treatises.

While late Renaissance codifiers of social conduct such as Della Casa and Guazzo engaged in the process of defining the boundaries of civil society and civil conversation, Baroque masters of etiquette took as a given the connections between social and learned conventions. In the early seventeenth century, academicians such as Matteo Pellegrini refined the relationship between courtly civility and learning in treatises such as his *Courtiership is Suitable to the Scholar* (1624), reflecting lingering uncertainties about the pairing of these activities. By mid-century, Jesuit authors based their own

[28]Ambr., ms. D.332 inf., ff. 68–69 (Bologna, 17 November 1597); BUB, *Aldrovandi*, ms. 21, Vol. IV, c 347r; CL, Vol. I, p. 403; Vol. III, pp. 1046, 1076.

[29]*Lettera dell'ecc. Cavallara all'ecc. Girolamo Conforto* (1586), in Dario Franchini et al., *La scienza a corte*, p. 49. Emphasis mine.

[30]Fragnito, *In museo e in villa*, p. 174. I believe this phrase originated with Galileo. For more on Costa, see Dario Franchini et al., *La scienza a corte*, esp. pp. 41–51.

advice on scholarly conduct upon this premise. When Daniello Bartoli wrote his *Recreation of the Scholar* (1659), he did not consider it necessary to inform readers that they would have to create museums to enjoy the pleasures of snails, shells, and the many delightful "little things" of nature.[31] Reinvigorating civil discourse by aligning it with the ethical norms of the Aristotelian curriculum taught in the Jesuit classrooms, Baroque philosophers such as Emanuele Tesauro stressed sociability as a distinct attribute of the elites, depicting it as a product of the training that Catholic educators could offer.[32]

Tesauro's *Moral Philosophy* (1670) was standard reading in the *seminaria nobilium*.[33] The norms of Aristotle's *Rhetoric* and *Ethics*, and the prescripts of the *Galateo*, became the basis of his philosophy of social discourse. Such rules, Tesauro advised, were intended only for civil persons, capable of moral discipline." Recall Malpighi's praise of the papal legate who visited Bologna in 1690 and exercised "judicious curiosity" in viewing the objects in the *Studio Aldrovandi*. "Prudent men hate excessive curiosity," counseled one contemporary.[34] Control of the appetites and emotions distinguished the collector and his audience from the "vile and plebeian people . . . not subject to the censure of the *Galateo*."[35] It made them, implicitly or explicitly, followers of the rules of conduct laid out in the etiquette books of the age.

Baroque culture completed the social elevation of learning that first appeared in the writings of sixteenth-century humanists.[36] By the time Daniel Georg Morhof wrote his *Polyhistor* (1688), the summa of this genre, he could reflect on an entire tradition of civil discourse that shaped the humanist republic of letters in the sixteenth and seventeenth centuries. "Among all well-mannered people," he wrote in a chapter on museums and colleges, "there have been public assemblies of learned men who discuss in common natural things, the arts and every kind of study." Morhof further specified the nature of the relationship between sodality and learning in a chapter entitled "On Learned Conversation." Here he remarked, "Muse-

---

[31]Sections of this treatise are reproduced in Daniello Bartoli, *Scritti*, Ezio Raimondi, ed. (Milan, 1960).

[32]For the relationship between religion and civility, see Jacques Revel, "The Uses of Civility," in Ariès and Duby, *A History of Private Life*, III, *Passions of the Renaissance*, Roger Chartier, ed., p. 182; Carlo Ossola, *Dal "cortegiano" all' "uomo di mondo"* (Turin, 1987), p. 139.

[33]Denise Aricò, *Il Tesauro in Europa. Studi sulle tradizione della Filosofia Morale* (Bologna, 1987), p. 8. See also her "Retorica barocca come comportamento: buona creanza e civil conversazione," *Intersezioni* 1 (1981): 317–349.

[34]Daniel Georg Morhof, *Polyhistor literarius, philosophicus et praticus* (Lubeck, 1747; 1688), Vol. I, p. 165.

[35]Emanuele Tesauro, *La filosofia morale* (Venice, 1729 ed.), p. 287.

[36]See particularly, Manfred Beetz, "Der anständige Gelehrte," and Emilio Bonfatti, "Vir aulicus, vir eruditus," both in *Res Publica Litteraria: Die Institutionen der Gelehrsamkeit in der frühen Neuzeit* (Weisbaden, 1987), Vol. I, pp. 155–191.

ums and Colleges . . . usually are established so that the increase of mutual learning occurs from the conversation of scholars."[37] A keen observer of the increase in Jesuit colleges, scientific societies, and museums that characterized the seventeenth-century scholarly world, Morhof explicitly linked the encyclopedic programs of polymaths such as Kircher to the sociable practices in which they engaged. "Truly nothing is more conducive to gathering knowledge than frequent conversation with learned men," he wrote.[38] Encyclopedism, as Morhof detailed at endless length in some 2000 pages of Latinate erudition, was the ultimate civil discourse because it provided an infinite number of things to converse about. Surely Kircher, whom Morhof held up as a model of learning, would have appreciated the wisdom of this statement.

By the seventeenth century, the relationship between museums and the *ars conversandi* provided an opportunity to produce new technological *mirabilia* to incorporate in the museum. In the Roman College, Kircher installed not only "speaking trumpets" that seemingly animated statues such as his famous "Delphic Oracle" but also a speaking tube that allowed him to converse with society at large. This acoustic device, measuring thirty span in length, connected the museum to Kircher's private quarters. Through it the custodian announced the arrival of visitors, contacting Kircher "in the remote recess of my bedroom."[39] While the Jesuit chose to separate his private study from the museum, after its transferral to a gallery on the third floor of the Roman College in 1651, he did not do so to isolate himself from the throngs of visitors who came to see his curiosities. Just as Beccadelli's gallery of famous men allowed him to continue to participate in humanist conversation from a distance, and Giganti's museum eased his entry into the humanist culture of Bologna, Kircher's speaking trumpet allowed civil society to intrude constantly in his life. While his contemporary Boyle developed elaborate ceremonies to keep visitors out of his laboratory, Kircher invented technologies that facilitated their entry into the museum.

The scientific civility nurtured by Italian patricians soon found a market abroad. By the mid-seventeenth century, aristocrats in Paris and other cities were eager to learn how to converse about nature. Home of the newly founded Paris Academy of Sciences and in proximity to the greatest court in Europe, Louis XIV's Versailles, Paris soon replaced Rome and Florence as the center of *politesse*, the French equivalent of *civiltà* and *buona creanza*.[40] The rising status of Paris in the civil world soon attracted naturalists and col-

---

[37]Morhof, *Polyhistor*, Vol. I, pp. 123, 151. For a brief discussion of this text, see Arpad Steiner, "A Mirror for Scholars of the Baroque," *Journal of the History of Ideas* 1 (1940): 320–334.

[38]Ibid., Vol. I, p. 165.

[39]*Phonurgia nova* (Campidonae, 1673), p. 112; also his *Musurgia Universalis* (Rome, 1650). For brief discussions and illustrations of these devices, see Reilly, *Athanasius Kircher*, p. 141; Godwin, *Athanasius Kircher*, pp. 70–71.

[40]The fundamental studies of this world are Norbert Elias, *The Court Society*, Edmund Jephcott, trans. (New York, 1983), and Maurice Magendie, *La politesse mondaine et les theories de l'hon-*

lectors from Italy, most notably Paolo Boccone. In his *Natural Researches and Observations on the Formation of Many Stones* (1671), Boccone advertised his arrival in Paris and his ability to make the French as adroit in the collecting of nature as the Italians. His work advertised lessons in natural history, in the form of "conferences in your home," and advised how a "French courtier" (*Seigneur de la Cour de France*) could form a cabinet of nature. In the former instance, Boccone stressed his experience as a teacher of Italian patricians: "In Italy I taught some Gentlemen of the first quality."[41] Boccone subsequently assured his French readers that it was possible for people other than the king to amass a "cabinet of curiosities," remarking, "If private persons such as Imperato and Calzolari could make a rather nice collection of these exquisite things, you *Monseigneur,* in Paris where politeness (*politesse*) and the abundance of rare and curious things have reigned for a long time, can do even better."[42] Paris, an urbane extension of the courtly world of Versailles that now superseded the courts in Florence and Rome, was a setting even more conducive to the collecting of nature than the culture Boccone had left behind.

Boccone's statements reveal several facets of the social expectations that defined the collecting milieu. For example, he carefully emphasized his experience in initiating gentlemen and exceptional women into the mysteries of nature, in other words, in establishing his courtly credentials as the botanist to the Grand Duke of Tuscany. He also underscored the study of nature as a civil pursuit, worthy of the Parisian upper classes who had not previously delighted in natural history to the extent that the Italians had done. While acknowledging the success of Italian collectors such as Calzolari and Imperato in forming theaters of nature, Boccone underscored the superior abilities of a French courtier who, situated in a more urbane setting than sleepy Verona or tumultuous Naples, had access to a plenitude of collectibles and possessed the civility that deemed them worthy of pursuit. Collecting nature was no longer important solely because it linked men of learning and curiosity in a common endeavor; it had become a reflection of the emerging absolutist centers of power and the new civility that they instituted.[43]

---

nêteté en France, au XVIIe siècle, de 1600 à 1660, 2 vols. (Paris, 1925). For a comparison of scientific culture in seventeenth-century France and Italy, see Mario Biagioli, "Scientific Revolution, Social Bricolage and Etiquette," and the forthcoming work by Jay Tribby.

[41]Boccone, *Recherches et observations naturelles sur la production de plusieurs pierres* (Paris, 1671), p. 97.

[42]Ibid., p. 110.

[43]On the role of collecting in the Ancien Regime, see Antoine Schnapper, "The King of France as Collector in the Seventeenth Century," *Journal of Interdisciplinary History* 17 (1986): 185–202; idem, *Le géant, la licorne, la tulipe: Collections françaises au XVIIe siècle. I. Histoire et histoire naturelle* (Paris, 1988); Pomian, *Collectors and Curiosities.*

FROM *STUDIO* TO *GALLERIA*

The transformation of the museum from *studio* to *galleria* parallels its transition from solitude to sound. Just as the conversability of the museum points to its function as a civil space, in a society obsessed with the ends and means of civility, its movement from the inner recesses of a house to a more accessible setting signaled a new publicity for the activity of collecting. The process of evolution was by no means straightforward, nor necessarily self-evident to the collectors who made it possible. While Aldrovandi proclaimed that his *studio* was "for the utility of every scholar in all of Christendom," borne out by its accessibility during his lifetime and by the donation of the museum to the Bologna Senate in 1603, he had nothing but praise for the more self-serving activities of the Grand Duke Francesco I, ensconced in his windowless *studiolo* in Florence.[44] Both were acceptable models of collecting, worthy of emulation.

As Claudio Franzoni suggests in his study of antiquarian collecting, one of the most important linguistic divisions within the vocabulary of collecting concerns the distinction between terms that defined a collection spatially and those that alluded to its philosophical configuration.[45] The words *stanza, casa, casino, guardaroba, studiolo, tribuna,* and *galleria* organized the domestic terrain of the museum. "One can truly call your *Casino* a house of nature, where so many miraculous experiments are made," wrote Aldrovandi to Francesco I, alluding to the Grand Duke's domestication of nature in his alchemical laboratory at San Marco.[46] Contemporaries described the collection of Cardinal Flavio Chigi in Baroque Rome as a "room of curiosities," linking the space that it inhabited to the nature of its contents.[47] While Renaissance collectors frequently labeled the museum a *studio, studiolo, stanzino* (little room), or occasionally *guardaroba* (wardrobe), Baroque collectors preferred terms like *tribuna,* frequently used to refer to the Uffizi Galleries, and *galleria.* The former reflected a sense of containment and privacy; the latter openness and sociability.

In his will of 5 March 1604, the apothecary Calzolari left "the *studio di antichità* that is in my house in Verona" to his nephew, underscoring his conception of the museum as familial possession, to be passed on to relatives rather than donated to the local government.[48] Initially the museum, as *studio,* distinguished the space of the collector from the space of the society that he collected. As Benedetto Cotrugli observed in 1573:

[44]BUB, *Aldrovandi,* ms. 34, Vol. I, c. 6r.

[45]Franzoni, "Rimembranze d'infinite cose," p. 358.

[46]BMV, *Archivio Morelliano,* ms.. 103 (= *Marciana* 12609), f. 29.

[47]Incisa della Rocchetta, "Il museo di curiosità del card. Flavio Chigi," p. 141.

[48]Umberto Tergolina-Gislanzoni-Brasco, "Francesco Calzolari speziali veronese," *Bolletino storico italiano dell'arte sanitaria* 33, f. 6 (1934): 15.

And he who delights in letters must not keep his books in the public study
(*scrittoio comune*), but must have a *studiolo* apart, in the most remote corner of
the house. It is best and healthy if it can be near the bedroom, so that one can
more easily study.[49]

Earlier in the century, the humanist Cortesi advised that the study "be
placed in the inner parts of the house" so that it would "be especially safe
from intrusion."[50] Surviving plans for late Renaissance museums support
such an organization. For example, Giganti's *studio* in Bologna testified to
the conscious placement of a collection within the interior space of a house.
A small library connecting the bedroom and the museum was the sole point
of access (Museum Plan 2). The collector, called by the Muses, retired to his
study in the same way that he withdrew to the bedroom; presumably the inspection of books and objects fueled the dreams to which philosophers such
as Cardano attached great significance.[51] The inventory of Della Porta's museum reveals the presence of a bed in the second room, furthering the confusion between these two different potentially intimate settings.[52]

As Carlo Dionisotti points out, the distinctions of public and private need
to be problematized to understand their relevance for earlier periods; a bedroom, theoretically the most intimate of spaces in elite houses, was not fully
private, nor was a museum.[53] By circumscribing the *studio* with the most private rooms in the house, a collector signaled the magnitude of the privilege
that he offered when showing visitors his museum. They had crossed one of
the few thresholds of intimacy that elite culture acknowledged during this
period—not the "intimacy" of the bedroom, probably one of the most public spaces in a house, but that of the study. Through repeated transgressions
of this implicit threshold, the study was transformed into a museum.

Advice to construct museums, libraries, and studies in proximity to the
most "personal" space in the home drew not only on contemporary experience with the arrangement of such rooms but also on Alberti's classically inspired designs. Describing the layout of a country house in his *Ten Books on
Architecture* (1415), Alberti specified, "The Wife's Chamber should go into

[49]Benedetto Cotrugli, *Della mercatura e del mercato perfetto* (1573), p. 86, in Franzoni, "Rimembranze d'infinite cose," p. 307; Du Cange's definition of *studiolum*—"Cellula, museum, con-
clave, ubi studetur. Gall, *Cabinet d'Etude:* museolum, scrinia, *Estudiole* dicimus"—also gives several examples that locate the museum next to the bedroom; Du Cange, *Glossarium Mediae et
Infimae Latinitatis*, Vol. VI, p. 395.

[50]Cortesi, *The Renaissance Cardinal's Ideal Palace*, p. 85

[51]Girolamo Cardano, *Sul sonno e sul sognare*, Mauro Mancia and Agnese Grieco, eds., and Silvia Montiglio and Agnese Grieco, trans. (Venice, 1989).

[52]Fulco, "Il museo dei fratelli Della Porta," p. 26 (item 61). *Cabinet*, as it evolved in seventeenth-century French, connoted the closet beyond the main bedchamber.

[53]Dionisotti, "La galleria degli uomini illustri," p. 452. For a more general overview of this
subject, see Orest Ranum, "The Refuges of Intimacy," in Ariès and Duby, *A History of Private
Life*, Vol. III, pp. 207–263.

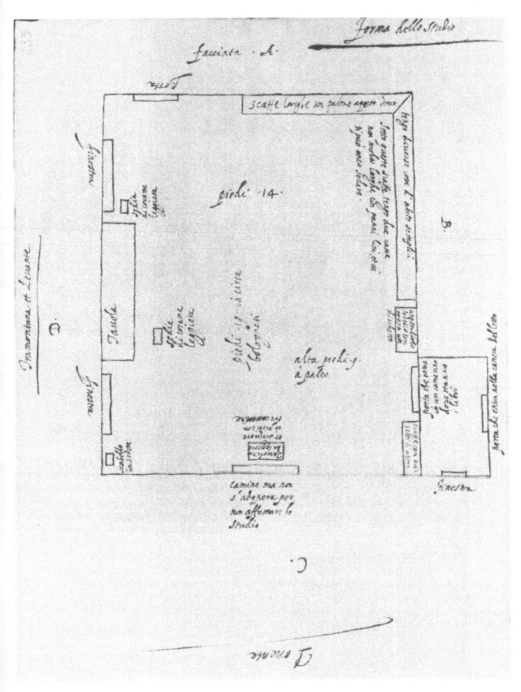

Museum Plan 2. The *studio* of Antonio Giganti (1586). From Ambr. S. 85 sup., f. 235r.

the Wardrobe; the Husband's into the Library."[54] This spatial arrangement
is borne out by the location of collections ranging from the *studio* of
Aldrovandi to the *studiolo* of Francesco I. The interlocuter of Alberti's treatise on family life, Giannozzo, recalled, "I kept my records at all times . . .
locked up and arranged in order in my study, almost like sacred and religious objects. I never gave my wife permission to enter that place, with me
or alone."[55] In this fashion, mercantile privacy was preserved and the affairs
of men left to men. By the seventeenth century, this gendered division of
space was a standard practice in most noble and patrician homes.[56]

While Alberti sharply defined the *studio* as an exclusively masculine
space, an image borne out by the relative absence of women in the spheres
of collecting and science, we can point to several noteworthy exceptions—
Isabella d'Este's collecting activities in late fifteenth-century Mantua and
Christina of Sweden's singular role as patron of learning in Baroque Rome.[57]
For the most part, however, collecting emerged out of a private and domestic culture that was almost exclusively male. The museum was a space
reserved within the home for scholarly activity, analogous to the contemplative space of the private family chapel, whose purpose was not entirely
divested from public life. Thus, it implicitly denied most women access. The
space of collecting was at once public and private, masculine space within
the domicile, and therefore civil in the most important sense.[58]

The museum, as *orbis in domo*, mediated between public and private by
literally attempting to bring the world into the home. The endless flow of
goods, information, and visitors that appeared on the doorsteps of the most
well-known museums determined that they could no longer be the hidden
worlds that monastic and eremitic images of *studium* suggested.[59] "If after the
arrival of my scribe, Giovan Corneglio, I have not responded to your letter

[54]Leon Battista Alberti, *Ten Books on Architecture*, James Leoni, trans. (1715), Joseph Ryck-
wert, ed. (London, 1965), vol. 17, p. 107; Pliny the Younger's description of his villa at Lau-
rentium (*Epist.* II.17) places his own library/study near the bedrooms.

[55]Leon Battista Alberti, *The Family in Renaissance Florence*, Renée Neu Watkins, ed. and trans.
(Columbia, SC, 1969), p. 209. This passage is discussed by Liebenwein, *Studiolo*,
pp. 41–42.

[56]Elias, *The Court Society*, p. 49.

[57]See C. M. Brown, " 'Lo insaciabilie desiderio nostro di cose antique': New Documents on
Isabella d'Este's Collection of Antiquities," in *Cultural Aspects of the Italian Renaissance*, Cecil H.
Clough, ed. (New York, 1976), pp. 324–353. For Christina of Sweden, see Gilbert Burnet, *Some
Letters Containing an Account of what seemed most Remarkable in Travelling through Switzerland, Italy,
Some Parts of Germany, &c. In the Years 1685 and 1686* (London, 1689), p. 244.

[58]Jean Bethke Elshtain, *Public Man, Private Woman: Women in Social and Political Thought*
(Princeton, 1981), pp. 3–16.

[59]The inscription above Pierre Borel's museum in Castres read, "Stop at this place (curious
one) for here you behold a world in the home, indeed a Museum, that is a microcosm or Compendium of everything rare"; *Catalogue des choses rares de Maistre Pierre Borel* in his *Les Antiquitez,
Raretz, Plantes, Mineraux, & autres choses considerables de la Ville, & Comte de Castres d'Albigeois* (Castres, 1649), p. 132.

as quickly as you wished," wrote Aldrovandi to Pinelli from his museum, "Your Most Illustrious Signor will excuse me for having been continuously occupied in various negotiations, public as well as private."[60] The antiquarian Giovan Vincenzo della Porta, "a man no less learned than unusual for the vast knowledge that he possesses" was singled out for "having through his own efforts created a most noble Museum to which scholars come from the farthest corners of Europe, drawn by its fame."[61] As we know from the 1615 inventory of his brother Giovan Battista's home, the Della Porta collections were indeed private yet open spaces, publicized through the informal networks of correspondence that formed the basis of the humanist scientific community. They reflected the transition of the museum from *studio*, a place of study and reflection, to *galleria*, a space for display and conversation. The reorganization of space reflected an equally significant cultural shift, that made both humanist sodality (societies of men) and courtly sociability (societies of both sexes) acceptable models for the production of knowledge. While scientific culture privileged the former, it did not entirely reject the later.

The evolution of princely collections perfectly illustrates the ways in which late Renaissance museums continued to incorporate both private and public notions of space in their conception and utilization. While the *studiolo* of Federico da Montefeltro at Urbino served largely personal functions, the *studiolo* of Francesco I, used by the Grand Duke from 1569 to 1587, operated in both contexts. Situated in the Palazzo Vecchio in Florence, both the seat of government and the Medici domicile, off the *Sala Grande* where affairs of state were conducted and leading into the private family chambers, it was a striking transition point: a room in which the Grand Duke could seclude himself without entirely leaving the realm of public affairs.[62] Francesco's study was more private than public; very few descriptions of it exist because few people were ever allowed access to it, with the exception of the court humanist Borghini who designed the original iconographic program of its *invenzioni*, Giorgio Vasari, and the other artists who worked on the room. Surrounded by the political intrigues of the Tuscan court, the *studiolo* and its contents were for the Grand Duke's eyes alone.

While Francesco I's *studiolo* was a personal space, his *Casino*, in which he fabricated everything from elegant glassware to plague remedies, was touted for its public benefits. The utility of the Grand Duke's personal laboratory paralleled the publicity surrounding the installation of a gallery in the botanical garden in Pisa in the late 1580s. Initially designed to instruct students in medicinal simples and other aspects of natural history considered

[60]Forlì, Biblioteca Comunale, *Autografi Piancastelli*, vol. 51, c. 486r (15 June 1595). "Giovan Corneglio" is Johannes Cornelius Uterwer.

[61]BUB, *Aldrovandi*, ms. 43, Vol. X, c. 284r.

[62]Both Lina Bolzoni and Luciano Berti concur on the ambiguity of the *studiolo's* position; Bolzoni, "L'invenzione dello stanzino," p. 264; Berti, *Il Principe dello studiolo*, p. 83.

useful to medicine, it soon competed with other collections as a noteworthy sight on the traveler's itinerary. When Robert Dallington visited Tuscany in 1596, he linked the appearance of the museum to the construction of the Grand Duke's residential palace in Pisa, describing the gallery as a "lesser house . . . adioyning a Garden of Simples, not much inferiour to that of *Padoa*"[63] (Museum Plan 3). A manifestation of the expansion of the Tuscan state, which now encompassed Pisa, the gallery was a symbol of Medici power. Its openness, in contrast to the closure of the family's personal collections in Florence, served to remind visitors of the benefits of a political dynasty that encouraged the increase in knowledge through the funding of botanical gardens and museums of natural history. Francesco I's decision to transform the family collections into the public galleries of the Uffizi signaled the final stage in the process of making collecting synonymous with the political and public identity of the ruling family of Florence.[64] By the end of the sixteenth century, the Medici, just like their client Aldrovandi, acknowledged the growing interest in museums as places of learning and display by opening their collections to all patricians.

The political maneuvers of the Grand Duke of Tuscany found their parallel in the growing rhetoric of public utility surrounding scientific space. Explicitly contrasting his own civic designs for a chemical laboratory with Tycho Brahe's princely laboratory at Uraniborg, Aldrovandi's contemporary, the chemical philosopher Andreas Libavius, describing the Danish astronomer's laboratory as "a private study and hideaway in order that his practice will be more distinguished than anyone else's" and the ideal space for philosophizing as a "dwelling suitable for decorous participation in society." Libavius' attack on the private *studio* indicated his participation in, and more importantly awareness of, the debate on secrecy versus openness that entered a wide range of discursive practices in the early modern period.[65] The laboratory, argued Libavius, was a civic and not a princely construct; it served to foster a sense of scientific community rather than exclusivity. Like the museum, it had to answer to the humanistic and later Baconian notions of utility that placed knowledge within the public sphere through its service to society. Thus, natural philosophers as well as naturalists such as Aldrovandi transformed the site of knowledge through a critical

[63]Sir Robert Dallington, *Survey of the Great Dukes State of Tuscany. In the yeare of our Lord 1596* (London, 1605), p. 24.

[64]For more on the transformation of the Uffizi, see Paolo Barocchi and Giovanna Ragionieri, eds., *Gli Uffizi. Quattro secoli di una galleria* (Florence, 1983), 2 vols.

[65]Andreas Libavius, *Commentarium . . . pars prima* (1606), Vol. I, p. 92, in Owen Hannaway, "Laboratory Design and the Aim of Science: Andreas Libavius and Tycho Brahe," *Isis* 77 (1986): 599. As Hannaway notes, citing Du Cange, *laboratorium* is a postclassical term, probably of monastic origin, that developed its more modern connotations from the sixteenth century on, paralleling the expansion of museum (p. 585). On the issue of secrecy versus openness, see William Eamon, "From the Secrets of Nature to Public Knowledge: The Origins of the Concept of Openness in Science," *Minerva* 23 (1985): 321–347.

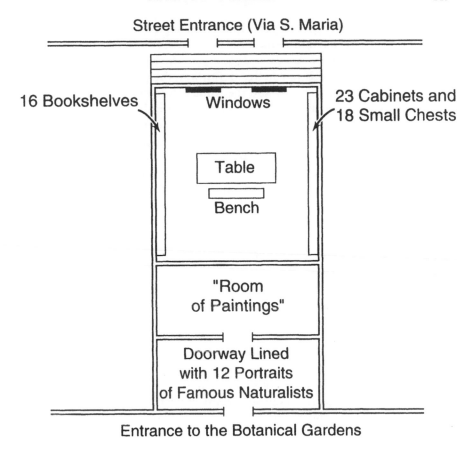

Museum Plan 3. The gallery of the botanical garden in Pisa (1626).

The custodian also was housed in the building, most likely on the first floor. The inventory does not indicate how one went from the "room of paintings" *(stanza dei quadri)* to the second-floor museum *(stanza grande)*, presumably via a staircase.

Sources: ASP, Università, 531, serie 5, cc. 1-31; Livorno e Pisa, pp. 516-517.

inspection of its intended uses. They reflected the idealism of late sixteenth- and early seventeenth-century scientific culture, which had only recently discovered the value of these new forms of communication.

By the seventeenth century, the museum had become more of a *galleria* than a *studio:* a space through which one passed. It reflected both the aristocratic ideal of collecting as a publicized activity and the humanist vision of the museum as a space in which learned men congregated, as Libavius put it, "suitable for decorous participation in society."The 1691 edition of the Florentine Accademia della Crusca's *Vocabulary* defined *galleria* as a "room of promenade (*stanza da passeggiare*) in which one keeps pictures and things of value."[66] In this normative definition, the gallery was noteworthy for its ability to contain society and the things elites most prized. In Naples, aristocrats began to include galleries as part of the public space of the palace to offset their more private wardrobes.[67] That careful observer of architectural convention, Vincenzo Scamozzi, noted the appearance of galleries in Venice during the same period: "nonetheless, following the example of Rome, for some time [galleries] have been introduced here in the houses of many Senators, Gentlemen, and Virtuosi for collecting."[68] While senators were obliged to have galleries due to their prominence in the city and gentlemen due to their social standing, Scamozzi's distinction of "virtuosi" as a separate category revealed his awareness of the importance of curiosity in remaking the study into a gallery. Curiosity, a most eminently civil discourse, transformed the museum even *before* architectural convention reflected this change. Describing the importance of Aldrovandi's collecting projects to the *gonfaloniere* of Bologna, Vincenzo Campeggi, in 1574, Fra Giovanni Volura praised "his Theater of Nature, visited continuously by all of the scholars that pass through here."[69] The civic notion of museum placed it in motion; forever opening its doors to visitors, the museum as *galleria* was the antithesis of the *studio,* which continued to reflect the earlier humanist ambivalence about the virtues of learning in society.

By the mid-seventeenth century, when collectors such as Kircher and Settala labeled their museums "galleries," the term was widely used. As Galileo observed, through his caustic labeling of Tasso's *Gerusalemme Liberata* as an "insignificant study" (*studietto*) compared to the "royal gallery" (*galleria regia*) formed by Ariosto's *Orlando Furioso,* the gallery had replaced the study as the normative form of the museum.[70] It conveyed the nobility and

---

[66]In Wolfram Prinz, *Galleria. Storia e tipologia di uno spazio architettonico,* Claudia Cieri Via, ed. (Modena, 1988), p. vii; also Nencioni, "La galleria delle lingue," in Barocchi, *Gli Uffizi,* Vol. 1, p. 17. Dionisotti, Franzoni, Nencioni, and Prinz note how the entry of *galleria* into the language of collecting from French in the late sixteenth century heralded a new spatial framework for the museum, as according to one contemporary description, "un luogo da passaggiare"; Franzoni, "Rimembranze d'infinite cose," p. 335. Cellini describes *galleria* as a *loggia* or *androne* in his autobiography; Dionisotti, "La galleria degli uomini illustri," p. 449. In his *The Fall of Public Man,* Richard Sennett defines "public" as motion, literally space to be moved through—a definition certainly in keeping with the historical development of *galleria* (p. 14).

[67]Gerard Lebrot, *Baroni in città. Residenze e comportamenti dell'aristocrazia napoletana 1530–1714* (Naples, 1979), p. 11.

[68]Vincenzo Scamozzi, *Dell'Idea dell'architettura universale* (Venice, 1615), in Liebenwein, *Studiolo,* p. 305 (n. 248).

[69]BUB, *Aldrovandi,* ms. 25, c. 304r (8 April 1574).

[70]This famous line has been discussed by many scholars, particularly Liebenwein, *Studiolo,* p. 135; Nencioni, "La galleria delle lingue," pp. 18–19; Lina Bolzoni, "Teatro, pittura e fisiognomica nell'arte della memoria di Giovan Battista Della Porta," *Intersezioni* 8 (1988): 486.

civility that Baroque collectors hoped to cultivate. Notably, Legati described Cospi's personal collection as a "gallery," but referred to the collection of Cospi's objects in the Palazzo Pubblico as the "museum." His distinction bears out Wolfgang Liebenwein's observation that *galleria* referred primarily to a spacious collection that added luster to a noble palace, while *museo* connoted a collection of nature or antiquities, reflecting a sustained engagement with the past.[71] Possibly this explains Bonanni's vehement rejection of the term "gallery" as a description of his restoration of the Roman College Museum in the 1690s. He intended the collection to be a place of serious study *in addition to* it being open to patrician visitors. Refinements in terminology also indicated a growing self-consciousness about the distinctions between civic and private museums. Ovidio Montalbani, a custodian of the *Studio Aldrovandi,* differentiated the public Aldrovandi collection he oversaw (*museum*) from his personal collection through the use of the diminutive (*privatum museolum; museolum meum*).[72] Using a multitude of different words, collectors explored the new conceptual possibilities that the growth of museums forced them to confront and explain.

Representations of museums revealed the trend toward openness, sociability, and publicity. The frontispiece of Imperato's *Natural History,* accompanying both the 1599 and 1672 editions, portrayed the apothecary's museum as a space in which gentlemen congregated to see and discuss the wonders of nature (see fig. 2). The attire of the three visitors leaves no doubt of their patrician status; wearing swords, plumed hats, and capes, they bear all the signs of their social class. Two civil gestures mark the action of the museum: the arm of Ferrante's son, Francesco, that directs the gaze of the visitors—the viewer of the catalogue has been implicitly included in this category—and the hand of the middle gentleman that registers his response to seeing such marvels.[73] Every detail in the illustration confirms the civil status of Imperato's museum, from the graceful stances of the visitors, right out of Castiglione's *Book of the Courtier* or Della Casa's *Galateo,* to the presence of the little dogs in the foreground, popular pets of the Italian upper classes. Enclosed in one room, indicating its origins as a study, Imperato's museum had all the social virtues of a gallery. In contrast, the 1622 engraving of Calzolari's museum in Verona was bereft of people, focusing only on objects (fig. 8). Despite this, the message above the image invited the "spectator" to admire the museum by "inserting an eye" into it. Such instructions evoked

[71]Legati, *Museo Cospiano,* p. 111; Liebenwein, *Studiolo,* p. 134.

[72]Ovidio Montalbani, *Curae analyticae* (Bologna, 1671), pp. 5, 15. The use of the diminutive, a linguistic device that played with the macrocosmic potential of the museum, a world in miniature, appears in other texts as well. Giovan Battista Cavallara, for example, described the collection of the Mantovan physician Filippo Costa as his *Studiolino;* "Lettera dell'eccell.mo Cavallara," in *Discorsi di M. Filippo Costa* (Mantova, 2d ed., 1586), sig. Ee. 3v.

[73]On reading gestures, see Peter Burke, "The Language of Gesture in Early Modern Italy," and Joneath Spicer, "The Renaissance Elbow," both in *A Cultural History of Gestures,* Jan Bremmer and Herman Roodenburg, eds. (Ithaca, NY, 1992), pp. 71–128.

SPECTATOR
OCVLOS INSERITO
CALCEOLARI
MVSAEI ADMIRANDA
CONTEMPLATOR
ET VOLVP ANIMO TVO
FACITO.

Figure 8. The museum of Francesco Calzolari. From Benedetto Ceruti and Andrea Chiocco, *Musaeum Francisci Calceolari Iunioris Veronensis* (Verona, 1622).

the prescription of Giulio Camillo in his *Art of Memory:* "let's turn scholars into spectators."[74] Even in museum illustrations where collectors and visitors were absent, humanist sociability intruded.

Images of Baroque museums only heightened this sensation. The urbane portrayals of Cospi and Kircher's museums, in 1677 and 1678, respectively,

[74]Camillo, *Opere* (Venice, 1560), Vol. I, pp. 66–67.

made both galleries implicitly accessible to members of the upper classes. In the former, the dynamic between Cospi and the custodian holds our attention (fig. 9). In the latter, the physical grandeur of the museum is juxtaposed to the diminutive civility with which Kircher greets visitors at the entrance (see fig. 5). Again, the gestures and appearance of the people represented in the museum draw the viewer into the picture. While the dwarf custodian, Sebastiano Biavati, displays one item in the Cospi collection, in the act of cleaning it, Cospi's hand holds no single object. Gesturing grandly across the expanse of the museum, he is clearly the *possessor* of all these objects.[75] Biavati's gesture is capable of containing only one artifact at a time; Cospi's sweeping arm encompasses the entire museum. Dressed in the noble fashion of the mid-seventeenth century rather than the late sixteenth century— note the sartorial contrast with Imperato's visitors—he invites other patricians to achieve what he already has accomplished: mastery of nature, appreciation of culture, social prestige, and potential immortality.

The De Sepi frontispiece presented a more classic image of a gallery. Like the image of the Settala gallery, it portrayed the museum as a space composed of long, narrow corridors receding backward into infinity. While the Roman College museum ascends upward into the heavens, it also extends outward into society. In the foreground, dwarfed by the objects in the collection, stands Kircher in the act of greeting visitors. Focusing on this part of the frontispiece, the importance of the Jesuit's civil gesture becomes apparent (see fig. 5). His hand extended out to the elder gentleman, he receives a letter. The younger man is either the visitor's companion, one of the many aspiring scholars eager to meet the famous polymath, or his custodian, monitoring the entry of every guest. The letter, perhaps the smallest artifact represented in the frontispiece, is as much a key to the museum as the colossal obelisks and mysterious loadstones that organize the philosophical display of knowledge. It represents the quintessential humanist gesture of access—communication and exchange—that defined the gallery as a "room to walk through." At the front of the museum, scholars engaged in the initial exchanges that allowed them to see the gallery; at the back, Kircher provided them with letters to send to friends and patrons that, in turn, permitted the bearer entry to yet another museum. This gesture, perhaps more than any other represented in the images of early modern Italian museums, marked the transformation of the *studio* to a *galleria*.

The etiquette of collecting was not only apparent in the language and gestures of its practitioners and their subsequent representations; it also influenced spatial organization. Museums, after all, were a sign of the increased investment in leisure activities by patricians who perceived the arrangement and location of a museum to be a reflection of the civil ethos they cultivated. The majority of collections were located on the second or

---

[75]This frontispiece and the collection are analyzed in greater detail in Tribby, "Body/ Building."

Figure 9. The museum of Ferdinando Cospi. From Lorenzo Legati, *Museo Cospiano annesso a quello del famoso Ulisse Aldrovandi e donato alla sua patria dall'illustrissimo Signor Ferdinando Cospi* (Bologna, 1677).

third floor of the house, a reflection of their earlier association with the private study and a sign of the rising status of their owners, who, over time, had removed themselves from the ground floor to the more distant reaches of the palace. By the mid-seventeenth century, however, collectors had begun to move the most visible pieces to the ground floor, normally reserved for commercial activities, to make them more easily accessible. Initially these ground floor galleries included a ceremonial bedroom to give visitors the illusion of intimacy they formerly had enjoyed upon entering a *studio*.[76] The increased sociability of collecting demanded greater attention to matters of presentation as collectors became more self-conscious of the effect their museums had upon visitors.

By the seventeenth century, particularly in a status conscious environment such as Baroque Rome, architectural etiquette was formalized in a series of proscriptive rules about the design of apartments. As Patricia Waddy details in her study of Roman palaces, the use of space denoted status. While rooms in sixteenth-century apartments were often arranged in clusters, reflecting a certain ambiguity about the distinction between private and public space, seventeenth-century apartments emphasized a linear progression of rooms from the most public *sala*, opening out to the stairs, to the private rooms behind the bedroom, which no visitor would enter.[77] The extent to which the host would venture into the more public parts of the house to receive a visitor was a measure of their relative status; a high-ranking visitor might be greeted on the stairs, while a less consequential one would be received in the antechambers.[78] Kircher's use of a speaking tube, and the employment of a custodian in the Cospi and Roman College museums, reflected similar social distancing mechanisms.

The spatial evolution of the museum further confirmed its role in the progressive aristocratization of Italian society. When John Evelyn visited the Imperato museum in 1645, he noted that it was in "one of the most observable Palaces in the City."[79] Certainly the strategic placement of the museum in one of the magnificent buildings that had brought the Neapolitan aristocracy into the city, away from their lands, only confirmed the notion that the Imperato family had used the museum to increase their status. The collection, a cultural novelty, was housed in an aristocratic palace, a social and architectural novelty in seventeenth-century Naples.[80] Galleries such as the one housing Imperato's theater of nature reflected the increased emphasis

[76]Patricia Waddy, *Seventeenth-Century Roman Palaces: Use and the Art of the Plan* (Cambridge, MA, 1990), p. 59. Here she is discussing sculpture and painting galleries specifically, but the point seems more broadly applicable.

[77]This type of organization is evident in the *gonfaloniere's* quarters in the Palazzo Pubblico in Bologna (see Museum Plan 5). Note, however, that the room beyond the bedroom is a "gallery," undoubtedly a reflection of the public nature of these apartments.

[78]Waddy, *Seventeenth-Century Roman Palaces*, pp. 3–10.

[79]E. S. de Beer, ed., *The Diary of John Evelyn* (Oxford, 1955), Vol. II, p. 330.

[80]See Lebrot, *Baroni in città*.

on hierarchy. The changing circumstances of the Italian upper classes transformed the image of the museum, accentuating its function as a visible, theatrical space. A close look at the transformation of Aldrovandi's collection
into the *Studio Aldrovandi* illustrates the process by which this occurred.

During Aldrovandi's lifetime, contemporaries described his personal
museum, situated in the family palace near Piazza Santo Stefano, as located
predominantly in one room. Known as the "museum," it was attached to but
distinct from his more private *studio*.[81] Undoubtedly, the collection began in
his study but eventually became autonomous, as its size and publicity grew.
Aldrovandi's decision to maintain the museum *next to* his study reminds us
of the organic link between these two spaces. In 1603, Aldrovandi had the
opportunity to envision a new organization for his collection, as part of his
instructions to the Senate of Bologna about its installment in the Palazzo
Pubblico. The expansion of the museum into a gallery offered a chance to
reflect on the use of space in classifying the objects it contained. Aldrovandi
differentiated the rooms of the new museum according to the nature of
the materials, separating manuscripts, covered with words, images, and
dried plants, from printed books, and larger natural objects—for example,
stuffed and articulated specimens—from smaller ones. He placed the fossils, stones, gems, minerals, and seeds that filled the sixty-six boxes of his
specimen cabinets, subdivided into 7000 compartments, in the last room of
his imagined museum (Museum Plan 4). The cabinets were to be accompanied by the "synoptic tables" that situated the objects in the vast field of
knowledge. The organization reflected a scholarly ideal that Aldrovandi
hoped the transformation of his collection into a civic museum would continue to enhance beyond his own lifetime.

While Aldrovandi organized the space of his museum with an encyclopedist's passion, later inventories of the *Studio Aldrovandi* reveal a slightly
different purpose. The 1696 inventory and Ghiselli's 1729 description, both
encompassing the Aldrovandi and Cospi collections, highlighted the political consequences of designating the museum as a public monument to the
scientific culture of Bologna.[82] The Palazzo Pubblico, like many civic buildings throughout Italy, was a palace whose partitions reflected the vicissitudes
of local government. As John Ray noted on his visit in February 1664,
Aldrovandi's collection "is kept in the cardinal legate's palace, commonly
called the *Palazzo del Confaloniero*."[83] Since the fall of the Bentivoglio family

---

[81]Unfortunately, documents reveal little about the organization of the museum before Aldrovandi's death; tantalizing clues come from two sources: a letter to his brother Teseo describing a ceremonial visit by Caterina Sforza in 1576 (BUB, *Aldrovandi*, ms. 35, cc. 203v–204r
[164v–165r]) and Pompeo Viziano's description of the museum to the Duke of Savoy in 1604
(BCAB. B.164, ff. 301–302r).

[82]BUB, Cod. 394 (408), Busta VI, f. II; Cod. 559 (770), XXI, f. 406.

[83]John Ray, *Travels through the Low Countries, Germany, Italy and France, With curious observations* (London, 1738), Vol. I, p. 200.

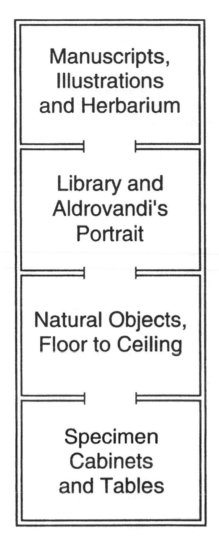

| Manuscripts, Illustrations and Herbarium |
| Library and Aldrovandi's Portrait |
| Natural Objects, Floor to Ceiling |
| Specimen Cabinets and Tables |

Source: Fantuzzi, p.78

Museum Plan 4. Aldrovandi's proposed plan of his museum (1603).

and the absorption of Bologna in the papal state in the early sixteenth century, the Palazzo Pubblico had two main occupants: the *gonfaloniere,* the civic head of government elected every two months by the Senate, and the papal legate, appointed in Rome to oversee the administration of the region.[84]

[84]The only overarching discussion of Bologna's political structure during this period remains Paolo Colliva, "Bologna dal XIV al XVIII secolo: 'governo misto' o signoria senatoria," in *Storia della Emilia Romagna,* Aldo Berselli, ed. (Bologna, 1977), pp. 13–34.

After some discussion, the museum was installed in the legate's quarters (Museum Plan 5). Possibly the continued absence of the legate, who increasingly left the Senate and the vice-legate to tend to the mundane affairs of government while he remained at the papal court in Rome, made his rooms the most logical place to locate the museum. Perhaps the civic and ecclesiastic governors insisted on this placement as a gesture of good faith as to what defined the "city" to which Aldrovandi had donated his museum. Aldrovandi himself may have recalled the role of several legates in his attempts to finance the publication of his natural histories. However it occurred, Pompeo Aldrovandi and Galeazzo Paleotti, the senators entrusted with the fulfillment of Aldrovandi's bequest, chose the spot in 1607 and calculated the expenses for the building of "new rooms" in 1609. The museum was formally installed there in 1617.[85] From this point on, it ceased to be Aldrovandi's personal possession and became incorporated into the highly politicized ritual space of Bologna.

The social and political pressures put upon this space inevitably modified Aldrovandi's original design. The 1696 inventory, for example, divided the museum into four "contiguous rooms,"[86] containing Aldrovandi's collection, a library, the vice-legate's study, and the former *gonfaloniere* Cospi's addition. Ghiselli notes that the custodian's study was the last room in the museum, possibly the *museolum* in which curators such as Ovidio Montalbani kept their own natural collectibles with which they embellished their editions of the Aldrovandian corpus. William Bromley's precise description of the museum in the early 1690s conforms well to this plan: the first room contained Aldrovandi's specimens, among them "712 Natural Rarities in little Glass-Bottles, with their Names"; the second, his "Study of Books"; the third, manuscripts and the woodblocks for his engravings; and the fourth, "more Curiosities, some natural, others artificial." By contrast, the descriptions given by Philip Skippon and John Ray, companion travelers through Italy in 1664, show a remarkable degree of conformity with Aldrovandi's original plan.[87] Clearly the installation of a museum as large as Aldrovandi's occurred gradually. By the time everything was in place, the senators and the custodian entrusted with its care had forgotten most of Aldrovandi's original suggestions or found them less relevant to their own proposed uses of the museum. They were more concerned with appeasing the legate, vice-legate, and *gonfaloniere* than the dead naturalist.

The donation of Cospi's collection precipitated a general reorganization of the museum. For a brief time, there were even two custodians: the pro-

[85]ASB, *Assunteria di Studio. Diversorum.* 10, n. 6 (*Carte relative allo Studio Aldrovandi*); Ferdinando Rodriguez, "Il Museo Aldrovandiano della Biblioteca Universitaria," *Archiginnasio* 49 (1954–1955): 207.

[86]This is how Legati described them in his *Museo Cospiano,* preface.

[87]William Bromley, *Remarks Made in Travels Through France and Italy* (London, 1693), pp. 122–124; Ray, *Travels,* Vol. I, p. 200; Skippon, *An Account of a Journey,* pp. 572–573.

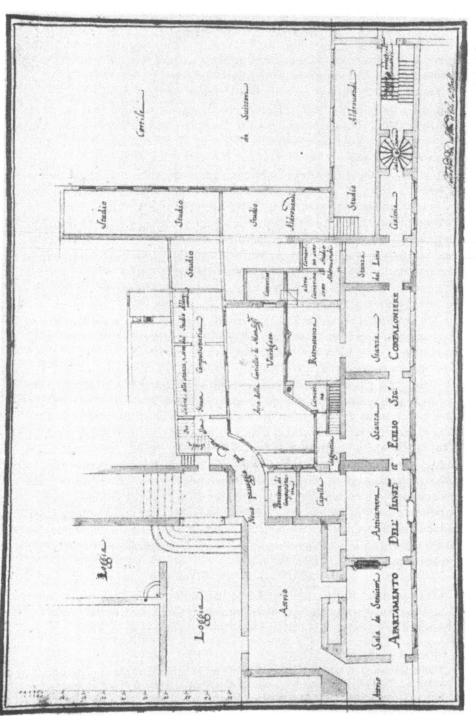

Museum Plan 5. The *Stadio Aldrovandi* in the early eighteenth century. From ASB, *Assunteria di Munizione. Recapiti* 2, f. 8.

fessor of natural history assigned to the *Studio Aldrovandi* and Cospi's beloved dwarf Sebastiano Biavati who was awarded the post of *Custode del Museo* (see fig. 9). By this time, the vice-legate's study now intruded into the space of the museum. Rather than being a map of the different parts of knowledge, as Aldrovandi had intended, the organization of the museum reflected the conflicting jurisdictional claims upon it. While Aldrovandi originally had donated the collection to Bologna, "wishing that my many labors be continued after my death, for the honor and utility of the City, and so that they may not have been for nothing,"[88] the commitment to these goals soon dissolved in the face of more pressing social demands, namely, the desire of one of the most prominent members of the city to add his own collection to the *Studio Aldrovandi.* Hardly a hallowed hall of science, the museum had become a sprawling, highly socialized affair which naturalists, virtuosi, dwarves, senators, the legate, and his entourage inhabited simultaneously. In the space of the *Studio Aldrovandi,* politics, religion, curiosity, and scholarship happily comingled. Without the presence of Aldrovandi, the museum no longer functioned primarily as the center of a community of learned naturalists, although it continued to fulfill some portion of this image. It had become a public showpiece for the city.

The evolution of Aldrovandi's museum from *studio* to *galleria* occurred over most of a century. Less than fifteen years sufficed to effect the same transition in the case of Kircher. Upon his arrival in Rome, the Jesuit began to create a museum in his private quarters at the Roman College. The donation of the patrician Alfonso Donnino's collection in 1651, and the decision to place Kircher in charge of its installation in the Roman College, precipitated the movement of Kircher's objects from his study into the gallery formed out of the corridor adjacent to the college library (Museum Plan 6). The gallery appeared in three different locations in the Roman College during the course of its history. From 1651 to 1672, it occupied a corridor of the third floor. The completion of the church of San Ignazio made it necessary to free the corridor so that the choir, rather than museum visitors, could pass through it. From 1672 until 1698, it was in a small, dark corridor near the second-floor infirmary; certainly the poor choice of location helps to explain why it had become so "mangled and dismember'd" when Misson observed it in 1687. During that period, the gallery had virtually ceased to be a sociable space, largely due to the inability of the Society of Jesus to find an appropriate successor to Kircher as its curator. In 1698, Bonanni reinvigorated the gallery by moving it to the fourth floor, near the chapel of Saint Louis.[89] He decorated the ceilings, added his own natural

---

[88]Fantuzzi, pp. 76, 84.

[89]Roberta Rezzi, "Il Kircheriano, da museo d'arte e di meraviglie a museo archeologico," in Casciato, pp. 296–300. Rezzi also summarizes the material from Bonanni's description of the gallery and provides a series of interesting reconstructions of the location of the museum in the Roman College.

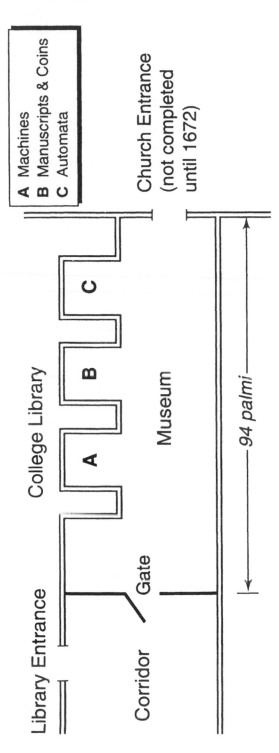

The gate was installed in the corridor in 1651 to prevent "free access" to the museum. The contents of the three side galleries are as described in the 1709 catalogue of the reorganized museum, which may have varied somewhat from the original.

Sources: ARSI. Rom. Historia (1704-29). 138. XVI, ff. 174^V-175^r; Filippo Bonanni, Musaeum Kircherianum (Rome, 1709), p.3.

Museum Plan 6. The Roman College museum (1651–1672).

history collection, and brought visitors back into the museum until the dissolution of the Society of Jesus in 1773. Despite the diminished scientific value of the collection in the eyes of eighteenth-century visitors, the Roman College museum continued to maintain its role as a sociable space for almost a century after Kircher's death.

The placement of the museum within the college served to reinforce its special status. An English visitor in 1675 noted that the museum was one of the "great rooms" off the courtyard.[90] The courtyards of the Jesuit colleges had become particularly important settings for the education of young patricians. By the mid-seventeenth century, they were the site of numerous theatrical productions in which the students at the *seminaria nobilium* participated.[91] There, young nobles learned to practice the civil conduct learned in the classroom. The proximity of the museum to the courtyard further underscored its role in the social world of Italian elites that the Jesuit educational system fostered. Notably, there were only two "spectacles" to which visitors, that is, nonmembers of the college, were invited: the theater and the museum.

Bonanni's 1716 description of the gallery elicits an interesting detail about Kircher's arrangement of the museum. After the assignment of the museum to "the part of the corridor contiguous to the library towards the unfinished Church," a gate was added by Kircher to more clearly demarcate the space of the museum and to prohibit "free access."[92] The gate and Kircher's famous speaking trumpet worked at seemingly opposite yet complementary purposes; while the former established distance, the latter bridged the divide between the now public museum and the private study of Kircher, where visitors no longer came. Lying beyond the space of the museum, as it is portrayed in the De Sepi frontispiece, the gate served as a tangible reminder of the social barriers that multiplied with increasing frequency in the course of the seventeenth century, an age in which, as one contemporary noted, "the world is governed almost entirely by princes."[93] The speaking tube, ingeniously providing access as well as distance, similarly testified to the intricacies of civil conversation in Baroque Rome. Both, in different ways, represented the new threshold of knowledge.[94]

[90]*A Tour in France and Italy made by an English Gentleman, 1675,* in *A Collection of Voyages and Travels* (London, 1745), Vol. I, p. 428.

[91]Gian Paolo Brizzi, "Caratteri ed evoluzione del teatro di collegio italiano (secc. XVII–XVIII)," in Mario Rosa, ed., *Cattolicesimo e lumi nel Settecento italiano* (Rome, 1981), pp. 177–204.

[92]ARSI, *Rom. Historia* (1704–1729), 138, XVI, f. 174v.

[93]In Benzoni, *Gli affani della cultura,* p. 25.

[94]The image of "the threshold of experiment" is introduced by Steven Shapin in his "House of Experiment," p. 374ff. I would like to broaden his question by asking where the "threshold of knowledge" was.

FRIENDS, FOREIGNERS, VISITORS

Access to the museum was regulated by more than the simple installation of a gate, such as the one stretching across the corridor of the Roman College, or the relative privacy or publicity of its location. A product of courtly and urban civility, conduct within the museum was strictly regulated by adherence to certain codes of behavior. An intricate series of social maneuvers that firmly aligned the practice of collecting with the gentlemanly arts of conduct conditioned entrance. Even in the early eighteenth century, one of the last visitors to the Aldrovandi museum, before its absorption into the collections of the Institute for Sciences, noted "that it is never opened but in the presence of a senator."[95] While the custodian assured visitors of its scientific worth, the senator reinforced its social and political value to the city of Bologna. Both monitored the conduct of visitors in the museum.

The rules of access were not uniform but generally operated according to principles of sodality and sociability. Since the quantity of visitors defined the success of a museum, naturalists could not afford to exclude anyone whose conduct was civil, thereby defining the museum as an intrinsically more open space than the court or the academy. Calzolari proudly described his museum as a place "where an infinite [number of] virtuosi converge every day." As an apothecary, this was a matter of business as well as pleasure. In contrast, Imperato complained in 1590 about the "scarcity of men" who frequented his museum. "Here in Naples I seem to be at the ends of the earth."[96] Undoubtedly his connections with the Accademia dei Lincei in the following decade ameliorated this situation; until then, he competed with Della Porta, more well known in the republic of letters, for scarce resources. Then, as now, southern Italy attracted fewer visitors than the northern and central regions where Calzolari and Aldrovandi resided; the Spanish presence did not compensate for the lack of prestigious universities that attracted foreign scholars.

About the same time, Aldrovandi wrote contentedly to the German naturalist Joachim Camerarius that a small group of scholars met nightly in his museum. Located in Bologna, a city through which many travelers passed on their way to and from Rome, Aldrovandi found it easier to cultivate the sort of scientific sodality that all collectors desired. Having encountered "many honorable gentleman and learned scholars" in his own travels, Aldrovandi repaid their courtesy through reciprocal displays of hospitality

[95]Johann Georg Keysler, *Travels through Germany, Bohemia, Hungary, Switzerland, Italy and Lorrain,* English trans. (London, 1760), Vol. III, p. 259.

[96]Both letters appear in Giuseppe Olmi, " 'Molti amici in varii luoghi': studio della natura e rapporti epistolari nel secolo XVI," *Nuncius* 6 (1991): 13, and can be found in the UBE, *Briefsammlung Trew,* Calzolari, 4 and Imperato, 10. This paragraph paraphrases the argument made in this excellent article.

in his own city.[97] By the 1570s, he could remark with great pride that his museum was "seen by many different gentlemen passing through this city, who visit my *Pandechio di natura,* like an eighth miracle of the world."[98] A century later, Kircher boasted that his museum was the centerpiece of a visit to the Eternal City. "No foreign visitor who has not seen the museum of the Roman College can claim that he has truly been in Rome," he exclaimed.[99] While collectors initially began with the idea of the museum as a nexus for the local scholarly community, the most successful ones eventually perceived it to be a repository of the entire patrician world.

The protocols governing the display of objects derived entirely from the rhetoric of friendship and community that shaped the humanist republic of letters and urban social relations.[100] The former determined the boundaries of the learned world; the latter, the milieu in which it thrived. Collecting and displaying nature refined the meaning of categories such as "friends" and "foreigners," both of whom visited museums. When the Venetian noble Pier Antonio Michiel refused to show Aldrovandi his botanical garden, Mattioli fueled the latter's indignation, writing, "if he were a gentleman, he would not have been so discourteous to a gentleman like you, who is not unlearned." Michiel, he implied, must be a man of low birth—"the son of some slave or a Greek or some gentleman's servant"—to have been so rude.[101] Since the incident occurred in 1554, before Aldrovandi had begun to make a reputation for himself, perhaps we can conclude that Michiel did not yet perceive the young Bolognese patrician to be a full-fledged member of the community of naturalists. He was not yet a "friend," in the eyes of the Venetian, but still a "foreigner."

As Ronald Weissman observes in his study of Renaissance Florence, "Renaissance moralists invitably divided the social world into two camps— friends and strangers—and avoidance of the latter was considered as important as cultivation of the former."[102] This generalized advice described the normative world of the Italian city in the fourteenth and fifteenth centuries. Writers such as Alberti counseled frequently against trusting people who were not, in some way, "known." The appearance of museums in private homes and public institutions during the sixteenth century reflected a relaxing of these standards. While Guazzo defined public conversation as that undertaken "abrode with strangers" and private conversa-

---

[97]Raimondi, "Le lettere di P. A. Mattioli," p. 38 (Prague, 29 January 1558).

[98]BAV, *Vat. lat.* 6192, Vol. II, f. 657r (Bologna, 23 July 1577).

[99]PUG, *Kircher*, ms. 560 (VI), f. 111 (Rome, 23 October 1671), in Rivosecchi, *Esotismo in Roma barocca*, p. 141.

[100]For a general overview, see Guy Fitch Lytle, "Friendship and Patronage in Renaissance Europe," in *Patronage, Art and Society in Renaissance Italy*, F. W. Kent and Patricia Simons, eds. (Oxford, 1987), pp. 47–62.

[101]Raimondi, "Le lettere di P. A. Mattioli," p. 27 (Goritia, 19 September 1554).

[102]Ronald F. E. Weissman, *Ritual Brotherhood in Renaissance Florence* (New York, 1982), p. 27. On friendship and access to personal space, see p. 32ff.

tion as that which occurred "at home in the house," this distinction held little meaning for collectors who frequently invited strangers into their homes.[103] The expanded realm of sociability had widened the moral meaning of friendship.

"Networks of friendship were the building block of social discourse and of politics," writes Richard Trexler.[104] This statement also holds true for scientific culture, where such networks expanded beyond the city to include men from different regions and occupations whose passion for natural history led them to search for the friendship of, to paraphrase Guazzo, those like themselves. Naturalists "honored" each other by writing letters, exchanging specimens and paying visits; such actions reinforced the identity they strove to create. "How dear to me is the friendship of Maranta," wrote Mattioli to Aldrovandi, while engaged in mortal battle with this rival over who deserved to possess the papers and herbaria of Luca Ghini.[105] Like most educated elites, naturalists used the rhetoric of friendship to facilitate mutual communication and exchange, in search of a secure berth in a world of perpetually uncertain relationships. Friendship, the security of knowing another, emerged from the cycle of letters and visits that organized the life of a collector. Tesauro compared conversation and friendship in the following terms: "civil conversation is a reciprocal communication of thoughts, just as friendship is a communication of emotions." In the letters of early modern naturalists, these two practices converged. When visitors remarked on the vast quantities of correspondence preserved in Aldrovandi's and Kircher's museums, they were noting the prominence of these two collectors in the republic of letters that existed only through the epistolary and face-to-face rituals in which its members participated and through the cultural institutions that emerged from such practices.

By the seventeenth century, increased travel and the omnipresence of galleries virtually dissolved the divisions between "friends" and "foreigners," at least among patricians and scholars who recognized themselves as belonging to the same community. "Father Kircher is my particular friend, and I visit him and his gallery daily . . . ," wrote Robert Southwell to Boyle in 1661. "He is likewise one of the most naked and good men that I have seen, and is very easy to communicate whatever he knows; doing it, as it were, by a maxim he has."[106] Southwell was one of many Protestants who spent their time in Rome in Kircher's museum. In this instance, the friendship between virtuosi transcended confessional differences (although Oldenburg and Boyle continued to be suspicious of the motives behind the Jesuit's hospitality). When Boccone arrived in Rome in 1678, he reported back to Antonio Magliabecchi: "I saw . . . Father Kircher . . . and many other

[103]Guazzo, *Civil Conversation*, Vol. II, p. 114.

[104]Richard C. Trexler, *Public Life in Renaissance Florence* (Ithaca, NY, 1991, 1980), p. 139.

[105]Raimondi, "Lettere di P. A. Mattioli," p. 51 (Prague, 16 September 1560).

[106]*The Works of the Honourable Robert Boyle*, Vol. VI, p. 299, in Pighetti, *L'influsso scientifico di Robert Boyle*, p. 95.

virtuosi."[107] Boccone's philosophical differences with the program of natural history instituted by the Jesuits did not prevent him from enjoying their conversation. Their shared interest in collecting and mutual commitment to the culture of curiosity allowed them to maintain civil relations. Indeed, the Roman College museum was recommended by writers of travel guides as one of the most "conversable" spaces in Baroque Rome. "Does the conference of learned persons please you? See Father Kircher for unknown languages and mathematics," advised Jacob Spon.[108]

If admission to one's home or museum signaled the conferral of friendship, then this was the tie that bound together collectors and visitors all over Europe. Such exchanges reaffirmed the "reality" of the republic of letters by defining the social parameters of that world. When Leibniz arrived in Bologna in 1689, he carried with him a letter of introduction from Magliabecchi, librarian to the Medici in Florence and a central figure in the Italian republic of letters. This letter was addressed to the mathematician Domenico Guglielmini, who promptly introduced the German philosopher to the leading intellectuals of the city, including the current custodian of the *Studio Aldrovandi*, Silvestro Bonfiglioli, and Malpighi. As Leibniz reported to Magliabecchi, "Guglielmini led me to Malpighi, at least a distinguished man in whose home I spent many fruitful hours in pleasant conversation."[109] This singular episode, typical of many learned exchanges, juxtaposed several of the most important elements of the civil world of scientific discourse. Magliabecchi's letter of introduction, Guglielmini's courtesy, and Malpighi's hospitality all conferred upon Leibniz official recognition of his status as a traveler of no inconsiderable honor and reputation. This pattern was repeated over and over again in visits to museums.

The published travel journals of the seventeenth century, a genre perfected in particular by the British who came to Italy, are the best guide to relations between collectors and visitors. Designed for an audience of potential travelers who intended to follow the same path on the "Grand Tour," they initiated readers into a culture of seeing, requesting, and receiving that further defined the image of the museum as one of the most desirable spaces for the "foreigner" to enter.[110] During a trip to Italy in the late 1660s, the Scottish physician Sir Andrew Balfour wrote a number of let-

[107]BNF, *Magl.* VIII. 496, f.4v (Rome, 11 August 1678).

[108]Jacob Spon, *Voyage d'Italie, de Dalmatie, de Grece, et du Lévant, fait aux années 1675 & 1676* (The Hague, 1724 ed.), Vol. I, p. 26.

[109]In André Robinet, *G. W. Leibniz Iter Italicum (Mars 1689–Mars 1690). La dynamique de la République des lettres. Nombreux textes inédits,* Studia dell'Accademia Toscana di Scienze e Lettere "La Colombaria" (Florence, 1987), Vol. 90, p. 309 (31 December 1689).

[110]On the role of travelers to Italy, see Peter Burke, *The Historical Anthropology of Early Modern Italy* (Cambridge, U.K., 1987). Early modern travel guides to Italy are comprehensively listed and analyzed in Ludwig Schudt, *Italienreisen im 17. und 18. Jahrhundert* (Vienna, 1959). For a more theoretical discussion of travel, see Georges Van Den Abbeele, *Travel as Metaphor from Montaigne to Rousseau* (Minneapolis, 1992).

ters to an anonymous virtuoso, offering him advice on how to negotiate the etiquette of travel successfully. Foremost in his mind was the ability to see natural curiosities. Urging his friend to sample the exotica of the Pisan botanical garden, Balfour recommended the following measures as a means of gaining entry:

> . . . you can have no access to it, except by the recommendation of the Physitian, that is Professor of *Botany* for the time, therefore I think it will be worth your while to make your address to him, for a Libertie, first, to see the Garden and Gallery; secondly, to get from the Gardener or himself, the Seeds . . . and a Peece of the Plant for drying. . . .

In contrast to the formality necessary in Pisa, the physician recommended that in Padua, "You may easily make your address to the Professor that keeps the garden . . . but the Gardener will be sufficient to do your turn."[111] Through the writing of travel journals, a genre as popular and novel as the books of courtesy, virtuosi publicized their success in traversing the social terrain of other regions, thus allowing us to see the learned world of Italy through their eyes. They measured their success in the number of entries and the quality of access to the leading cultural institutions—museums, academies, courts, gardens, and homes of learned men. As Balfour counseled, admission to a museum was predicated on *who* rather than *what* one knew and how one deployed that knowledge. Following some of the most interesting and articulate visitors to early modern Italy—the English naturalist John Ray, the physician Philip Skippon, and the theologian Gilbert Burnet, all affiliated with the Royal Society—we can see these mechanisms at work.

Their varied experiences in Bologna highlight the tissue of conventions woven around museum-going. Arriving in the city in February 1664, both Ray and Skippon noted—using the exact same phrase—that they had toured the Aldrovandi collection "by the favor of Dr. Ovidio Montalbano," its current custodian.[112] Most likely the introduction to Montalbani had been facilitated by Bolognese natural philosophers in contact with the Royal Society, for example, Malpighi. Following the trail from the *Studio Aldrovandi,* we see our two travelers introduced to Giacomo Zanoni, custodian of the botanical gardens founded by Aldrovandi. "Dr. Montalbani very civilly brought us to the house of Jacopus Zenoni, an apothecary, a skilful herbarist, and a collector of rarities."[113] The affable Zanoni showed them his garden and natural history collection, apologizing for its depletion since the recent sale of its contents to the Duke of Modena. In compensation, he provided them with a letter, addressed to Father Ganzia, a Theatine in

[111]Sir Andrew Balfour, *Letters Write to a Friend, . . . Containing Excellent Direction and Advices For Travelling thro' France and Italy* (Edinburgh, 1700), pp. 96, 230–231.
[112]Ray, *Travels,* Vol. I, p. 220; Skippon, *An Account of a Journey,* p. 572.
[113]Ray, *Op. laud.,* Vol. i, p. 200, in Murray, *Museum,* Vol. I, pp. 80–81.

Modena, that ensured them access to the Duke's collection. All in all, it was a very pleasant trip.

Traveling a similar route in 1685–1686, Gilbert Burnet had a completely different experience. The importance of personal and written introductions as a means of entry into civil society is borne out by his account, composed in the form of a letter to Boyle, who was always curious about what the Italians were doing.[114] What is most noteworthy about Burnet's description is the relative *absence* of remarks about famous galleries and botanical gardens. Lest we think that Burnet was uninterested or disdainful, we need only consider his wistful remarks on his stay in Bologna: "I saw not one of the chief Glories of this place," he complained, "for the famous *Malphigius* was out of Town while I was there."[115] As a member of the Royal Society, the second Italian to have been admitted, Malpighi facilitated the visits of numerous English travelers who passed through his native city. Little wonder that his absence caused Burnet so much consternation. Burnet probably had come bearing letters from Oldenburg and Boyle, only to find the recipient not at home. His inability to compensate for this unanticipated turn of events only reinforces our sense of the rigidity and formality of the rules of conduct.

While travel accounts establish the pattern of behavior, they do not tell us how the system of introductions worked, only what it accomplished (or prevented). Instead, the letters of introduction themselves reveal how the protocol unfolded in practice. Writing to Aldrovandi in October 1561, Alfonso Cataneo, professor of medicine and natural philosophy at the University of Ferrara, warmly recommended Ridolfo Argento to his colleague. "This [man] is my dear friend," he explained, "whom I have directed to Your Excellence upon his arrival in Bologna, since he is a doctor and a gentleman, worthy of seeing certain little things (*cosette*) that interest him. I know that you will not neglect to show him the usual courtesy for love of me."[116] Cataneo's letter established several important criteria. First, the visitor was "a doctor and a gentleman," both scientifically and socially worthy of the sight of Aldrovandi's marvels. Given the inordinate number of doctors who traveled and collected and their steady rise in status throughout the early modern period, this phrase summarized the social attributes that came with entry into this particular profession.[117] Second, social status conditioned the ability to see the "little things" of nature. By definition, people of worth

---

[114]For more on this subject, see Pighetti, *L'influsso scientifico di Robert Boyle,* and Steven Shapin's forthcoming study of Boyle and gentlemanly discourse in England, *The Social History of Truth.*

[115]Burnet, *Some Letters,* p. 244.

[116]BUB, *Aldrovandi,* ms. 38(2), Vol. II, c. 37 (Modena, 12 October 1561).

[117]This phrase wonderfully parallels the image of "a scholar and a gentleman" discussed by Steven Shapin in his article under that same title. Apparently the gentleman physicians of Italy were more successful than the gentleman scholars of England in making the association stick.

rather than those excluded from the universe of the *Galateo* better understood the intricacies of finely woven arguments and the subtleties of the scientific paradoxes that abounded in a museum. Finally, the etiquette of access was a two-way street. Having established Argento's credentials, Catanio reminded Aldrovandi that, since he had followed the prescribed course of conduct, the naturalist was "obliged" to admit his protegé.

While the arrival of a visitor from Ferrara was a discrete, indeed unremarkable, event, the ritual surrounding his arrival was not. Repeated constantly among men of quality, it was a defining feature of civil behavior. Other bearers of letters were introduced to Aldrovandi as "learned and polite," men with "virtues and good manners," "courteous," and "worthy."[118] Letters were the only way to determine that visitors were "men like oneself" and therefore admissible; surely this explains why so many recommenders described the bearer as someone who should be treated as an extension of themselves. In 1600, when Peiresc decided to go to Rome from Padua, he wrote his patron Pinelli, requesting letters of introduction; later in life, he performed the same service for numerous scholars on their way to the papal city.[119] Shortly afterward, upon the death of Pinelli in 1601, he fulfilled the ideal of humanist mentorship by taking over Pinelli's responsibilities as one of the principal coordinators of the republic of letters. In a sense, the persona of Peiresc *was* an extension of Pinelli's identity.

By the mid-seventeenth century, the humanist courtesy of letter writing had become a formalized system of introduction. Without demonstrable proof of one's connection to the republic of letters, as the case of Burnet already has shown, a scholar had few ways of entering the museums and academies in the cities he visited. Robert Boyle reputedly missed the opportunity to see the fledging museum at the Roman College on his one trip to Italy in 1641 because he declined the attentions of the English Jesuits, fearing they would try to convert him. His friend John Evelyn, however, made good use of contacts at the English College in Rome to meet the incomparable Kircher.[120] For British Catholics, the Jesuit networks were even more supportive. Writing to Father Andrew Leslie at the Scottish College in Rome in 1652, a compatriot in Paris conveyed the news that a group of young "men of good condition and hopes, . . . [had] gone to Rome. . . ." He continued, "I pray your R[everend] let al kindness be ourse their shouen to them, not only in our hous, bot acquent with F[ather] Athanasio."[121] Through such introductions, visitors gained access to men of learning and reputation such as Kircher. He, in turn, expressed a willingness to use his connections to

[118]BUB, *Aldrovandi*, ms. 38(2), Vol. II, c. 106; Vol. III, c. 121r; Vol. IV, c. 212r.

[119]Gassendi, *The Mirrour of True Nobility and Gentility*, Vol. I, p. 25; Vol. II, p. 41.

[120]Reilly, *Athanasius Kircher*, pp. 148–149.

[121]Father W. Christie to Father Andrew Leslie, Paris, 20 January 1652, in Edward Chaney, *The Grand Tour and the Great Rebellion: Richard Lassels and "The Voyage of Italy" in the Seventeenth Century*, Biblioteca del viaggio in Italia (Geneva, 1985), Vol. 9, p. 104.

facilitate the travel of other scholars, just as the De Sepi frontispiece suggested. When the Danish naturalist Nicolaus Steno went to Vienna in 1669, he asked Kircher to introduce him to the right people at the Habsburg court. "I now supplicate Your Reverence for the kindness that you offered me when I studied with you in Rome, promising me by means of your letters to procure for me the facility to be able to see the curiosities of Vienna and surrounding places."[122] As an eminent naturalist and a prominent convert, Steno surely received an enthusiastic introduction from Kircher to the Catholic princes and scholars of Central Europe.

Enumerating upon the virtues that he hoped a young nobleman under his tutelage would acquire after visiting Italy, Richard Lassels remarked, "I would have him learne of the Italians, to receive those that visite him, with great civility and respect." The sensitivity to customs of reception was a leit motif of Lassels' guide. Arriving at the Medici court to see the galleries and gardens, Lassels commented that the Grand Duke "admits willingly of the visits of strangers, if they be men of condition." Generalizing about this behavior on the part of the Italians, he explained, "they are extremely civil to one another . . . not knowing whose turne it may be next, to come to the highest honors."[123] Decorum, as Tesauro outlined in his advice to young noblemen, was learned only by "practicing with civil persons."[124] This occurred first at home and in the Jesuit colleges, where such skills as "accompanying and receiving"[125] were inculcated, and second, in social settings such as the museum, laboratories of civility as well as nature. Access to a museum was not simply about scientific credentials, though surely membership in various academies and societies facilitated one's entrance, nor was it solely conditioned by curiosity. In a culture defined by the vicissitudes of urban and court hierarchies, collectors, like other members of society, were hedging their bets when they opened their doors.

Rarely do we have a precise idea of who visited a museum. We can extrapolate from travel journals and letters, but it is difficult to imagine the aggregate, let alone understand its significance. The case of Aldrovandi's museum, however, offers a unique point of departure: the visitor's books kept meticulously by the naturalist and his assistants in the late sixteenth century. Aldrovandi belonged to a small group of collectors for whom we can document the presence of visitors' books, although his is probably the most extensive of the ones that have survived. His Swiss rival, Conrad Gesner, created a *Book of Friends* in the last ten years of his life (1555–1565)

[122]Nicolaus Steno, *Epistolae et epistolae ad eum datum*, Gustav Scherz, ed. (Hafniae, 1952), Vol. I, p. 208 (Innsbruck, 12 May 1669).

[123]Lassels, *The Voyage of Italy*, Vol. I, sig. f.r, pp. 14, 217.

[124]Tesauro, *La filosofia morale*, pp. 276–277.

[125]I have taken this directly from the 1640 statutes of a Jesuit college in Bologna, as published in Brizzi, *La formazione della classe dirigente*, p. 238.

that contained 227 autographs.[126] Some entries even duplicated those in Aldrovandi's book, for instance, the museographer Samuel Quiccheberg who visited both naturalists. Francesco Barberini and Settala also possessed catalogues similar to Gesner's *Book of Friends*. Settala's "little book" (*libriccuolo*), for instance, "indicated the rare names of his many correspondents and friends throughout the world."[127] Barberini had a "Catalogue of . . . learned men" in which he recorded the names of *viri illustres* who came to his palace to see the library and museum.[128] The Danish collector Olaf Worm's catalogue of visitors in his Copenhagen museum mirrored this pattern: "many royal persons and envoys visiting Copenhagen ask to see the museum on account of its great fame and what it relates from foreign lands, and they wonder and marvel at what they see," one contemporary reported. "As evidence of having seen it, they testify with their own hand in a book remaining with him."[129] Such visitors' books immortalized the fame of a museum and its creator by recording their connection to the social, political, and intellectual centers of power.

While Gesner collected 227 autographs, Aldrovandi acquired almost seven times that number (1,579), expanding the use of the "book of friends" and "catalogue of illustrious men" to define the social world of the museum in its broadest sense. As Aldrovandi and several observers tell us, there were two books, one for the most illustrious visitors and the other for the majority of people who toured the museum.[130] Upon seeing the museum in 1604, shortly before Aldrovandi's death, Pompeo Viziano marveled at the number of people who had visited the naturalist's *studio*, describing the catalogues of visitors in the following terms:

> . . . in two large books, that he conserves among the other [books], an infinite number of Princes, Cardinals, Prelates, Knights, and other people of note (*alto affare, et di elevato ingegno*) that have passed through Bologna, attest in their own hand to having seen and diligently considered [the museum] with great satisfaction.[131]

The first book, spanning the period from 1566 to 1644, recorded only signal visits: the arrival of the archbishop of Bologna (Gabriele Paleotti), a Gonzaga duke, the archbishop of Ravenna, and several papal legates, in

[126]Gesner's *Liber amicorum* is owned by the National Library of Medicine; the contents are discussed by Richard J. Durling, "Conrad Gesner's *Liber amicorum* 1555–1565," *Gesnerus* 22 (1965): 134–157.

[127]Giambattista Pastori, *Orazione funebre* (1668), in Folgari, "Il Museo Settala," p. 91.

[128]Gassendi, *The Mirrour of True Nobility*, Vol. II, p. 41.

[129]In Arthur MacGregor, "Collectors and Collections of Rarities in the Sixteenth and Seventeenth Centuries," in *Tradescant's Rarities*, MacGregor, ed., p. 80.

[130]BUB, *Aldrovandi*, mss. 41 and 110.

[131]BCAB, B. 164, f. 301r (Pompeo Vizano, *Del museo del S.r Dottore Aldrovandi*, 21 April 1604).

whose quarters the museum ultimately resided. "Cardinal Enrico Gaetano, legate to Bologna, saw the *mirabilia* of nature in the *studio* of doctor Ulisse Aldrovandi," read one entry for 1587 in the *Book in Which Men of Extraordinary Nobility, Honor and Virtue Who Have Seen the Museum That the Most Excellent Ulisse Aldrovandi Gave to the Most Illustrious Senate of Bologna Write Their Own Name in Perpetual Memory of the Thing.*[132] Nobility, honor, and virtue, the traits cultivated by all Italian patricians, were the criteria for appearing in either of the two books. What distinguished the few visitors who recorded their names in the gilt-edged book, covered in expensive red fabric (fig. 10), was their unusual level of visibility in the local political culture. Undoubtedly, the signing of the book was the culmination of an important ceremonial visit, distinguished by virtue of the activity surrounding it from the more routine tours of the museum.

The second book survives in the form of a cross-referenced index of names compiled by Aldrovandi and his assistants over several decades. A project as grand and as infinite as the *pandechion* of nature, it created a "museum" of the urban patriciate and scholarly community, recording and classifying all the people interested in natural history. Aldrovandi's arrangement of the second book reflected his vision of the catalogue as a form of social taxonomy. The first section of his *Catalogue of Men Who Have Visited Our Museum* was arranged "according to the order of dignity, study and profession"; the second, by city or region. It was composed mainly of signatures, written on scraps of paper by Aldrovandi, his assistants, and the visitors themselves, and later pasted into different sections and organized alphabetically. The sheer number of visitors recorded in this way testifies to the Bolognese naturalist's willingness to open up his theater of nature to the world.

With patience and meticulous care, Aldrovandi and his assistants extracted the details of each visitor's circumstances (tables 1 and 2). Some were so well known that no questions needed to be asked: the unspecified pope—possibly Gregory XIII, as part of the contingent to view the monstrous dragon in 1572?—various princes, cardinals, bishops, and famous men of learning such as Carlo Sigonio who, among all the professors and university graduates, merited a singular entry, neither as *doctor* nor *lector* but as a man "most learned, all-sustaining and extraordinary." Perhaps Aldrovandi later regretted not assigning him to the more prestigious book, though admittedly, Sigonio's virtues were more intellectual than social. Sigonio was one of a number of humanists who came to the museum; in fact, few people who claimed association with the learned world of late Renaissance Italy had *not* seen it. Scholars and naturalists, known personally to

[132]BUB, *Aldrovandi*, ms. 41, c. 2r (*Liber in quo viri nobilitate, honore et virtute insignes, viso musaeo quod Excellentissimus Ulyssis Aldrovandus Illustriss. Senatui Bononiensi dono dedit, propria nomina ad perpetuam rei memoriam scribunt*). The book, however, was started in Aldrovandi's lifetime, since the entries date from 1566—significantly, the first signature was Gabriele Paleotti's—to March 1644.

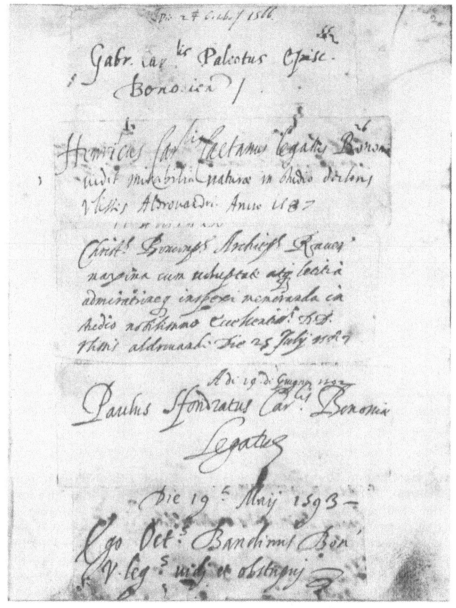

Figure 10. Aldrovandi's catalogue of visitors. From BUB, *Aldrovandi,* ms. 41.

Aldrovandi by reputation, correspondence, or study at the University of Bologna, graced the pages: Joachim Camerarius,[133] Caspar Bauhin, the Dutch collector Bernard Paludanus, Fra Evangelista Quattrami (botanist

[133]"The extraordinary physician, Doctor Joachim Camerarius who has written many works on various natural things, especially plants."

TABLE 1 Visitors to Ulisse Aldrovandi's Museum Until 1605

Visitors	*Number*
Popes	1
Legates and Vice-Legates	7
Papal Governors	4
Cardinals	10
Archbishops	11
Bishops	21
Theologians	15
Inquisitors	3
Other Clergy	190
Princes	6
Nobles	118
Courtiers	4
Diplomats and Princely Servants	7
Famous Men	26
Appointed and Elected Officials	26
Lawyers	43
Secretaries	8
Notaries	1
Military Leaders	6
Professors	21
Philosophers	6
University Graduates	21
Scholars	907*

to the d'Este in Modena), Giovan Antoni Bertioli (ducal pharmacist in Mantua), the Florentine court painter Jacopo Ligozzi, Aldrovandi's one-time colleague Girolamo Cardano, and that benevolent humanist of humanists, Giovan Vincenzo Pinelli. Scattered throughout the lists of professors, doctors, and scholars were numerous students of Aldrovandi, described as *discipulus* or *scolaris meus*. In certain cases, Aldrovandi proudly recorded their current occupation: "Doctor Andreas Langnerus of Magdeburg, my disciple and scribe, whom I advanced to the College of Philosophy and Medicine"; Pietro di Wittendel "who served for eight years at the University of Padua, Bologna and Pisa," as *amanuensis* for the physician Mercuriale; and so on.[134] The majority received only simple entries, a reflection of their more ephemeral relationship to Aldrovandi's pursuit of knowledge.

---

[134]For a sampling of entries, see Alessandro Tosi, ed., *Ulisse Aldrovandi e la Toscana* (Florence, 1989), pp. 439–442.

TABLE 1   *(Continued)*

Visitors	Number
Doctors	87
Anatomists	1
Surgeons	1
Apothecaries	6
Distillers	1
Mathematicians	2
Cosmographers	1
Antiquarians	1
Printers	2
Booksellers	2
Painters	6
Poets	1
Jewelers	1
Women	1**
Unknown	4

SOURCE: BUB, *Aldrovandi*, ms. 110 (*Catalogus virorum, qui visitarunt Musaeum nostrum, et manu propria subscripserun in nostris libris Musaei. Secundum ordinem dignitatum, studiorum, et profes-sionum*). Aldrovandi and his assistants recorded 1579 visitors to the museum.
*The category *studiosus* included one entry of indeterminate number, which I have discounted from the tally.
**The only woman visitor listed was Ippolita Paleotti, listed as *studiosa*.

Certain striking features emerge from an analysis of the method of entry. It is impossible to ignore the level of precision about those characteristics which the title of the catalogue revealed in sequential importance: "dignity, study and profession." In fulfillment of this premise, the lists recorded singular entries such as "Coadjustor of Bologna," "Cosmographer," "Jeweler," "Cleric of the Apostolic Chambers," "Councillor for the Holy Roman Emperor Rudolf and *Nuncius* to the Pope and Princes of Italy," and most delightfully, "Carver for the King of Poland." Truly the book was a collection of the most noteworthy, and in some instances unusual, occupations held by early modern elites. Here we should pause momentarily to contemplate the amassing of such detail. What sort of questions must Aldrovandi's assistants have put to visitors upon entrance and exit to the museum? We can imagine a version of the following: "From whence do you come?" "By whom are you recommended?" "What is your occupation?" The need to ask and answer such questions indicates a high level of consciousness among patricians about the nature of identity. What Aldrovandi essentially demanded of his visitors was the means by which they wished to be remembered in his

TABLE 2 Aldrovandi's Visitors

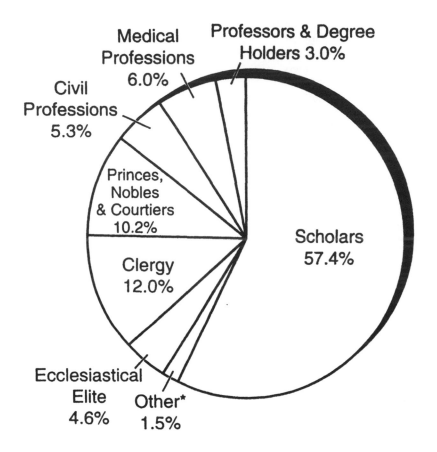

Medical Professions 6.0%

Professors & Degree Holders 3.0%

Civil Professions 5.3%

Princes, Nobles & Courtiers 10.2%

Clergy 12.0%

Scholars 57.4%

Ecclesiastical Elite 4.6%

Other* 1.5%

* "Other" encompasses groups too small to be displayed separately: Antiquarians (0.06%), Cosmographers (0.06%), Mathematicians (0.1%), Painters (0.4%), Poets (0.06%), Distillers (0.06%), Jewellers (0.06%), and Women (0.06%). It also includes four entries which are undecipherable (0.25%).

classification of all the professions and identities in the world. Not unlike Tommaso Garzoni's *Universal Piazza of All the Professions in the World* (1585), which went into numerous editions precisely during the years in which Aldrovandi created his books, the catalogues of visitors formed a map of society. Within their pages, memory, status, and identity came together, and the lists continued to grow.

Some answers reflected well-known aspects of social norms and conventions. Among the multitude of dukes, barons, and counts who signed the book, four entries stand out. They recorded men of high standing who— perhaps disdainful of the queries about "study" and "profession"—replied that they were *aulicus* ("courtiers") or, even better, *aulicus et familiaris* ("courtier and servant"). The latter reminds us that, in addition to the 1,579 visitors dignified with an entry in the catalogue, there were many other "invisible" visitors[135]—people who saw Aldrovandi's *mirabilia* but whose seeing did not count. Only one woman—the learned Ippolita Paleotti—received an entry. Even Caterina Sforza, who arrived in 1576 with "fourteen of fifteen coaches and carriages containing fifty Gentlewomen, the flower of the first families of this city, accompanied by more than 150 Gentlemen" was not asked to sign the visitor's book.[136] While Aldrovandi wrote a lengthy report of this important ceremonial visit to his brother Teseo and hoped that many good things would come from his hospitality toward "such honored company," he did not equate the honor of the visit of an important female patron with the honor of the visitors whose names appeared in his catalogue.

Unlike the Baroque galleries, in which women gradually made an appearance—for example, note the importance of Queen Christina of Sweden's visit to the Roman College museum in 1656[137]—Aldrovandi's museum, as portrayed in his visitors' books, reflected the ideal of humanist sodality more than courtly sociability. Following the lead of social commentators such as Della Casa and Guazzo, who advertised civility as a form of discourse best practiced with other men, Aldrovandi and many of his contemporaries perceived women to be the true "foreigners" in the museum.[138] Servants

[135]Here I am mirroring some remarks made by Steven Shapin on laboratory technicians in his "The Invisible Technician," *American Scientist* 77 (1989). 554–563.

[136]BUB, *Aldrovandi,* ms. 35, cc. 203v (Bologna, 29 May 1576).

[137]PUG, *Kircher,* ms. 556 (VII), f. 40; Villoslada, *Storia del Collegio Romano,* pp. 276–277. Kircher subsequently dedicated his *Iter Exstaticum* (1656) to Christina. Admittedly, it is unclear exactly how contemporaries perceived Christina; certainly her gender was ambiguous if not problematic in the eyes of her critics. Yet Christina's success in forming a scientific salon and a library visited by the Republic of Letters, and in patronizing Ciampini's Accademia Fisico-Matematica, signaled a change in attitudes toward learned women from the period of the Accademia dei Lincei, when no woman had participated formally in the scientific culture of the academies and museums. For more on these circumstances, see Susanna Åkerman, *Queen Christina of Sweden and Her Circle: The Transformation of a Seventeenth-Century Philosophical Libertine* (Leiden, 1991).

[138]On conversation with women, see Guazzo, *Civil Conversation,* Vol. I, pp. 231–249.

could be ignored since they were not of the same social class. Women had to be excluded. Such practices, revealed implicitly in the absence of letters and gifts from women, were formalized in the regulations governing scientific sodalities such as the Accademia dei Lincei and in the homosocial organization of the Jesuit colleges as societies of men.[139] Certainly no woman ever entered Cesi's museum. And despite Christina's celebrated appearance at the Roman College, few graced the portals of the Jesuit institutions.

As Aldrovandi's catalogue of visitors suggests, more strongly than any other document emerging from the collecting culture of the period, the emerging scientific culture largely excluded women from its official representations. On the one hand, the strength of courtly models and the importance of courtiers as naturalists, collectors, and visitors allowed a limited place for women, not as participants in this discourse but as an audience whose presence facilitated the civil conversation of men. This explains Aldrovandi's pride in the visit of Caterina Sforza, whose presence allowed him to imagine his museum as a space in which conversations such as those depicted in Castiglione's *Book of the Courtier* could be enacted. In honor of this special visit, he even had his wife and sister greet visitors at the door, thus completing his side of the courtly conversation about nature. On the other hand, the very organization of his catalogue of visitors allowed no place in which to record such encounters, nor would they have added to his standing in the republic of letters. Caught between two conflicting models, like most of his contemporaries, Aldrovandi was ambivalent about their function in practice, but decisive in his attempts to set the record straight.[140] No woman would be included if she could not be placed in a category such as *studiosa* that called her gender into doubt.[141] Only one, a relative of his most important local patron Paleotti, met this standard.

The invisibility of women and servants underscores the social function of Aldrovandi's catalogue, which did not indiscriminantly record all visitors. Each entry was a calculated attempt to enhance the reputation of the museum and its creator, by collecting people of worth within its pages. Next to the nameless servants and women who did not warrant the label *studiosa*, we should note the presence of numerous scholars, who added bulk but not substance to the catalogue. They at least merited cursory entries, as marginal members of the humanist world of learning. As a statistical breakdown of the

[139]Lombardo, *With the Eyes of a Lynx*, pp. 42–43. I also am indebted to Richard Lombardo's unpublished essay entitled "Representing the Accademia dei Lincei: The Luca Valerio Affair." For more on this general subject, see David F. Noble, *A World Without Women: The Christian Clerical Culture of Western Science* (New York, 1992).

[140]In a different context, see Jed, "Making History Straight: Collecting and Recording in Sixteenth-Century Italy," in Jonathan Crew, ed., *Reconfiguring the Renaissance: Essays in Critical Materialism*, Bucknell Review, vol. 35, no. 2 (Lewisburg, PA, 1992), pp. 104–120.

[141]Margaret King, *Women of the Renaissance* (Chicago, 1992), pp. 194–198ff.

contents indicates, scholars—a category referring primarily to students or visitors who offered no other means of identification—comprised well over half (57.4 percent) of the visitors (see table 2). While many scholars merited individual entries, a large portion, particularly foreigners, were assigned anonymous headings: "six German scholars," "two Spanish scholars," "four French scholars," or simply "another scholar." They added quantity, not quality, to the catalogue. Undistinguished by reputation yet worthy of admission on intellectual grounds—the mutual *curiositas* that all shared—they testified to the strong pedagogical function of Aldrovandi's collection and to his desire to magnify, within the bounds of social convention, the size of his list.

The overall pattern of visitors to Aldrovandi's museum reveals several interesting trends. Approximately two-thirds held or were studying toward university degrees, suggesting a fairly close association between the worlds of learning and collecting. Numerous members of the medical profession also appeared—doctors, anatomists, and surgeons (89) and a few apothecaries (6). After scholars, clergy comprised the largest group of visitors; 262 of them graced Aldrovandi's portal. Ecclesiastics traveled more frequently than most people as part of their spiritual duties, hence the appearance of inquisitors, visitors, legates, and governors, all making their appointed rounds. Next to the clergy, nobles, diplomats, and courtiers, the third largest group (161), enjoyed a similar mobility. Aldrovandi's success in bringing so many prominent clergy, courtiers, and elected officials into his theater of nature gives us a glimpse of the extent to which these types of museums had made the viewing and possession of nature an attractive pastime for the educated elite. By the late seventeenth century, when the botanist Paolo Boccone gathered together his letters on various "natural observations," it was no longer worthy of note that so many of the recipients were nobles.[142]

Aldrovandi not only kept records throughout his lifetime but specified that the names continue to be recorded after his death. "It would also please me," he specified in his gift of 1603, "if the Gentlemen and Men of Letters who have visited and will visit the Museum after my death will continue to write their names in my two books designated for this purpose."[143] Unfortunately, the custodians continued the practice only haphazardly in the early seventeenth century, until it was finally discontinued around 1644. Unlike Aldrovandi, who perceived the recording of all information to be an infinite, continuous process, the custodians of the museum, lacking the level

[142]Paolo Boccone, "Indice delle Materie, delle Osservazioni Naturali, e de' Cavalieri, Letterati, a' quali sono indirizzate, e dedicate," *Osservazioni naturali* (Bologna, 1684). Of the twenty-six letters, ten are addressed to nobles and *signori*, six to counts and senators, two to *marchesi*, and one each to a lawyer and a canon. Only six are addressed to people with a "professional" interest in nature, that is, four professors of medicine, one professor of mathematics, and one *protomedico*.

[143]Fantuzzi, p. 84.

of personal investment in the museum that the creator had had, quickly forgot the reasons for this exercise. Like the museum itself, the catalogues of visitors collected people as a means of eternalizing anyone of social worth who had seen the collection; they were portraits of the "best" that this particular world had to offer. Entering the museum in 1604, a year before Aldrovandi's death when the list was almost complete, Pompeo Viziano marveled at the fact that he briefly inhabited a space in which popes, prelates, princes, and the famous scholars of bygone days had stood. Adding his own name to the catalogue, he enjoyed the sensation that he was part of this most eminently civil enterprise.

THE VIEW FROM BEYOND

The space of collecting, marked by noteworthy artifacts such as Aldrovandi's catalogue of visitors and Kircher's gate and speaking trumpet, reflected the complexities of the early modern social world. Gender, privilege, and status defined the culture of learning. As sites of knowledge, museums reveal some of the fundamental underpinnings of scientific culture. In settings such as Aldrovandi's *teatro di natura,* new models of sociability, the changing status of intellectuals, and the expanding material culture of science forged the discipline of natural history. While most collectors, to borrow Gassendi's description of Peiresc, "shunned the society of Women," they occasionally relaxed the principle of sodality to allow women of exceptional learning and high social standing a glimpse of the wonders that the natural world offered.[144] Knowing the advantages of sociability, they occasionally retreated into their museums just as Montaigne enjoyed his *solitarium.* Faced with a variety of competing models that defined the site of knowledge—solitary and conversable, a company of male humanists and a theater for courtiers— Italian naturalists preferred to allow it to exist in all these different forms, sometimes in the same museum, rather than choosing one definitive pattern. Certainly Italy was not the only region in which the rearticulation of the space for philosophizing occurred. Since we have spent a great deal of this chapter following the travels of English virtuosi in Italy, an epilogue is the appropriate place to reverse the viewpoint, considering briefly how England appeared to foreign scholars.

Restoration England, home of Baconianism and the Royal Society that made Bacon's pronouncements about natural history the foundation for a new form of philosophical inquiry, created a scientific culture that placed the collecting of nature at its center.[145] There, as Peter Dear, Steven Shapin, and Simon Schaffer have detailed, a natural philosophy based on gentle-

---

[144]Gassendi, *The Mirrour of True Nobility and Gentility,* Vol. II, p. 176.

[145]Ken Arnold, *Cabinets for the Curious: Practicing Science in Early Modern English Museums* (Ph.D. diss., Princeton University, 1991).

manly conduct emerged.[146] Men such as Oldenburg created epistolary networks as far reaching as those of Aldrovandi and Kircher, and the members
of the early Royal Society, following the advice laid out in treatises such as
Bacon's *New Atlantis,* perceived the collecting of information and the possession of objects to be essential features of the creation of a new encyclopedia of knowledge, a goal that was simultaneously idealistic and utilitarian.
Implicitly, gentlemen were the best qualified individuals to conduct the exchanges that facilitated the Royal Society's many projects; only they had the
credentials to validate the authenticity of the knowledge presented to the
fledgling scientific organization. Despite the general similarity in the trajectory, upon closer inspection, certain essential differences emerge, a reflection of the varied social and political climates in England and Italy.

Two episodes, both occurring in the mid-seventeenth century, highlight
the contrast. In 1683, the Ashmolean Museum at Oxford opened its doors.
A gift of Elias Ashmole, member of the Royal Society, alchemist, magus,
and natural philosopher, it was heralded as a sign of the revivification of scientific culture at the University. Like many of the science museums we have
already encountered, it was designed to facilitate new forms of scientific
inquiry. Yet it diverged in one important respect: the criteria of admission. Throughout the late seventeenth and eighteenth century, visitors
from the continent expressed their dismay and sometimes horror over the
Ashmolean's public status. "On 23 August we wished to go to the Ashmolean
Museum," wrote the German traveler Zacharias Conrad von Uffenbach in
1710, "but it was market day and all sorts of country-folk, men and women,
were up there (for the *leges* that hang upon the door *parum honeste & liber-
aliter* allow everyone to go in). So as we could have seen nothing well for
the crowd, we went down-stairs again and saved it for another day." The
Ashmolean, as Von Uffenbach sourly noted, was a new sort of public institution, open to the very people whose exclusion from the world of learning
defined the civil boundaries of the scholarly community and implicitly determined the principles of admission to Renaissance and Baroque museums. It was a highly uncivil space.

Von Uffenbach's displeasure at the literal openness of the Ashmolean
translated into pointed comments about the general definition of "public"
institutions in England. Not only did the open admission standards disinte-
grate the gender and class barriers that defined the private, hence exclusive,
nature of the museum—"even the women are allowed up here for a six-
pence"—but the establishment of the price of admission commodified the
experience of scholarship. His experience in the "world-famed public

---

[146]Peter Dear, " 'Totius in verba': Rhetoric and Authority in the Early Royal Society," *Isis* 76
(1985): 145–161; Steven Shapin and Simon Schaffer, *Leviathan and the Air-Pump: Hobbes, Boyle
and the Experimental Life* (Princeton, 1985). I have already cited many of Shapin's more recent
articles earlier in this chapter.

library of this University," the Bodleian, only confirmed his worst fears about the dangers of the public in a scholarly setting:

> But as it costs about eight shillings and some trouble to gain an entrance, most strangers content themselves with a casual inspection. Every moment brings fresh spectators of this description and, surprisingly enough, amongst them peasants and women-folk, who gaze at the library as a cow might gaze at a new gate with such noise and trampling of feet that others are much disturbed.[147]

The pinnacle of his trip to England, a visit to the famed Royal Society, proved to be equally disillusioning. Finding the society and its museum to be in complete disarray, Von Uffenbach commented on the inevitability of its state.

> But that is the way with all public societies. For a short time they flourish, while the founder and original members are there to set the standard; then come all kinds of setbacks, partly from envy and lack of unanimity and partly because all kinds of people of no account become members; their final state is one of indifference and sloth.[148]

Despite attempts to painfully distinguish the gentlemanly members of the Royal Society from society at large, many continental visitors perceived it as reflective of the ambiguous social hierarchies of late seventeenth- and early eighteenth-century England that did not differentiate categories of people according to the standards used in the urban and courtly world of continental Europe. Nobility, honor, and virtue—to invoke Aldrovandi's categories—all were terms in use in England, but they did not have quite the same meaning. If members of the Royal Society created museums that demanded payment as the criterion for admission, a mercantile rather than a gentlemanly request, and imagined women and peasants, people clearly excluded from Morhof's *Polyhistor* (which Von Uffenbach surely had read), as an appropriate audience, what other incivilities were they capable of?

The discomfort of Von Uffenbach and other visitors with the public agenda of Baconian science, as practiced in a less courtly and increasingly mercantile society, only reinforces the perception that the republic of

---

[147]Zacharias Conrad von Uffenbach, *Oxford in 1710*, W. H. Quarrell and W. J. C. Quarrell, eds. (Oxford, 1928), pp. 2–3, 24, 31; see Martin Welch, "The Foundation of the Ashmolean Museum" and "The Ashmolean as Described by its Earliest Visitors," in *Tradescant's Rarities*, MacGregor, ed., pp. 41–69. The image of the Ashmolean as the prototype for the "public" museum appears in the emphasis on this word in contemporary descriptions of it. Borel singled it out as "Le Cabinet publique" in his list of museums in 1649, alluding to the earlier Tradescant collection at Oxford that was subsumed under Ashmole's in 1683. The Chevalier de Jaucourt described it as the museum "that the University had built for the progress and the perfection of the different branches of knowledge" ("Musée," *Encyclopédie*, Vol. X, p. 894). The entry in the *OED*, Vol. VI, p. 781, under "museum," like the Crusca reference to Aldrovandi, underscores the normative function of the Ashmolean in shaping the use of *museum* in English.

[148]Zacharias Conrad von Uffenbach, *London in 1710*, W. H. Quarrell and Margaret Mare, trans. (London, 1934), p. 98.

letters, for all its success in forging bonds among scholars in different regions, had its limits. Von Uffenbach, not unlike many Italian patricians, understood the category "public" to define a relatively closed world. This undoubtedly explains the remark of his compatriot, the museographer Michael Bernhard Valentini, in his *Museum of Museums* (1714): "In Italy one finds hardly any fully public museums."[149] While English travelers marveled at the intricacies of Italian civil behavior, so different from their own, they often enjoyed the added distinctions it conferred. In contrast, foreign visitors to England were bewildered by the relative absence of protocol and the lack of identifiable connections between the Royal Society and English court culture. Familiar with the workings of absolutist courts such as those in Florence, Madrid, Versailles, and Vienna, they found it difficult to understand why a scholar could not be a courtier.

In 1668, Lorenzo Magalotti, the would-be secretary of the Accademia del Cimento, voyaged to England. Among other things, he planned to present Charles II and the Royal Society with a copy of the academy's *Essays on Natural Experiences* (1667). A man who considered himself both "courtier and philosopher," Magalotti was the epitome of the scientific culture then flourishing in Italy.[150] Educated at the Roman College and the University of Pisa, he was learned as well as civil. With his credentials, he anticipated a flurry of visits to museums, academies, and the court. Arriving in England, Magalotti dutifully visited the two museums of note in London: the collection of John Tradescant in Lambeth, not yet subsumed into the Ashmolean Museum, and the repository of the Royal Society, then in the house of Robert Hooke. Neither left much of an impression on him. Having just left a court at which Francesco Redi was busy experimenting on nature in every corner of the Medici palace and grand ducal laboratory and pharmacy, Magalotti undoubtedly expected more extravagant displays of nature. Continuing his itinerary, he appeared on the doorstep of the Royal Society on 13 February 1668 for one of their Thursday meetings, yet did not enter. As Magalotti explained to Leopoldo de' Medici, prince of the Accademia del Cimento, "having found that entry was not permitted to a simple passerby; I did not want to get a place for myself as a scholar, firstly because I am not one, and secondly because even if I were I should not consider it the most advantageous character for getting into courts."[151]

Even as a virtual newcomer, Magalotti understood that the category of "scholar" in England conferred a different status than the image of a "courtier and philosopher" in Italy. Four months later, now experienced in

---

[149]Valentini, *Museum museorum, oder der allgemeiner Kunst- und Naturalien Kammer* (Frankfort, 1714), sig. xx2r.

[150]Cochrane, *Florence in the Forgotten Centuries*, p. 255.

[151]BNF, *Ms. Gal.* 278, pp. 145r–146r, in W. E. Knowles Middleton, ed. and trans., *Lorenzo Magalotti at the Court of Charles II. His Relazione d'Inghilterra of 1668* (Waterloo, Ontario, 1980). The discussion of the Tradescant and Royal Society collections can be found on p. 140 of this work.

the ways of English gentlefolk, he elaborated on his dilemma. "I could never tell Your Highness how prejudicial it is for a man of fashion from that side of the mountains to pass for a philosopher and mathematician. The ladies at once believe that he must be enamoured of the moon, or Venus, or some silly thing like that."[152] Magalotti, more committed to his image as courtier than philosopher, preferred to keep his distance from the Royal Society, at least in a scholarly capacity. While admiring their philosophy, he distrusted their social base, instinctively recognizing it as something alien to his courtly sensibilities. The Royal Society repository might contribute to the advancement of knowledge, in some abstract sense, but it lacked the refinement that seventeenth-century Italians associated with the experiencing of going to the museum.

Magalotti, a product of the social world of Baroque Italy, a courtly and patrician culture in all respects, could not help but be responsive to the pressures that his society exerted upon him, even far from home. For him, like all the naturalists and virtuosi we have been studying, civil society determined where the site of knowledge would be and what shape it would take. After touring the *Musaeum Tradescantium,* he remarked, "to tell the truth there is nothing there that can be called rare nowadays and that is worth going across the river to see, as one has to do."[153] Tradescant's museum contained objects, but it did not contain knowledge. Von Uffenbach would have applied this statement to the Ashmolean Museum as well.

[152]Ibid., p. 62.
[153]Ibid., p. 140.

PART TWO

Laboratories of Nature

A site of encyclopedic dreams and humanist sociability, the museum also was a setting in which to examine nature. Nature, in its broadest sense, was the object that collectors strove to contain and display; through their actions, they created new ways of perceiving nature that marked the emergence of natural history as a discipline. At the beginning of the sixteenth century, natural history was hardly distinguishable from other subjects worthy of humanist inquiry. " Nature" was not the initial point of departure for the humanists who discussed the merits of Pliny's *Natural History* in the 1490s, inspecting it like any other text that demanded the attention of trained philologists. Discussions of Pliny, however, opened up the possibilities of a relatively unexplored material world, fast expanding in the wake of Columbus's arrival in the Americas in 1492. By the middle of the sixteenth century, the accumulation of natural objects in the humanist *studio,* initially collected to resolve problems with words, had transformed the site of humanist inquiry into a laboratory of nature. Not all naturalists immediately embraced this new form of inquiry. For some, books continued to take precedence over nature, now a competing " book." Even naturalists who did not revere Aristotle and Pliny felt that engaging in a critical dialogue with the past was the only way to dismantle outdated theories. As collectors became more comfortable with the idea of studying nature as a material object rather than a textual subject, they gave equivalent weight to the activities surrounding the gathering of nature, while continuing to emphasize the importance of reading old texts and creating new ones. Rather than allowing texts to define nature, they used nature to form new texts.

Part II offers a selected reading of the role of collecting in transforming the discipline of natural history. It explores the rituals that evolved around the study of nature, both in and out of the museum. Besides the appearance of botanical gardens and lectureships in natural history, signs of institutional success that we readily associate with the formation of a discipline, the emergence of natural history can also be traced through the shared experiences that defined the community of collectors: the symbolic and scientific meaning of travel, a growing desire to perceive nature as an object of experiment, and a persistent interest in the medicinal uses of nature, the field Galen, Dioscorides, and Avicenna had defined as *materia medica*. Chapter 4 outlines the process by which nature entered the museum. Objects entered a collection through many different channels. Not always the gifts of favored patrons and associates, they regularly were the results of herborizing trips, contact with groups such as fishermen and falconers, and daily inspection of nature's wares in the marketplace. All of these different activities, repeated over and over by several generations of naturalists, marked the pilgrimages of nature. Naturalistic excursions identified nature as the first *musaeum,* a site worthy of contemplation and a setting in which naturalists could commune. There they further refined the parameters of humanist sociability, determining who had the ability to collect nature and the credentials to interpret it.

Such activities set the stage for the experimental life of the museum. The experiments and medicines produced in the museum transformed it into a laboratory in which collectors gradually came to see the testing, dissection, and distillation of nature as one of their primary goals, as important to the reconstitution of natural history as their textual emendations and classifying projects. Collecting became a prelude to manipulating nature, and knowledge became demonstrable. Chapters 5 and 6 detail a series of scientific debates that occurred in the museum. Chapter 5 looks at the experimental culture of the museum, considering its role in debates about fossils, spontaneous generation, and other paradoxes of nature. Chapter 6 investigates the use of the museum as a medical arena, as seen particularly in the involvement of prominent collectors in debates about antidotes such as theriac and the proper composition of medicines. Both chapters underscore the role of the museum as an *instrument*—as powerful in its ability to magnify nature as the microscope and one as fraught with conflict. The museum certainly provided a new space for scientific activity, in all of its forms, and collectors were quick to acknowledge this. The significance of its novelty, however, was more ambiguous. Did the museum serve to defend orthodoxy or demolish it? Through the empirical programs developed in this setting, naturalists seemed to have fulfilled both possibilities, creating a location that can only be described as a contested site. What remained uncontested, however, was the authority that collectors derived from their control of these laboratories of nature.

FOUR

Pilgrimages of Science

If reading gives so much utility to scholars, travel gives them ten times more.
—ULISSE ALDROVANDI

Collecting did not begin in the museum. Rather than seeing the museum as a point of departure, we should imagine it as an end: the resolution of a long and complex voyage beginning at the moment an object was possessed by human hands, continuing to the point at which it was designated a collectible specimen, and ending with its arrival and display in the museum. In the interim stages, a variety of different activities resulted in the domestication of nature. This chapter is concerned with the means by which nature was collected. What impulses led Renaissance naturalists to experience nature through the prism of their senses with an intensity unmatched by their predecessors? How were they able to accumulate and accommodate their knowledge? The first "theater of nature," after all, was not the museum of natural history but nature herself. To create their own theaters of nature, collectors initially had to come to terms with the original *musaeum*. They drew upon all the cultural resources at their disposal, particularly their humanist training, to establish and often personalize their relationship with nature. Through such actions, naturalists not only collected nature but made this knowledge meaningful.

Travel, both near and far, was a precondition to the domestication of nature; it bridged the distance between nature and the museum, between the site of the journey and its result. The desire to travel led scholars out of their studies and into nature, where its rediscovery as a *locus amoenus* was celebrated. "Nature is not silent but speaks to us everywhere and teaches the observant man many things if she finds him attentive and receptive," wrote Erasmus in "The Godly Feast" (1522).[1] Naturalists reveled in their ability to

[1] Erasmus, "The Godly Feast," in *The Colloquies of Erasmus*, Craig R. Thompson, trans. (Chicago, 1965), p. 48. The image of the *locus amoenus* ("pleasant site") is a particularly important one to humanistic inquiry. Places identified with this label facilitated the sort of philosophical speculations that led to knowledge.

make natural history a form of intimate knowledge, engaging nature (as well as each other) in dialogue, just as Erasmus had suggested. As Carlo Ossola and Gigliola Fragnito have observed, the museum simultaneously enhanced the mystery of exotic nature, through its concentration of rare and miraculous objects, and proclaimed the ability of the collector to manage nature's more proximate parts.[2] In the museum, the unknown became knowable, and the known showable.

Increased contact with other parts of the world undoubtedly contributed to the greater openness on the part of naturalists regarding the origins of authoritative information about nature. Yet natural history continued to be a haven for philologists, who were as fascinated with the etymologies of living creatures as they were with the origins of the word *musaeum*. Even such well-traveled scholars as the botanist Melchior Wieland often placed linguistic skills above the powers of observation.[3] However, the textual foundation of natural history fueled a sort of empiricism precisely because questions of language, and the virtual absence of illustrations in ancient natural histories, pushed naturalists to observe as a way of resolving humanists' debates about the proper names and descriptions of living things. "One cannot know simples by reading books," admonished Calzolari, "unless this reading is accompanied by direct observation."[4] As the model of a physician who included observation in his medical activities, Aldrovandi professed to have described only those things

> that I have seen with my own eyes, touched with my hands, dissected, and likewise conserved one by one in my little world of nature, so that everyone may see and contemplate them daily. They are preserved in image and in example (*in pitture et al vivo*) in our museum for the utility of scholars, collected in travel by me not only while I was a student but, for the most part, after I took my degree.[5]

Natural history had become a matter of doing in conjunction with learning, a practice with specific operations and procedures that formed the groundwork of a philosophical system of understanding. In the world of sixteenth-century natural history, authority and knowledge were gained through the

---

[2]Carlo Ossola, *Autunno del Rinascimento: "Idea del Tempio" dell'arte nell'ultimo Cinquecento* (Florence, 1971), pp. 243–263; Fragnito, *In museo e in villa*, pp. 174–175.

[3]Richard Palmer, "Medical Botany in Northern Italy in the Renaissance," *Journal of the Royal Society of Medicine* 78 (1985): 154; Anthony Grafton, "Rhetoric, Philology and Egyptomania in the 1570s: J. J. Scaliger's Invective against M. Guilandinus's *Papyrus*," *Journal of the Warburg and Courtauld Institutes* 42 (1979): 167–194. For a comparable study of humanism and natural history in Northern Europe, see Peter Dilg, "*Studia humanitatis et res herbaria:* Euricius Cordus als Humanist und Botaniker," *Rete* 1 (1971): 71–85.

[4]*Lettera di M. Francesco Calzolari spetiale* (Cremona, 1566), sig. Cr; "la cognitione de' semplici non può haversi dal legger libri, quando insieme non vi sia conguinta la sperienza de gli occhi stessi . . . ."

[5]Aldrovandi, *Discorso*, p. 180.

accumulation of "facts"—pre-Baconian particulars revising the scientific universals that were the cornerstone of late Renaissance natural philosophy.[6] Experience now played a greater role in the constitution of scientific authority, and the naturalists who claimed the greatest level of "experience" subsequently came to possess the highest degree of knowledge.

In their decision to expand the uses of observation as a technique of investigation, academically trained naturalists found themselves more frequently rubbing shoulders with their less well schooled contemporaries. They brought their learned tomes into the fields, making apothecaries and herbalists more knowledgeable about the writings of Pliny, Theophrastus, and Dioscorides through commentaries and translation. In exchange, men with more intimate knowledge of living nature showed physicians and professors where and when to gather specimens and proved themselves to be an invaluable source of information about the local names and uses of each item. An apothecary such as Calzolari possessed neither a university degree nor much skill in Latin; yet his knowledge of the plants of Northern Italy was praised by scholars such as Ghini, Mattioli, and Aldrovandi, who all herborized with him on various occasions. Naturalists, firmly entrenched within the courts, the universities, and the medical profession, were careful to reinforce the social hierarchies that shaped their relationships. But they nonetheless acknowledged that information per se had no boundaries. In different contexts it simply took on different forms and meanings. Knowledge itself was heterodox, but those authorized to possess it were not. While natural history allowed men from many different walks of life to contribute to its material basis, only humanistically trained naturalists of high social standing were considered qualified to define the field of study. In imitation of physicians, who cast themselves as philosophers to distinguish their work from that of surgeons, apothecaries, midwives, and empirics, naturalists understood their form of inquiry to be superior to all others because it was grounded in the tenets of natural philosophy. Experience was not adequate unto itself, but required the proper intellectual framework to make it meaningful knowledge.

While museums were imbedded within patrician culture, collecting itself was neither a "high" nor a "low" practice. It depended on a network of communication and exchange that stretched from the fisherman's barges of the Italian ports and the medical and scientific lore perpetuated by street mountebanks in the piazza; to the letters of the virtuosi who assiduously hoarded such information; and ultimately to the libraries, pharmacies, and museums of early modern Italy, where knowledge was put into place. The process by which knowledge was made, through the movement of objects across all of these different arenas, is the primary concern of this chapter. Enhancing certain practices—the field trip, the inspection of wares in the

---

[6]Daston, "Baconian Facts"; idem, "The Factual Sensibility," *Isis* 79 (1989): 452–470.

marketplace, the exchange of specimens—naturalists made collecting central to the reinvention of nature as an object of humanist inquiry. Strong cultural predispositions guided the naturalists who voyaged into nature. Pilgrimage, the primary mode in which travel occurred, was an activity laden with all sorts of spiritual and symbolic meaning. Put most generally, it was a search for truth, a truth now found in the contemplation of nature. Claiming that travel was essential to the study of nature, collectors such as Aldrovandi and Kircher felt compelled to leave the cities they inhabited as an exercise in self-knowledge. Nature provided them with the perspective that they lacked as long as they stayed at home. Only through travel could they come to know themselves.

NATURE'S PILGRIMS

The desire to experience nature in her original form turned collectors into travelers. As the Ferrarese professor of medicine and critic of Pliny, Nicolò Leoniceno, queried in 1493, "Why has nature provided us with eyes and other organs of sense but that we might discern, investigate, and of ourselves arrive at knowledge?"[7] In their eagerness to seek out nature *in situ*, naturalists scaled mountains, herborized in the woods and fields, and peered into all of nature's crevasses in search of specimens. As Aldrovandi observed, a museum was "collected in travel," and this basic precept shaped the practice of natural history more intensely from the sixteenth century onward. In their collective reminiscences, travel seemed to become a virtual precondition for becoming the new sort of naturalist. Autobiographical reflections returned constantly to this theme. The German naturalist Leonhard Rauwolf remarked in 1581, "From my youth I had the strong desire to go to foreign lands, especially those of the Orient, as these were more famous and more fruitful than the others . . . but also much more to discover and to learn to know the beautiful plants and herbs described by Theophrastus, Dioscorides, Avicenna, Serapion, etc. in the location and places where they grow."[8] Such sentiments were evident in naturalists of markedly different backgrounds, from the patrician Aldrovandi to the herbalist Anguillara, who increasingly saw nature as the potential *resolution* of their quest for knowledge rather than simply the object of their desire.

A collection was the product of numerous voyages into nature. "With great pleasure I received your last letter, informing me of your happy return from the mountains, and of the magnificent treasure of simples you brought back," wrote Mattioli to Aldrovandi after his trip to Monte Baldo in 1554.[9] By the end of the sixteenth century, travel had become an essential rite of

[7]In Edward Lee Greene, *Landmarks of Botanical History*, Frank N. Egerton, ed. (Stanford, 1983), Vol. II, p. 543.
[8]In Dannenfeldt, *Leonhard Raulwolf*, p. 31.
[9]Raimondi, "Lettere di P. A. Mattioli," p. 23.

passage for the aspiring naturalist. "There is no man, I thinke that wyl deny, but that the searchinge out of the nature of thinges . . . is performed by no meanes more effectually, then by traveill," wrote Jerome Turler in 1575.[10] Cardano echoed the same advice in his autobiography: "An acquaintance with other lands is . . . especially helpful . . . to anyone interested in the nature and productive usefulness of plants and animals."[11] Cardano's appraisal of travel revealed the medical origins of this advice. Galen had created the prototype of the physician–traveler through his voyages to Palestine, Cyprus, Alexandria, and the island of Lemnos, all taken to enhance his knowledge of nature as a source of medicine. This topos appeared repeatedly in the writings of early modern naturalists who claimed to travel "in the guise of Galen."[12] Reading works such as Dioscorides' *Materials of Medicine* and Galen's *On the Properties of Medicinal Simples* inspired naturalists across Europe to seek out nature in her own habitat. Following both his namesake Ulysses, his mentor Ghini, and the revered authority of Galen, Aldrovandi, like many naturalists, was the prototype of the medical humanist as traveler.[13] His decision to observe nature directly was the product of his humanist education, the current excitement about natural history in his student years, and an appropriate sense of manifest destiny, as the new Ulysses for a new age.

Within a century, the model first put into practice by Renaissance naturalists was formalized in prescriptive treatises such as the Danish physician Thomas Bartholin's *On Medical Travel* (1674). Writing for his sons Caspar and Christopher and his nephew Holgar Jacobsen on the eve of their departure for medical study abroad, Bartholin remarked, "Today there are many travelers; indeed it seems as if the whole of Europe is on the move."[14] Advising them to follow "the examples of ancient wise men," Bartholin described the physician as the consummate "visitor"—one who by profession was destined to see and know many things outside the boundaries of local experience. Amplifying a theme introduced in the writings of Cardano and Turler, he proclaimed, "no one puts much faith in the authority of a physician who has not set foot outside of his native land." Travel, in other words, had become a credential. The individual experiences of Renaissance

[10]Jerome Turler, *The Traveiller* (London, 1575), p. 33.

[11]Girolamo Cardano, *The Book of My Life*, Jean Stoner, trans. (New York, 1962, 1930), p. 101.

[12]Aldrovandi, *Discorso naturale*, p. 216; other references to Galen include Balfour, *Letters*, ii; Bartholin, *On Medical Travel*, p. 49; Dannenfeldt, *Leonhard Rauwolf*, p. 4; and Raimondi, "Lettere di P. A. Mattioli," p. 28. The imitation of the ancients is discussed by Reeds, "Renaissance Humanism and Botany," pp. 524–528.

[13]For a discussion of the relationship between humanism and travel, see Jonathan Haynes, *The Humanist as Traveler: George Sandys's Relation of a Journey begun An. Dom. 1610* (Rutherford, NJ, 1986).

[14]Bartholin, *On Medical Travel*, pp. vi, 47. The quotations that follow are taken from pp. 47 and 50.

naturalists were now a form of collective initiation for seventeenth-century physicians, the product of a new medical curriculum that stressed *all* of nature as the object of study. Given these circumstances, the reappearance of Holgar Jacobsen as the curator to the Royal Danish collections at the end of the century was fairly predictable; physicians like him were trained to be collectors.[15] Perhaps the designation of physicians as "visitors" also explains their prominence in the catalogue of visitors housed in the *Studio Aldrovandi*.

Most naturalists cited some form of early "apprenticeship" as their initial introduction to the experience of collecting as an empirical practice. During his studies at the Jesuit college in Ancona in the mid-seventeenth century, Filippo Bonanni recalled his proximity to the sea as formative in the development of his interest in natural history, in particular, conchology. He initially had gone to the port of Ancona to look for samples to compare with those he had seen in the museum of Camillo Pichi, and he became so intoxicated with the process of collecting that he immediately began to form his own museum.[16] A century earlier, Aldrovandi confessed that beachcombing was one of his most successful methods of collecting:

> Often I have collected in similar [places], a number of times finding fantastic things, worthy of admiration. Thirty years ago, while fishing at Ravenna, three miles from where my brother was then Abbot, by the luck of the sea I found many bizarre marine creatures on the beach, among which was a dolphin's skeleton . . . . I had it brought to Bologna and placed it in my Theater of nature, where it still remains intact.[17]

The voyage into nature involved the discovery of proximate as well as distant wonders. As a group, naturalists learned that the act of travel was more important than the distance of the journey. Like the experiences that occurred within the museum, it reinforced the humanist sense of sodality that defined the community of naturalists. Travel was an experience they all could share, initiating succeeding generations of naturalists into the mysteries of nature through their participation in this collective rite of passage.

Naturalists described themselves not only as travelers, propelled by curiosity and a sense of duty toward the betterment of mankind, but also as pilgrims. In his *Simples* (1561), Luigi Anguillara, director of the botanical garden in Padua, explained, "I have made many pilgrimages to various lands."[18] Anguillara was not the only self-described "pilgrim" into nature. Aldrovandi's mentor Francesco Petrollini also spoke of his desire to go "in pil-

[15]Holgar Jacobaeus, *Museum Regium, seu catalogus rerum tam naturalium quam artificialium* (Hafnia, 1696).

[16]"Elogio di P. Philippo Buonanni," *Giornale de' letterati d'Italia* 37 (1725): 365–366.

[17]BUB, *Aldrovandi*, ms. 21, Vol. IV, c. 91r.

[18]Anguillara, *Semplici*, p. 15, in Greene, *Landmarks of Botanical History*, Vol. II, p. 734.

grimage."[19] Predictably, Aldrovandi elaborated on this image in his autobiography. At age 16, upon the invitation of a Sicilian pilgrim, Aldrovandi left Bologna "dressed as a pilgrim" to see the world.[20] This was not the first time that he had wandered away from home, but his second adventure took him farther than Rome: from Genoa and Savona into southern France and over the Pyrenees into Spain, where he communicated at several holy sites. Despite a near drowning on the voyage between Marseilles and Genoa—an event, curiously enough, repeated on Kircher's arrival to Italy in 1634[21]—the young Aldrovandi was invigorated by his experiences. "When they finally arrived in Genoa, he wished to travel the world over for several more months, having accustomed himself to a life so curious and delightful for the variety of things observed." Specifying what had made travel pleasurable, Aldrovandi recalled "not only . . . the holy relics of holy bodies that he visited with devotion, but also many other natural things that he desired to know at that time." Fired with enthusiasm, he planned to leave immediately for Jerusalem after stopping in Bologna.

While Aldrovandi never again traveled outside of Italy after 1538, his early experiences determined the course of his life. His self-described transition from merchant—a profession he had begun to learn after his first taste of travel—to pilgrim to naturalist marked the choices of a man tantalized by the vision of a wider world. "Despite [my] continual studies, while I was a student, every year during vacations I wandered in various parts of Italy, at great expense, scaling the highest mountains in the Apennines, Monte Baldo . . . and the Sybilline mountains in order to find natural things."[22] From the 1550s to the 1570s, he attempted to interest various patrons in financing an expedition to the Indies, the Near East, or more modestly, Crete. Confiding his plans to Mattioli, the elder naturalist responded encouragingly, "when I was young like you, nothing would have kept me from seeing Egypt, Syria, Constantinople, Lemnos, Cyprus, Crete, and all the islands of the archipelago, if that were my desire. But now it is impossible. How much it pleases me that you wish to go further than you have gone in the past."[23] Inevitably Aldrovandi found himself offering Giuseppe Casabona the same advice some forty years later, as the grand ducal botanist was about to depart for Crete, when age had finally made it impossible for him to travel. With the example of cosmopolitan naturalist–explorers such as Gonzalo Fernández de Oviedo, author of the *General and Natural History*

[19]Emilio Chiovenda, "Francesco Petrollini botanico del secolo XVI," *Annali di botanica* 7 (1909): 443.

[20]The following quotations are taken from Aldrovandi, *Vita*, pp. 6–8.

[21]Kircher, *Vita admodum P. Athanasii Kircher Societatis Iesu Viri toto orbe celebratissimi* (Augsburg, 1684), pp. 46–47.

[22]Aldrovandi, *Vita*, p. 9.

[23]Raimondi, "Lettere di P. A. Mattioli," p. 28.

of the Indies (1535–1557) in mind, late sixteenth-century naturalists were well aware of the limitless potential of nature to yield further treasures, beyond those catalogued by the ancients, who had confined their travels—and therefore descriptions—to the Mediterranean basin.

Like Kircher, frustrated in his attempts to become a missionary to China,[24] Aldrovandi eventually contented himself with training future scholars in the techniques necessary for the new history of nature, while reaping the benefits of others' travel. If neither became real pilgrims, visiting the revered holy sites and cataloging the natural wonders of distant lands in person, as they hoped to do in their youth, they were instrumental in establishing nature as a site worthy of pilgrimage. Ironically, the best collectors were not often the most well traveled.[25] Instead, they were men whose taste for travel, and inability to go as far as they would have liked, led them to accumulate tangible reminders of natural and human diversity. Travelers established first contact for the Europeans with the wider world. Collectors formalized the meaning of the encounter by sifting through the reports and objects that entered their museums, offering the sort of critical synthesis that few travelers could hope to achieve. And they did this in a religious framework that privileged pilgrimage as the preferred metaphor for travel.

As a good Catholic—despite his brush with the Inquisition—and a man with certain spiritual leanings, Aldrovandi could not help but see travel as a form of pilgrimage. For Kircher, it was a religious calling enhanced by his encyclopedic and speculative propensities. Kircher actively searched out ways to combine pilgrimage, ecstatic visions, and the study of nature. Explorations of the hills near Marino in 1661 revealed "the place where Saint Eustace saw his vision of Christ in a stag's horns."[26] Kircher's rediscovery of an important *locus conversionis* and his subsequent transformation of it into a site of pilgrimage indicate the extent to which the Jesuit collector saw the landscape, and any object he encountered, in terms of its theological significance. This effect was brought to its pinnacle in his *Ecstatic Voyage* (1656) and *Second Ecstatic Voyage* (1657), two imaginary voyages into nature written in the form of dialogues. In the former, Kircher (renamed Theodidactus), guided by the angel Cosmiel, ascended to the heavens to see the workings of the supralunar world; in the latter, guided by Hydriel and Cosmiel, he explored the sublunar world. In both instances, the accompaniment of an angel allowed Kircher unique perspective on the universe; with Hydriel and Cosmiel at his side he could soar to unheard of heights, visit all the planets, circle the poles,

---

[24]Kircher was bypassed as a missionary in favor of another German Jesuit, Adam Schall, who became the astronomer and advisor to the Emperor of China.

[25]One exception to this rule is Settala, who traveled for seven years in the 1620s, particularly in the Levant. Despite the scope of his voyage, he nonetheless conforms to the pattern of traveling in one's youth.

[26]Kircher, *Historia Eustachio-Mariana* (Rome, 1665); idem, *Vita*, p. 63; described in Godwin, *Athanasius Kircher*, pp. 13–14.

and enter the volcanoes that he perceived as the key to his thesis concerning the balance of elements in the world. Unhampered by the physical and temporal constraints of earthbound naturalists, Kircher was a pilgrim who made all nature his playground. While Aldrovandi regretted that he had never seen Crete or the Indies with his own eyes, Kircher felt that his faith had allowed him to circumvent the obstacles that kept him in Rome by expanding the boundaries of his experience through divinely inspired imagination. Travel, as he illustrated better than anyone, was a cognitive activity; like collecting, it needed to be experienced with the "eye of the mind" to achieve its fullest potential. At the end of Kircher's "ecstatic voyage" lay the truth about the cosmic structures that he spent his life uncovering.

Through actual and imaginary voyages, naturalists made observation and exploration increasingly meaningful activities. Returning to Jan van Kessel's portrayal of *Europa,* we should note the inscription scrawled across the pages of a book of naturalistic illustrations in the lower right-hand corner: "Pilgrims are those who look in the cities for shells for their staffs" (see fig. 3). Recalling the practices of pilgrims who arrived in Santiago de Compostela laden with shells to mark the places they had visited, Van Kessel imagined the collector to be engaged in a similar quest, patiently accumulating mementos to mark the progress of his voyage. Like pilgrims, whose staffs bore their weight for countless miles, wordlessly recording the holy sites, the naturalist's "staff"—the museum—was a testimony to the extent of his devotion to nature and the quest for knowledge that could only be gained in pilgrimage.

LOCAL KNOWLEDGE

One of the most important lessons that emerged in this changing climate was the necessity of local knowledge. Although the new passion for observation was formalized in the scientific expeditions undertaken by various monarchs and religious orders to all parts of the world, the image of the natural philosopher as traveler was also evident in the daily practices of naturalists. Collectors, dreaming of places they could not see, mapped out the coordinates of their immediate environment. Thomas Bartholin described the medical traveler, the prototype of the naturalist, as an Argonaut bringing back the Golden Fleece. Certainly this must have been the case for men such as the Spanish physician Francisco Hernandez, whose explorations of Mexican flora and fauna at the behest of Philip II so excited the community of naturalists, or the Augsburg physician Leonhard Rauwolf whose description of plants in the Levant surpassed the works of Pierre Belon, Melchior Wieland, and Prospero Alpino, all travelers to the Near East.[27] Others,

[27]Bartholin, *On Medical Travel,* p. 57; on Hernandez, see David Goodman, *Power and Penury: Government, Technology and Science in Philip II's Spain* (Cambridge, 1988), pp. 22–29; on Rauwolf and other travelers to the Levant, see Dannenfeldt, *Leonhard Rauwolf.*

unable to fund such extraordinary voyages, contented themselves with local travel, dependent upon friends to bring them more unusual collectibles. While firmly committed to the collection of the exotic, the majority of naturalists subsequently exalted local nature due to their necessary rediscovery of its appeal.

After the fires of youthful ambition had waned, many naturalists, perhaps recalling Luca Ghini's advice about the importance of ordinary nature, concentrated upon the development of more detailed regional natural histories. Aldrovandi, for all his encyclopedic curiosity, conducted the majority of his summer excursions within the vicinity of Bologna, venturing as far north as Verona and Trent and as far south as Rome, but little more. This localization was borne out in the composition of his herbarium, which contained more domestic than foreign specimens. With the help of students and colleagues, Aldrovandi identified 565 different plants growing near his home—an admirable feat when compared with the 84 specimens that Prospero Alpino and Onorio Belli identified on Crete, or the 350 plants that Calzolari listed between Verona and Monte Baldo.[28] Even Mattioli, a man who prided himself on his disdain for common botanizing, preferring to have others collect for him, betrayed his own origins by favoring descriptions of plants that he had seen in the vicinity of Trent in 1527.[29] Despite their attempts to recapture the global nature of Pliny's *Natural History,* most collectors discovered that it was better to know a small portion of nature in its entirety than to claim to know all of nature imperfectly.

The tendency to build outward from one's immediate environment was apparent even in the writings of naturalists who traveled extensively. Prospero Alpino's three years in Cairo, as physician to the Venetian consul there from 1581 to 1584, made him the uncontested expert on the medicine and natural history of Egypt, when he returned to Padua to describe the marvels of coffee and numerous other unknowns and to debate the feasibility of such plants growing in Europe.[30] His friend and correspondent, Onorio Belli, physician to the Venetian consul in Crete from 1583 to 1599, became a similar resource for anyone interested in obtaining specimens from that island. The services provided by the Republic of Venice's physicians were relatively limited in comparison to what the Jesuits could offer. The Venetian community of naturalists extended their reach as far as the limits of the Venetian state and its diplomatic networks allowed: the Jesuit missionaries commanded a vast stretch of territory that encompassed most parts of the world.

[28]Oreste Mattirolo, *L'opera botanica di Ulisse Aldrovandi (1549–1605)* (Bologna, 1897), pp. 64, 90–91; A. Baldacci and P. A. Saccardo, "Onorio Belli e Prospero Alpino e la Flora dell'isola di Creta," *Malpighia* 14 (1900): 143; Tergolina-Gislanzoni-Brasco, "Francesco Calzolari," p. 10.

[29]F. Ambrosi, "Di Pietro Andrea Mattioli e del suo soggiorno nel Trentino," *Archivio Trentino* 1 (1882): 49–61; Palmer, "Medical botany in Northern Italy," p. 152.

[30]See Prospero Alpino, *De medicina Aegyptiorum* (Venice, 1591) and *De plantis Aegypti* (Venice, 1592).

In China, Jesuits such as Martin Martini and Michael de Boym proved indispensable for Kircher's attempts to collect the knowledge of this land and its culture. Kircher cited their respective books—the *New Chinese Atlas* (1654) and *Chinese Flora* (1656)—frequently in his *China Illustrated* (1667). Martini, Kircher's former pupil, was praised for having "communicated to me many things, his keen insight having been well trained for this by his mathematical studies." [31] Boym returned to Rome in 1664, laden with manuscripts and specimens for Kircher to include in his publication. Like his predecessors, Kircher acknowledged the importance of establishing contact with naturalists who had seen what they claimed to describe so as to lend credibility to his own work through the inclusion of their testimony. Increasingly, naturalists understood the value of accumulated experience, borne of prolonged familiarity with a single region, and they adjusted their patterns of collecting accordingly. But if a naturalist could not travel to the regions he wished to study, then the reports of reliable, on-site witnesses might adequately substitute for personal experience. Other Jesuits, trained in Rome before their departure to various missions, met the criterion of reliability.

Behind the increased emphasis on personal experience lay an even greater transformation: the evolution of natural history from the study of universal Nature to the investigation of specific nature. What factors contributed to this shift? In the 1490s, critics and defenders of Pliny called for greater sophistication in the study of nature. While disagreeing violently on the merits of Pliny's *Natural History*, all concurred that *new observations* would provide the tools with which to resolve the debates. [32] Nicolò Leoniceno's adversary, that vocal defender of Pliny, Pandolfo Collenuccio, best summarized the consensus that emerged from this initial episode in the renaissance of natural history: "In my opinion, he who is to write about herbs, and hand down knowledge of them, ought to study not only books but also the face of the earth, not letters alone, but also the open fields," he wrote in his *Plinian Defense* (1493):

> For fitness to give instruction in botany, it does not suffice that a man read authors, look at plant pictures, and peer into Greek vocabularies . . . . He ought to ask questions of rustics and moutaineers, closely examine the plants themselves, note the distinction between one plant and another; and if need be he should even incur danger in testing the properties of them and ascertaining their remedial value. [33]

[31] Kircher, *China Illustrata,* Charles D. Van Tuyl, trans. (Muskegee, OK, 1987), p. iv.

[32] The principal studies of these interestings debates are as follows: Lynn Thorndike, "The Attack on Pliny," in his *A History of Magic and Experimental Science,* Vol. IV, pp. 593–610; Castiglioni, "The School of Ferrara and the Controversy on Pliny"; Karen Reeds, "Renaissance Humanism and Botany," pp. 523–525; and Charles G. Nauert, Jr., "Humanists, Scientists and Pliny: Changing Approaches to a Classical Author," *American Historical Review* 84 (1979): 72–85.

[33] In Greene, *Landmarks of Botanical History,* Vol. II, p. 551. I have modified the translation slightly.

Within a half-century, such techniques had become standard practice. Treatises such as Euricus Cordus's *Botanologicon* (1534)—a student of Leoniceno—and Antonio Musa Brasavola's *Examination of All Simples* (1536), both dialogues on the proper method of botanizing, advised naturalists on the skills necessary to gather nature's bounty. Heeding the advice of Dioscorides that the observation of nature was a *continuous* rather than a singular experience, naturalists returned frequently to the same places, observing seasonal changes and collecting different parts of specimens at different times of the year. Such practices gave special meaning to the image of nature as a *locus amoenus*. Rather than haphazardly observing nature, collectors prided themselves on their intimate knowledge of a particularly fertile corner of the world, chosen partly by accident of birth or circumstance but explained as a setting that one was "destined" to know.

At the same time, university professors such as Luca Ghini, founder of the academic study of nature in Bologna and Pisa, established the field trip as a standard feature of student training; by the 1540s, professors and students were regularly scouring the countryside for bits of unobserved nature. In imitation of Ghini, Aldrovandi frequently took his students on summer excursions, while students of Anguillara in Padua made pilgrimages to Mattioli in Goritia to learn from the acknowledged master of Dioscorides. Botanizing further strengthened the ties between mentors and disciples, as an essential rite of inclusion. The students invited to participate in these trips gained the benefit of closer contact with a prominent naturalist, in a setting removed from the formality of the university lecture hall. While earlier naturalists had recognized the value of studying nature *in situ*, Ghini and his students were among the first to collect specimens. Rather than leaving nature where they found it, returning only with their books and notes, they brought home samples.[34] The growth of herbaria at the hands of such scholars as Aldrovandi and Cesalpino gives testimony to the facility with which these techniques spread; Aldrovandi had more than 14,500 specimens and 2,000 illustrations of plants by 1570. The speed and thoroughness with which he acquired them led Mattioli to remark, "if any man in Europe today has seen and collected the greatest number of plants, you are the first among them all."[35] Aldrovandi was one of the first students exposed to the new curriculum in natural history who reaped its benefits.

The activities of a relatively unknown naturalist, Gherardo Cibo, illustrate well this tendency. Like Ghini, Cibo (1512–1600) never published any of his work, except via the pen of Mattioli, who praised him in later editions of his translation of Dioscorides. Yet this gentleman–scholar, living quietly in Rocca Contrada, produced some of the most detailed studies of nature of

[34]Ibid., p. 709; and Agnes Arber, *Herbals, Their Origin and Evolution* (Cambridge, 1986, 1912), pp. 138–143.

[35]Mattirolo, *L'opera botanica di Ulisse Aldrovandi*, pp. 90–91; Raimondi, "Lettere di P. A. Mattioli," pp. 32–33.

this period, collecting and illustrating thousands of plants.[36] Like Aldrovandi, he had traveled in his youth, primarily between Rome and the courts of Charles V in Germany, Spain, and the Low Countries. He probably had listened to Ghini's lectures on medicinal simples in Bologna. Despite Andrea Bacci's description of him as one "not content" to confine his botanizing to Italy alone, after 1540 he left his home only once, to return to Rome. For most of his life, his studies of nature were confined to the Marches, the region overlooking the Adriatic that surrounded his home. Yet Cibo was praised by Mattioli, Aldrovandi, and Bacci, professor of natural history at *La Sapienza* in Rome, as an astute observer of nature. As Lucia Tongiorgi Tomasi has recently demonstrated, Cibo was admired particularly for his precise sketches and detailed illuminations of plants. While many other naturalists brought artists along on their excursions, or employed them in their museums where they captured nature post facto, Cibo combined both skills—observing and illustrating—as he went.[37] After receiving samples of his work from Cibo's brother Scipione in 1563, Mattioli described it as "a mirror of purest crystal," so well did he capture nature.[38] But Cibo did not simply depict nature; he also developed the image of the naturalist in the garden of nature, leaving behind a detailed visual record of the activities composing a field trip. Through his illustrations, we can follow a typical Renaissance naturalist at work.

In the background of the plants Cibo chose to illustrate, with all of their parts carefully articulated, the naturalist repeatedly appears in the act of discovering, depicting, and collecting nature. In one image, two men are portrayed contemplating a specimen of hellebore (fig. 11). The plant is magnificently displayed in the foreground from leaf to root. To the left, sit two naturalists in the act of creating the entry that accompanied this page in Cibo's herbarium. One holds up the specimen, while the other consults a passage from a book. Such a scene accords well with textual descriptions of field trips. It is highly reminiscent, for example, of Brasavola's imagined conversation with an old apothecary, Senex, and his assistant Herbarius in the Ferrarese naturalist's *Examination of All Simples*. In this dialogue, designed to demonstrate the superiority of knowledge derived from reading *and* observing, Brasavola had the unlearned apothecary hold up a plant while he read aloud from Dioscorides's *Materials of Medicine:*

[36]For more on Cibo, see E. Celani, "Sopra un Erbario di Gherardo Cibo conservato nella R. Biblioteca Angelica di Roma," *Malpighia* 16 (1902): 181–226; *Gherardo Cibo alias "Ulisse Severino da Cingoli"* (Florence, 1989); Lucia Tongiorgi Tomasi, "Gherardo Cibo: Visions of Landscape and the Botanical Sciences in a Sixteenth-Century Artist," *Journal of Garden History* 9 (1989): 199–216.

[37]The following paragraph essentially paraphrases the observations of Lucia Tongiorgi Tomasi, "Gherardo Cibo," passim. Her observations on Cibo's development of a "self-portrait" in the recurring image of the herborist evident in many of his illustrations are particularly interesting (pp. 202–205).

[38]The letter is reproduced in Celani, "Sopra un Erbario di Gherardo Cibo," p. 216.

Figure 11. Cibo's Renaissance botanizers. From British Library, Ms. Add. 22332, f. 94.

Senex: Well here I have it, and read you the text of Dioscorides to me.
Brasavola: Listen, then; for here are his words.[39]

Cibo captured with his paintbrush what Brasavola recorded in words: the constant interplay between text and artifact central to the study of nature. His illustrations also articulated the image of nature as a *musaeum:* verdant, teeming with infinite specimens, it was in nature rather than in the museum that he felt most at home. Unlike his contemporary Imperato, Cibo chose not to portray himself in the museum. Instead, he preferred to depict the act of collecting nature as the image he most associated with the identity of the naturalist.

Cibo's herbarium was not a static record. He constantly revised it as later trips yielded new results.[40] Many entries were crossed out, old names replaced with new ones, as greater familiarity with his environment and further discussions with other naturalists yielded greater precision. For Cibo, the herbarium became a dynamic investigative tool that recorded the process of observing nature and established the meaning of what he had seen. Like most of his contemporaries, Cibo scrupulously followed advice about the need to localize descriptions. Each entry recorded "the location, the day and the hour in which it was gathered, as well as the names of the persons who had accompanied him on that particular herb gathering expedition."[41] Such care reflected the increased concern with accuracy, as naturalists better understood the importance of regional differences. As Brasavola advised, "the collector must not be indifferent to the matter of the province, and its climatic peculiarities."[42] Nature was no longer the universal unknown, as it had appeared to humanists in the 1490s, but a map of the universe whose features sharpened with time.

While botanists took the lead in reviving and modifying observational procedures, these techniques were not confined to the study of plants. Building upon the work of his mentor Ghini, Aldrovandi developed a pattern for investigating all of nature that was imitated by the most ambitious naturalists. In his *On Animal Insects* (1602), Aldrovandi explained his method of observing nature. "When I reflect on the many days I have given to this study, and what expenses I have incurred, I cannot but wonder how I have been able to obtain possession of, and to examine, and to describe such a number of minute creatures," he wrote.

> For the obtainment of my object, I was in the habit of going into the country for months during the summer and autumn, not for relaxation, like others; for at these times I employed all my influence, as well as money, to induce the country-people to bring me such insects, whether winged or creeping, as they

[39]In Greene, *Landmarks of Botanical History,* Vol. II, p. 672.
[40]Tongiorgi Tomasi, "Gherardo Cibo," pp. 205–206.
[41]Ibid., p. 205.
[42]In Greene, *Landmarks of Botanical History,* Vol. II, p. 695.

could procure in the fields or underground, and in the rivers and ponds. When any were brought to me, I made inquiries about its name, habit, locality, etc. I often, too, wandered over the marshes and mountains, accompanied by my draughtman and amanuenses, he carrying his pencil, and they their notebooks. The former took a drawing if expedient, the latter noted down to my dictation what occurred to me, and in this way we collected a vast variety of specimens.[45]

In this passage, Aldrovandi provided a complete description of the activities and personnel necessary to collect and catalogue nature. Observing nature was no haphazard affair, but entailed the complicated management of artists, scribes, and assistants under the watchful eye of the naturalist in charge. By the late sixteenth century, it had become an enterprise of monumental proportions. In many respects, the bucolic images provided by Cibo had little bearing on the activities undertaken by more prominent naturalists such as Aldrovandi. While Aldrovandi depended on men such as Cibo to add to his theater of nature, it had been a long time since he ventured out into the fields with a simple sack and a single companion. In fact, it probably was a sign of great prestige that he did *not* do any of the manual work involved in the field trip but had artists and scribes record what he "saw."

As a result of the efforts made by Ghini, Mattioli, Aldrovandi, and their colleagues, by the end of the sixteenth century naturalists no longer searched for universal specimens, assuming that American or African creatures were interchangeable with their European counterparts, but understood that each belonged to a local context whose natural history had to be charted. Printed works such as the Roman naturalist Castor Durante's *New Herbarium* (1585), which systematically illustrated each plant and standardized the categories of description—names, form, place, qualities, and virtues—were the product of almost a century of debate about the proper means of knowing and describing nature. Regardless of where one studied nature or what one collected, the localization of description had become an underlying principle. Such work created the foundation for more grandiose projects such as Cesi's *Phytosophical Tables,* which praised Durante's herbarium. Only by knowing nature in all of her particular forms could a new encyclopedia of nature emerge.

NATURE IN THE MARKETPLACE

The desire to experience nature brought naturalists in closer contact with empirics and unlettered craftsmen since it was often not the scholar who first laid eyes on such objects. Local knowledge was a product of regional communities of collectors having a wide range of social and professional

---

[45]Aldrovandi, *De animalibus insectis libri septum* (Bologna, 1602), in Willy Ley, *Dawn of Zoology* (Englewood Cliffs, NJ, 1968), p. 158.

identities. While naturalists saw themselves as members of the republic of letters, they were dependent upon men who knew nature by profession to complete the unwritten chapters in their new histories of nature. Initially, the "book of nature" was far removed from the humanist's *studio*. The men who possessed it were neither physicians nor philologists but tradespeople who understood nature to be a commodity upon which their economic survival depended. "Nature" was not only courtly and humanist knowledge but also artisanal knowledge; it had been a commodity long before it became a curiosity. Thus, the marketplace was an arena that a naturalist had to enter before he could truly claim to know nature. The tradespeople who gathered and sold different parts of nature for their livelihood possessed the experience that naturalists claimed to value. Entering the marketplace, collectors transformed empirical know-how into humanist knowledge. They sought out nature in the marketplace but were not contaminated by its uncourtly and potentially dishonorable associations. The butchers, fishermen, and street vendors who kept naturalists apprised of the latest natural oddity would never formally enter their museums, let alone sign the visitors' book. They were the "invisible assistants" who made the collection possible, but only through their anonymity.[44] Despite the persistent erasure of these individuals from the collecting milieu, we can reconstruct some aspects of their impact on natural history through the conversations that naturalists memorialized in the form of letters, explaining to one another where they obtained their artifacts.

Most obviously, empirical observers infused the academic study of nature with new terminologies. By the middle of the sixteenth century, natural history was "on the boundaries of two languages."[45] On one side, humanistic scholarship had revived the academic discipline of natural history by purifying its classical idiom. Closer attention to philology and etymology had produced more accurate editions of the writings of Aristotle, Theophrastus, and Dioscorides; no longer faced with the linguistic corruptions and mistranslations of Pliny, naturalists could compare words and things, secure in their knowledge that they had the right word in the right language. At the same time, increased dependence on experience broached a different problem for Renaissance naturalists: the language of experience was neither Latin nor Greek but a bewildering array of vernacular languages and dialects that identified objects within a local cosmology. In the mid-sixteenth century, Calzolari's friend, Prospero Borgarucci, attempted "to interview local herbalists in Welch" to know more about plants in that region of the world.[46] He was just one of many naturalists who understood that proper

[44]See Shapin, "Invisible Technician."

[45]I have appropriated this phrase from Mikhail Bakhtin, *Rabelais and His World*, Helene Iswolsky, trans. (Bloomington, IN, 1984), p. 465.

[46]Prospero Borgarucci, *La fabrica de gli speziali* (Venice, 1567), p. 183, in Palmer, "Medical Botany in Northern Italy," p. 149.

communications was essential to good natural history. Identification was one of the most important goals in the collecting of specimens; to be ordered, they had to be named. "Thursday morning, the 20th of May, above Malamaco, these fishermen took a fish weighing more than 1000 lbs. No one knows what it is called."[47] The presence or absence of a name was usually the first thing noted; when a new specimen was collected, a naturalist's first instinct was "to ask the circumstances of the name."[48] Gradually, regional names crept into the official lexicon of nature—sometimes turned into Greek and Latin neologisms, sometimes left as they were—as naturalists found existing terminology inadequate for the expanded history of nature.

In his remarkable study of the world of Rabelais, Mikhail Bakhtin offers an interesting opening into the problems of language during this period. Contending that Rabelais drew upon oral sources for his encyclopedic writings, Bakhtin introduces the example of the sixty different names for fish recorded in the fourth book of *Gargantua and Pantagruel*. Noting that Rabelais wrote his work decades before the publication of the ichthylogies of Pierre Belon and Guillaume Rondelet, Bakhtin concludes that Rabelais acquired his knowledge through conversations with fishermen in Brittany, Normandy, and Marseilles: "All these were absolutely fresh words, as fresh as the fish Rabelais must have seen in the markets; they had never been used in written or printed form, they had not as yet been processed with an abstract, systematic context."[49] Like Rabelais, trained in the medical culture of Montpellier that shared many characteristics with Bologna, naturalists facilitated the interplay between oral and written languages through their appropriation of "fresh words." In the inventory of Antonio Giganti's museum, for example, were

> illustrations of two New World fish, the so-called upside-down fish. One may call them fishermen's fish because one pierces other fish with the spines with which it is armed while the other catches them with a sack that it has on its tail, extending and then contracting it. Men make use of these fish in the sea as they would arms and sparrow hawks on land.[50]

Rabelais' use of vernacular idioms on a thousand subjects found its parallel in his contemporaries' decision to incorporate common descriptions, such as that of the "upside-down fish," into their learned treatises on nature. Undoubtedly, many naturalists perceived themselves to be marketplace philologists, or proto-anthropologists, as they explored the dialects and descrip-

[47]BUB, *Aldrovandi*, ms. 143, Vol. III, c. 127v.

[48]BUB, *Aldrovandi*, ms. 136, Vol. VIII, c. 183r.

[49]Bakhtin, *Rabelais and His World*, pp. 456–457. Another interesting work to consult in this regard is Roy Porter, "The Language of Quackery in England 1660–1800" in *The Social History of Language*, Peter Burke and Roy Porter, eds. (Cambridge, 1987), pp. 73–103.

[50]Ambr., ms. S.85 sup, c. 250r, in Fragnito, *In museo e in villa*, p. 197.

tive language of their less learned contemporaries. By doing this, they reaffirmed the importance of humanist education in studying nature.

Correspondence between naturalists documents the translating mechanisms at work. Letters often contained more of this material than published treatises because they reflected the process of identification rather than the decisive result that appeared in print. "The plant that I sent you . . . is called *lunaria tonda* by some poor herbalists," wrote Costanzo Felici to Aldrovandi. Likewise, Alfonso Cataneo identified a specimen—"a hammerfish as it is vulgarly called by fishermen"—based upon its local name.[51] Given their sensitivity to language, Renaissance naturalists were particularly concerned with the coordination of vernacular names and classical labels. "Here they call it a waterbird . . . ," wrote Annibale Camillo to Aldrovandi; "I want to have your opinion and the name under which it was described by the ancients."[52] The process of naming nature tells us something about how a museum was created. A collector's sources of information were not confined to the immediate social and intellectual circles in which he normally circulated. Instead, they reflected the fruitful interaction of high and low culture.[53] Aldrovandi's letters, notebooks, and travel journals[54] reveal numerous instances in which naturalists, perplexed by the inability of texts to offer definitive answers to medical and scientific problems that arose in daily practice, relied upon the comments of untrained observers. Contacts such as Bernardo Castelletti in Genoa, the apothecary Giulio Cesare Moderati, and the doctor Costanzo Felici, the latter two in Rimini, offered Aldrovandi valuable information about the abundance of marine life that washed up on the shores of the Mediterranean and the Adriatic. "So that you will not marvel at my facts when I send you the most ordinary things, I warn you that my intention is to send you almost every sort of fish, rare or common, that one catches in our port," wrote Felici.[55] When a monstrous whale arrived in the port of Ancona in 1584, Antonio Giganti immediately alerted Aldrovandi of this event. Francesco Petrollini promised to pass on whatever he found at "that marina in Ravenna."[56] But naturalists did not confine themselves to a knowledge of local flora and fauna arrived through trips to the beaches and

[51]BUB, Cod. 526 (688), c. 1r; BUB, *Aldrovandi*, ms. 136, Vol. XIX, c. 154r.

[52]BUB, ms. 136, Vol. XVI, c. 214.

[53]This point is made by Marco Ferrari about the place of the books of secrets in early modern culture in his "I secreti medicinali," in *Cultura popolare nell'Emilia Romagna. Medicina, erbe e magia* (Milan, 1981), p. 86.

[54]I am referring particularly to mss. 136 and 143 of the Aldrovandi collection in Bologna. These notebooks were organized chronologically and geographically, spanning the period between 1566 and 1601.

[55]Costanzo Felici, *Lettere a Ulisse Aldrovandi*, Giorgio Nonni, ed. (Urbino, 1982), p. 39 (Rimini, 20 October 1557).

[56]BUB, *Aldrovandi*, ms. 136, Vol. X, c. 221 (Giganti to Aldrovandi, 29 July 1584); Chiovenda, "Francesco Petrollini," p. 445 (13 April 1553).

botanical excursions into the mountains. "I go often to the fishmarket," confessed Antonio de Campagnoni, "and find diverse fish."[57]

In 1644, Sforza Pallavicino remarked, "philosophy lives in the shops and the countryside, as well as in books and in academies."[58] The association between "shops and countryside" as new sites of scientific inquiry is borne out in the places that naturalists found amenable for the study of nature. Thomas Bartholin, advising physicians that they could cull medical information from many different environments—pharmacies, alchemical laboratories, hospitals, and the marketplace—remarked, "from all [we learn] something."[59] Naturalists were not altogether unlike Paracelsus who recorded "the memory of the people" in the hope that folklore might yield a greater truth than the classical texts he despised.[60] Offering a more modified critique of the perceived scholastic tendency to invest the ancients with unimpeachable authority, they suggested that empirical observation might sharpen the truths already found in texts, making them respond to the needs of the early modern society rather than the world of ancient Greece and Rome.

In search of an oral history of nature, collectors found themselves in contact with a world unacknowledged by the universities and courtly academies. "The Marketplace was the center of all that is unofficial," wrote Bakhtin; "it enjoyed a certain extraterritoriality in a world of official order and ideology."[61] The decision to enter the marketplace in search of new materials with which to study nature forged a new relationship between knowledge and experience. Naturalists' new attentiveness to the word of a fish vendor or a seller of secrets suggested a slightly different role for him than the traditional image of a natural philosopher allowed; his interpretations were derived from observation as well as commentary, from experience as well as erudition. While broadening the definition of what constituted natural philosophy, naturalists rigorously maintained the intellectual hierarchy, which established physicians and philosophers as the sole interpreters of nature. We need only recall one botanist's description of gardeners as "an enemy of [natural history] and friend of edible things" to understand how wide the perceived chasm between trained naturalists and unlettered practitioners continued to be.[62]

[57]BUB, *Aldrovandi*, ms. 38, Vol. I, c. 225r (Venice, 9 May 1554).

[58]Sforza Pallavicino, *Del bene*, pp. 346–347, in Carlo Ginzburg, "High and Low: The Theme of Forbidden Knowledge in the Sixteenth and Seventeenth Centuries," *Past and Present* 73 (1976): 37.

[59]Bartholin, *On Medical Travel*, pp. 151–152.

[60]Charles Webster, "Paracelsus and Demons: Science as a Synthesis of Popular Belief," in *Scienze, credenze occulte, livelli di cultura*, pp. 13–14.

[61]Bakhtin, *Rabelais and His World*, pp. 153–154; see also Peter Burke, *Popular Culture in Early Modern Europe* (New York, 1978).

[62]BUB, *Aldrovandi*, ms. 136, Vol. XXV, c. 59r (Evangelista Quattrami to Aldrovandi, Ferrara, 20 December 1595).

The idea of gathering scientific information in the marketplace was recommended not only by Rabelais but also by classical authorities. Aldrovandi
tells us that Aristotle employed thousands of unrelated practitioners in writing his own history of nature,

> commissioning various fishermen, hunters, bird-catchers, herbalists and other
> scrutinizers of mineral things in various parts of Africa, Asia and Europe to col-
> lect all the plants and every sort of animal on land and sea, and bring them to
> Aristotle in Alexandria, so that he could dissect them and write their nature
> and history, considering the use of the internal and external parts.[63]

What distinguished the sixteenth-century use of empirical evidence from
the classic notion of experience was the ability of late Renaissance naturalists to let such information actively shape their writings. As Bartolomeo
Maranta put it, no one could "advance knowledge of simples . . . without
seeing different places and talking to diverse men [who are] experts in their
profession."[64] While the framework within which collectors operated was
still wholly Aristotelian, their histories incorporated direct observation and
oral tradition. They acknowledged the authority of "professional experts" as
essential to the fact-gathering process, just as Bacon would do in his outline
for a new natural history. Classical wisdom was increasingly tempered by the
untutored descriptions of apothecaries, gardeners, fishermen, and hunters
whose occupations gave them privileged access to the secrets of nature.

In his 1586 autobiography, Aldrovandi represented a visit to the fishmarkets of Rome in 1549–1550 as a signal event that sparked his interest in
natural history. "In that same time that [Guillaume Rondelet] was in Rome,
I began to be interested in the sensory knowledge of plants, and also of dried
animals, particularly the fish that I saw often in the fishmarkets."[65] Even
more interesting, Aldrovandi suggested that reading the humanist Paolo
Giovio's *On Roman Fish* (1524), a work based primarily on observations
made in the same setting, had prompted his own desire to enter the marketplace. Just as Descartes observed and dissected animals in the butcher
shops of Paris a century later, Aldrovandi went to the fish vendors to learn
about fish.[66] For all his humanist learning, he knew that they had more information about this particular natural phenomenon than he could ever
hope to collect. Hence, he would draw upon their expertise to complete his
own education as a naturalist.

In the fishmarket, naturalists hunted through the stalls in search of rare
and unusual specimens that might be more profitably collected or dissected
than eaten. Subsequently, the rhythm of the marketplace was integrated

[63]Aldrovandi, *Discorso naturale*, pp. 180–181.
[64]Vallieri, "Le 22 lettere di Bartolomeo Maranta," p. 749 (Naples, 5 August 1554).
[65]Aldrovandi, *Vita*, pp. 28–29.
[66]See René Descartes, *Treatise on Man*, Thomas Steele Hall, trans. (Cambridge, MA, 1972),
pp. xii–xiii.

into the pattern of collecting. Just as botanizing had its seasons, so did the gathering of fish. "Right now there are not as many fish in the market as there will be at Lent," warned the Roman naturalist Ippolito Salviani, author of the *History of Aquatic Animals* (1554), in a letter to Aldrovandi. "And I must tell you that, of the forty fish that you requested, several cannot be found in Rome."[67] At home as well as abroad, Aldrovandi had friends and assistants willing to help him in his endless quest for specimens. "I will begin to frequent the market and the bridge with your list in hand," promised Pietro Fumagalli in Bologna, "and I will purchase everything that I find, having the fishermen inform me of the type of fish and the time at which it was caught to the extent that they can. Thus I will slowly acquire most, if not all, of those fish that you requested."[68] All of this material found its way into Aldrovandi's voluminous publications, where he thanked regular informants for their generosity. As with his list of visitors that singled out the socially prominent from the ordinary mass of scholars, men of similar social standing were acknowledged by name, while the empirical "experts" receded into virtual anonymity, as absent from his publications as they were from his museum. Most of these tradesmen, after all, lacked a skill essential to even the most marginal participation in the community of naturalists as defined by the republic of letters: basic literacy. The ability to read and write marked the essential division between those who provided information and those who created knowledge.

Beyond the marketplace, collectors also could be found haunting the docks and beaches to view the daily catch as it came off the boats, in search of an unusual find. "You have here a fish that is one of the most rare and extravagant parts of the sea," wrote Bernardo Castelletti to Aldrovandi in 1578. "It was given to me dried, as I send it to you, by the fisherman who caught it, who, upon discovering such a strange thing, did not throw it back into the sea, as fisherman usually do with useless fish. Indeed, he kept it alive for as long as he could, and then dried it to show as a wonder." As the circumstances of the monstrous fish illustrate, the priorities of the naturalist in acquiring beautiful and unusual specimens were often quite different than those of the fisherman, with an eye for the edibility or inedibility of the day's catch. Yet the two different ways of observing nature—one alimentary, the other aesthetic—were not incompatible. The collector, in fact, became a different market for the empirical "experts" in nature, who recognized curiosity as another commodity. Some months later, Castelletti reluctantly admitted that he had been tricked by the same fisherman, who had invented the monster, knowing the new market for natural rarities.[69] By designating even the lowliest fisherman or herbalist as a potential source of information, nat-

[67]BUB, *Aldrovandi*, ms. 38(2), Vol. II, c. 4 (Rome, 8 February 1560).

[68]BUB, *Aldrovandi*, ms. 38(2), Vol. III, c. 2r (Bologna, 7 September 1558).

[69]BUB, *Aldrovandi*, ms. 136, Vol. IX, c. 5v (Genoa, 22 February 1579); c. 129r (n.d.).

uralists learned to their chagrin that what nature could not produce, the market could invent. In time, they developed a certain caution toward their new "assistants" as their own observational skills sharpened.

The role of empirical practitioners was not confined to procurement. They often advised naturalists on the identification and classification of specimens. Giulio Cesare Moderati invoked the testimony of a Rimini fisherman regarding the ability of a dried sponge "to grow as if it were full of little creatures" when put in the water. Collectors such as Ottavio Ferri of Milan and the physician Antonio Anguisciola in Piacenza brought taxonomic problems to local fishermen as well as owners of the fishmarkets. "I will tell you what I know of the names of the fish that you asked about," wrote Ferri to Aldrovandi, "and I got my information from those who sell fish in the market."[70] Identifying the marketplace as the source of one's information relieved many naturalists from the burden of authenticating it. If an oral description conformed well with written natural histories, it reinforced the primacy of the text, even when modifying its particulars. If a specific phenomenon could not be located in any text, it was designated as uncertain information, not yet divested from its ambiguous social origins, the marketplace, and not yet transformed into humanist knowledge. The marketplace became an imaginary site in which collectors could locate all the knowledge that, in various ways, resisted the humanist synthesis they stubbornly attempted to enact on every natural object that came their way. Like nature, it was a repository of knowledge that preceded the transferral of objects into the museum.

Besides fishermen, what other groups can we point to that provided naturalists with regular supplies of artifacts and information? Apothecaries, distillers, herbalists, miners, falconers, and vendors of scientific "secrets" were all likely sources of information. Visitors compared their experiences in Calzolari's pharmacy to those in the public fishmarket, suggesting the extent to which pharmacies also participated in the marketplace culture of science.[71] While many apothecaries enjoyed some degree of social prominence, their interest in nature proceeded from their livelihood, making them somewhat like fishermen in the eyes of university-educated naturalists. Apothecaries collected and displayed nature, but did so primarily to advertise the materials for numerous powders, pills, oils, and elixirs that their elite customers craved. Yet by the sixteenth century, many pharmacies became sites in which to search for potentially collectible items. "I was not able to find anything for you," wrote Giulio Mascarelli to Aldrovandi, "except for a truly wonderous piece of petrified nutmeg in the shop of an apothecary, which I was un-

[70]BUB, *Aldrovandi*, ms. 136, Vol. XII, c. 2r (Rimini, 9 September 1586); ms. 136, Vol. XI, c. 94v (1585–1586); ms. 136, Vol. IX, cc. 3v–4 (Anguisciola to Aldrovandi, Piacenza, 9 July 1579).

[71]Olivi, *De reconditis et praecipius collectaneis*, sig. +3r. (9 April 1581).

able to obtain."[72] The conflicting demands between objects for use and objects for display sometimes determined that collectors would go away empty handed. During epidemics, for example, all the exotic items that might normally be sold to collectors were transformed into medicines that allegedly warded off the latest fatal disease.

A similar process of negotiation can be observed in the relations between naturalists and various empirical specialists. "I have befriended some falconers," wrote Alfonso Cataneo from the Estense court in Ferrara, "who promise me wonderful things."[73] The information and specimens relayed by Cataneo and other friends at the Northern Italian courts, particularly in Mantua and Ferrara where falconers were plentiful, contributed materially to the success of Aldrovandi's three-volume *Ornithology* (1599–1603). Courts were the best place to find falconers since hunting was essential to court life, so Aldrovandi wrote to court physicians and naturalists such as Cataneo when he commenced work on his study of birds. Similarly, he turned to his friends in Central Europe such as Mattioli when he wished to know more about minerals. Like Paracelsus and Georg Agricola, Italian naturalists peered down shafts in search of mineralogical information. On a trip to Trent in 1562 with Camillo Paleotti, Aldrovandi visited the mines of Peregrine in the nearby mountains.[74] When Mattioli became Imperial physician at the court of Maximilian II, he repaid friends in Italy by plying them with new specimens. "If we stay in Prague this year," he confided to Aldrovandi, "I will undoubtedly visit the mines to see if I can find something worthy of you there."[75] Mattioli was distraught to hear of the death of Agricola in 1555, shortly after his arrival in Prague; he had hoped to benefit from the German physician's own experience in studying the rock and mineral formations of Central Europe and was disappointed when this did not occur. Despite this setback, he nonetheless was able to make use of less illustrious contacts in various mining towns to obtain the information he desired.

Within every city, collectors such as Aldrovandi and Della Porta depended on the cooperation of distillers, glassmakers, and metal workers to provide them with valuable information about chemical and alchemical processes. Like apothecaries, these craftsmen worked with the materials of nature out of which a collection was formed, creating new objects of worth that also were deemed collectible. "Messer Giorgio dal Stuzzo in Marzaria . . . is a great distiller," reported Leonardo Fioravanti, "and he makes many beautiful things in his shop."[76] Della Porta employed a distiller in his academy in Naples in the 1560s and frequently visited craftsmen's workshops to compile observations on the uses of glass and the transmutation of metals that formed the

[72]BUB, *Aldrovandi*, ms. 137, Vol. I, c. 34r (n.d.).

[73]BUB, *Aldrovandi*, ms. 38(2), Vol. II, c. 19 (Ferrara, 23 April 1559).

[74]Aldrovandi, *Discorso naturale*, p. 194.

[75]Raimondi, "Lettere di P. A. Mattioli," p. 32 (Ratisbona, 19 January 1557).

[76]Fioravanti, *Dello specchio di scientia universale* (Venice, 1564), p. 38r.

nucleus of several sections of his *Natural Magic.*[77] Such unlearned sources provided the moment of *experience* that the naturalist refashioned into *experiment* in the museum. The natural philosopher, as Thomas Bartholin aptly put it, sought out "merchandise in its own market"; thus, collectors appeared wherever *experientia* could be acquired.[78]

Slaves were an equally valuable source of information, for they brought news of other cultures and their customs. Bernardo Castelletti, Aldrovandi's main contact in Genoa, procured valuable information about the medicinal properties of a plant he had recently identified on Elba by interviewing the men in the ports:

> I mainly learned about it from a native, who had been a slave for many years in Barbary, where he told me it is greatly prized by the Moors for various diseases, particularly syphilis, having miraculously cured desperate cases.... Here [in Genoa] I have gathered information on this plant from several Moorish galley-slaves, who told me that it purges the lower regions of the body well.[79]

Castelletti's recognition that people from all walks of life might contribute to the identification of particular specimens was borne out by the introduction of such information into numerous scientific texts, an instance of oral culture crystalized through the medium of print. Herbalists such as Anguillara and "professors of secrets" such as Fioravanti also described the properties of exotic plants based on the authority of slaves.[80] Like everything derived from the marketplace, this information did not count as knowledge until some recognized form of authority confirmed it. While the primary location of experimental life was the *studio,* the process of acquiring the artifacts and information that formed the museum knew no such bounds. Stepping outside of the museum and into the piazza, collectors accumulated their data through an intricate and fruitful juxtaposition of reading, travel, and observation that constituted the "experimental life" of early modern natural history.

THE ASCENT OF MONTE BALDO

All voyages into nature were not equal. While Ghini might admonish students that even the lowliest weed was worthy of study, they nonetheless were inclined to favor certain theaters of nature over others. Within Europe, Italy was considered a particularly fertile region for naturalistic excursions, and

---

[77]William Eamon, "Books of Secrets in Medieval and Early Modern Europe," *Sudhoffs Archiv* 69 (1985): 43–44; idem, "Science and Popular Culture in Sixteenth-Century Italy," *Sixteenth Century Journal* 16 (1985): 478, 480. Aldrovandi also corresponded with several *distillatori,* as a quick glance at BUB, *Aldrovandi,* ms. 38(2), reveals.

[78]Bartholin, *On Medical Travel,* p. 58

[79]BUB, *Aldrovandi,* ms. 76, c. 8v (Genoa, 20 November 1598).

[80]Anguillara, *Semplici,* p. 39; Psuedo-Falloppius, *Secreti diversi et miracolosi* (Venice, 1563), p. 6r.

many foreign scholars—Jean and Caspar Bauhin, Joachim Camerarius, Carolus Clusius, Conrad Gesner, François Rabelais, Guillaume Rondelet, Leonhard Rauwolf, and William Turner, to name only a few—studied and traveled there. In the late fifteenth century, the humanist Rudolf Agricola visited Italy specifically "to see first-hand the plants that Pliny had known."[81] Certain regions emerged as particularly fruitful ground: the mountains near Verona, rich in flora, the lands surrounding Trent, known for their mineral deposits, and Sicily, which Kircher described as a "theater of nature" after a visit in 1638. One particularly celebrated site was Monte Baldo. Towering majestically above Verona, it was—and still is—a favorite spot for naturalists to visit. "*Baldus* is famous for the great variety of choice simples growing therein . . . ," wrote John Ray in 1663. "The Paduan herbarists making simple voyages yearly hither, hath gotten *Baldus* its reputation." In the same decade, Andrew Balfour also remarked on the fecundity of the region surrounding Verona: "There is excellent Herborizing upon the Hils near to the City, but especially upon the *Monte Baldo,* which is about 20 miles distant."[82] While the rare naturalist experienced the mysteries of Egypt, Crete, or the New World first-hand, all could share in the pleasures this local Garden of Eden offered.

The wonders of Monte Baldo were publicized in two treatises: Francesco Calzolari's *Voyage to Monte Baldo* (1565) and Giovanni Pona's *Monte Baldo Described* (1617), originally published in Latin in 1601. Not coincidentally, Calzolari and Pona were apothecaries in Verona who had rich museums of natural history, though the former achieved greater fame.[83] In both instances, proximity to Monte Baldo enhanced their reputations as apothecaries and naturalists. When illustrious visitors arrived, they guided them up its slopes. Calzolari led annual summer jaunts there, except when illness prevented him from enduring the ardors of the climb. As a tribute to Calzolari, Mattioli named a previously uncatalogued violet indigenous to Monte Baldo, *Calceolaria,* one of the first instances in which naming of a plant commemorated a living person.[84] By the time of Calzolari's death in 1604, there were few naturalists of note who had not enjoyed the sight of Monte Baldo under Calzolari's guidance.

The most famous excursion occurred in June 1554. In 1551, Luigi Anguillara herborized with Calzolari on the mountain. Three years later, he traveled from Padua to join Calzolari, Aldrovandi, the physician Andrea Alpago, and a group of local Veronese botanists for a grander expedition. Quite possibly this was the trip that Aldrovandi had tried to persuade Francesco Petrollini to join; Petrollini declined because his duties as a

[81]Reeds, "Renaissance Humanism and Botany," p. 528.

[82]Ray, *Travels,* p. 187; Balfour, *Letters,* p. 235.

[83]On Pona's museum, see his *Index multarum rerum quae repositorio suo [Johannis Ponae] adservantur* (Verona, 1601).

[84]Tergolina-Gislanzoni-Brasco, "Francesco Calzolari," p. 6.

physician in Codignola made it impossible for him to take more than two weeks off.[85] By all accounts, the trip was a great success. Aldrovandi returned to Bologna laden with specimens for his herbarium, which he generously shared with Ghini, and feeling that he had begun to establish himself in the community of naturalists. It was an important turning point in his career. For years afterward, the participants in this singular excursion recalled the trip with a degree of nostalgia and precision that suggests how greatly it impacted their lives. It became a common point of reference for the development of the late Renaissance community of naturalists, an object lesson in the success of their enterprise and a parable of the failure of certain participants to properly appreciate the invention and implementation of new techniques of investigating nature. While Calzolari and Aldrovandi probably were acquainted from the latter's student days in Padua, the 1554 trip was the first opportunity to measure each other's skills and establish a friendship. Their mutual delight in the encounter signaled the beginnings of a lifelong relationship. Already in July 1555, Calzolari sighed when he recalled "the happy voyage that we took together to Monte Baldo."[86]

Aldrovandi first met Anguillara in 1554, although he had probably heard of him by reputation prior to their encounter in Verona. More traveler and herbalist than humanist, Anguillara was prefect of the botanical garden in Padua (1546–1561), but not a university professor. He had already incurred the wrath of Mattioli, who seemed to enjoy making enemies more than friends and loudly proclaimed the ignorance of Anguillara's students—in his opinion, they no more knew the "plants of Dioscorides" than they could distinguish the difference between "basil and lettuce."[87] Primed by Mattioli, Aldrovandi's encounter with Anguillara on Monte Baldo was an opportunity to unmask his ignorance. "I was most pleased that the voyage in the Mountains allowed you to know the ignorance of Luigi the Eel Skinner [a pun on Anguillara's name] . . . ," wrote Mattioli contentedly. "I am most sorry that they were honored with company like yours, because they will have learned an infinite number of wonderful things from you, while you will have extracted almost nothing from them."[88] Even Wieland, no friend of Mattioli, chortled over Anguillara's exposure. "I read the [letter] addressed to messer Gabrielle [Falloppia]," he wrote to Aldrovandi in September 1554, "in which everything about Micheli and that ass, Luigi the Roman, is recounted. I think you now are aware of their laziness."[89]

Despite Calzolari's glowing portrayal of the trip, the two naturalists nearly came to blows, over what exactly no one specified. Most likely, Anguillara,

[85]Chiovenda, "Francesco Petrollini," pp. 443–444.

[86]Cermenati, "Francesco Calzolari, "p. 102.

[87]Raimondi, "Le lettere di P. A. Mattioli," p. 26; also Palmer, "Medical botany in Northern Italy," p. 150.

[88]Raimondi, "Le lettere di P. A. Mattioli," p. 24.

[89]De Toni, *Spigolature aldrovandiane XI*, p. 10.

unlike Brasavola's imagined herbalist Senex, did not take Aldrovandi's criticisms of his knowledge of ancient sources with good grace. Both men represented two different images of the naturalist. Anguillara was a man whose knowledge was based primarily on experience, while Aldrovandi had little experience—this is what he hoped to gain on the trip—but a great deal of learning. Almost six months later, Calzolari was still distraught about "the discord between you and messer Aluise." "Of course it displeases me greatly . . . ," he wrote to Aldrovandi in November 1554. "I wish that you would restore the good faith between you (*voria cum il proprio sangue potervi quietar fra noi*), but you do not possess the patience."[90] In retaliation, Anguillara neglected to mention Aldrovandi a single time in his *Simples,* earning the epithet of "The Beast" from Mattioli for this latest infamy.[91]

Diplomatically, Calzolari preferred to recall only the high points of their trip. His elegant description of the voyage to Monte Baldo in 1565 captured all the exuberance and none of the discord that this gathering of naturalists in one of nature's finest settings produced, presenting the excursion as the epitome of humanist sodality. Beyond its own diversity, the soil and climate of Monte Baldo allowed foreign plants to take root with remarkable facility, and this made it a particularly unusual theater of experimentation as well as observation. "Our Mountain, no other than a beautiful garden in which many rare plants are brought by cultivators from various and diverse parts, is naturally most fecund," proclaimed Calzolari.[92] In this particular locale, he assured his readers, one could find "hundreds of each species in little space" and countless "most delightful spots" (*amenessimi luoghi*).

In spare but elegant prose, Calzolari retraced the path that he and many others took up the mountain. Describing plants in the order in which they appeared, as well as "the notable sights necessary to recognize the way," the apothecary led readers through "fertile hills and shady little valleys" to the summit, from which one could see the Veronese plains, dotted with castles and cities, "as if it were a most beautiful Flemish painting." From Verona to the base of the mountain, one encountered fields of geraniums, sunflowers, orchids, rhododendrums, and heliotropes, as well as samples of digitalis, belladonna, myrrh, and catnip. Ascending to the *valle d'Artilone,* one found "all the species of the most beautiful and rarest plants born, I will say, not only in Italy but perhaps in all of Europe," among them yews, violets, *noli mi tangere,* and samples of *costo,* a precious ingredient in antidotes such as theriac. Concluding the pamphlet, Calzolari compared Monte Baldo to Crete, Cyprus, Syria, and the Indies—in short, all the most desirous places to travel—exclaiming "why shouldn't Monte Baldo be numbered one of the

[90]Cermenati, "Francesco Calzolari," p. 100.
[91]Raimondi, "Le lettere di P. A. Mattioli," p. 56.
[92]Calzolari, *Il viaggio di Monte Baldo,* p. 4. The passages that follow are taken from pp. 4, 6, 11, 16, 15 in sequential order.

principal [spots] for herbalists?" To bolster this claim, he enumerated the famous naturalists, Venetian nobles, and Veronese gentlemen who "several times expended great effort with me in order to research it."

The language used by Calzolari to recapture the sensation of experiencing this privileged theater of nature was highly reminiscent of the descriptive strategies used by Petrarch in *The Ascent of Mount Ventoux*. Like Petrarch's letter, Calzolari's *Voyage to Monte Baldo* discussed the preparation for the climb, the ascent, the events of the summit, and the descent.[93] Perhaps he had read one of the numerous Italian editions of Petrarch's work such as the one that appeared in Venice in 1549, or more generally imbibed the image of the mountain as a place of moral virtue through contact with local humanists. In 1336, Petrarch climbed Mount Ventoux, also known for its "conspicuous height."[94] It was a journey he had anticipated for years and he chose his companions—his younger brother and a servant—wisely. In a passage curiously reminiscent of Calzolari's difficulties with his own companions, Petrarch observed, "It will sound strange to you that hardly a single one of all my friends seemed to me suitable in every respect, so rare a thing is absolute congeniality in every attitude and habit even among dear friends." Attempting to circumvent the ardors of the ascent, Petrarch was frustrated in his attempts to find an easier path. This led him to reflect, "But nature is not overcome by man's devices; a corporeal thing cannot reach the heights by descending." As Mattioli instructed the young Aldrovandi after he had returned from his summer travels in 1553, "Although you suffered in climbing high mountains, this will increase your content, knowing much better than before how arduous the path ascending to virtue is."[95] Like Petrarch, Renaissance naturalists imagined *in via* to be one of the fundamental metaphors of existence. Ascending a mountain was a form of moral progress.[96] At its summit lay knowledge, for Petrarch the product of a solitary and frustrating voyage, for Renaissance naturalists the result of a communal experience.

Upon reaching the summit of Ventoux, Petrarch randomly opened his copy of Augustine's *Confessions*—his version of a pocket Dioscorides—and read the following passage (X.8, 15):

> And men go to admire the high mountains, the vast floods of the sea, the huge streams of the rivers, the circumference of the ocean, and the revolutions of the stars—and the desert themselves.

[93]See Robert M. Durling, "The Ascent of Mount Ventoux and the Crisis of Allegory," *Italian Quarterly* 18 (1974): 7–28. Thanks to Ken Gouwens for bringing this article to my attention.

[94]The following passages come from Francesco Petrarca, "The Ascent of Mount Ventoux," in *The Renaissance Philosophy of Man*, Ernst Cassirer, Paul Oskar Kristeller, and John Herman Randall, Jr., eds. (Chicago, 1948), pp. 36–46.

[95]Raimondi, "Le lettere di P. A. Mattioli," p. 15.

[96]I borrow these terms from Durling, "The Ascent of Mount Ventoux," pp. 11, 13.

Returning home, Petrarch repeatedly looked back at the summit to remind himself of the lessons he had learned on its heights. While Renaissance naturalists equated their voyages into nature with spiritual quests, they would not have agreed with the moral that Augustine offered Petrarch. Nature was the site of self-discovery, not of loss. As the events surrounding the 1554 trip to Monte Baldo demonstrate, it was an opportunity for naturalists to test their powers against one another, pitting new methods of textual criticism against empirical practices. Ultimately it was a battle waged against nature herself. While Petrarch's ascent was a solitary, unrepeatable journey, illuminating the frustrations of spiritual fulfillment in an earthly life, the naturalist's ascent incrementally yielded success through repetition. Standing on the top of Monte Baldo, in command of the names for the plants covering its slopes, Calzolari must have felt that he could almost know God through the wonder of nature.

For both Petrarch and Calzolari, the mountain summit provided a unique vantage point. On the top of Mount Ventoux, Petrarch contemplated the difficulties of self-knowledge and the problems of revering the ancients in a Christian world. On the top of Monte Baldo, Calzolari gained new perspective on the study of nature. But he also pondered the difficulties of keeping the community of naturalists intact, as his distress over Aldrovandi's and Anguillara's quarrel illustrates. Just as Petrarch had put his revelations into writing, Calzolari chose to publicize the significance of the mountain that had been so formative in determining the course of his own life, and perhaps the direction of an entire community of scholars. While the "new fullness of knowledge and perspective" that Petrarch gained was spiritual, Calzolari's was both spiritual and material.[97] Through repeated trips to Monte Baldo, he transformed it from a *locus amoenus*, known to few, into a "theater of nature" to which generations of naturalists went in pilgrimage.

"VULCAN'S FURNACE"

Like the protagonist of Dante's *Divine Comedy*, who experienced the terrors of Inferno before the joys of Paradise, naturalists also perceived their voyages as a descent into nature. While travelers to north-central Italy sought out the gardenspots of the Alps and Apennines, travelers to the south, taking "the steep and savage path,"[98] explored the gurgling sulfur fields in Pozzuoli, gazed at the swirling currents of the straits of Messina, and clambored up the slopes of Etna and Vesuvius to peer into the open mouth of hell. Had Virgil repeated his role as guide in the seventeenth century, he could well have asked the same question of naturalists that he put to Dante: "Why

[97]Durling, "The Ascent of Mount Ventoux," p. 25.
[98]*The Divine Comedy of Dante Alighieri: Inferno*, Allen Mandelbaum, trans. (Berkeley and Los Angeles, 1980; Toronto, 1982), p. 19 (II, 142).

does your vision linger there below among the lost and mutilated shadows?"[99] Appropriately, the lands south of Rome were full of danger and promise for travelers. Naturalists remained unsurprised, and even pleased, that the topography of Italy conformed to an imagined moral logic. The path from Rome to Naples to Sicily, as culled from the natural histories of the day, led straight to perdition. Mixing classical and religious motifs, naturalists imagined Sicily as both the gateway to Hell and the entrance to the Underworld. Their physical coexistence imitated their textual conflation in the writings of humanists.

Like Aldrovandi, Kircher's greatest adventures occurred in his youth. After his trip to Sicily in 1637–1638, he confined his excursions to Lazio and Umbria, the regions adjacent to his adopted city, Rome. Kircher displayed little of the active interest he subsequently took in natural history before his southern travels.[100] Called to Rome at the behest of Francesco Barberini, the Jesuit was hard at work on his Coptic vocabulary and interpretations of the Egyptian hieroglyphs, when diplomatic duties intervened. Frederick of Hesse, recently converted to Catholicism, had expressed an interest in seeing Sicily. Kircher, as the most luminous German Jesuit philosopher in Rome, was assigned to his entourage. Between Rome and Calabria, he dallied in the sulfur fields near Naples, examined the peculiar formation known as the "wood-fossil-mineral" first identified by Cesi a decade earlier, and eventually arrived in Sicily. In Catania, he caught his first glimpse of Etna, looming ominously above the city, and climbed the slopes. There he commented on the remarkable fertility of volcanic ash, which produced enormous chestnut trees.[101] From a distance he watched Etna erupt, producing an illustration of the sight "observed by the author in the year 1637"[102] (fig. 12). This was his first taste of the powerful forces at work in that region of the world.

In March 1638, as Kircher and his companions attempted to navigate the straits of Messina, the skies grew dark. Surely this was Charon's ferry that took them across the waters, navigating the treacherous path between Scylla and Charybdis? They arrived on the shores of Calabria just in time for an earthquake to commence. Nearby Stromboli, part of the Ischian islands, erupted. In the distance, Vesuvius began to smoke and rumble. Curiosity overcame fear, and Kircher climbed to the summit. Staring down into the crater through the smoke and flames, he described the sight as "the workshop of Vulcan's furnace."[103] His eyewitness report later appeared in his *Subterranean World*, as "seen by the author in 1638," and allegedly inspired the

[99]Ibid., p. 265 (XXIX, 5–6).

[100]The following narrative paraphrases material in Godwin, *Athanasius Kircher*, p. 13; Reilly, *Athanasius Kircher*, pp. 5–13; and Kircher, MS, Vol. I, sig. **2r–**4r.

[101]Kircher, *China illustrate*, p. 180.

[102]See Kircher's "Aetnae descriptio," in MS, Vol. I, pp. 200–205.

[103]Ibid., sig. **3v.

Figure 12. The eruption of Mount Etna, 1637. From Athanasius Kircher, *Mundus subterraneus* (Amsterdam, 1664).

publication of his magnificent encyclopedia of nature. Reading the description that he published almost thirty years after the eruption, we can still feel his excitement and awe:

> When finally I reached the crater, it was terrible to behold. The whole area was lit up by fires, and the glowing sulphur and bitumen produced an intolerable vapor. It was just like hell, only lacking the demons to complete the picture.[104]

Overcome by the power and fury of the volcano, Kircher cried out, "O God, . . . how incomprehensible are your ways!"[105]

---

[104]For the image of Vesuvius erupting, see the illustration between sig. **3v and **4r. The passage from *Mundus subterraneus* is translated in Reilly, *Athanasius Kircher,* p. 71.

[105]Kircher, MS, Vol. I, sig. **3v.

Vesuvius was a formidable sight during the early seventeenth century, when it experienced a period of intense volcanic activity. Many other naturalists were also struck by the grandeur and fury of this imposing volcano. While a student in Padua, Peiresc availed himself of the opportunity to travel to Naples, where he saw Vesuvius in 1601.[106] In December 1631, it spewed forth ashes as far as Venice. A decade later, when Thomas Bartholin laid eyes on it, he reported, "nor then did Vesuvius cease to smoke, and I saw the sulphur boiling in the base of the chasm when I climbed the mountain in November 1643." Like Kircher, Bartholin also witnessed the activity of Stromboli, remarking, "the devil seems to inhabit the place."[107] While botanists praised the fertility of Monte Baldo, would-be geologists flocked to Etna and Vesuvius in anticipation of a magnificent natural spectacle.

Kircher's description of his experiences in southern Italy bore an uncanny resemblance to well-known classical descriptions of earlier volcanic activity. His voyage was a spiritual test of a different order than Petrarch had experienced on Mount Ventoux, reminding him of the fragility of human existence by forcing him to confront his own mortality. In his autobiography, the Jesuit described his trip in the following terms: "Thus I endured new dangers to my life . . . the greatest dangers, from which the Lord thought I was worthy of rescue."[108] Similarly his adventures on the slopes of Etna and Vesuvius were undertaken "with great peril to life"[109] Describing these events, Kircher readily quoted classical writers alongside descriptions of more recent eruptions. Kircher's most compelling image of the volcano, as "Vulcan's furnace," was hardly original. In his *Civil War*, the Roman writer Lucan depicted an eruption in the following terms: "In Sicily, furious Vulcan opened the mouth of Etna and launched its fiery debris not straight up to the sky but showering downward to threaten Italy's shore." Lucretius described the source of volcanic eruption as "Etna's mighty furnaces" in his celebrated poem, *On the Nature of Things*.[110] In the *Aeneid*, Virgil chose to place Vulcan not beneath Etna, inhabited by a captive and restless giant, but in one of the Lipari islands north of Sicily, aptly named Vulcano. Virgil described it this way:

An island rises
Near the Sicanian coast and Lipare,
Aeolian land, steep over smoking rocks.
Below them roars a cavern, hollow vaults
Scooped out for forges, where the Cyclops pound
On the resounding anvils; lumps of steel

[106]Gassendi, *Mirror of True Nobility and Gentility*, p. 38.
[107]Bartholin, *On Medical Travel*, pp. 69, 73.
[108]Kircher, *Vita*, pp. 53–54.
[109]Kircher, MS, Vol. I, p. 201.
[110]*Lucan's Civil War*, P. F. Widdows, trans. (Bloomington, 1988), p. 18 (I, 603–606); Lucretius, *On the Nature of Things*, Palmer Bovie, trans. (New York, 1974), p. 204 (VI, 682).

Hiss in the waters, and the blasts of fire
Pant in the furnaces; here Vulcan dwells,
The place is called Vulcania, and here
The Lord of Fire comes down.[111]

Like the naturalists who saw the ascent of Monte Baldo as a recapitulation of Petrarch's discovery of nature as a *locus amoenus,* Kircher and many other visitors to southern Italy portrayed their encounters with the volcanoes as a voyage into the terrain of Virgil, Lucan, Lucretius, and occasionally Dante, the material confrontation with their humanist past.

The terrible conjuncture of events in 1637–1638—the eruptions of Etna, Vesuvius, and Stromboli and the Calabrian earthquake—was a turning point in Kircher's life. These experiences led him to temporarily put aside the study of hieroglyphs in favor of a more comprehensive investigation of the natural world. He had survived the very event that had killed Pliny in A.D. 79 and personally confronted the power of Vesuvius by climbing its slopes. Like Dante, he could explain, "His is no random journey to the deep: it has been willed on high."[112] The frontispiece of the second volume of the *Subterranean World* illustrated the reorientation of his philosophical interests (fig. 13). Here Mercury persuaded Kircher, represented as a female muse intent on the study of hieroglyphs, to leave behind the world of human objects, and implicitly the humanist *studio,* to study the secrets of nature, indicated by the presence of miners in the background. As the frontispiece reminded readers, destiny led Kircher to study nature, whose structures were as mysterious and impenetrable as the hieroglyphs themselves. His voyage into nature was a calling in the truest sense of the word, a pilgrimage shaped by the omnipresent images of volcanic eruptions in the writings of classical authors and by a Christian search for spiritual fulfillment in the earthly world. As the Jesuit explained in his autobiography, God had saved him from certain death for a reason. From that point on, confronting and understanding the very thing that almost had caused his demise was to become his vocation. His boundless curiosity and vast erudition had found a purpose at last.

Kircher's experiences led him to ask the following questions: What chain of events generated an earthquake or a volcanic eruption? How did terrestrial cataclysms figure in the delicate balance of elements in the universe? From these speculations, Kircher developed an image of the earth as a globe filled with a series of competing, internal processes, among them a system of subterranean fires that spread outward from the center, creating fissures and occasionally gurgling to the surface to form volcanoes. Such images accorded well with his theories of universal magnetism, which posited the

---

[111] *The Aeneid of Virgil,* Brian Wilkie, ed., and Rolfe Humphries, trans. (New York, 1987), p. 194 (VIII, 433–442).
[112] Dante, *Inferno,* p. 59 (VII, 10–11).

Figure 13. Kircher's subterranean world. From Athanasius Kircher, *Mundus subter-raneus* (Amsterdam, 1664).

existence of attractive and repulsive forces as an explanation for most things in the world.[113] In contrast to Renaissance naturalists, concerned primarily with problems of nomenclature, identification, and classification, Kircher preferred to study the *processes of nature*. Thus, he imagined Naples and Sicily over places such as Monte Baldo as preferred "theaters of nature" because they were open fissures, windows on the soul of the universe. Sicily, which Lucretius described as a "land of marvels," continued to be a site of wonder where the imagination could roam.[114]

NATURAL THEATERS

The tumultuous sights revealed on Etna and Vesuvius to visitors such as Kircher, Peiresc, and Bartholin and inhabitants such as Boccone, and the excitement generated by new observations and discoveries, represented the culmination of a tradition of investigating nature that began in the late fifteenth century and accelerated in the mid-sixteenth century. Certainly Monte Baldo, Etna, and Vesuvius were not the only sites that merited such intensive inspection, although they were probably the most well known. Different communities and generations of naturalists had their own *amenessimi luoghi* ("delightful spots"), to borrow Calzolari's phrase. In the early 1590s, all of Italy, or at least those who collected nature, awaited the results of Giuseppe Casabona's voyage to Crete.[115] Like Sicily, Crete was a site known for its remarkable fertility, celebrated by classical writers. Aldrovandi, always ready to taken advantage of an opportunity to increase his collection, praised the Grand Duke Ferdinando I for underwriting this excursion and advised his botanist Casabona on what to observe and collect. In November 1591, impatiently awaiting Casabona's return, he wrote again to Ferdinando, inquiring of his botanist "who, having been on that most fertile island of simples for an entire year, I assume will have brought back many beautiful things."[116] Undoubtedly, Aldrovandi hoped to add any new observations to his notebooks in which he had divided the study of nature geographically, entitled *Observations of Natural Things That Are Born in Different Parts of the World*.[117] Four years later, when Casabona wrote to inform Aldrovandi that the Grand Duke now planned to send him to Corsica, Aldrovandi responded by reminiscing about how reading Theophrastus had inspired him to undertake the same trip. Characteristically, he encouraged his young colleague to expand his itinerary to include Sardegna, Elba, Sicily,

[113]On Kircher's theories of magnetism, see his *Ars magnesia* (Würzburg, 1631); *Magnes, sive de Arte Magnetica* (Rome, 1641); and *Magneticum naturae regnum* (Rome, 1667).

[114]Lucretius, *On the Nature of Things*, p. 34 (I, 728).

[115]For a detailed discussion of this excursion, see Tongiorgi Tomasi, "L'isola dei semplici," *KOS* 1 (1984): 61–78; and Olmi, "Molti amici in varii luoghi," pp. 23–26.

[116]ASF, *Mediceo* 830, c. 293r, in Tosi, *Ulisse Aldrovandi e la Toscana*, p. 387.

[117]BUB, *Aldrovandi*, ms. 143. Here I use the title he gave it in a letter to Belisario Vinta.

and "all the other little islands of the Tyrrhenian sea."[118] Calzolari confined his image of nature as a *locus amoenus* to one specific location; Aldrovandi expanded it to include all that nature offered.

While Monte Baldo continued to attract naturalists, other mountains soon competed with it as botanical paradises. In the early seventeenth century, Cesi took members of the Accademia dei Lincei up Monte Gennaro, near his country residence above Tivoli, playground for the Roman nobility. In a letter to Galileo of 21 October 1611, he confessed to "having visited meticulously and researched my Monte Gennaro nearby, with four most learned botanists."[119] Much like Calzolari, Cesi preferred to return to his mountain frequently to know the seasonal variety of its indigenous plants. In June 1612, he informed Johann Faber, "Sunday I will ascend Monte Gennaro . . . and I anticipate seeing many plants." A year later, he promised his friend that he would send "whatever natural delights will be obtained from our Amphitheater."[120] In describing Monte Gennaro as *Amphitheatrum nostrum,* Cesi revealed his affinities with such naturalists as Calzolari, who saw botanizing as a form of humanist sodality. The Linceans, as a self-defined community of naturalists, needed to have certain "amenable places" in which to commune with nature, just as they required museums to study and dissect it. For Cesi, the hills behind his country villa provided the primary *locus amoenus* for his academy.

Second only to Monte Gennaro was the landscape that yielded the "wood-fossil-mineral," to which Cesi returned frequently in the 1620s to inspect this wonder of nature. While Monte Gennaro never gained any popularity beyond the immediate circle of the academy—it truly was the sort of personal space that a *locus amoenus* should be—the site where the metallophyte had been discovered gained some popularity, as word of it spread through the republic of letters. Peiresc became so excited about this paradox of nature, after reading Stelluti's published account of it, that he begged friends traveling to Italy to bring him a piece of the metallophyte. According to Peiresc, the Grand Duke Ferdinando II was so fascinated by it that he planned to launch a new expedition to Acquasparta to collect samples of this curious wood.[121] Whether the Tuscan courtiers ever really arrived there is inconsequential to the story. However, the desire that moved them to consider the voyage is not. Such symbolic gestures render visible an imaginary landscape in which naturalists and their patrons found comfort in the act of experiencing nature.

[118]BUB, *Aldrovandi,* ms. 21, Vol. IV, c. 170r.

[119]CL, Vol. II, p. 175 (Tivoli, 21 October 1611). This episode is also discussed in Gabriella Belloni Speciale, "La ricerca botanica dei Lincei a Napoli: corrispondenti e luoghi," in Lomonaco and Torrini, *Galileo e Napoli,* pp. 59–79.

[120]CL, Vol. II, pp. 243, 359 (San Polo, 29 June 1612; Monticelli, 1 June 1613).

[121]Peiresc, *Lettres à Cassiano dal Pozzo,* pp. 172, 176, 180, 185, 192. As Stelluti describes it, the metallophyte was found near Todi.

Kircher's image of the naturalist's voyage as a moral choice was not lost on his followers. Less than twenty years after his *Subterranean World* appeared, with its gripping description of a trip to Sicily, his disciple Bonanni published his study of mollusks. While little in the text established a relationship between the two treatises, other than their defense of Aristotelian doctrine, the frontispiece indicated a more direct connection (fig. 14). On the title page of Bonanni's *Recreation of the Eye and the Mind in the Observation of Snails,* a young noble—presumably Bonanni in his student days at Ancona—holds high the spiral shell of a snail. He offers it up to a muse who contemplates an open oyster shell containing two pearls. With his other hand, the shell collector gestures to the embarrassment of riches the sea produces. The muse, already initiated into the well-known mysteries of oysters, is about to experience the more heady contemplation of snail anatomy and generation. Clearly the image depicts Bonanni's own moment of self-discovery. Behind him, Neptune rises majestically out of the sea, heralding the arrival of even vaster riches. While Kircher imagined Vulcan to be his mythological guide in his voyage into the underworld, Bonanni portrayed Neptune as his mentor and beneficent provider. Accompanied by Neptune and "reading shells" with Cicero, Bonanni brings the facts of nature, represented by his shells, to the attention of philosophy, embodied as a muse. He makes the little things of nature significant. In a different fashion than Peiresc's musing about the "wood-fossil-mineral," Bonanni's frontispiece charted another facet of the imaginative landscape into which collectors voyaged. Through the deployment of powerful mythologies, he and his mentor Kircher situated the act of possessing nature within a cultural framework that their intended audience understood and appreciated. Like Calzolari's subtle invocation of Petrarch, these illustrations facilitated the image of natural history as a culturally acceptable pursuit.

Reflecting on the different "theaters of nature" that naturalists visited—bustling marketplaces, dimly lit mining shafts, abundant beaches, verdant mountain groves, and live volcanic craters—and the manner in which they subsequently portrayed them, one is struck again and again by their literal theatricality. "The last eruption of 1635 displayed a variety of spectacles," wrote Bartholin in his description of Etna.[122] In the eyes of early modern naturalists, nature was *drama personified.* Inviting them into her midst, she lured them with her seemingly infinite possibilities and the potential for heroic adventure. When collectors brought nature into the museum, as Thomas Bartholin did when he brought back "a piece of that solidified flow" from Etna "for the Museum Wormianum,"[123] they attempted to recapture not only the totality of nature but also the excitement, conflicts, and expectations that their initial voyages in the vast "natural" theater had produced.

[122]Bartholin, *On Medical Travel,* p. 79.
[123]Ibid.

Figure 14. Bonanni's shell collector. From Filippo Bonanni, *Recreatio mentis et oculi in observatione animalium testaceorum* (Rome, 1684 ed.).

FIVE

Fare Esperienza

They that search out the secrets of Nature, in cursory discourses, fall unfortunately upon the thorns of subtleties and snares of questions, and do nothing but weave and unweave them with a fine thread of controversies.

—JOHN JONSTONE

In the illustration adorning the 1671 Latin edition of the Tuscan court naturalist Francesco Redi's *Experiments on the Generation of Insects* (1668), nature is represented within the laboratory[1] (fig. 15). At the top, a reclining muse—possibly Nature herself—enjoys the attention of two putti, who playfully collect the bees, butterflies, and other winged creatures that she contemplates. In the philosophical and cultural universe that Redi inhabited, these bits of nature have now become objects of experimental worth, their generative mysteries—and Redi's disavowal of their spontaneous generation—unraveled within the pages of the text. At the bottom, Minerva, the embodiment of *scientia,* puts Nature to the test. The microscope through which she observes an insect, and the glass vials to the left alert us to the "laboratory-like" conditions in which she works, a representation of the activities that organize objects in the Medici naturalistic collections where Redi performed his experiments. To the right, the muse who brought nature into the laboratory overlooks the activities of her experimental sister as if to supervise them, establishing a bridge between the pastoral setting above and the scientific enclave below. While Minerva observes and experiments with her right hand, her left hand, holding open a book of unspecified authority, reminds us that observation is only a valid exercise when mediated by the words of the past.[2]

[1] I owe this point to William J. Ashworth's unpublished paper, "The Garden of Eden: Evolution of a Seventeenth-Century Image" (History of Science Society, Seattle, WA, 1990), which discusses the Redi illustrations in the larger context of changing images of nature. The illustrations are, respectively, from *Francisci Redi Patritii Aretini Experimenta circa generationem insectorum ad nobilissimum virum Carolum Data* (Amsterdam, 1671) and *Francisci Redi Nobilis Aretini, Experimenta circa res diversas naturales speciatim illas, quae ex Indiis adferuntur. Ex Italico Latinitate donata* (Amsterdam, 1675).

[2] This point is made more comprehensively by Jay Tribby in an analysis of Redi's *Osservazioni intorno alle vipere* (1664); see his "Cooking (with) Clio and Cleo: Eloquence and Experiment in Seventeenth-Century Florence," *Journal of the History of Ideas* 52 (1991): 417–439.

Figure 15. Francesco Redi, *Experiments on the Generation of Insects* (1668). From *Francisci Redi Patritii Aretini Experimenta circa generationem insectorum ad nobilissimum virum Carolum Dati* (Amsterdam, 1671 ed.).

The dialectic between ancient authority and contemporary scientific practice is further reinforced by the motto emblazoned on the frontispiece: *Optimi consultatores mortui* ("the dead are the best counselors"). Below the table, Minerva casually rests her foot on a turtle, reminding us not only of the delightful experimental results that their decapitation produced, but also of the repetitive experimental ethos—"testing and retesting" (*provando e riprovando*), as the motto of the Accademia del Cimento to which Redi belonged went—that the most avant garde natural philosophers cultivated by the mid-seventeenth century.[3]

Redi's image of nature in the laboratory is part of a diptych, completed by the illustration accompanying the 1675 Latin text of his *Experiments on Diverse Natural Things, Particularly Those Which Come from the Indies* (1671), a further response to the philosophical precepts imbedded in Kircher's *Subterranean World* (fig. 16). In this series of experiments, reported in the form of a letter to Kircher, Redi tackled the problem of the serpent's stone, fabled to possess remarkable curative powers through its magnetic action on any venomous wound. Kircher supported this view and conducted numerous experiments with it at the Roman College to publicize its success; Redi politely responded that no living creature had ever been cured of anything by the application of a serpent's stone. The frontispiece allegorically depicts Redi's confrontation with his opponent. Here Minerva, now seated at a richly bedecked experimental table, receives the natural wealth of the "Indies." Staring down upon a native carrying an armadillo and offering up a serpent's stone, she reproves him for not testing the phenomenon experimentally. As Redi constantly affirmed in his writings, "I do not put much faith in matters not made clear to me by experiment."[4] Pointing uncompromisingly toward the books and microscope in front of her, the tools of natural philosophy, Minerva challenges "Received Wisdom" to confront the Learned Past and the Experimental Present. She has become the symbol of the experimental dialectic that characterized natural history and natural philosophy in general by the mid-seventeenth century.

Behind the universal truths that Redi's frontispiece proclaimed lay the implied context that informed all of his work: the court of the Grand Dukes of Tuscany. The Medici court, represented by Redi's iconographically improvised laboratory, was not the only "theater of experiment" in which nature could be gathered and examined, although it was probably one of the most well known. As Paolo Rossi observes, different images of natural philosophy corresponded with different images of nature, and the allegorized

[3]The bibliography on early modern experimental culture is scattered throughout this chapter, although I refer the readers particularly to the work of Peter Dear, Simon Schaffer, and Steven Shapin. Besides the work of Jay Tribby, for more on Redi, see Bruno Basile, *L'invenzione del vero. La letteratura scientifica da Galilei ad Algarotti* (Rome, 1987), and Findlen, "Controlling the Experiment."

[4]*Francesco Redi on Vipers*, Peter K. Knoefel, trans. and ed. (Leiden, 1988), p. 47.

Figure 16. Francesco Redi, *Experiments on Diverse Natural Things, Particularly Those Which Come from the Indies* (1671). From *Francisci Redi Nobilis Aretini, Experimenta circa res diversas naturales speciatim illas, quae ex Indiis adferuntur. Ex Italico Latinitate donata* (Amsterdam, 1675 ed.).

depiction of Redi's work displayed his preferred vision of the proper means to attain knowledge.[5] The image of the laboratory—a space filled with objects and instruments whose use engendered knowledge—represented a certain style of philosophizing that the Tuscan naturalist wished to advertise. Redi was not alone in his choice, however. He represented the culmination of new ideas about "experiencing" nature that made the museum and the laboratory privileged sites of knowledge by the mid-seventeenth century. While naturalists expanded the parameters of "nature" through their travels, they also reconceptualized their view of what constituted knowledge, giving objects a greater role in the assessment of truth. As Adi Ophir writes, "The institutionalization of special places for the search for knowledge—where words are systematically linked to things, discursive authority is correlated with access and visibility, and the epistemological is coordinated with the social—was a crucial stage in the historical process that constituted science as an established cultural system."[6] These different activities came together in settings such as the museum, producing a substantially altered view of the proper means to know nature.

Images such as those populating Redi's natural histories further reinforced the function of the museum as a precise visual matrix for scientific practices. As William Ashworth has observed, by the mid-seventeenth century, the naturalist was more frequently represented as studying nature in the laboratory rather than *in situ*.[7] By identifying the study of nature with a specific physical site, noteworthy for its separation from nature, naturalists iconographically aligned the museum with the new techniques of natural history that put nature under the gaze of science in a setting that artificially isolated the parts form the whole. In contrast to late Renaissance portrayals of the naturalist as *homo viator*, best represented in the sketches of Gherardo Cibo, seventeenth-century artists preferred to imagine the naturalist first as a demonstrator, exemplified by the frontispiece of Imperato's 1599 catalogue (see fig. 2), and finally as an experimenter. The voyage into nature, precipitated by the act of reading about nature, eventually ended in the laboratory, where books and artifacts contributed to the production of humanist and experimental knowledge, and the shaping of natural history as an investigative enterprise.

The Redian theater of nature in Baroque Florence and the museums of Kircher in Rome and Settala in Milan were the culmination of almost a century and a half of intensive reconceptualization of the place of nature in the formation of *scientia* and the role of the collector as a manager of nature's resources. Collectors such as Aldrovandi, Calzolari, and Imperato initiated the passage of nature into the study. There they tested discrete portions of

[5] Paolo Rossi, "The Aristotelians and the 'Moderns': Hypothesis and Nature," *Annali dell'Istituto e Museo di Storia della Scienza di Firenze* 7 (1982): 5.

[6] Ophir, "A Place of Knowledge Recreated," p. 165.

[7] Ashworth, "The Garden of Eden: Evolution of a Seventeenth-Century Image," passim.

it to understand better the relationship among natural philosophy, medical knowledge, and sensory experience, culled from the pages of Aristotle, Galen, Dioscorides, and other ancient authorities. The museum, as Aldrovandi broadly observed, was a place in which one could "join theory with practice," a view that most Renaissance naturalists shared.[8] Their Baroque counterparts, operating in a scientific climate conditioned by the work of Galileo and his contemporaries, and the instrumental mediation of the senses, more boldly examined the process of nature itself. From their perspective, this could only be accomplished by repeated observation *and* repeated intervention. While the former was a skill cultivated by naturalists from the earliest time, the latter belonged more precisely to the experimental culture developed in the seventeenth century. Through the use of instruments such as the microscope, the air pump, and the numerous machines populating Baroque museums, naturalists refined the meaning of what it meant to experience nature. Such artificial interventions produced new forms of intimacy, as in the case of the microscope, and a new understanding of distance, as naturalists transformed the distinguished visitors to their museums into experimental witnesses, clarifying the meaning of demonstration in a scientific context.[9]

In his work on the Royal Society, Steven Shapin has asked where "the place of experiment" was.[10] The appearance of museums, laboratories, botanical gardens, and anatomy theaters in early modern Europe played a significant part in the transposition of knowledge from the discursive to the visual arena. All of these structures, which ultimately contributed to the dislodgement of *scientia* from its textual setting, shared the display and containment of knowledge as their common goal. As Andrew Wear perceptively notes in a discussion of William Harvey's method, "personal knowledge of Nature had to replace public knowledge embodied in books."[11] Thus, the formation of a museum was one way by which the subject of science, natural philosophy, could contain its object, nature. Knowledge, formerly embedded in texts, was created by a community of collectors, experimenters, and visitors whose viewing of nature established its authoritative image. Infusing classical prescriptions about the importance of observing nature with new

[8]BUB, *Aldrovandi*, ms. 6, Vol. I, c. 40r; BMV, *Archivio Morelliano*, ms. 103 (= *Marciana* 12609), f. 29.

[9]For a parallel study of the same subject, see Steven Shapin and Simon Schaffer, *Leviathan and the Air-Pump: Hobbes, Boyle and the Experimental Life* (Princeton, 1985).

[10]Shapin, "The House of Experiment," p. 373. See also Hannaway, "Laboratory Design and the Aim of Science," and Giovanna Ferrari, "Public Anatomy Lessons and the Carnival: The Anatomy Theatre of Bologna," *Past and Present* 117 (1987): 50–106.

[11]Andrew Wear, "William Harvey and the 'Way' of Anatomists," *History of Science* 21 (1983): 234. As Owen Hannaway observes, "science no longer was simply a *kind of knowledge* (one possessed *scientia*); it increasingly became a *form of activity* (one did science)." Hannaway, "Laboratory Design and the Aim of Science," p. 586. For the general intellectual parameters of this transition, see Ong, "System, Space and Intellect in Renaissance Symbolism," p. 69ff.

attitudes toward the descriptive potential of natural history, collectors dissected, compared, and described specimens, creating a demonstrative and experimental scientific culture. While this culture was deeply imbedded in the authoritative past—note Redi's respect for "dead counselors" even as late as the 1670s—it nonetheless gradually undermined, though never dissolved, the status of "weighted words" through the desire to establish their meaning artifactually. By the mid-seventeenth century, even disciples of Kircher such as Bonanni and Lana Terzi insisted that knowledge be authenticated in experiment *and* authority, an indication of the extent to which this mode of inquiry had penetrated all corners of the learned world.

How did the museums of late Renaissance and Baroque Italy compare with other "laboratory" spaces? In *Leviathan and the Air-Pump*, Steven Shapin and Simon Schaffer define experimental space in Restoration England as follows:

> The laboratory was therefore a disciplined space, where experimental discursive, and social practices were collectively controlled by competent members. In these respects, the experimental laboratory was a better space in which to generate authentic knowledge than the space outside it in which simple observations of nature could be made.[12]

The precisely articulated experimental etiquette circumscribing the laboratories of the Royal Society appears in marked contrast to the more fluid parameters of the museum, whose creators openly delighted in its ambiguities.[13] Quite often their commitments to other communities—the academic culture of the university, the social world of the courts, and the competing intellectual styles of the new scientific academies and Jesuit colleges—led them to contravene each other since these allegiances defined the different purposes of their museums. Though self-conscious about the novelty of their enterprise, collectors were less inclined to articulate rules of inclusion and exclusion than members of an academy, simply because they were never entirely sure who belonged to their community or what the criteria for membership was, beyond a certain degree of education and civility.

Yet despite their heterodoxy, common goals emerged that distinguished the collectors of nature as a group. While collectors disagreed strongly about the precise image of nature that their museums produced, all concurred that assembling and displaying nature was integral to the knowledge-making process (and assumed that scholars and gentlemen of worth were best prepared to perform these acts). Aldrovandi certainly saw his museum as the key to his ability to shape a more accurate natural history than any physician or natural philosopher who had preceded him, and other collectors followed suit. Their confidence was predicated not only on the encyclopedic supposition that more was better but also on the idea that a museum made nature uniquely proximate; like the telescope and the mi-

[12]Shapin and Schaffer, *Leviathan and the Air-Pump*, p. 36.
[13]I am indebted to Bill Eamon for clarifying this point for me.

croscope, it was an instrument that enhanced the senses. At the same time, collecting gave the community of naturalists a distinct social identity, one that was hardly as precise as that provided by membership in a society, but nonetheless viable and articulated. Unable to completely limit access to the museum, collectors were nonetheless highly conscious of whom they admitted and what they displayed.

Drawing upon the penchant for public display in early modern culture, collectors made the spectacle a technique of investigation. In search of patrons, naturalists fashioned their demonstrations in imitation of court spectacles, hoping to attract the attention of powerful princes and nobles whose support enhanced the prestige and importance of a museum. Naturalists emphasized the visual nature of their work in order to ennoble it. Experiencing nature became part of the aesthetic production of the late Renaissance and Baroque court, where knowledge was simply another form of display; concommitantly it also drew upon an urban and market culture that placed spectacle, in all its forms, at its public center.[14] In the museum, just as in many other scientific settings, credibility was a problem of social worth as well as philosophical acuity.

This chapter focuses on the role of the museum in experiential and experimental culture. "Experience" was the product of collecting nature; "experiment" the result of undoing nature in the laboratory, exposing her structures and replicating her operations. Such activities placed the museum at the heart of some of the most important scientific debates of the period. Within its walls, decisions were made about the nature of fossils, the effects of magnetism, the ability of living forms to generate spontaneously, and a variety of other paradoxes that confounded the scientific community. In the hands of naturalists, museums became *sites of competing meaning* as the objects in their possession were used to materially ground their philosophies of nature.

THE COLLECTING OF EXPERIENCE

In his *studio* in sixteenth-century Bologna, Antonio Giganti possessed several optical instruments "in order to experiment" (*per far l'esperienza*).[15] While we have no record of exactly what Giganti showed visitors when they viewed his collection, or whether he used these optical instruments for his own pleasure and edification, the experiments he performed, possibly with the help of his neighbor Aldrovandi, prefigured the more spectacular demonstrations practiced by Redi, Kircher, and Settala a century later.

[14]For a general overview of theatricality in early modern Italy, see Burke, *Historical Anthropology of Early Modern Italy*. It is discussed more specifically in the scientific realm in Ferrari, "Public Anatomy Lessons and the Carnival."

[15]Ambr., ms. S.85 sup., cc. 242r, 243r, in Fragnito, *In museo e in villa*, pp. 182, 184. The phrase *far l'esperienza* literally translates as "to make an experience" or "to experiment."

Possibly he had read Della Porta's *Natural Magic* and attempted to construct some of the objects the Neapolitan magus described. Giganti's optical instruments produced no results that later philosophers would savor, nor ones that his contemporaries felt worthy of recording. Yet his decision to signal their demonstrative role in the writing of a simple inventory of his museum suggests how interwoven the cultures of collecting and experimenting had become by the late sixteenth century. Not only well-known naturalists such as Aldrovandi and Della Porta but even ecclesiastic servants such as Giganti felt obliged to show visitors a few experiments when they came to see his *studio*.

The experimental culture of the museum was at best a heterogeneous affair. In the space of the *studio*, the classical problemata of natural philosophy were put to the test through the collection and comparison of data and the replication of recorded experiments. This process encompassed a wide range of phenomena from the testing of the quotidian to the display of the miraculous powers of natural substances such as asbestos and phosphorus to the spectacular confrontations between poisons and antidotes. It also included debates about the proper description and classification of specimens, particular problematic phenomena such as fossils, zoophytes, coral, bezoar stones, unicorn's horns, and previously undescribed *naturalia* that did not readily fit into classical categories. In all instances, naturalists used artifacts to confirm, refute, or extend commonly held truths about nature. Ovidio Montalbani was probably representative of most collectors when he begged Kircher to send him "anything and everything that will provide me with something to philosophize about,"[16] underscoring the indiscriminate nature of his interest in the artifacts that were necessary to the writing of natural history.

What did a collector hope to gain from these exercises? In social terms, prestige and patronage; in philosophical terms, "experience." Experience was something few naturalists possessed and many hoped to achieve. Frequently invoked as the soundest method with which to arrive at a better understanding of nature, *experientia,* "a knowledge of singular and particular things,"[17] was the foundation of good natural philosophy. From the botanical garden in Padua, Anguillara wrote that the ideal naturalist was one "who possessed theory as well as practice, and has experienced many things."[18] As writers from Aldrovandi to Boccone reminded their readers, experience was the cornerstone of the new "sensory natural history" to which they all hoped to contribute.[19] It initially reinvigorated canonical views of nature and ultimately called them into question. In 1544, the Florentine philosopher

---

[16]PUG, *Kircher,* ms. 563 (IX), f. 230v (Bologna, 26 September 1663).
[17]This is how the surgeon Ambroise Paré, paraphrasing Aristotle's *Metaphysics* 981a: 1–5, defined it; *Les Oeuvres d'Ambroise Paré* (Paris, 1585), p. 1214.
[18]Anguillara, *Semplici,* pp. 14–15.
[19]BUB, ms. 21, Vol. IV, c. 36.

Benedetto Varchi observed the early appearance of *experientia* in Renaissance scientific discourse:

> Although the custom of modern philosophers is to always believe and never prove anything that one finds written in good authors, most particularly in Aristotle, there is no doubt that it would be more certain and delightful to do otherwise and descend every now and then to experience.[20]

Varchi particularly cited the lectures of Luca Ghini that he had heard as a student in Bologna as an example of this new approach to nature. Renaissance naturalists, while largely men of texts, gradually introduced *experientia* into their lectures on natural history. While Aldrovandi and other humanistically trained naturalists perceived their interest in experience to be a measure of their devotion to ancient authority, they set the stage for the more empirical, less textually based studies of nature by colleagues such as Calzolari and Imperato, whose humanist credentials were marginal but whose experience of nature was great. The interest of Renaissance naturalists in the meaning of experience contributed to a more critical inspection of this category in the seventeenth century.

The results of humanist curiosity can be seen in the work of naturalists such as Boccone, who could imagine no other method by which to proceed. "The most sensory Philosphers (*sensati Filosofi*) agree that one cannot come to an understanding of the properties of new species of plants without experiences," he wrote in his *Museum of Rare Plants* (1697). Boccone addressed his collected essays specifically to "botanists and students of experimental philosophy," suggesting his self-conscious cultivation of natural history as a component of the new philosophy.[21] While Renaissance naturalists begrudgingly allowed experience a foothold in the learned world as necessary to the revitalization of natural history, by the end of the seventeenth century, the leading naturalists advocated it as the primary means of establishing truth. What factors led naturalists to accord *experientia* such an central place in the study of nature within a century and a half?

Experience encompassed a wide range of activities in scientific discourse. In the language of medieval natural philosophy, *experientia* and *experimentum* described "tried and proven remedies"[22] that authenticated knowledge through proof. *Experientia* was a formal category describing a means of attaining knowledge, frequently advocated by Aristotle in his own studies of nature. In the words of Galileo, Aristotle "always put sensory experience

[20]Benedetto Varchi, *Questione sull'alchemia* (1544), Domenico Moreni, ed. (Florence, 1827), p. 34.

[21]Boccone, *Museo di piante rare della Sicilia, Malta, Corsica, Italia, Piemonte e Germania* (Venice, 1697), pp. 1–2.

[22]Luke Demaitre, "Theory and Practice in Medical Education at the University of Montpellier in the Thirteenth and Fourteenth Centuries," *Journal of the History of Medicine* 30 (1975): 114; Thorndike, *A History of Magic and Experimental Science*, Vol. VI, p. 21.

(*sensata esperienza*) before natural discourse," and this was a guiding principle for later naturalists who drew inspiration from the sagacities of the
Philosopher.[23] As Peter Dear convincingly points out in his study of Jesuit
mathematics, the meaning of experience changed dramatically from the
medieval to the early modern period. While the scholastic notion of experience emphasized experience as common knowledge that supported the
universal laws of nature without any necessary verification, early modern
philosophers defined experience as the specific description of the behavior and appearance of natural phenomena. Experience became *evidence,*
thereby taking an active role in the construction of axiomatic principles.[24]
In many respects, the experimental culture of collecting fulfilled the heightened desire for knowledge of particulars that naturalists since Aristotle had
coveted. While the repository of the Royal Society became a locus for the
production and containment of Baconian "facts," the museums of early
modern Italy exhibited an abundance of Aristotelian data.[25] Both philosophical frameworks lent great significance to these nuggets of wisdom and
the knowledge produced from their manipulation, reinforcing the importance of collecting as a foundational activity for natural history.

Experientia encompassed not only "experience" but "experiment." Studying the language of books of secrets, William Eamon has distinguished "*experience* as a guide to scientific discovery and *experiment* as the test of truth."[26]
Yet the linguistic ambiguity of these two words should alert us to their pro-
cedural ambiguity. In a world of mutable categories, an *experientia* was never
a fixed entity, never a product of commonly agreed upon procedures. Instead, it conveniently summarized a multitude of different practices unified
by a certain engagement with objects and a heightened sense of audience.
An "experience" could just as well occur in the piazza as in the museum.
"For many years there was a charlatan, also a viper-catcher, who made sev-

[23]These words are put in the mouth of Salviati in the *Dialogi,* in Neil W. Gilbert, *Renaissance
Concepts of Method* (New York, 1960), p. 230. For a broader discussion of Galileo's use of this
term, see ibid., p. 230ff, and Gabriele Baroncini, "Sulla Galileiana 'esperienza sensata,' " *Studi
secenteschi* 25 (1984): 147–172. The classic study of this subject remains Charles Schmitt, "Experience and Experiment: A Comparison of Zabarella's View with Galileo's in *De Motu,*" in his
Studies in Renaissance Philosophy and Science (London, 1981), VIII, pp. 80–138. The comparison
between Zabarella and Galileo is also pursued by Rossi in his "The Aristotelians and the 'Moderns,'" esp. pp. 19–20.

[24]Peter Dear, "Jesuit Mathematical Science and the Reconstitution of Experience in the
Early Seventeenth Century," *Studies in the History and Philosophy of Science* 18 (1987): 134. See
also idem, "Totius in verba," pp. 152–153. Note that the division between medieval and early
modern notions of experience was by no means absolute. Even Boccone, who used the term
esperienza repeatedly in his work in a Galilean sense, at one point defined it in Aristotelian
terms: "In nature many causes are unknown to us, and therefore it is necessary to content oneself with experiences alone"; *Museo di piante rare,* p. 12.

[25]For a broader discussion of Baconian "facts" and collecting, see Arnold, *Cabinets for the Curious;* Daston, "The Factual Sensibility," pp. 452–467; and idem, "Baconian Facts."

[26]Eamon, "Science and Popular Culture in Sixteenth-Century Italy," p. 484.

eral experiments with vipers . . . in Piazza del Duomo," wrote Settala to Redi from Milan.[27] While we know something of the experiences that collectors shared in their museums, we can never really be certain what they meant when they retrospectively labeled their actions "experimental." At best, we can imagine that the use of this label made them significant in the eyes of contemporaries, as examples of a way of philosophizing that suggested an intermixing of things with words, physical demonstrations with textual proofs, and a certain novelty of approach.

The new emphasis on experience gave sensory evidence a larger place in the predominantly textual world of learning. "One delights in practical natural philosophy, which one learns from ocular testimony," wrote Gabriele Falloppia to Alfonso II d'Este in Ferrara in 1560.[28] Naturalists activated this sort of experiential curiosity by testing a vast array of phenomena in front of a learned and appreciative audience. "Tomorrow I place cold acqua vitae and coriander in the fire," reported Anguillara to the Duke of Ferrara from his laboratory at the botanical garden in Padua.[29] Smells, sensations, and sights all found a greater place in the description of natural phenomena, and naturalists began to pay closer attention to these aspects of classical descriptions of plants, animals, and minerals to identify and classify them better. Olivi listed the smells of the medicinal simples that Calzolari collected in his museum, praising the apothecary for his decision to include sensory information in his descriptions.[30] Imperato also highlighted the smell, taste, and feel of various waters in his *Natural History*, reinforcing the importance of physical contact with every phenomenon as a form of essential knowledge.[31]

A visitor entered a natural history museum with his senses alert, registering its sights, sounds, smells, and tactile sensations as part of the *experientia* that could be gained there. The skills cultivated during botanical excursions became a sort of virtuoso display in the museum. During a trip to Rome, for example, Anguillara identified a true example of *aspalatho*, a valuable ingredient for antidotes, by "reading" nature directly in the home of another naturalist:

> . . . finding myself at the house of the most excellent physician Messer Ioseppe Cincio in Rome in 1545, in the company of the most excellent and learned Cesare Odone, I came across a piece of wood on a table in his *studio* that . . . smelled like a rose but without any bitterness, from which I quickly determined that it was *aspalatho* and communicated this with many.[32]

[27]Laur., *Redi* 222, c. 244r (Milan, 4 January 1671).
[28]In Pericle di Pietro, "Epistolario di Gabriele Falloppia," *Quaderni di storia della scienze e della medicina* 10 (1970): 53 (Padua, 15 November 1560).
[29]BEst., *Est. It.*, ms. 833 (Alpha G I, 15) (Padua, 9 October 1561).
[30]Olivi, *De reconditis et praecipuis collectaneis*, sig. +4v, p. 13.
[31]Imperato, *Dell'historia naturale*, pp. 166–176.
[32]Anguillara, *Semplici*, pp. 37–38.

To test a sample of the plant *antitora,* in one form an antidote for plague and snakebites, in another form a poison, Aldrovandi "tasted the tip of the root." In privileging his personal experience of an object, he was not unlike his contemporary Fioravanti who also claimed to try all his miraculous remedies on himself.[33] Through sharpened faculties such as Falloppia's ocular sensibilities, Anguillara's acute sense of smell, and Aldrovandi's discriminating palate, Renaissance naturalists refined their histories of nature in a manner sanctioned by classical authorities. "I call this sensory [philosophy] the mother of universal philosophy, from which it derived its origins," explained Aldrovandi, paraphrasing Aristotle. "If this particular is taken away, the universal does not remain, since memories are born from sensory experiences, and universals from memories . . . for there is nothing in the intellect that is not first in the senses."[34] Through his collecting of experience, Aldrovandi became known as "a true sensory Philosopher."[35] Nature, as he and his contemporaries learned, could be read as easily through sensory data as through the pages of a book. This was the experience that naturalists most coveted, as they translated textual images into experiential practices.

While Renaissance naturalists gave equal weight to all sensory information, their followers prioritized the role of sight in the study of nature.[36] The Linceans gradually emphasized the visual characteristics of objects over other descriptors as the microscope played a larger role in their anatomical studies. Like their Baroque successors Kircher, Redi, and Boccone, they offered little discussion of taste, feel, and smell. Even when describing the testing of venom in his *Observations on Vipers* (1664), Redi presented it more as a visual spectacle than as a tactile experience in the manner of Aldrovandi. He did not test poisons on himself but watched his viper catcher Jacopo Sozzi, who clearly was *not* a natural philosopher, engage in this sort of perilous experiment.[37] The emphasis on sight reflected the increased entry of natural history into court culture. Demonstrations became less the personal revelations that complete intimacy with an object yielded and more a technique of persuasion for an audience distant from the phenomenon itself.

[33]De Toni, *Spigolature aldrovandiane III,* p. 10 (Bologna, 22 November 1576). See, for example, Leonardo Fioravanti, *De' capricci medicinali* (Venice, 1573), p. 156v.

[34]BUB, *Aldrovandi,* ms. 21, Vol. IV, c. 72, in Aldo Andreoli, "Un inedito breve di Gregorio XIII a Ulisse Aldrovandi," *Atti e memorie dell'Accademia nazionale di scienze, lettere e arti di Modena,* ser. 6, 4 (1962): 138. This passage is a post-Aristotelian rendering of material found in *De anima* 432a7 and generally in *De sensu et sensatum.* It also appears in the writings of William Harvey; see Wear, "William Harvey and the 'Way of Anatomists,' " p. 237.

[35]BUB, *Aldrovandi,* ms. 25, Vol. XV, c. 308v (Della Nuntiata, 8 April 1574).

[36]Ezio Raimondi, "La nuova scienza e la visione degli oggetti," *Lettere italiane* 21 (1969): 265ff. Raimondi draws upon the writings of Lucien Febvre in making this distinction. Both suggest that Renaissance naturalists gave more weight to smell and sound than to sight. Instead, I would put all sensations on relatively equal footing.

[37]Tribby, "Cooking (with) Clio and Cleo," pp. 438–439.

Calzolari and Imperato may have tasted and smelled every object they possessed in their capacity as apothecaries; Kircher, Redi, and Settala better understood how to turn artifacts into spectacles that filled the leisure hours of princes and virtuosi. Situated within a scientific culture that perceived experiments as spectacles rather than demonstrations, they imagined a group of spectators who were less impressed with the general *experientia* that the possession of artifacts entailed and more interested in the elaborate "experiences" that could be generated from each item.

The origins of experimental culture lay in the emergence of experience as a form of authority. In the 1610s, the Linceans, heavily influenced by Imperato's natural history and attracted to the new philosophy developed by Galileo, attempted to reverse the traditional relationship between experience and authority. "And therefore we hold that one ought to believe more in the observation of natural things than in imagined objects and suppositions, derived from one sole observable principle without the means and end of the thing itself, from which one can construct a rule," wrote Colonna in 1618:

> The observable thing perfectly gives a method. It is not method that makes an object conform to its presuppositions. The nature of things cannot mold itself to the caprice of man, imagining how he thinks; rather man should mold his caprice according to natural things exactly observed. From this a Method can emerge.[38]

The Linceans contributed to the erosion of textual authority by suggesting that scientific method was found in nature. Through such maneuvers, they attempted to strip objects of the philosophical accoutrements that gave them significance in the world of Aristotelian and Plinian natural history, a technique that later naturalists would use to great effect. For the Linceans, artifacts did not recall the language of traditional learning but trumpeted the sounds of the new philosophy, contributing to experimental knowledge rather than Aristotelian *experientia*. The establishment of new techniques to generate knowledge from the experience of nature as an essential step in this formation of this cultural divide.

In the 1570s, when Aldrovandi suggested that the presentation of artifacts in museums "placed [philosophy] clearly in front of the eyes," he alluded to the explicit relationship between seeing and philosophizing that used objects to *reinforce* rather than negate the scientific canon.[39] By the 1610s, naturalists such as Colonna criticized this widely accepted Renaissance practice, arguing that such techniques engendered "imagined objects" rather than revealing the actual material contours of the natural world. It was left to followers of *both* styles of philosophizing—the Jesuits who upheld the

---

[38]Fabio Colonna, *La sambuca lincea* (Naples, 1618), in Giuseppe Olmi, "La colonia lincea di Napoli," in Lomonaco and Torrini, *Galileo e Napoli*, p. 54.

[39]Aldrovandi, *Discorso naturale*, p. 193.

Aristotelian tradition and the Tuscan court naturalists who championed the place of natural history in the new experimental philosophy—to continue the debates about the use of objects in the writing of natural history. Ultimately, the former mode of inquiry gave way to the latter, but until the beginning of the eighteenth century, it was by no means clear that this would come to pass. In the interim, naturalists challenged one another's claims to knowledge by invoking their common possession of experience.

ANATOMIES IN THE MUSEUM

While naturalists debated the significance of experience, they also refined their conception of the setting in which it became meaningful. When Giulio Camillo suggested that scholars should become spectators, he acknowledged the important relationship between the reform of knowledge and the redefinition of scientific practice that made the museum a theatrical site of inquiry. Camillo's imagined transformation of humanist pedagogues was borne out by the sociable practices of collectors who made every object visually pleasurable. They achieved this aesthetic by manipulating artifacts, making knowledge itself a form of display. Dissecting and testing phenomena were the two primary means of creating this effect. The former revealed the secrets hidden within each object, while the latter measured it against a set of canonical truths that gave it philosophical definition. Both techniques demanded the presence of spectators to affirm their validity.

The experimental culture of the museum was a product of the mutual expectations of collectors and spectators. Objects in a collection became the focal point for a series of discursive exercises that defined the culture of reading and observing. The humanist understanding of experiment took it to be a dialogue with the past, a product of historical time.[40] This attitude conditioned numerous tests of natural phenomena, from Aldrovandi's, Imperato's, and Redi's experiments with vipers to Kircher's and Settala's attempts to recreate the famous experiments of Archimedes with burning mirrors. Aristotle, for example, had declared that bears gave birth to a lump of flesh and then licked it into the shape of a cub. By the seventeenth century, one could find the fetus of a cub displayed in the *Studio Aldrovandi* in contradiction of this belief. "In the Museum of the Senate of Bologna a bear cub is preserved, enclosed within a jar, which has been extracted from the womb, but no part appears imperfect," wrote Gottfried Voigt in his *Physical Delights* (1671).[41] Lining the shelves of most museums were hundreds of objects which, like the bear cub whose presence demolished classical myth, directly responding to ancient scientific beliefs. They existed to do battle with

[40]This is made particularly evident in the recent work of Jay Tribby, "Cooking (with) Cleo and Clio."

[41]Gottfried Voigt, *Deliciae physicae* (Rostock, 1671), p. 143.

their textual representations, challenging the products of historical time to confront the material realities of the present. In 1633, Peiresc informed Cassiano dal Pozzo that he had dissected a chameleon to prove, contrary to Pliny, that it had teeth and therefore did *not* survive on air alone.[42] Next to these discredited objects lay a number of equally contentious items that continued to support ancient truths, for example, the remains of a siren, examples of unicorn's horn, or the popular *remora*. Whether unmasking or upholding the scientific canon, naturalists felt increasingly obliged to demonstrate what they were already certain they knew.

Anatomical demonstrations were an important part of the experimental life of the museum, attracting the same people who crowded into the anatomy theaters in search of a scientific spectacle. While dissecting had been widely used in medical teaching since the late thirteenth century, the appearance of anatomy theaters and the elevation of anatomy as a discipline made it an increasingly popular technique of display. By the mid-sixteenth century, wherever learned patricians who professed an interest in nature converged, a dissection was sure to occur. In 1548, Benedetto Varchi compared the appreciation of anatomy to the study of art. Both exhibited the remarkable artifice to which men of culture were innately drawn. Addressing his lectures on human anatomy to "the most noble Academicians and all you most virtuous Listeners" gathered in the garden of Palla Rucellai, Varchi identified the learned audience that appreciated anatomy as a cultural aesthetic.[43] Congregated in the famous *Orti Oricellari*, where numerous political conversations among the leading Florentine patricians had taken place, the physicians, painters, and virtuosi who listened to Varchi's lecture on two local monsters participated in the process of making anatomy a suitable subject for civil conversation and therefore a necessary part of the scientific culture of the museum.

The collective decision of a group of Florentine patricians to dissect two monsters in the garden of one of the city's leading families serves as background to the anatomical demonstrations that occurred in museums with increasing regularity. Two different models of learning reinforced the importance of dissecting as a form of "experience." Within the context of the university, professors who were accustomed to giving private lessons in their homes continued to perform private dissections for students. The anatomies *in studio* enhanced the pedagogical function of collecting. Lessons begun in public anatomy theaters and botanical gardens were often refined, if not

---

[42]Peiresc, *Lettres à Cassiano dal Pozzo*, n. 6, p. 127.

[43]Varchi, *La prima parte delle lezzioni di M. Benedetto Varchi nella quale si tratta della generazione del corpo humano, e de' mostri*, "Proemio," n.p., and *Lezzione di M. Benedetto Varchi, sopra la generazione de' mostri . . . fatta da lui publicamente nell'Accademia Fiorentina la prima, & seconda domenica di luglio, l'anno 1548* (Florence, 1560), p. 94r. For a more detailed discussion of this text, see Hanafi, *Matters of Monstrosity in the Seicento*, ch. 2.

concluded, in the museum. The German naturalist Volcher Coiter recalled the anatomies he had observed in Aldrovandi's house as a student in Bologna.[44] The didactic model of display underscored the principle of limited access. When Gassendi wrote that Peiresc frequented the private as well as the public anatomies of Fabricius in Padua. "who out of the singular good will he bore to *Peireskius*, did admit him to be present," he signaled the privilege accorded to the young French law student whose knowledge of nature was sufficient to merit entry into Fabricius's *studio*. Private dissections, such as the ones performed by Fabricius in front of Peiresc in the 1590s, created new observations that provided material for new natural histories. Therefore, only *legitimately knowledgeable* observers—physicians, surgeons, naturalists, and men of high social standing and great learning—could contribute to the meaning of the dissection, as it subsequently was recorded.[45] Their collective participation in the anatomy reinforced the connections between humanist sodality and experiential curiosity.

The second model of display drew upon the social conventions outlined in Varchi's Sunday afternoon lessons for his fellow academicians and their illustrious visitors. In this context, the act of dissection contributed less to philosophical *scientia* than to social knowledge. The public dissections performed in museums, like the anatomies undertaken in the university dissecting theaters during Carnival in many Italian cities, catered to a more heterogeneous audience.[46] The most important members of the local government, distinguished visitors, courtiers, virtuosi, and occasionally women might be present, all underscoring the ceremonial importance of dissecting. For signal events such as the anatomy of the papal serpent or the opening of the vipers that entered into Aldrovandi's theriac, both the Archbishop of Bologna and the current *gonfaloniere* were present.[47] The appearance of local notables in the museum validated the form of knowledge that it represented by making dissecting and observing socially acceptable pursuits. Anatomy, as Aldrovandi wrote in 1592, was a spectacle worthy of a "learned and discerning eye."[48] Varchi's decision to give two anatomical lessons to the Florentine patriciate conformed well to the general climate in which displaying nature was fast becoming a popular pastime for the upper classes. Humanists and university-trained scholars provided the erudition that transformed good observations into good natural philosophy; courtiers, their senses sharpened by their charged social environment, supplied the discernment that made dissecting a tasteful enterprise.

[44]Olmi, "Molti amici in varii luoghi," n. 77, p. 28.

[45]For a similar discussion of "competent observers" within the Royal Society, see Shapin and Schaffer, *Leviathan and the Air-Pump*, esp. pp. 55–60.

[46]Ferrari, "Public Anatomy Lessons and the Carnival."

[47]See BUB, *Aldrovandi*, ms. 21, Vol. III, c. 167r.

[48]In Olmi, "Molti amici in varii luoghi," p. 29.

Of all the sixteenth-century naturalists, Aldrovandi was the most articulate about the protocols surrounding dissection and the knowledge he wished to gain from it. His museum was not only a *hortus siccus* but also an anatomical theater and chemical laboratory in which nature's interior and exterior were scrutinized. The active role of objects in the process of reading nature was reiterated constantly by Aldrovandi in his writings on natural history:

> Who wishes to judge these natural things, beyond theory, must have practice, not only in the description of the exterior parts, but also in the particular anatomy of plants and animals, as is evident in several birds [described] in my *Ornithology*, now in press.[49]

Elsewhere Aldrovandi affirmed his commitment to the dissection and observation of all specimens that came under his roof. "This is the true philosophy," he continued, aptly summarizing the Aristotelian image of *experientia*, "knowing openly the generation, temperature, nature and faculty of each thing by means of experience."[50]

For Aldrovandi, each specimen had the potential to produce multiple forms of knowledge. While living, he could observe its habits and illustrate its movements; dead, he could understand its generation, articulate its skeletal structure, and preserve it for all to see. From nature he gained insight into the processes he studied; from books he learned the questions to ask of every artifact he encountered. Aldrovandi did not just collect vipers but also employed the surgeon Gaspare Tagliacozzi to dissect them in front of the local medical establishment. These investigations not only proved his views about the proper time to kill them for insertion in the Bolognese theriac but also demonstrated that vipers possessed sexual organs "and therefore did not conceive by the female's biting off the head of the male."[51] Few living creatures that entered Aldrovandi's museum remained intact. Under his watchful eye, his assistant Antonio Ulmo dissected several impregnated hens to observe the descent of the egg from the septum into the uterus "for the common benefit of students." Uncertain about whether Aristotle or Galen was right as to which organ—the heart or the liver—emerged first in generation, Aldrovandi personally opened twenty-two incubated eggs to observe chick embryos, ultimately siding with Aristotle who championed the heart.[52] When a captured eagle was brought to his museum, he observed it for eight days before dissecting it. Perhaps the only animal in his possession

---

[49]BUB, *Aldrovandi*, ms. 70, c. 8r.

[50]BUB, *Aldrovandi*, ms. 21, Vol. IV, in "Un inedito breve di Gregorio XIII," p. 138.

[51]Aldrovandi, *Ornithologiae* (1599), Vol. I, p. 27, in Thorndike, *A History of Magic and Experimental Science*, Vol. VI, p. 281. For more on Tagliacozzi's role in the theriac disputes, see chapter 6, present volume.

[52]*Aldrovandi on Chickens*, pp. 76, 91; Alessandro Simili, "Spigolature mediche fra gli inediti aldrovandiani," *Archiginnasio* 63–65 (1968–1970): 35.

spared these attentions was a pet monkey. Instead, Aldrovandi tested numerous wines on it to see what effect they would have; the result was a simian alcoholic whose antics no doubt greatly amused his guests.[53]

While ancient and contemporary histories of nature provided Aldrovandi with a philosophical framework for his anatomies, the more popular "books of secrets" gave him precise instructions for experiments to perform on phenomena. The number of annotations and underlined passages in Aldrovandi's personal copies of various books of secrets, with particular reference to the sort of exotic ingredients that collectors prized as well as the ever-present plague remedies, demonstrates the importance of this sort of medical literature to his experimental work. "I, Ulisse Aldrovandi, finished reading the entire [book] on the 28th day of November 1568," noted the naturalist at the bottom of his copy of the *Secrets of Alexis*. He gave the same attention to the popular medical works of his fellow Bolognese, Leonardo Fioravanti.[54] It is not hard to imagine Aldrovandi, with one of Fioravanti's numerous treatises in hand, testing plague antidotes, potions guaranteed to ensure longevity, and numerous other recipes that used the exotic ingredients he carefully amassed. What other audience could such literature have hoped to reach that would have the understanding, the interest, and most importantly, the ingredients to activate each recipe, removing it from the printed page to the laboratory? In the museums of Aldrovandi and his contemporaries, textbook *experimenta* and spectacular *experientia* converged under the rubric of "secrets." In the hierarchy of operations, however, anatomies preceded all other forms of display since they provided the surest method of penetrating the structure of nature that Aldrovandi, as a committed Aristotelian, wished to reveal.

In 1592, Alfonso Cataneo sent Aldrovandi a present of three birds, with the request that his illustrious friend "share that which you will have observed by your divine genius in dissecting these birds." Aldrovandi responded with a letter detailing the anatomy of a swan. Giovan Battista Cortesi, Tagliacozzi's student, performed the dissection; two other university professors, the philosopher Federico Pendasio and Girolamo Mercuriale, then professor of theoretical medicine at Bologna, witnessed it. Aldrovandi scrupulously followed the procedure he had outlined in his *Natural Discourse*, immediately making an "external description" and subsequently inviting Cortesi to dissect the swan in front of the three professors that very evening. "We observed every part . . . that neither by Aristotle . . .

[53]On the eagle dissection, see Thorndike, *A History of Magic and Experimental Science*, Vol. VI, p. 269. On the monkey, Ricc., Cod. 2438, Pt. I, lett. 91 (Bologna, 27 June 1587).

[54]See the BUB's copy of *De' secreti del Reverendo Donno Alessio Piemontese* (Venice, 1564–1568), Pt. I, p. 157r. For more on Fioravanti, see Eamon, " 'With the Rules of Life and an Enema': Leonardo Fioravanti's Medical Primitivism," in *Renaissance and Revolution: Humanists, Scholars, Craftsmen and Natural Philosophers in Early Modern Europe*, J. V. Field and Frank A. J. L. James, eds. (Cambridge, U.K., in press).

nor by other writers has been memorialized." Aldrovandi particularly lingered over the trachea, whose "artifice" was so "miraculous and worthy of spectacle" that Pendasio and Mercuriale discoursed spontaneously upon it. The naturalist concluded by assuring Cataneo that an image and full description of the swan would appear in his *Ornithology*.[55]

Several features mark this dissection, typical of the procedures enacted on museum artifacts. First, Aldrovandi chose not to dissect the object himself but relied upon the skills of a surgeon. Tagliacozzi performed this service for him in the 1570s; the famous surgeon's disciples continued their association with Aldrovandi's museum into the 1590s. Like many naturalists, Aldrovandi often claimed the experience that more specialized hands placed before him. Cortesi could open the swan, but only he and other natural philosophers could discourse upon it. Rather than eroding the division between *ars* and *scientia*, the anatomies in the museum reinforced the perceived differences between these two skills, while allowing the collector to claim both as attributes. Aldrovandi's role as the facilitator of anatomical spectacle allowed him to take credit for surpassing Aristotle in the knowledge produced from the viewing of the swan. His colleagues, both university philosophers, could bear witness to what he had seen, lending authority to any report he would publish. Lastly, the descriptive language used by Aldrovandi in detailing the anatomy of the swan signified its cultural worth: ingeniously crafted, it was an object worthy of its spectators (*degno di spettacolo*). Brought to Bologna from a court known for its exotic birds—earlier Cataneo had supplied Aldrovandi with information from the Estense falconers—the swan ennobled the museum through its presence.

The anatomical studies performed within the homes of university professors such as Aldrovandi and Fabricius, where learned men gathered to endorse the Aristotelian idea of *experientia*, paved the way for more sociable uses of dissection. Led by the Linceans, anatomizing nature soon became the passion of the Roman nobility. Johann Faber kept a special table in his *studio* in Rome "to hold the said Anatomies at home."[56] Dissections were not only for the benefit of his medical students at *La Sapienza* but also for patrons and fellow academicians. Faber described a demonstration that he performed on a specimen from Francesco Barberini's museum in 1624, at the cardinal's request:

> Signor Cardinal Barberini did me the honor of requesting that I perform a small anatomical study of a monster that had been presented to him, that is, a calf with two heads, which I did in my house in the space of four hours in the presence of my scholars two days ago, observing everything and making illustrations.[57]

[55] The entire letter is reproduced in Olmi, "Molti amici in varii luoghi," pp. 29–30.

[56] BANL, *Archivio di S. Maria in Aquiro*, ms. 412, f. 84v.

[57] CL, Vol. II, pp. 869–870 (Rome, 20 April 1624).

As Faber's description of the event indicates, by the 1620s, the sort of anatomical demonstration performed in Palla Rucellai's garden eighty years earlier had found its way into a cardinąl's museum. Both were key settings controlled, respectively, by the leading cultural brokers of Renaissance Florence and Baroque Rome. Their association with the act of anatomizing nature marks the permutations of this scientific spectacle, as it traveled among the gardens, museums, and other sociable spaces of the Italian elite. Faber's willingness to perform the demonstration in Barberini's museum rather than in his personal *studio* indicates the extent to which the Lincean embedded the act of dissection within the social hierarchies of the papal court, where all knowledge was funneled through the papal nephew, an extension of Urban VIII. Years later, when a dragon appeared in Rome in 1660, Kircher hurried to the prefect of Barberini's museum, Hieronymo Lancia, to obtain permission to examine it.[58] At that point, Barberini was no longer the papal nephew, but he continued to be a powerful figure in Rome, a patron for whom naturalists dissected.

Even as Faber rushed to Barberini's museum to establish the anatomical particulars of his two-headed calf, his fellow Lincean Galileo was crafting a new instrument that would transform the meaning of dissection. In September 1624, Galileo presented Cesi with a "little eyeglass (*occhialino*) in order to see little things upclose." By May, Faber was recounting to Cesi the wonder he had experienced upon viewing a fly under Galileo's "beautiful little eyeglass." A year later, he coined the term "microscope" to contrast its effects with the other Lincean instrument that Della Porta and Galileo had made famous.[59] In the remaining years of the academy's existence, the Linceans completely reconfigured the procedures they used to examine objects and expanded the scope of the artifacts they deemed worthy of investigation. The invention of the microscope in the same year that Maffeo Barberini was elected pope seemed to be yet another "marvelous conjuncture" that punctuated the lives of aspiring virtuosi. Galileo's latest instrument allowed the Linceans to describe the anatomy of the bee, the Barberini family emblem, more thoroughly than anyone had done before.

Cesi, Stelluti, and Faber immediately began to observe Roman bees and encouraged Colonna to compare them to the Neapolitan variety. The results of their work appeared in Cesi's *Apiarium* and in a five-page footnote to Stelluti's translation of Persius's *Satires,* published in 1630. As Stelluti wrote, "With the Microscope I minutely observed all the parts of the Bee, whose form I esteemed well enough to represent here because it is worthy of knowledge and [deserves] to be seen by everyone" (fig. 17). While Aldrovandi publicized his anatomical skills as a central feature of his image as the new Aristotle, Stelluti proclaimed that "neither Aristotle nor any

[58]Kircher, MS, Vol. II, p. 90.
[59]CL, Vol. II, p. 924 (Bellosguardo, 23 September 1624); Kidwell, *The Accademia dei Lincei and the "Apiarium,"* pp. 82–84.

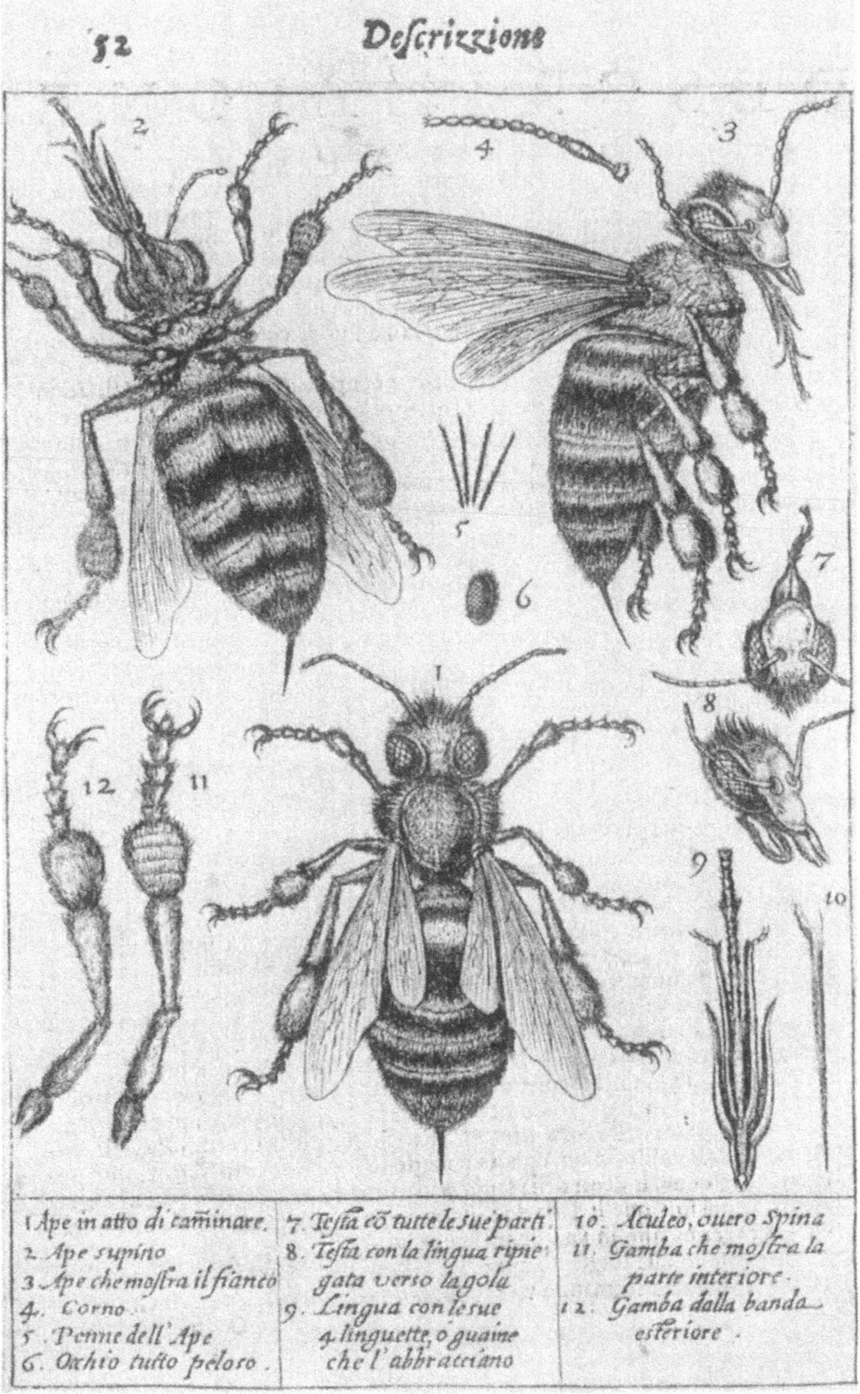

Figure 17. The Lincean bee. From Francesco Stelluti, *Persio tradotto* (Rome, 1630).

other philosopher and ancient or modern naturalist has ever observed or known [these things]."[60] Like Cataneo's swan, the bee was an object of anatomical worth whose significance extended beyond the museum in which it was preserved, intersecting with the political culture of the papal court. Urban VIII's pleasure at Cesi's *Apiarium* deemed their new technique of investigation a success. Enthusiasts such as Stelluti and Colonna soon moved on to other creatures—dust mites, flies "with eyelids larger than ours," and so forth.[61] Microscopic observations fueled the imaginative possibilities of natural history in ways that dissecting alone no longer seemed to achieve. In the 1620s, the microscope offered a means of laying claim to the novelty that always seemed to elude the grasp of the Lincean naturalists since it produced "experiences" that no ancient and few moderns could claim.

At the same time, one of the more well-traveled academicians, Cassiano dal Pozzo, did not neglect the larger, equally unstudied objects that added to the luster of their edition of Hernandez's "Mexican Treasure." Throughout the 1630 and 1640s, he produced anatomies of flamingos and other exotic creatures to delight his patron Francesco Barberini.[62] By the time of his death in 1657, Dal Pozzo was well known for his passion for dissecting and illustrating nature. As Carlo Dati, member of the Accademia del Cimento, wrote in his eulogy of the Roman collector, "and with a truly Lincean eye he wished to see anatomy because, in my opinion, besides chemistry nothing is more delightful than anatomizing nature."[63] Through their dissections, illustrations, and microscopic observations, the Linceans initiated the process of transforming anatomy from an Aristotelian exercise into an emblem of experimental philosophy.

Virtuosi, quick to imitate the latest fashion in Rome, soon spread the techniques used by the Linceans to other cities. By the 1660s, anatomizing "with a Lincean eye" infused the scientific culture of Florence, which competed with Rome as the center of Italian court culture. Microscopes had become a standard feature of all scientific investigations of nature and a staple in museums of natural history. As Dati observed, the Accademia del Cimento as a group prioritized chemical operations above anatomies, preferring to watch animals painfully expire in Torricelli's vacuum than to dissect them. One notable exception to this rule was Redi, who reinvigorated the culture of dissecting at the Tuscan court. While Kircher used the microscopes crafted by Jesuit instrument makers and displayed in the Roman Col-

---

[60]Francesco Stelluti, *Persio tradotto* (Rome, 1630), in Giuseppe Gabrieli, *Contributi alla storia della Accademia dei Lincei* (Rome, 1989), Vol. I, p. 354.

[61]Ibid., pp. 354, 361.

[62]Solinas, "Percorsi puteani: note naturalistiche ed inediti apppunti antiquari," in his *Cassiano dal Pozzo*, pp. 103–106.

[63]Dati, *Delle lodi del Commendatore Cassiano dal Pozzo*, n.p.

lege museum to detail the features of "subterranean animals," Redi drew upon the Grand Duke of Tuscany's instrument collection to demystify the smallest, most difficult parts of nature. Both belonged to a culture that made dissecting and experimenting paradigmatic ways of interacting with natural phenomena. For Kircher, the techniques of the new philosophy provided new proofs of ancient tenets. For Redi, they offered an opportunity to create a form of scientific knowledge in opposition to classical authority, allowing him to continue the program of investigation initiated by the Linceans.

By the mid-seventeenth century, anatomizing nature had become not only a preferred social skill but an example of courtly *sprezzatura,* the "nonchalance" favored by Castiglione and his followers.[64] As court physician and naturalist to Ferdinando II and Cosimo III, Redi studied nature amidst one of the most noteworthy gatherings of scientific virtuosi in Baroque Italy. In 1667, the year the Accademia del Cimento dissolved, Redi was only beginning to achieve the fame that made him one of the foremost naturalists of his day. He had recently published his *Observations on Vipers* and was in the process of completing his *Experiments on the Generation of Insects.* Both publications memorialized the culture of experimentation that privileged anatomy as a fashionable pursuit among Tuscan courtiers. While Aldrovandi confined his dissections to the space of the *studio,* Redi made the entire court his laboratory. Rather than inviting visitors to see his experiments by bringing them into the museum, Redi brought nature *to* his audience by performing demonstrations throughout the court—in the Medici pharmacy and laboratory, during the hunt, after dinner, in short, anywhere suitable to the leisurely perusal of nature. Anatomy was no longer a neo-Aristotelian prerogative but a courtly imperative. Microscope in hand, surrounded by eager assistants, Redi further refined the experimental etiquette that governed the study of nature.

With the help of his assistants, Redi decapitated 250 vipers to determine the origins of their venom, placed scorpions under the microscope to examine their stinger, and opened more than 20,000 oak-galls to see if trees actually gave birth to the animals contained within them. Most notably, Redi concerned himself with the paradoxes that arose, in his opinion, from either refusing to observe nature or allowing *a priori* assumptions to influence the act of seeing. Kircher had placed dead flies in honey water and had seen a new batch emerge spontaneously; Redi promptly collected his own specimens to counter that more flies appeared only when the dish was left uncovered. Kircher affirmed that oxen's dung produced caterpillar-like worms that became bees; Redi scattered samples of dung around the

---

[64]For more on anatomical nonchalance, see Biagioli, "Scientific Revolution, Social Bricolage and Etiquette," p. 19.

Tuscan court, "following Father Kircher's directions," to observe that indeed gnats and flies, but not bees, eventually swarmed around the exposed specimens, though never around the covered ones.[65] When Bonanni affirmed that snails did not have hearts and used his microscopes to prove this Aristotelian dictum, Redi scornfully replies, "To see the heart of terrestrial snails, one doesn't need to help one's sight with the microscope, nor is it necessary to sharpen one's eyelashes."[66] The technology of the new philosophy was inadequate unless accompanied by sound experimental method. This could only be produced in the courtly culture of repeated demonstration in which men of high social standing—most obviously the Grand Duke of Tuscany Ferdinando II and his brother Leopoldo—provided unimpeachable testimony.

As Redi constantly reminded his critics, anatomizing nature had become an experimental rather than a purely demonstrative science. Courtly rather than scholastic credentials determined the quality of witnesses as well as the status of the experimenter, making it impossible for a Jesuit pedagogue such as Bonanni to *ever* produce a better anatomy of a snail than the courtier Redi. Puzzling over the statements of naturalists such as Kircher and Bonanni who continued to uphold Aristotelian ideas about nature while claiming to experiment, Redi imagined that they had been "blinded by inexperience."[67] Describing them as "uncourtly" would be another way to put it. Both Redi and his Jesuit adversaries shaped their anatomical demonstrations and experiments to conform to the expectations of their local culture. Visitors anticipated Aristotelian spectacles when they entered the Roman College museum and anti-Aristotelian displays when they observed Redi at work in the Medici naturalistic collections. For an audience that perceived debate as the highest form of rhetorical enjoyment, the contrasts between these two modes of inquiry only heightened the pleasure of the spectacle.

Like Aldrovandi, Redi did not always perform the operations with his own hands. His standing at court demanded that he allow assistants to complete the more mundane aspects of dissection, intervening only when a display of virtuosity was required. Frequently, the viper catcher Sozzi performed more ordinary experiments. In 1681, Sozzi dissected (*dal taglio anatomico*) a number of vipers that had been labeled male to prove that they were female. Redi subsequently placed them under the Grand Duke's microscopes to confirm Sozzi's assertion, allowing the new technology to provide the final arbitration. That same year, Redi, Giovanni Alfonso Borelli, and two English visitors watched Tilmanno Trutuino dissect a poisoned dog.[68] When

[65]Kircher, MS, Vol. II, pp. 358, 360–361, 367–368; Redi, *Experiments on the Generation of Insects*, Mab Bigelow, trans. (Chicago, 1909), pp. 34, 43.

[66]Redi, *Osservazioni intorno agli animali viventi*, p. 58.

[67]Redi, *Experiments on the Generation of Insects*, p. 81.

[68]Wanda Bacchi, "Su alcune note sperimentali di Francesco Redi," *Annali dell'Istituto e Museo di Storia della Scienze di Firenze* 7 (1982): 52–54.

Redi chose to dissect something ignoble—for example, doormice and squir-
rels—he felt obliged to explain it as a necessity occasioned by excess leisure.
"I have taken up the pastime of dissecting them," he wrote to Jacopo dal
Lapo from the hunting grounds of Artimino in 1682.[69] The anatomical *sprezzatura* necessary for dissecting the more exotic animals in the Grand Duke's
zoo—for example, a bear presented to Redi after its death—was developed
only by practicing on less noble but anatomically useful creatures. By the
1670s, Redi was almost as well known as his court associate Steno as a naturalist who had restored integrity to the problematic act of seeing through
his constant dissection of nature. When visitors such as Francesco d'Andrea
came to Florence, they aspired to being allowed "the honor of seeing some
animal dissection by Signor Redi."[70]

The arrival of Steno in Tuscany in 1666 precipitated a flurry of dissections, as Redi and the Danish anatomist engaged in furious slaughter of
every living creature within reach of the court to impress the Grand Duke
and his courtiers with their mutual virtuosity. Steno's reputation as an
"anatomist of great fame" preceded him. The previous year he had amazed
scholars in Paris with his skillful dissection of a brain. For Steno, anatomizing nature was the primary means of advancing knowledge. His installation
at the Medici court confirmed the popularity of dissecting as a courtly
ethos. Criticizing common techniques of dissection that only superficially
scratched the surface of nature, Steno advocated a new anatomy that did
not use "Ancient Laws as inviolable Rules in Dissection."[71] He was precisely
the sort of experimental anatomist who felt at home in the anti-Aristotelian
scientific culture of Tuscany. Only months after his arrival, Steno was already displaying "various wonderful experiences" before the Grand Duke
and had become an intimate of Redi. As Redi recounted to Vincenzo Viviani
in 1667,

> Signor Steno does me the honor of favoring my table day and night, and I am
> content to enjoy his most learned and amiable conversation. What's more we
> are never at leisure, dissecting every day and making wonderful observations
> around the court in Florence.[72]

Collaborating with Steno, Redi discovered even further opportunities to discredit ancient beliefs and dismantle the erroneous statements of naturalists
from Aldrovandi to Kircher, who had neither Redi's experimental propensities nor his anatomical wit. Like Redi, Steno praised Ferdinando II as a

---

[69] *Lettere di Francesco Redi patrizio aretino* (Florence, 1779), Vol. I, p. 81.

[70] Antonio Borelli, "Francesco d'Andrea nella corrispondenza inedita con Francesco Redi,"
Filologia e critica 7 (1982): 168.

[71] Nicolaus Steno, *A Dissertation on the Anatomy of the Brain*, Edv. Gotfredsen, ed. (Copenhagen, 1950), p. 35.

[72] In Lionello Negri, Nicoletta Morello, and Paolo Galluzzi, *Niccolò Stenone e la scienza in
Toscana alla fine del '600* (Florence, 1986), pp. 25, 27.

generous patron of the new anatomy. "Animal description has been advanced to no small extent by the various animals which you allowed me to open," he wrote in 1667.[73] While Steno increasingly chose to put aside scientific activities for spiritual duties after his conversion to Catholicism, his presence in Florence strengthened Redi's position as one of the leading anatomists of nature. Together they filled the leisure hours of the Grand Duke with elaborate and costly spectacles.

The courtly games cultivated by Redi and Steno in Florence were replicated in other Baroque theaters of experiment. In Rome, Ciampini's Accademia Fisico-Matematica (Physico-Mathematical Academy) repeated Redi's insect experiments in 1678.[74] Despite the fact that the academy included several prominent Jesuits, they did not advocate the idea of spontaneous generation, siding with Redi based on the results of their experiments. Most contemporaries saw little conflict between the Jesuit program of experimentation and Redi's more methodologically sophisticated investigations of nature. What they noted instead were the commonalities: the display of objects, use of instruments, and advocacy of experimentation. In Bologna, Legati praised Kircher and Redi for demonstrating to virtuosi how the microscope, a "receptacle and Theater of the most marvelous Works of Nature," should be used.[75] In Milan, Settala enjoyed the friendship of Redi, Steno, and Kircher. Steno's visit to Settala's museum allowed the physician Paolo Terzago an opportunity to display his anatomical skills: "among the other anatomical demonstrations, [he performed one on] a heifer's ovary in the house of Signor Settala, with Manfredo present, extracting the egg and diligently removing its cortex."[76]

Steno, as the originator of this sort of anatomical *sprezzatura*, was the most fitting audience for Terzago's virtuosity. Had Kircher ever visited Milan, Settala surely would have demonstrated his burning mirrors and magnetic instruments since he owed the success of their construction to numerous scholarly exchanges with his friend in Rome. Regardless of the philosophical position that one took, in the Baroque museum, all nature was on display, and the performance of a particularly difficult anatomy, such as the dissection Terzago did for Settala and Steno, fulfilled the criterion of virtuosity that defined experimental culture. Anatomizing nature, a practice begun as a means of shoring up the Aristotelian program in the universities and the humanist *studio*, evolved into a form of theater, performed in the courts, academies, and museums, where it took its place alongside all the other natural spectacles that collectors offered their patrons.

[73]Steno, *Elementorum myologia specimen*, in *Steno. Geological Papers*, p. 69.
[74]BNF, *Magl.* VIII. 496, f. 5r (Rome, 11 August 1678).
[75]Legati, *Museo Cospiano*, p. 215.
[76]Bartolomeo Corte, *Notizie istoriche*, p. 189, in Belloni, "La medicina a Milano fino al Seicento," in *Storia di Milano* (Milan, 1958), Vol. XI, pp. 640–641.

NATURAL SPECTACLES

While naturalists used the objects in their museums to refine the uses of anatomy as a form of inquiry, they also gravitated toward artifacts noteworthy for their *inherent spectacularity*. Dissecting yielded the meaning of an object through dismemberment. At the end of an anatomical demonstration, a living creature became a series of observations, illustrations, and disembodied parts that bore little resemblance to the whole it once had been. Anatomizing nature produced knowledge but did not preserve nature. For this reason, many collectors often hesitated to dissect the rarest specimens they acquired. Surely it was more delightful to display an entire crocodile, armadillo, or bird of paradise than to dismember it? Even naturalists such as Aldrovandi, who imagined anatomy as the primary means of acquiring knowledge, preferred to preserve unusual specimens. In the case of a completely unknown species, they argued, the contemplation of the exterior provided enough new information, at least until it became more commonly available in Europe. Torn between dissecting and preserving the animate objects that entered their museums, naturalists turned readily to the contemplation of inanimate objects that could be tested without being dissected. The possession of rare animals led to memorable dissections, detailed in the individual anatomies of each creature, but such events increasingly failed to meet the emerging criteria of *experientia* as communal knowledge. Only a court naturalist such as Redi had access to a sufficient quantity of specimens to create a truly experimental form of anatomy. "Some natural observations need repeated experiences," counseled Boccone, offering a definition of experimental practice.[77] Inanimate objects that possessed remarkable transformative powers yielded the spectacular repeatable "experiences" that defined another facet of the experimental culture of the museum, highly indebted to the tradition of natural magic.

Natural magic provided an appropriate framework in which to develop a demonstrative culture of science. In his *Forerunner to the Great Art* (1670), the Jesuit Lana Terzi divided "experiences" into three different categories, in imitation of Francis Bacon's tripartite division of natural history. The first concerned the generation of "all material and sensible things," the second praeternatural phenomena, and the third "artificial experiences." Under the latter heading, Lana Terzi subsumed natural magic, "that part of natural science that . . . works extraordinary and marvelous effects."[78] Naturalists were as fascinated with the hidden powers of natural objects as they were with uncovering their anatomical structure. In their collecting habits, they

[77]Boccone, *Museo di piante rare*, p. 51.
[78]Francesco Lana Terzi, *Prodromo all'Arte Maestra*, Andrea Battistini, ed. (Milan, 1977), pp. 54, 56.

gravitated toward objects reputed to have special properties. Like the guests of the Royal Society, who saw only the successes and none of the failures of their experimental program,[79] the majority of museum visitors were treated to the most predictable tests that the canon of natural magic could offer. Tastefully woven asbestos garments stubbornly refused to burn; luminescent stones glowed in the dark, magnets hidden within elaborate machines caused untraceable movements, lenses produced flames from remarkable distances, and distorting mirrors made the process of seeing a problematic enterprise. Through their mutual participation in these activities, naturalists and spectators forged a common culture of expectations about the activities conducted within the interior space of a museum.

Experimenting with nature was a test of social power as well as philosophical knowledge. "If I were to have the power or the force, Most Serene Signor, I would experiment with an infinite number of wonderful secrets, since I have noted many of them," wrote Aldrovandi to Francesco I in 1577, acknowledging the Grand Duke's superiority as an experimenter.[80] The courtly and patrician understanding of experiment presented it as a product of leisure. Faced with an audience who wanted to see something new, or at least something predictably marvelous, naturalists created experiments that made wonder a primary form of *experientia*. The revival of natural magic traditions in the context of the museum, as seen most notably in the work of Della Porta, Kircher, Schott, and Settala, fulfilled the courtly love of paradox and enhanced the power of display by giving it philosophical legitimacy; it also contributed to the image of the naturalist as a *manipulator* of resources. Used to princes and nobles who could arbitrarily rearrange the order of things, museum goers delighted in the image of the collector who could similarly command nature at will.

Contemporary books of secrets portrayed naturalists as inveterate experimenters, "experimenting continuously with his own hands," as the publisher of the anonymous *Diverse and Miraculous Secrets* (1563) described the Paduan anatomist Falloppia.[81] "Secrets" defined the sort of natural knowledge the collector activated in the museum. Natural philosophers such as Della Porta based their reputation upon the quantity of secrets they amassed and experienced. Writing to his patron, Cardinal Luigi d'Este, Della Porta boasted that he had collected "more than 2000 secrets of medicine, and other wonderful things," impoverishing himself, friends, and patrons to test them.[82] Della Porta's *Natural Magic* quickly disseminated this image throughout Italy. Collectors such as Giganti were constantly in search of more "secrets of nature" to adorn their museums, in the hope of

[79]Shapin, "The Invisible Technician."
[80]BUB, *Aldrovandi*, ms. 6, Vol. I, c. 47r.
[81]Pseudo-Falloppius, *Secreti diversi et miracolosi*, n.p.
[82]BEst, *Est. It.*, ms. 835 (Alpha G 1, 17) (Naples, 27 June 1586).

achieving the absolute knowledge and control of nature that Della Porta claimed to possess.[83]

In the case of Aldrovandi, contemporaries acknowledged the level of his success by imagining his museum as a complete repository of arcane knowledge. "Just as the waters run to the sea, all the marvelous things and the most beautiful and recondite secrets of nature should assemble in your famous Museum," exclaimed the Jesuit Gioseffo Biancano. Fulfilling his prophecy, Biancano offered the naturalist "two beautiful secrets," known only to him.[84] Secrets, as Girolamo Ruscelli noted, were "little things" and therefore as tiny, precious, and seemingly trivial as the objects that cluttered museums.[85] Quite often they represented difficult, exotic, or dubious knowledge aimed at a literate audience who enjoyed the spectacularity and theatricality of scientific production. While dissections revealed secrets by making knowledge public, natural spectacles heightened the mystery of nature and the power of the collector who activated nature's hidden properties. When Renaissance scholars invoked the opinion of Aldrovandi as the "highest truth," they gave testimony to the power that came from possessing secrets.[86]

In late Renaissance Florence, Francesco I and his illegitimate son Don Antonio gave experimental culture the social prestige necessary to ensure its success through their activities in the laboratory of the *Casino di San Marco,* making natural magic a form of courtly knowledge. Don Antonio possessed an enormous manuscript collection of secrets, among them "the diverse secrets of Giovan Battista Porta Neapolitan" from which he and his father derived many of their experiments.[87] In 1577, Aldrovandi traveled to Florence with the explicit purpose of seeing the Grand Duke's laboratory. His description of the Grand Duke, "working day and night, investigating various secrets of nature . . . that he most generously and courteously makes available to the sick after discovering and experimenting upon them," vividly confirms the image of Francesco as a prince–practitioner who modeled himself on Della Porta. "In the space of eight hours, during two visits, Your Highness deigned to show me so many wonderful, virtuous and attractive secrets experimented on by yourself," wrote Aldrovandi.[88] Knowing that the Medici prince did not always perform his own experiments but often had "several masters working on different projects in different rooms,"

[83]BUB, *Aldrovandi,* ms. 136, Vol. VIII (12 May 1580).

[84]BUB, *Aldrovandi,* ms. 136, Vol. XXXI (Piacenza, 10 November 1602).

[85]Ruscelli, *De' secreti del Reverendo Donno Alessio Piemontese,* pt. IV, sig. *2r.

[86]BUB, *Aldrovandi,* ms. 136, Vol. XIX, c. 155v.

[87]Don Antonio's books of secrets are located in the BNF, Magl. XVI, 63, I. The only detailed study of Don Antonio is P. Covoni's, *Don Antonio de' Medici e il Casino di San Marco* (Florence, 1892), but see the more recent study by Paolo Galluzzi, "Motivi paracelsiani nella Toscana di Cosimo II e di Don Antonio dei Medici: Alchimia, medicina, 'chimica' e riforma del sapere," in *Scienza, credenze occulte, livelli di cultura* (Florence, 1982), pp. 32–62.

[88]BUB, *Aldrovandi,* ms. 6, Vol. I, cc. 30v–31r, 33r.

Aldrovandi understood the privilege accorded to him as a naturalist worthy of witnessing the production of princely knowledge.[89] Like the surgeons who saved Aldrovandi from performing the manual operations of anatomy, Francesco I employed craftsmen to turn his ideas into material realities. The credit for all the objects produced in his *Casino,* however, devolved to the Grand Duke. Like Aldrovandi, he was a natural philosopher, an experimenter and not a craftsman.

As Aldrovandi's visit to Florence indicates, the Grand Duke's laboratory was a place in which to initiate fellow virtuosi into experimental culture. Secrets were no longer buried within the textbooks of physicians and natural philosophers. Through their integration with the world of scientific collecting, they had become a part of the theatrical culture of science. The Grand Duke's investment in the experimental practices of natural magic and pharmaceutical chemistry only heightened the status of such collectors as Calzolari and Imperato. As apothecaries, they possessed the skills and equipment necessary to instruct other naturalists in these arts. Calzolari proudly displayed his collection of distilling devices (see Museum Plan 1); Imperato cultivated a reputation as a naturalist who was assiduous in the chemical investigation of nature. Praised by Fioravanti as the "patron of many rare secrets and noble experiments,"[90] Calzolari not only guided naturalists up the slopes of Monte Baldo but brought them into his laboratory to test what they collected. Mattioli was so dependent upon Calzolari's endless array of specimens that he made two extended visits to the apothecary's museum, where he stayed for two months in 1572. "Calzolari in Verona has written me that you were with him this year and saw all the beautiful things in his Theater," noted Mattioli in a letter to Aldrovandi, "I was still there only a short while ago."[91] While the Grand Duke of Tuscany had the financial resources to fund an entire laboratory of costly glassware and exotic ingredients, most naturalists found it necessary to frequent pharmacies, distilleries, and the museums of apothecaries to gain chemical knowledge of nature.

One of the highlights of Aldrovandi's visit to Calzolari's museum in Verona in 1572 was an asbestos experiment in which he participated. Aldrovandi recorded the event in his notebook:

> We experimented [*habbiam' fatto esperienza*] in his house, placing it in the flame of a burning candle. It lit up as if in flames, so that everyone thought that it had turned to ashes. Nonetheless, once cooled, its substance and appearance remained the same as they were before being placed in the fire. The Excellent Marcantonio Menochi of Lucca, doctor in medicine and philosophy, who was with me in the *studio* . . . can attest to it.[92]

[89]In Berti, *Il Principe dello studiolo,* p. 58.
[90]Fioravanti. I can't find the reference.
[91]*Lettere di Pietro Andrea Mattioli,* p. 59 (Trent, 20 March 1572); Tergolina-Gislanzoni-Brasco, "Francesco Calzolari," p. 12.
[92]BUB, *Aldrovandi,* ms. 136, Vol. V, c. 181r (1569–1570).

In the mid-sixteenth century, asbestos was a relatively novel substance whose properties were little understood and much admired. Paradoxical, malleable, and infinitely spectacular, it fit well within the culture of demonstration then emerging. Aldrovandi himself kept a large quantity of asbestos in his museum "to make spectacles for people."[93] Imperato found it so compelling that he requested samples of it from Northern correspondents and included an illustration of an asbestos experiment in his *Natural History*[94] (fig. 18). In fact, this was the *only* experiment he depicted in the entire work.

By the seventeenth century, asbestos experiments evolved from simple demonstrations into elaborate entertainments for the Italian patriciate and foreign visitors. Most well known were Settala's asbestos experiments in Milan. Displaying his skills as an inventor, Settala fashioned an asbestos purse that

> . . . was thrown many times on a great quantity of lit charcoal in the presence of many signors and princes, in particular the Most Serene Archduke of Innsbruck and the Most Serene Grand Duchess of Tuscany, curious to see the experiment with their own eyes. Nor was anyone wounded.[95]

Collectors frequently made little trinkets woven out of asbestos that they presented to visitors as souvenirs and as further proofs of their virtuosity. Kircher had an asbestos map he displayed in the Roman College, while the *Studio Aldrovandi* contained a feathered garment woven of the same substance.[96] The number of references to asbestos experiments, beginning with Aldrovandi's visit to Calzolari, suggests how well this sort of *experientia* defined the experimental life of the museum. Initiated as an example of the curious effects produced by nature, it soon became an eminently civil experiment that exhibited artifice and paradox, qualities worthy of its noble audience. While the natural knowledge asbestos produced by the seventeenth century was minimal—Baroque collectors had little more to say about the causes of its inflammability than did Aldrovandi—its social value was great. Asbestos exemplified the objects that made natural magic an appropriate framework for displaying nature.

While Calzolari enjoyed the popularity that came with being the only collector of repute in Verona, Imperato competed with Della Porta as the preeminent naturalist and experimenter in Naples. Constantly occupied by his investigations of mineralogy, metallurgy, and the composition of exotic

[93]BUB, *Aldrovandi*, ms. 51, Vol. I, c. 17v.

[94]E. Migliorato-Garavini, "Appunti di storia della scienza del Seicento: tre lettere di Ferrante Imperato ed alcune notizie sul suo erbario," *Rendiconti dell'Accademia Nazionale dei Lincei. Morali*, ser. 8, VII (1952): 37 (Naples, 10 March 1596).

[95]In Folgari, "Il Museo Settala," p. 114; see also Terzago, *Museo o galleria adunata dal sapere*, p. 232. For other descriptions of asbestos experiments, see Lassels, *The Voyage of Italy*, Vol. I, pp. 125–126; *The Diary of John Evelyn*, Vol. II, p. 502; anon., *A Tour in France and Italy Made by an English Gentleman, 1675*, p. 419; Mission, *A New Voyage to Italy*, Vol. I, p. 121.

[96]Kircher, MS, Vol. II, p. 67; *The Correspondence of Henry Oldenburg*, Vol. V, pp. 299–300.

AMIANTO PIETRA FIBROSA, DALLE CVI FIBRE
si fan lacci, e tele, che stanno à fuoco.

Figure 18. Imperato's asbestos experiments. From Ferrante Imperato, *Dell'historia naturale* (Naples, 1599).

plague remedies, Imperato was instrumental in establishing an experimental culture that moved away from the humanist view of knowledge as a textual entity and toward a more artifactual understanding of nature. Mattioli lauded Imperato for "his great diligence and agility in investigating the secrets of nature."[97] Physicians such as Bartolomeo Maranta and the Lincean Nicola Antonio Stelliola collaborated closely with Imperato in their research on theriac and other medicines, while Colonna depended on the apothecary's collection for his investigations of plants and fossils. In Imperato's museum, Maranta performed a series of tests to determine the right proportion of wine necessary to dissolve theriac ingredients: "and thus I established it, helped by the experience diligently repeated many times by Messer Ferrante Imperato."[98] The experiences that occurred within Imperato's museum were not dissimilar to those in Aldrovandi's *studio*, although Imperato more often personally demonstrated artifacts for his visitors whose social status, as physicians, was higher than his own. Both belonged to a far-ranging epistolary network of naturalists who shared the results of

[97]Mattioli, *Discorsi attorno ai libri di Dioscoride,* in Neviani, "Ferrante Imperato," p. 36.
[98]Maranta, *Della theriaca e del mithridato* (Venice, 1572), pp. 211, 218.

their observations and experiments with one another.

In the course of experimenting with nature to make better medicines from it, Imperato developed an interest in the composition of nature and used his museum as an instrument through which to inspect, catalogue, and categorize all its different parts. After capturing a pregnant viper, Imperato observed it hourly to find out more about its generation. He and his companions gathered large quantities of toads to demonstrate that "toadstone" (*pietra di rospo*) was not formed between the toad's cranium and his skin but was in fact a stone like any other. The unmasking of the toadstone, like the disavowal of fossils as inorganic creations, directly opposed Della Porta's view of nature as infinitely metamorphic, full of meaningful and mysterious resemblances. Instead, Imperato argued that nature's causes did not exceed the powers of ordinary observers. He performed experiments, such as one that involved placing different clays in water to see if they would float, that underscored nature as a knowable entity, full of ordinary rather than magical wonder.[99] As an apothecary particularly concerned with the chemical properties of substances, Imperato heated and distilled almost every object that came within his possession. Yet he did not embark on the search for the philosopher's stone, a subject that occupied many of his contemporaries. Like many naturalists in the early seventeenth century, Imperato gradually came to see experimentation as a means of completing the anatomization of the world that began on the dissecting table and ended in the chemical laboratory. From his perspective, dissecting and testing nature were equivalent forms of inquiry, both producing utilitarian knowledge.

Despite their shared interest in collecting and distilling nature, Imperato and Della Porta typified two different trends among late Renaissance naturalists. The rivalry between the two arose from philosophical as well as political differences.[100] Imperato professed to study nature *only from nature* and for the betterment of medicine, while Della Porta perceived natural history as a prelude to natural magic. Visitors to their museums underscored the differences between these two forms of inquiry. In 1601, Peiresc traveled to Naples, where letters from Pinelli gained him entry to both museums. Della Porta enjoyed Peiresc's company so much that he invited the French scholar to observe a few experiments. "Nor did he only see what ever they kept in their studies, and precious treasuries," wrote Gassendi, "but he was present at their Experiments of all kinds." The initial contact in Della Porta's museum, sealed by Peiresc's witnessing some of his more extravagant spectacles, led to great mutual "familiarity" between the

[99]Bruno Accordi provides a concise summary of these activities in his "Ferrante Imperato (Napoli, 1550–1625) e il suo contributo alla storia della geologia," *Geologica Romana* 20 (1981): 47, 51, 54.

[100]This is a point made well by Giuseppe Olmi, "La colonia lincea di Napoli," in Lomonaco and Torrini, *Galileo e Napoli*, pp. 40–41. It is worth remembering that Imperato received visits from the Spanish Viceroy, while Della Porta was known for his suspect (anti-Spanish) political activities.

two naturalists.[101] Unlike the "safe" tests such as asbestos, available for all to
see, Peiresc was made privy to discussions and demonstrations of uncertain
processes with indeterminate ends—perhaps a few magnetic demonstrations, transmutations of base into precious metals, or experiments with mirrors? Crossing the boundary between the social world of collecting and the
scientific culture of exchange that made friendship and virtuosity a requisite for entrance, Della Porta allowed Peiresc to observe the intimate workings of his experimental theater.

Conversely, Peiresc's visit to Imperato's museum elicited no special
memories. As a humanist fascinated by arcane knowledge, Peiresc inevitably felt more drawn to Della Porta than Imperato. The former shared
his appreciation for recondite secrets, just as Kircher later did, and enjoyed
Peiresc's encyclopedic erudition. Instead, Imperato could converse only
about nature with his learned visitor. He had no classical statues, magic
lanterns, speaking tubes, distorting mirrors, or other objects of humanist
erudition in his museum, and therefore possessed none of the artifacts that
mediated *experientia* within the republic of letters. Imperato collected objects but did not possess "wisdom," in the humanist sense of the term, and
therefore could not provide the learned conversation that initiated the
sharing of secrets.

Changing attitudes toward the value of ancient authority in the early seventeenth century brought new visitors to Imperato's museum who did not
perceive his lack of humanist embellishments to be detrimental to his ability to philosophize. Comparing Della Porta's attitude toward observation
with Imperato's empiricism, Tommaso Campanella remarked, "Nonetheless the most studious Della Porta forces himself to recall this science, but
only historically, without explanation; and the *studio* of Imperato can be a
foundation for uncovering it."[102] Distinguishing Imperato's active notion of
experientia from Della Porta's more formal use of it as a philosophical category, Campanella indicated his own preference for a collector who read directly from the book of nature. While Della Porta used the objects in his possession to demonstrate historical truths, Imperato saw his museum as a
space in which to create knowledge directly from artifacts rather than
around them. Della Porta more closely resembled Girolamo Ruscelli, who
"continually experimented on all the secrets that we could recover from
printed books or from ancient and modern manuscripts" in his Accademia
Segreta.[103] Ruscelli was the author of the pseudonymously published *Secrets*

[101]Gassendi, *Mirrour of True Nobility and Gentility*, Vol. I, pp. 28, 45.

[102]Tommaso Campanella, *Del senso delle cose e della magia*, Antonio Bruers, ed. (Bari, 1925),
pp. 221–222. Campanella wrote this treatise in the context of a public dispute with Della Porta
(ca. 1590) on the sympathy and antipathy of things, shortly after the publication of Della
Porta's *Phytognomia*.

[103]In William Eamon and Françoise Peheau, "The Accademia Segreta of Girolamo Ruscelli:
A Sixteenth-Century Italian Scientific Society," *Isis* 75 (1984): 339–340.

of Alexis (1555) that appeared in the libraries of naturalists such as Aldrovandi, Cesi, and Jan Eck[104] and competed with Della Porta's *Natural Magic* in popularity. He typified the humanist community of natural philosophers that gave natural magic and textual knowledge a privileged place in late Renaissance experimental culture. With the formation of a Lincean colony in Naples, however, Imperato's museum began to play a greater role in the experimental culture of the city. Della Porta's museum was the official center of the Lincean colony, but the majority of Neapolitan members, most notably Stelliola and Colonna, preferred the atmosphere of the apothecary's museum to the *studio* of a Renaissance magus.

Cesi first encountered Della Porta and Imperato during a trip to Naples in 1604. Both introduced the Roman noble to the local scientific culture through conferences in their museums. "I have learned much from discussions with them, and I have and will have the most wonderful secrets," reported Cesi to Stelluti upon his return to Rome. "And with these two I spent a good part of my time in Naples, to great profit."[105] Cesi's decision to include both Della Porta and Imperato within the orbit of his academy reflected his own conflicted feelings about where philosophical novelty lay. Della Porta was the most renowned experimenter in Italy and a philosopher of great reputation; there was no question about his becoming a key member of the Accademia dei Lincei. And, at least in the early stages of the academy's formation, Cesi genuinely believed that natural magic provided an essential foundation to the restructuring of the natural sciences. It was the most comprehensive form of sensory knowledge then available.

Cesi also delighted in the technology that accompanied natural magic. During a second trip to Naples in 1610, Della Porta invited the Roman prince and his traveling companion Stelluti to witness one of his famous experiments with burning mirrors. "While we were in Naples at Della Porta's house, we observed many noteworthy things . . . ," wrote Stelluti:

> He had a crystal lens that burned every wood, even green, from a distance of four feet. And the said Porta taught Signor Cesi about the parabolic section that burns infinitely, which he discussed obscurely in his *Natural Magic*. But because there is no one in Rome who knows how to make similar metal mirrors, [Cesi] has not been able to try it nor is it clear that it works.[106]

Most of the demonstrations enacted by Della Porta for Cesi and other visitors were experiments with optical illusions, loadstones, and burning mirrors—the very secrets that filled his *Natural Magic*. Like Galileo's micro-

---

[104]BUB, *Aldrovandi*, ms. 148; BANL, *Archivio Linceo*, ms. 32; BANL, *Archivio di Santa Maria in Aquiro*, ms. 412, ff. 62r, 74v. Most of the books of secrets available in the BUB were originally in Aldrovandi's possession, as can be discerned by his inscriptions and notations in many of them.

[105]CL, Vol. I, p. 41 (Rome, 17 July 1604).

[106]CL, Vol. I, p. 155 (Fabriano, 18 February 1611).

scope, they were the *mirabilia* of the academy and reflected the patrician tastes of its members who equated philosophical novelty with the possession of instruments.

Yet as Linceans such as Colonna began to develop a more radical epistemology that divested the study of nature from its humanistic origins and therefore the tradition of natural magic, Cesi increasingly felt uncomfortable about his association with Della Porta. By the 1620s, the model for Lincean experimentalism had shifted slightly. While continuing to enjoy the culture of secrets and magical demonstrations, Cesi hesitated to admit any alchemists to the academy and persevered in the writing of the "new" natural history. Other Linceans also followed this path. In his 1626 eulogy for Virginio Cesarini, Cesi's cousin and fellow academician, Monsignor Agostino Mascardi praised the fact that the Roman noble "even performed chemical distillations and with exquisite diligence experimented in order to see with his own eyes those transmutations both in simples and minerals which are difficult for the speculative intellect to comprehend."[107] Cesarini, in other words, modeled himself after naturalists such as Imperato rather than Della Porta.

Until his death in 1615, Della Porta continued to inform Cesi about experiments of interest in Naples and shared his best secrets with the Roman prince. "Some time ago I sent the secret of speaking from afar to you," wrote Della Porta to Cesi in 1609, inquiring whether or not Cesi had experimented with it yet (*havesse fatto l'esperienza*). In 1612 he announced the successful creation of the philosopher's stone by one of his associates.[108] By this time, Cesi had become relatively unreceptive to the exchange of secrets that structured all of Della Porta's communications. A new spectacle had been placed before him, one more tangible than the elusive philosopher's stone and more emblematic of the direction in which he was heading. In the spring of 1611, Galileo demonstrated to the Linceans the luminescent powers of a stone discovered by some Bolognese alchemists, subsequently dubbed the "Bologna stone" or, more poetically, the "solar sponge."[109] By June, Faber informed Imperato about the *lapis Bononiensis*. Eager to see this latest natural curiosity but suspicious of an object whose powers fell so clearly within the realm of natural magic, Imperato cautiously predicted, "I believe that it is not natural but artificial."[110] Similar to phosphorus and offering all sorts of rich comparison with the ubiquitous loadstone, the "so-

[107]Agostino Mascardi, *Orazioni e discorsi* (Milan, 1626), p. 9, in Redondi, *Galileo Heretic,* p. 95.
[108]Ibid., Vol. I, p. 115 (Naples, 28 August 1609); Vol. II, p. 300 (Naples, 16 December 1612).
[109]The philosophical significance of the "solar sponge" is discussed in Redondi, *Galileo Heretic,* pp. 5–27. I have drawn my discussion of the controversy surrounding it from material in this chapter. Redondi identifies it as barium sulfide.
[110]CL, Vol. II, p. 163 (Naples, 10 June 1611).

lar sponge" became both an object of the new philosophy and a proof that Aristotelian science could still find ways to absorb new objects into traditional natural philosophy.

Over the next two years, the Linceans beseeched Galileo to send them samples of the stone for their museums. By the spring of 1613, a "box of luciferous stones" arrived at the Cesi palace. As Cesi responded to Galileo, "I thank you in every way, for truly this is most precious, and soon I will enjoy the spectacle that, until now, absence from Rome has not permitted me."[111] Like asbestos, the *lapis Bononiensis* provided multiple experiments for naturalists to enjoy, producing endlessly engaging paradoxes. While committed Aristotelians such as the Jesuit Niccolò Cabeo and the Paduan professor Fortunio Liceti found ways of subsuming the luminescent stone within traditional philosophy, by declaring its ability to retain light to be a form of magnetism, others invoked more daring hypotheses, raising the problematic issue of atomism. For a group of philosophical dilettantes such as the Linceans, viewing the "solar sponge" tingled their speculative sensibilities, drawing them closer to the "dangerous" (anti-Catholic) knowledge of the new philosophy.

The activities in Imperato's museum more comfortably fell within these new guidelines for experimenting. The apothecary, in fact, proved to be more helpful than Della Porta in deciphering difficult phenomena such as the *lapis Bononiensis*. Imperato's critical stance toward the properties of most natural objects won him the support of Colonna, who replaced Della Porta as the head of the Lincean colony in 1615. Cesi, Colonna, Faber, and Imperato all shared a common interest in plants and minerals, particularly fossils and other "ambiguous bodies." When the German Lincean Theofilo Müller traveled to Naples, Faber supplied him with a letter of introduction to Imperato. "As soon as he arrived, I showed him my natural Theater and also all the living plants among my simples," the apothecary reported to Faber.[112]

On Cesi's second visit to Naples, Imperato presented him with an unusual mixed specimen that eventually found its way into the prince's research on metallophytes.[113] After Cesi's discovery of the metallophyte near Acquasparta—an object he deemed equivalent in subtlety to Galileo's "solar sponge"—he and Faber frequently consulted Imperato about its unusual properties. Possibly Imperato advised them on an experiment, recorded in Stelluti's *Treatise on the Wood-Fossil-Mineral.* A piece of metallophyte left "in a room in Signor Duke Cesi's palace in Acquasparta was found, after several

[111]Ibid., p. 357 (Monticelli, 30 May 1613).

[112]BANL, *Archivio di S. Maria in Aquiro*, ms. 420, f. 362 (Naples, 6 April 1612).

[113]Stelluti, *Persio tradotto*, p. 170, in Gabriele, "L'orizzonte intellettuale e morale di Federico Cesi illustrato da uno suo zibaldone inedite," pp. 678–679.

months, all converted into wood, not without marvel by [Cesi] and others who saw it."[114] Through frequent communications between Rome and Naples, the Linceans, with the help of Imperato, gradually compiled evidence that supported the organic origins of fossils. Cesi's metallophyte provided further proof that organic and inorganic objects were not always distinct. Despite Cesi's admiration for the Neapolitan apothecary, he never once considered him a candidate for membership in the academy. Imperato was an assiduous experimenter, worthy of admiration by all proponents of the new philosophy. But he was also an apothecary, not a gentleman, and therefore had to content himself with the glory of assisting the Linceans in producing and explaining the latest natural spectacles that delighted the Italian patriciate.

PROBLEMATIC OBJECTS

Collaborating with Colonna, Imperato participated in one of the most heated controversies of the early modern period: the debate over fossils. As many scholars have noted, the definition of a "fossil object" changed dramatically between the sixteenth and seventeenth centuries.[115] By the seventeenth century, the inorganic nature of fossils was openly contested. Naturalists such as Colonna and Steno suggested that fossils were not simply the product of nature's play, inorganic analogues of the animals and plants whose appearance they inexplicably replicated, but the organic remains or imprints of living creatures. These arguments were expanded and refined in the work of English naturalists such as Robert Hooke, John Ray, and John Woodward, leaving it to a later period to resolve the scientific and theological quagmires that they unleashed. "To find out the truth of this question," wrote Edward Lhwyd, curator of the Ashmolean Museum, to John Ray in 1690, "nothing would conduce more than a very copious collection of shells, of skeletons, of fish, of corals, pori, etc. and of these supposed petrifications."[116] The museums of Renaissance naturalists such as Aldrovandi, Mercati, and Imperato initiated the relationship between the fossil debates and collecting practices that became a standard feature of geology and paleontology in later periods.

In his utopian definition of the laboratory, Johann Valentin Andreae described a space that embodied the ambiguity between textual and experi-

[114]Stelluti, *Trattato del legno fossile minerale*, p. 8.

[115]See, for example, Rachel Laudan, *From Mineralogy to Geology: The Foundations of a Science, 1650–1830* (Chicago, 1987); Nicoletta Morello, *La nascita della paleontologia: Colonna, Stenone e Scilla* (Milan, 1979); Roy Porter, *The Making of Geology: Earth Science in Britain 1660–1815* (Cambridge, U.K., 1977); Paolo Rossi, *The Dark Abyss of Time: The History of the Earth and the History of the Nations from Hooke to Vico*, Lydia G. Cochrane, trans. (Chicago, 1984); and Rudwick, *The Meaning of Fossils*.

[116]Lhywd to Ray, 25 November 1690, in Joseph M. Levine, *Dr. Woodward's Shield: History, Science and Satire in Augustan England* (Berkeley, 1977), p. 29.

ential practices: "Whatever has been dug out and extracted from the bowels of nature by the industry of the ancients, is here subjected to close examination, that we may know whether nature has been truly and faithfully opened to us."[117] The confrontation between text and artifact that collecting engendered opened up numerous problems about representation, replication, and the nature of scientific authority. Fossils are the most well-known example of a natural artifact whose meaning changed because of its collectibility. As Kircher put it, fossils frequently were found "in the museums of curious men," whose curiosity was aroused by the paradox of their resemblance to animate forms.[118] They contained spectacular but problematic knowledge.

In his *Dissertation on Tonguestones* (1616), Colonna advertised Imperato's collection as a place in which proper philosophizing about nature could be undertaken. Describing the "petrified objects" that he had unmasked as actual impressions of plants and animals rather than "jokes of nature," which was the common term for fossils and other paradoxical phenomena, Colonna advised, "and if you have no means of possessing them, go to the Museum of my friend Imperato."[119] Colonna was hardly the first naturalist to perceive the museum as a space in which to resolve paradoxes. In the late sixteenth century, Aldrovandi and Mercati, also aided by Imperato, underscored the image of the museum as a setting conducive to investigating the problem of resemblance in nature. Aldrovandi recalled a trip to Rome during which Mercati "meticulously showed me all the mineral things of his *Metallotheca*, and likewise all the plants of the [papal] garden, and he gave me part of all the things I did not have."[120] Aldrovandi subsequently incorporated the results of this and many other exchanges into his *Museum of Metals*, not published until 1648, in which he cautiously suggested that some fossils were incomplete organisms, an hypothesis that acknowledged that not all "figured stones" were images made by nature's pen.[121] However, Aldrovandi tactfully chose to remain silent about the *cause* that led a creature to form halfway in stone, preferring to imagine a fossil as a form of inanimate monstrosity.

To formulate his conclusions, Aldrovandi perused the hundreds of tiny drawers filled with stones, precious gems, amber-encased objects, and actual fossils that comprised Mercati's *metallotheca*, in the hope of receiving some insight into their origins by noting the various similarities and differences among them (fig. 19). Despite their increased skepticism about the origin

[117]Johann Valentin Andreae, *Christianopolis*, Felix Emil Held, trans. (New York, 1916), p. 197.

[118]Kircher, MS, Vol. II, p. 38.

[119]Fabio Colonna, *De glossopetris dissertatio* (1616), in Morello, *La nascita della paleontologia*, p. 83. For more on *lusus naturae*, see Findlen, "Jokes of Nature and Jokes of Knowledge."

[120]Tosi, *Ulisse Aldrovandi e la Toscana*, p. 415 (Bologna, 29 August 1595).

[121]Schnapper, *Le géant, la licorne, la tulipe*, p. 19.

Figure 19. Cabinet of fossils in Mercati's *metallotheca*. From Michele Mercati, *Metallotheca* (Rome, 1717).

of fossils, Aldrovandi's and Mercati's commitment to Aristotelianism precluded them from completely overturning accepted theories. Instead, they chose to consolidate the classic discourse on fossils by attaching it to the material reality displayed in their museums. Sensory experience of fossils increasingly suggested that the resemblances between different objects in their museum indicated common origins. Yet there were many other mysterious affinities between objects that eluded the powers of human reason. Rather than dismantling ancient authority with the evidence in the Vatican museum, Aldrovandi and Mercati chose to critique certain specifics to strengthen their faith in the system of understanding they espoused.

By the seventeenth century, Colonna, Steno, and the Sicilian painter Agostino Scilla mustered the evidence in various museums to argue for a new theory of fossils. Careful comparisons between shark's teeth and "tonguestones" (*glossopetrae*) created the first crack in the system. Following Colonna's lead, Steno and Scilla published the results of their comparisons between fossils and living creatures, both advocating the theory of organic origin. Like Colonna, they relied on the work of various collectors to support their hypothesis. In his *Dissection of a Shark's Head* (1667), Steno hesitated to attack common beliefs about fossil formation without some material confirmation. However, a visit from Settala allayed his fears:

> This digression was quite ready to go to press when Manfredo Settala, Canon of Milan, who is known to everybody for his singular knowledge of natural history and his indefatigable zeal for enriching his museum, on a short visit to this place told me that he possessed among his rarer objects many things that favour my conjectures rather clearly, which I was very glad to hear, since I know quite well how much they gain in weight by this man's concurrence.[122]

Steno's postscript confirms the importance of collecting to the resolution of paradox by the seventeenth century. Possession had become a form of proof, although even for Steno it took second place to the "conjectures" that led him to speculate about a different explanation for fossils. Surely Steno's careful attention to evidence was one of the reasons that Boccone praised his work as "so well grounded in experience."[123] When Steno returned briefly to Copenhagen in 1672, he visited the museum of his friend Ole Worm, reporting to Leopoldo de' Medici that Worm had shown him a fish embedded in a stone.[124] Five years after he published his treatise, Steno was still wondering if he had drawn the right conclusion.

Kircher partially deserved the credit for Steno's oscillation. As the most famous Catholic natural philosopher in Rome, he served as a role model for

---

[122]Nicholas Steno, *The Earliest Geological Treatise*, Avel Garboe, trans. (London, 1958), pp. 44–45. I have translated the title freely to make it more understandable; in Latin it is entitled *Canis carchariae dissectum caput*.

[123]Boccone, *Museo di fisica* (Venice, 1697), p. 181.

[124]Steno, *Epistolae*, p. 278 (Copenhagen, 19 November 1672).

Steno during his conversion. Despite the work of Colonna and Steno, Kircher firmly believed that fossils were inorganic, produced by the mysterious lapidifying juices that coursed through the veins of the earth. While Robert Hooke and John Ray lamented their inability to replicate fossils experimentally, thereby resolving the fact of the matter,[125] Kircher conducted numerous experiments in the Roman College museum to prove the existence of the *vis spermatica*, the universal generative principle responsible for the appearance of all natural creations. To this end, he calcified urine, produced samples of *arbor metallica*, performed "crystallogenesis," and mixed chemicals to create a marbling effect in stone, all experimental "proofs" of nature's plastic virtues.[126] While none of these activities literally produced a fossil, they represented many of the ways in which nature could shape herself, independent of any human intervention. Combining and recombining objects, Kircher offered a form of proof by analogy. If chemicals could "paint" stone of their own volition, he argued, then why couldn't nature do the same at the behest of God, forming rocks that seemed to mirror the images of plants and animals?

Kircher's approach to the fossil controversy as a problem of experimental knowledge reflected his own engagement in the debate about the proper relationship between natural philosophy and sense experience. Like the members of the Accademia del Cimento, Kircher enthusiastically subscribed to the program of experimentation developed during the mid-seventeenth century as a means of establishing proof. Following the guidelines of the new philosophy, his experiments tested hypotheses and decided between alternatives. Kircher, as his protegé Petrucci affirmed, did not wish to be one of those "unexperimented persons who asserts too easily the belief that they have seen [something] with their own eyes."[127] In his *Experimental Kircherian Physiology* (1680), a summary of 300 of the Jesuit's most successful experiments, Johann Kestler confirmed this image, describing his master as "the prodigious miracle of our age who has excited the admiration of the whole world by the innumerable experiments on which he has based his universal sciences."[128] Unable to reproduce many of the experiments in Della Porta's *Natural Magic*, Kircher labeled his predecessor a charlatan, impious as well as experimentally unsound. Instead, he presented himself as a Catholic magus who could discern truth from falsehood and

[125]Rossi, *The Dark Abyss of Time*, p. 14.

[126]Kircher, MS, Vol. II, pp. 27, 42, 52–53, 335–336.

[127]Petrucci, *Prodomo apologetico*, p. 79. This is a paraphrase of Peter Dear's description of the Accademia del Cimento's experimentalism in his "Narratives, Anecdotes and Experience: Turning Experience into Science in the Seventeenth Century," in Dear, ed., *The Literary Structure of a Scientific Argument* (Philadelphia, 1991), p. 133. Dear's studies of the experimental work of Jesuit mathematicians parallel my own discussion of natural history and the natural magic traditions on many points.

[128]Johann Kestler, *Physiologia Kircheriana Experimentalis*, in Thorndike, *A History of Magic and Experimental Science*, Vol. VI, p. 569.

reason from unreason by experimentally testing everything that entered the Roman College museum. He had "seen" the processes that spontaneously formed fossils. What further proof was needed?

When his *Subterranean World* appeared, naturalists from Steno to the members of the Royal Society eagerly read it and attempted to reproduce the experiments Kircher had performed. In 1665, Oldenburg reported, not without some disappointment, to Boyle that the "very first Experiment singled out by us out of Kircher" had failed, "and 'tis likely the next will doe so too."[129] Kircher's fossil experiments, like numerous other entries in his *Subterranean World,* possessed the remarkable ability of succeeding in Jesuit colleges and other Catholic centers of learning but nowhere else. Bound to his own *a priori* assumptions about the shape of the natural world, Kircher's experiments confirmed the tenets of Aristotelian natural history with the techniques of the new philosophy.

By the time Scilla published his *Vain Speculation Undeceived by Sense* (1670), the latest assault on ancient fossil theories, he had eradicated all the doubts that came from being a traditional natural philosopher by declaring sense experience to be the *only* basis for natural history. Boccone, always a barometer of opinion about the latest trends in studying nature, waxed enthusiastic about Scilla. In 1671, he urged Redi to search "the museums of curious men in Florence" to develop "the comparison of marine *Echini* with petrified ones," no doubt in response to reading Scilla's treatise. In 1678, he finally met his fellow Sicilian during a stay in Rome. Two years later, he reported to Magliabecchi that he had seen Scilla's "Observation on the parts of altered animals."[130] Steno continued to hesitate about the implications of his findings, attempting to take the middle ground due to his allegiances to Kircher—"The same phenomena can be explained in many ways, and even Nature pursues the same ends through diverse means in her operations," he remarked diplomatically.[131] Scilla, however, presented himself as an observer like Colonna, who derived all knowledge from objects. For Scilla, objects rather than texts spoke authoritatively. Exhibiting the same decisiveness in his investigation of coral, considered a plant–mineral until the eighteenth century, Boccone announced that his own research had produced similarly conclusive results in the classification of yet another ambiguous body. After comparing the anatomy and physiology of plants to coral, he proclaimed that coral "must be" considered a mineral because it exhibited the external form but not the internal structure of a plant.[132] Like Scilla, Boccone envisioned the resolution of paradoxes as his primary goal.

[129] *The Correspondence of Henry Oldenburg,* Vol. II, p. 615 (21 November 1665).

[130] Laur., *Redi* 209, f. 217 (Paris, 8 September 1671); BNF, *Magl.* VIII.496, ff. 7, 19 (Rome, 17 December 1678 and 10 May 1680).

[131] In Morello, *La nascita della paleontologia,* p. 139.

[132] Boccone, *Recherches et observations curieuses sur la nature du corail blanc & verge, vray de Dioscoride* (Paris, 1671), p. 2.

Fossils were only one of the many problematic objects that shaped the experimental culture of the museum. Collecting did not produce neutral knowledge nor was the museum an "objective" space for philosophizing. As naturalists connected *experientia* to the possession of objects, they increasingly found themselves forced to make decisions about the meaning of this new form of knowledge. What was the purpose of displaying nature? When Kircher responded that he intended his magnetic experiments for the "investigation of the learned," "admiration of the ignorant and uncultured," and "relaxation of Princes and Magnates," he underscored the ambiguity of the experimental enterprise.[133] Patrons such as the Grand Duke Ferdinando II supported the experiments of the Accademia del Cimento while eagerly clamoring for Kircher's latest spectacles. "I await the curiosities and experiments that you were to send me at your convenience," wrote Ferdinando II to Kircher in 1659.[134]

The indecisiveness of naturalists and their patrons toward experimentation was also reflected in the display of objects. In certain contexts, loadstones represented the choice between two competing cosmologies: as artifacts that alternately reinforced geocentrism or supplied a necessary ingredient in the argument for heliocentrism. But in the demonstrations of artificial magic that Kircher and Schott performed in the Roman College, they were playful objects that decided little.[135] Like the Society of Jesus's decision to advocate the Tychonic compromise, which allowed both the earth and the sun to be centers of the universe, loadstones embodied the paradoxical attractiveness of the new philosophy even for the most traditional Aristotelians. Certainly it would have been safer to refuse to display them or bury them beneath the plenitude of other objects. Instead, the Jesuits who collected loadstones invented progressively elaborate ways to make them the central spectacle of their museums.

While choosing to take the middle ground on issues such as magnetism, Kircher displayed other objects whose presence constituted a material defense of traditional natural philosophy. In 1645, Kircher witnessed a vacuum experiment with Torricelli's new machine at the house of a Medici cardinal. Quickly he constructed a counterexperiment, prominently displayed in his museum, to disprove the latest claim of the new philosophy. As a good Aristotelian, he *knew* that nature abhorred a vacuum. Next to the failed air pump, unsuccessful perpetual motion machines gave testimony to the impossibility of infinite movement, again in accordance with Aristotelian physics. "Kircher exhibits different experiments on perpetual motion in his

[133]Kircher, *Magneticum naturae regnum*, p. 346.

[134]PUG, *Kircher*, ms. 556 (II), f. 63 (Florence, 18 November 1659).

[135]In his *Universal Magic of Nature and Art* (1657–1659), Schott described "those matters that Father Kircher showed in his museum in his lessons on artificial magic and that I demonstrated in that very place for three years by using a magnet," quoted in Baldwin, "Magnetism and the Anti-Copernican Polemic," p. 170.

museum, which he reproves rather than assists," wrote De Sepi.[136] For Kircher, perpetual motion was the philosopher's stone of the new science, as elusive as anything that ancient alchemists had envisioned. Yet this only heightened his desire to construct a machine that replicated its effects. In the demonstrative culture of Baroque natural philosophy, all opinions had to be displayed, removed from the pages of textbooks and turned into museum exhibits. Seeing and believing became one.

Continuing along this judicious path, Kircher also condemned alchemists who "corrupted" the art of chemistry by claiming to effect fantastic transmutations and improbable generations, for example, the *homunculus*. Instead, he developed a program of "Hermetic Experiments" that defined the limits of probability as well as piety. Real transmutations did exist in nature and Kircher demonstrated as many of these as he could collect, making fossils the most important proof of nature's metamorphic capabilities. When Christina of Sweden visited the Roman College museum in 1657, she watched Kircher make the "Vegetable Phoenix," a plant palingenesis.[137] Kircher strictly differentiated his hermetic experiments from the impious alchemy that could not be experimentally proven and therefore not publicly displayed. By using the objects in his museum to dissociate himself from the overly speculative propensities of *both* ancient and new philosophies, Kircher strove to create a synthesis that would provide common middle ground for all natural philosophers. Given this premise, he was persistently puzzled by the suspicion and derision with which Redi, other members of the Accademia del Cimento, and the Royal Society greeted his work.

In contrast, his disciple Lana Terzi received a warm welcome from the English natural philosophers when he published his *Forerunner to the Great Art*. Unlike Kircher, Lana Terzi believed in the reality of the philosopher's stone and based his belief partially on the viewing of a nail "in the Gallery of the Grand Duke of Tuscany, of which one part is all iron, and the other . . . shows itself to be purest gold." In imitation of Kircher's experimentalism, Lana Terzi advocated alchemy as a form of knowledge only after he had "tried and seen this and other experiments with my own eyes, from which no doubt remains about the possibility of transmuting metals."[138] In the work

[136]Middleton, "Science in Rome, 1675–1700," p. 139; De Sepi, pp. 47, 57. I owe the image of perpetual motion as a philosopher's stone to Pamela Smith.

[137]For a general overview of the Jesuit attitude toward alchemy, see Martha Baldwin, "Alchemy in the Society of Jesus," in Z. R. W. M. von Martels, ed., *Alchemy Revisited: Proceedings of the International Conference on the History of Alchemy at the University of Groningen 17–19 April 1989* (Leiden, 1990), pp. 182–187. On Kircher's hermetic experiments, see De Sepi, p. 45; and Schott, *Ioco-seriorum naturae et artis, sive magiae naturalis centuriae tres* (Würzburg, 1666), Century II, p. 138ff.

[138]Lana Terzi, *Prodromo all'Arte Maestra*, pp. 193, 196. For more on Lana Terzi, see Pighetti, "Francesco Lana Terzi e la scienza barocca"; and Vasoli, "Sperimentalismo e tradizione negli 'schemi' enciclopedici di uno scienziato gesuita del Seicento," *Critica storica* 17 (1980–1981): 101–127.

of Lana Terzi, the experimental curiosity fostered in the course of the seventeenth century no longer was held in check by the strictures of traditional philosophy. Lana Terzi's willingness to put every belief to the test led him to admire and incorporate the work of Bacon and Boyle in his encyclopedia as well as recapitulate the alchemical program that his master Kircher abhorred. Eroding the boundaries of the Jesuit Aristotelian program, Lana Terzi participated in its eventual dissolution, all in the name of *experientia*.

By the mid-seventeenth century, collecting had become an accepted practice for Aristotelian naturalists, Jesuit experimentalists, and self-professed Baconians and Galileans, all of whom claimed it as an integral feature of their scientific activities. Museum objects were not authoritative in themselves but rather served as touchstones for varying claims to produce truth. Authority was no longer grounded only in a canonical text but also in the personal testimony of naturalists who performed experiments.[139] No longer unified by the scholastic community of discourse that gave medieval natural philosophers a common scientific culture in which to operate, early modern naturalists faced the problem of creating new social groupings as old ones dissolved. Experiencing nature offered a new basis for communication, linking naturalists who philosophically had little in common but who shared a belief in a set of common procedures and who cultivated the same patrician audience. Experiments were the most dramatic way to place the contents of a museum "under the gaze of the spectators."[140] Their display in front of the ruling elite, who also claimed to possess privileged knowledge, socially legitimated the new configuration of scientific culture. Making their artifacts of interest for courtly patrons and transforming them into evidence for whatever philosophical position they hoped to maintain, collectors established *experientia* as socially desirable and philosophical necessary. The museum, as Kircher proclaimed in the preface to this *Great Art of Light and Shadow*, existed to replicate the "sensory Theater of the World," in all of its infinity. Unable to associate it successfully with any one philosophical position, naturalists used *experientia* to provide common ground on which to conduct struggles for intellectual authority and used the same practices to support different conclusions.

[139] Peter Dear also makes this point in discussing the rhetorical strategies of the early members of the Royal Society throughout his "Totius in verba," pp. 146, 152, 157.

[140] Scarabelli, *Museo o galleria adunata dal sapere*, sig. +3r.

SIX

Museums of Medicine

I most firmly believe that the science of medicine little by little, will land in the whorehouse . . . because one day everyone will be a doctor.

—LEONARDO FIORAVANTI

Like the dragons dissected in Aldrovandi's *studio* in 1572 and Barberini's *studio* in 1660, or the loadstone that Kircher gave special prominence, vipers enjoyed the status of favored objects among collectors. They appeared frequently in the museums of physicians and apothecaries, whose possession of nature was predicated upon the idea of medical utility, and in the museums of Italian patricians, who perceived medical knowledge to be socially advantageous. Like many of the artifacts in early museums of natural history, vipers were essential ingredients for the most popular and costly medicines then in use. "And these vipers are the foundation of theriac," affirmed Aldrovandi. Collecting and dissecting them preceded their inclusion in theriac, the "royal antidote of antidotes" first described by Galen and used in western Europe and the Islamic world from antiquity to the eighteenth century.[1] In treatises such as *On Theriac to Piso, On Theriac to Pamphilius,* and *On Antidotes,* Galen identified theriac as a sixty-four-ingredient compound, able to cure any ill known to mankind. In accordance with the macrocosmic principles of premodern medicine, theriac was designed to mirror man's physiological complexity; each ingredient corresponded to a particular part and function of the human body. The combination of so many simples in one

[1]BUB, *Aldrovandi,* ms. 6, Vol. II, c. 153r (Bologna, 13 August 1573); ms. 21, Vol. III, c. 134. The two main studies of theriac are Gilbert Watson, *Theriac and Mithridatium: A Study in Therapeutics* (London, 1966); and Thomas Holste, *Der Theriakkrämer: Ein Beitrag zur Frühgeschichte der Arzneimittelwerbung,* Würzburger medizinhistorische Forschungen, 5 (Hannover, 1976). For discussions of medieval theriac, see Michael McVaugh, "Theriac at Montpellier 1285–1325," *Sudhoffs Archiv* 56 (1972): 113–144; and William Eamon and Gundolf Keil, " 'Plebs amat empirica': Nicholas of Poland and His Critique of the Mediaeval Medical Establishment," *Sudhoffs Archiv* 71 (1987): 180–196. For a discussion of theriac in Islamic medicine, see Gary Leiser and Michael Dohls, "Evliya Chelebi's Description of Medicine in Seventeenth-Century Egypt," *Sudhoffs Archiv* 71 (1987): 197–216. I will refer to other specific studies of sixteenth-century debates about theriac throughout the chapter, particularly the work of Olmi, Palmer, and Rosa.

antidote literally produced a pharmacy in miniature, assuring the person who took theriac that the compound would anticipate all ills. "Theriac has two principle virtues," wrote Maranta in his *On Theriac and Mithridatum* (1572), "one is that . . . it preserves the healthy, the other that it cures the sick."[2] Throughout the early modern period, every Italian city large enough to have some level of organized medical practice transformed the annual production of theriac into an elaborate ceremonial that began in May with the capture and killing of vipers and ended in June with the concoction of this state medicine, usually in a public setting approved by the College of Physicians. Ambassadors as well as physicians and apothecaries attended the making of this antidote; it was an event no gentleman with any pretensions to scientific learning could afford to neglect. "I am killing several vipers in order to make theriac," noted the apothecary Giovan Battista Fulcheri in Lucca.[3] Renewed emphasis on contagion theory also contributed to the popularity of theriac in the sixteenth century since it was a medicinal compound whose innate virtue literally pulled disease from the body in contrast to humoral medicines that cured a patient by restoring equilibrium.[4] From the 1570s to the 1650s, theriac and its effect on contagious diseases took on added meaning due to the virulent outbreaks of plague.[5]

Widely consumed by Italian patricians to preserve health and ward off the ill effects of various "poisons," theriac was a standard feature of elite medical practice. Patrician illnesses required costly ingredients—spices, exotic

[2]Maranta, *Della theriaca et del mithridato*, pp. 8, 163. Theriac was reputed to cure plague, syphilis, epilepsy, apoplexy, asthma, catarrh, and a variety of everyday ills that preyed upon the human body; see Girolamo Donzellino, *De natura, causis, et legitima curatione febris pestilentis . . . in qua etiam de theriacae natura ac viribus latius disputatur* (Venice, 1570), c. 14v.

[3]For a description of the ceremony in Venice, see Angelo Schwartz, ed., *Per una storia della farmacia e del farmacista in Italia. Venezia e Veneto* (Bologna, 1981), pp. 48–49; for Bologna, see Piero Camporesi, *La miniera del mondo. Artieri, inventori, impostori* (Milan, 1990), pp. 261–262; BUB, *Aldrovandi*, ms. 38(2), Vol. III, c. 119 (Lucca, 24 April 1571).

[4]Melissa P. Chase, "Fevers, Poisons and Apostemes: Authority and Experience in Montpellier Plague Treatises," in *Science and Technology in Medieval Society*, Pamela O. Long, ed. (New York, 1985), pp. 153–169; and Vivian Nutton, "The Seeds of Disease: An Explanation of Contagion and Infection from the Greeks to the Renaissance," *Medical History* 27 (1987): 1–34. Thanks to Katy Park and Nancy Siraisi for help in refining this argument.

[5]I am thinking particularly of the pandemics of 1575–1577, 1630–1631, and 1656–1657. For a statistical survey of the demography of these plagues, see Lorenzo del Panta, *Le epidemie nella storia demografica italiana (secoli XIV–XIX)* (Turin, 1980), pp. 138–178. The literature on these epidemics is large, so I will mention only a few works available in English: Carlo M. Cipolla, *Fighting the Plague in Seventeenth-Century Italy* (Madison, WI, 1981); Giulia Calvi, *Histories of a Plague Year: The Social and the Imaginary in Baroque Florence*, Dario Biocca and Bryant T. Ragan, Jr., trans. (Berkeley, 1989); and Richard Palmer, *The Control of Plague in Venice and Northern Italy 1348–1600* (Ph.D. diss., University of Kent at Canterbury, 1978). For recommendations of theriac as a plague remedy, see Mercati, *Instruttione sopra la peste* (Rome, 1576), p. 95; Giovan Filippo Ingrassi, *Informatione del pestifero e contagioso morbo* (Palermo, 1576), in Dollo, *Filosofia e scienza in Sicilia*, p. 26; and Girolamo Ruscelli, *De' secreti del donno reverendo Alessio* (Venice, 1564–1568 ed.), Pt. I, p. 49v.

animals, and precious minerals, in short, the very items contained in muse-ums—in contrast to the medicines used on less wealthy patients.[6] Vipers, whose inclusion in theriac distinguished it from mithridatum and other well-known compounds, provided precisely this sort of substance. As the source of poison as well as the cure, vipers were highly meaningful objects, particularly in a Christian culture that perceived them as the embodiment of the coexistence of good and evil in the world. Just as acknowledging one's sins cured the soul, swallowing the flesh of a viper, specially prepared for in-clusion in theriac, healed the body. When naturalists chose to highlight the collecting of vipers as an important feature of their activities, they signaled their adherence to the traditions of natural magic and sympathetic medi-cine that made the viper a privileged, albeit problematic, object. They also underscored the close relationship between medicine and natural history. Only by anatomizing vipers and inspecting the other parts of nature desig-nated as "simples"—the essential ingredients for medical compounds—could medicines such as theriac be restored to their original Galenic state. Only by testing the efficacy of medically useful objects such as vipers, uni-corn's horns, bezoar stones, and serpent's stones through numerous ex-periments could their potency be confirmed. Thus, the collecting habits of early modern naturalists were directly linked to the reform of medicine. Medicine, like natural philosophy, took its place within the public culture of science that made seeing and experimenting the criteria for truth.

In September 1577, Aldrovandi received two Libyan vipers from the Grand Duke of Tuscany, Francesco I. They appeared in two different forms: writhing in a box delivered by courier and immortalized in a watercolor by the Florentine court painter, Jacopo Ligozzi (fig. 20). The latter was done at Aldrovandi's request. As he explained to Francesco I, one of the vipers had died before his own illustrator could return, so he no longer was able to record accurately both creatures as they appeared in their natural state, something Ligozzi had done for the Grand Duke prior to their shipment to Bologna. Never one to waste an opportunity, however, he affirmed that he would preserve the dead viper "to put it in my Museum."[7] In 1580 he re-ported to Francesco I the results of an experiment he had performed with the grandducal gift. In a classic example of confrontation between effect and antidote, Aldrovandi attempted to determine how long venom was po-tent. Tasting the liquid extracted from the two vipers, both dead at this point, Aldrovandi discovered that it was indeed still active. Upon feeling the ill effects, he spit it out and immediately "took a little bit of a most perfect

[6]Lodovico Settala dubbed vegetable cures such as garlic and onions the "theriac of peasants and the poor," highlighting the social specificity of this medicine; Settala, *Preservatione dalla peste* (Milan, 1630), p. 37. See also Piero Camporesi, *Bread of Dreams: Food and Fantasy in Early Modern Europe,* David Gentilcore, trans. (Chicago, 1989), pp. 73, 103. As Camporesi observes, popular medicines for the lower classes were vegetable based.

[7]BUB, *Aldrovandi,* ms. 6, c. 3r (Bologna, 19 September 1577).

Figure 20. The grandducal vipers as depicted by Jacopo Ligozzi. From BUB, *Aldrovandi. Tavole di animali.* IV, 132.

theriac, twelve years old."[8] Like many objects in his museum, the vipers yielded productive medical knowledge.

Aldrovandi had two reasons to memorialize the Grand Duke's gift. First, it was a present from an illustrious patron, a testimony to their mutual interest in the study of nature. Aldrovandi only recently had returned from a trip to Florence, where he had visited the Grand Duke's collections and conversed with Francesco I about their contents. Undoubtedly, the gift of the vipers was a sign of Francesco I's willingness to continue his relationship with Aldrovandi. Second, the vipers arrived shortly after Aldrovandi had concluded a lengthy and acrimonious dispute with the College of Physicians in Bologna over the proper ingredients in theriac. The trip to Florence occurred almost as an afterthought—although Aldrovandi was never one to leave things to chance—as he returned from Rome in June 1577. Aldrovandi had spent most of the spring in the papal city, successfully persuading Gregory XIII to intercede with the College of Physicians on his behalf and resolve the dispute in his favor. Thus, the appearance of two vipers in his museum the following fall had an added significance: they testified to his unimpeachable authority as the arbiter of medical knowledge in Bologna, if not all Italy. The pope had confirmed Aldrovandi's primacy in this realm by forcing the College of Physicians to reinstate his cousin within its ranks; the Grand Duke added weight to this decision by offering Aldrovandi two of his best vipers, tangible reminders of the Bolognese naturalist's unceasing efforts to improve the study of nature as medically necessary knowledge.

Like many collectors of nature, Aldrovandi perceived his primary goal to be the reform of *materia medica,* the formal study of nature as a medical entity prescribed by Galen, Dioscorides, and medieval commentators such as Avicenna. The appearance of museums of natural history in Italy was closely tied to the transformation of the medical profession. While Aldrovandi's catalogue of visitors does not allow us to calculate precisely how many physicians, surgeons, apothecaries, medical students, and professors of medicine visited his museum, the number of them who appear in his correspondence confirms that medical practitioners were an important audience for his activities and one of the most significant group of collectors.[9] From a variety of different perspectives, physicians, apothecaries, and professors of medicine all perceived collecting to be integral to the advancement of medical knowledge and to the enhancement of their standing within the medical community.

Apothecaries collected specimens as a natural part of their professional activities; they were the ingredients for the medicines sold in pharmacies.

[8]BUB, *Aldrovandi,* ms. 6, Vol. II, c. 57V (Bologna, 15 August 1580).

[9]As table 2 indicates, 6.0 percent of his visitors were identified as members of the medical profession, but it is unclear how many of the professors, university graduates, and students who composed an additional 60.4 percent held or were studying toward medical degrees.

Collecting increased the status of men such as Calzolari and Imperato by publicizing their possession of the most exotic ingredients that nature could supply. When Imperato described his museum as "full of an infinite number of rare things for the apothecary's profession," he defined the professional goal behind collecting.[10] Each item Imperato displayed reinforced the authoritative nature of the medicines he sold to customers. Physicians and patients linked the efficacy of medicines to their rarity, the criterion by which objects also found their way into museums. Collecting further defined good medical practice by expanding the material possibilities of medicine, privileging physicians and apothecaries who mastered the sensual reality of nature. Apothecaries were the first collectors to limit their museums consciously to the natural world because of their professional interest in the subject; physicians only followed suit as the medical profession placed greater emphasis on the teaching and practice of *materia medica*. "My theater of nature consists of nothing but natural things," affirmed Imperato, "that is, Minerals, Animals and Plants, of which I have collected several thousand to date."[11] While many collectors combined the acquisition of nature with that of antiquity, paintings, and other approved humanistic subjects, apothecaries such as Imperato defined nature as medically necessary knowledge.

From a different perspective, physicians increasingly collected natural specimens to fit their new image as observers of nature and practitioners of *materia medica*. "We must gather those things which seem to enrich the medical storehouse from all sides, so that like Argonauts we may bring back the golden fleece from our journey," counseled Bartholin in his *On Medical Travel*. "Everything which occurs attracts the eye of the physician."[12] By the mid-sixteenth century, physicians commonly acknowledged the study of nature as the foundation of good medical and pharmaceutical practice. "[The physician] cannot be an expert, as Galen testifies, unless he really knows the true instruments of his profession, that is, the pharmaceutical aspect of medicines, as simples as well as compounds," affirmed Aldrovandi. "We cannot compose a medicine without first knowing the simples."[13] The desire to make natural history an important part of medical education translated into specific curricular reforms within the faculties of medicine at most Italian universities. Permanent lectureships in natural history were introduced, botanical gardens founded, and museums were given new prominence as centers of pedagogy. By the end of the sixteenth century, no candidate for a medical degree in Italy could graduate without knowing something about

[10]Imperato, *Dell'historia naturale*, sig. A.2. The Poitiers apothecary Paul Contant also described his cabinet as "an apothecary's source of material"; in Barbara Balsiger, *The 'Kunst- und Wunderkammern': A Catalogue Raisonné of Collecting in Germany, France and England 1565–1750* (Ph.D. diss., University of Pittsburgh, 1970), pp. 144–145.

[11]In Neviani, "Ferrante Imperato," p. 75 (Naples, 25 September 1597).

[12]Bartholin, *On Medical Travel*, p. 57.

[13]Aldrovandi, *Discorso naturale*, p. 203.

medicinal simples, a practice that quickly spread to other European universities. Natural history had become part of the "expertise" required to become a physician.

The transformation of the medical curriculum soon translated into new responsibilities for the practicing physician. Increased interest in collecting and studying nature led the Colleges of Physicians to redefine the scope of their authority to include the regulation of pharmacy. The new legitimacy of natural history as a field of study gave physicians and, to a lesser extent, apothecaries who possessed this knowledge added credentials to inspect medicines. Physicians used their training in medicinal simples to expand their regulation of apothecaries, just as the study of anatomy justified their increased supervision of surgeons and midwives. Collecting not only improved the visibility of various members of the medical profession but also gave them the tools to expunge the "errors" that had crept into medicine since the time of Galen. It reinforced the authority of humanistically trained physicians, whose primary goal was the restoration of ancient medical learning, and it increased the status of those members of the medical community who participated in the rebirth of an ancient discipline. In the controversies about medicines that occurred throughout the sixteenth and seventeenth centuries, the intellectual goals of the new medical learning and social objectives of an increasingly hierarchical and centralized medical profession coalesced.

Nowhere was this more apparent than in the debates about theriac which involved the entire community of naturalists in the second half of the sixteenth century. As the most elaborate and comprehensive of all the ancient medicines, theriac became a focal point for numerous debates about the quality and availability of ingredients and the efficacy of antidotes. Rather than questioning the validity of theriac as a panacea, naturalists—like the subjects who attacked the king's evil counselors rather than the monarch himself[14]—castigated the corrupt recipes and inept practices that had crept into the medical profession. Such an approach gave Renaissance physicians the impetus to reform rather than reject the existing system of medicine. By collecting nature, naturalists activated lists of ingredients, forcing themselves to compare Galen's canonical list of simples that entered into theriac against actual specimens. Increased travel, closer comparison of different samples with textual descriptions, and the development of a wide network of naturalists who exchanged information, specimens, and opinions created a dynamic medical culture committed to reviving theriac in its original form. By bringing medicinal simples such as the two Medici vipers into the *studio*, physicians and apothecaries used the objects in their museums to make decisions about the purity of ingredients and the validity of substitutes when ancient substances could not be found.

[14]Thanks to Anthony Grafton for this apt analogy.

As already discussed in chapters 4 and 5, identifying a specimen was not simply a matter of experience but also authority. In the emerging professional hierarchies of late Renaissance Italy, physicians who did not collect nature possessed neither the empirical knowledge nor the social prestige of their colleagues who followed this latest trend. Apothecaries possessed *experientia* but not knowledge due to their inferior academic credentials. They could support the conclusions of professors of medicine such as Aldrovandi but were incapable of establishing independent opinions about nature. Collecting nature demarcated the competing spheres of medical activity, as different sectors of the medical profession jealously guarded their privileges. Through the introduction of *materia medica* into the universities and through the central role of this form of knowledge and its associated practices—collecting, anatomizing, and experimenting—in shaping the debates over the preparation and dispersal of medicines, natural history emerged as an important feature of early modern medical culture.

RETHINKING THE ROLE OF *MATERIA MEDICA*

Before turning to the debates about theriac, we need to consider the institutional landscape in which they were conducted. Collecting and university reform both resulted from the impact of humanistic culture on the medical profession. A fresh look at old texts and the appearance of a few new ones engendered new practices both in the *studio* and in the university. Many prominent collectors of nature were professors of medicine, while others received their initial training in the universities. Thus, they were in a position to both benefit from and transform the medical curriculum to reflect the growing interest in *materia medica*. Until the middle of the sixteenth century, natural history played little or no part in the medical curriculum of the European university system. "Although in Ferrara there are many physicians in all other ways learned, very few of them have a knowledge of simples," remarked Alfonso Pancio in 1579.[15] Reflecting back upon the changes that had occurred in the study of natural history during his lifetime, Aldrovandi described his discipline as a "faculty buried for so many hundreds of years in the gloom of ignorance and silence."[16] The traditional teaching of medicine allowed little room for subjects such as anatomy, chemistry, and natural history that were not properly philosophical forms of inquiry. "He that would be an excellent physician must first be a philosopher," advised Giovan Battista Silvatico in his history of the College of Physicians in Milan.[17] Medicine was part of an unbroken plane of knowledge that moved imper-

---

[15]AMo, *Archivio per materie. Storia naturale. Naturalisti e semplicisti* (1579).

[16]AI, *Fondo Paleotti*, ms. 59 (F 30) 29/7 (Bologna, 11 December 1585).

[17]Giovan Battista Silvatico, *Collegio Mediolanensis Medicorum origo* (Milan, 1607), pp. 26–27, in Carlo Cipolla, *Public Health and the Medical Profession in the Renaissance* (Cambridge, 1976), p. 4.

ceptibly from the study of philosophy to the study of medicine, with natural history falling somewhere between the two.

Even with the introduction of more empirical components to the medical curriculum—anatomical dissections, botanical demonstrations, and chemical experiments—professors of medicine continued to exalt theory over practice. The museum itself reflected the ambiguous relationship between these two categories. "I know . . . that the things shown to me by Costa have inadvertently turned me away from medical issues to philosophical questions," reflected Giovan Battista Cavallara upon seeing the museum of the physician Marcello Donati in Mantua.[18] While increasingly committed to the importance of sensory information, physicians nonetheless mirrored the prejudices of their profession when they expressed a preference for discussions of natural philosophy over *materia medica*. Only with the attempts of naturalists such as Aldrovandi to elevate the study of nature to the status of philosophy, and with its success as courtly pastime for princes such as Francesco I, did it become a fashionable pursuit among physicians.

The introduction of natural history into the medical curriculum of the Italian universities arose from the growing demand for practical demonstrations of *materia medica* to train physicians to recognize the materials out of which medicinal compounds were made. "How can anyone exercise his art without knowing the means and the subsidies?" exclaimed Gasparc Gabrieli in 1543.[19] At the beginning of the sixteenth century, physicians considered the study of nature to be peripheral to their medical training; a half-century later it was no longer superfluous. "Giovan Angello must attend the practice of the composition of medicines," wrote Giovan Pietro Giudoli to Aldrovandi in 1561, reminding the naturalist of his promise "to host him in your home in order to make him understand the knowledge of simples and other things pertinent to pharmacy."[20] Less than a decade after his own graduation from the University of Bologna, Aldrovandi was already actively engaged in the process of transforming the medical curriculum, using his museum to train physicians in the art of medicinal simples.

Humanistic inquiry into medicine and natural history provided the impetus behind these changes. The rediscovery of Theophrastus's botanical writings and the publication of standard editions of Galen, Dioscorides, Pliny, and Avicenna gave *materia medica* a new prominence in the textual universe that most physicians inhabited. Professors of medicine and natural history such as Ghini, Falloppia, and Aldrovandi were instrumental in the revival of this aspect of medicine. Through their efforts, the humanist debates about classification, identification, and medical utility were brought into the classroom because of their decision to lecture on texts that integrated

[18]*Discorsi di M. Filippo Costa*, sig. Ee.4v.
[19]Gioelli, "Gaspare Gabrieli," p. 38.
[20]BUB, *Aldrovandi*, ms. 38(2), Vol. I, c. 216 (Masera, 31 August 1561).

materia medica into medicine. With the appearance of Mattioli's edition of Dioscorides's *Materials of Medicine* in 1544, reissued frequently throughout the second half of the sixteenth century, scholars turned their attention to this important and relatively neglected subject. That same year, Ghini read Dioscorides's book on minerals in his inaugural lecture at Pisa.[21] By the 1550s, professors throughout Italy incorporated it into their lectures. "Signor Falloppia has not yet begun to lecture on Dioscorides," wrote the medical student Annibale Terenti to Aldrovandi from Padua in the fall of 1558, "but I believe that he will begin soon. And if Your Excellence could write a few words to him recommending me, so that I could more easily attend to the faculty of simples, I would be most grateful."[22] The anatomist Falloppia was not the only professor to build his curriculum around the new editions of Dioscorides then rolling off of the presses. Prospero Alpino, who held the chair in medicinal simples at Padua from 1594 to 1616, devoted at least eight of his year-long lectures to Dioscorides.[23] Aldrovandi, who befriended Falloppia during a trip to Padua in 1554, lectured on all five books of the *Materials of Medicine* at Bologna between 1556 and 1566. Like his friend Giganti, who listed *il Dioscoride* as the only book worthy of note in his museum, Aldrovandi gave special prominence to the work of this ancient authority, writing commentaries on the *Materials of Medicine* that complemented and refined those published by Mattioli.[24] This was a technique that he and other professors used frequently, not just in their lectures on Dioscorides but also in the teaching of related texts such as Galen's *On Medicinal Simples* and the second book of Avicenna's *Canon of Medicine*.[25] Through teaching, translation, and commentary, they revived the ancient tradition of *materia medica*, both in the classroom and in the museum.

[21]Alberto Chiarugi, "Le date di fondazioni dei primi orti botanici del mondo," *Nuovo giornale botanico italiano*, n.s. 60, 4 (1953): 815.

[22]BUB, *Aldrovandi*, ms. 38(2), Vol. IV, c. 359 (Padua, 13 November 1558). John Riddle has counted 49 Latin editions and 43 vernacular translations of Dioscorides' *Materia medica* in the sixteenth century alone, along with 96 printings of 36 separate commentaries on this book; Riddle, *Dioscorides on Pharmacy and Medicine* (Austin, TX, 1985), p. xix.

[23]Giuseppe Ongaro, "Contributi alla biografia di Prospero Alpini," *Acta medicae historiae Patavina* 8–10 (1961–1963): 118. For a general overview of the importance of Dioscorides to Renaissance medicine, see Jerry Stannard, "Dioscorides and Renaissance Materia Medica," in M. Florkin, ed., *Materia Medica in the XVIth Century*, Analecta Medico-Historica, 1 (Oxford, 1966), pp. 1–21.

[24]Tugnoli Pattaro, *Metodo e sistema delle scienze nel pensiero di Ulisse Aldrovandi*, pp. 150–152; Fragnito, *In museo e in villa*, p. 198. For Aldrovandi's commentaries, see BUB, *Aldrovandi*, ms. 77, Vol. I–III; Vol. XVI, c. 98; Vol. I, cc. 92–148; Vol. IV, pp. 2–9r, 44–48r.

[25]For a broader discussion of these two other medical authorities, see Owsei Temkin, *Galenism: The Rise and Decline of a Medical Philosophy* (Ithaca, 1973); and Nancy Siraisi, *Avicenna in Renaissance Italy: The Canon and Medical Teachings in Italian Universities after 1500* (Princeton, 1987), esp. pp. 74–124. On the fortunes of Galen's texts, see Richard J. Durling, "A Chronological Census of Renaissance Editions and Translations of Galen," *Journal of the Warburg and Courtauld Institutes* 24 (1961): 230–305.

While apothecaries did not have direct access to the medical education enjoyed by physicians, they nonetheless absorbed and participated in the same humanistic culture. Imperato, for instance, collaborated with Maranta (who taught at Salerno) in the making of theriac according to the new, more authentically Galenic guidelines. The Mantuan physician Filippo Costa praised Imperato for making a medicine, following the guidelines outlined in Maranta's *On Theriac and Mithridatum*, that "differs from the rules of Mesue and conforms to those of Galen." Maranta himself assured readers that Imperato distilled all his ingredients according to the procedure outlined by Dioscorides.[26] With the guidance of physicians trained in the humanistic medical curriculum, apothecaries transformed the new philosophy behind *materia medica* into a medical reality.

In a lecture of 3 November 1543, Gaspare Gabrieli, the first professor of simples at the University of Ferrara, aptly summarized the need to revitalize the study of natural history as an academic discipline. Decrying the lack of interest among physicians in *materia medica,* he wrote

> In my opinion, this attitude derives solely from the belief that the part of medicine dealing with knowledge of plants does not concern them. They leave the entire study of this branch [of medicine] to chemists, apothecaries, and wise-women. Thus at present the entire medicine of herbs is in the hands of the unlearned, the foolish, and superstitious wise-women. Not surprisingly, infinite errors occur from this incompetence.[27]

The reform of the medical curriculum in the sixteenth century minimized the role of unlearned practitioners in all aspects of medicine. Close textual rereadings of Galen, Dioscorides, and Avicenna revealed the importance of *materia medica* to the study as well as the practice of medicine; as a result, physicians invoked the classical authorities when they usurped the knowledge of the apothecaries through their inclusion of *materia medica* in the academic curriculum.

Both Galen and Dioscorides justified the subordination of apothecaries to physicians through the argument that medicine, as a philosophical discipline, was the highest art and its practitioners the most skilled healers. Philosophy, wrote Galen, was the means to truth, and only physicians could claim to know philosophy by virtue of their education.[28] According to the classical model, apothecaries were simply "makers of remedies" (*pharmakopolai*), while physicians determined what those remedies would be. Naturalists reactivated a venerable intellectual tradition of incorporating natural history within the study of medicine when they brought it into the university curriculum. "It is, I suppose, obvious to everyone that pharmacology is a necessity," wrote Dioscorides in the preface to the *Materials of*

[26]*Discorsi di M. Filippo Costa*, p. 48; Maranta, *Della theriaca et mithridato*, p. 238.
[27]Gioelli, "Gaspare Gabrieli," pp. 37–38.
[28]Temkin, *Galenism*, p. 10.

Medicine, "closely linked to the whole art of medicine, and forging with its every part an invincible alliance." The first-century physician also offered a concise summary of the proper method for obtaining knowledge of medically useful objects. "For I have exercised the greatest precision in getting to know most of the subject through direct observation, and in checking what was universally accepted in the written records."[29] This statement formed the basis of the curricular reforms instituted in the faculties of medicine in late Renaissance Italy.

The increased circulation and critical inspection of other texts, such as Pliny's *Natural History,* also spurred the introduction of natural history to the medical curriculum. In Ferrara, site of the most virulent debates about Pliny, Antonio Musa Brasavola and his pupils Gaspare Gabrieli and Amato Lusitano (the famous Portuguese commentator on Dioscorides) made natural history an essential part of medical education. "I advise anyone who wishes to know botany well or learn the science of medicine to go to Ferrara," wrote Lusitano.[30] Falloppia, also a pupil of Brasavola, held the lectureship in simples at Ferrara in 1547–1548 before moving on to more illustrious positions at Pisa and Padua.[31] Under Brasavola's tutelage, physicians learned the techniques necessary for the critical inspection of medicinal simples—*experientia* combined with direct knowledge of classical texts. Like many of his contemporaries, Brasavola expressed contempt for the Latin translations of medieval Islamic commentaries, a staple of medieval and Renaissance academic life, preferring unmediated access to the words of Galen and Dioscorides. Greater familiarity with the words of the ancients was juxtaposed to increased intimacy with nature. Both, as Brasavola affirmed, made the physician qualified to supervise apothecaries in the preparation of medicines.

Ferrara was only one of many universities that participated in the resurgence of interest in natural history and *materia medica*. In 1513, Leo X appointed Giuliano da Foligno to the first chair in simples. While the closing of *La Sapienza* during the Sack of Rome in 1527 temporarily curtailed this lectureship, it was revived at mid-century, when natural history enjoyed a particular prominence in Rome, and was held by well-known naturalists such as the papal physician Andrea Bacci and the Lincean Johann Faber.[32] Italian universities with strong medical faculties soon followed the lead of

---

[29]In Riddle, *Dioscorides on Medicine and Pharmacy,* p. 21.

[30]Amato Lusitano, *Commentarium in Discoridem,* Vol. V, p. 372, in Gioelli, "Gaspare Gabrieli," p. 6. For more on Brasavola, see Thorndike, "Brasavola and Pharmacy," in his *A History of Magic and Experimental Science,* Vol. V, pp. 445–471.

[31]Giovanni Cascio Pratilli, *L'Università e il Principe: Gli Studi di Siena e di Pisa tra Rinascimento e Controriforma* (Florence, 1975), p. 149; Alessandro Visconti, *La storia dell'Università di Ferrara 1391–1950* (Bologna, 1950), p. 46.

[32]Filippo Maria Renazzi, *Storia dell'Università di Roma* (Rome, 1804), Vol. II, pp. 65–66. Bacci held the chair in simples from 1567 to 1600, followed by Faber until 1630; Alessandro Simili, "Alcune lettere inedite di Andrea Bacci a Ulisse Aldrovandi," *Atti del XXIV Congresso Nazionale di Storia della Medicina* (Rome, 1969), p. 428.

Rome. By 1533, the Venetian senate established a permanent chair in simples at the University of Padua; Bologna and Perugia followed suit in 1534 and 1537, respectively.[33] With the reopening of the University of Pisa in 1543, the Grand Duke Cosimo I wooed the eminent naturalist Ghini away from Bologna to hold its first chair in the subject. Varchi recalled Ghini's lectures during his student days in Bologna as evidencing "not only great knowledge but practice of all minerals."[34]

Ghini's students in Bologna and Pisa, along with Brasavola's in Ferrara, because the first generation of naturalists to make the teaching of *materia medica* a common feature of the medical curriculum. Aldrovandi, Bacci, Cesalpino, and Falloppia, among others, soon established their hegemony over this emerging subdiscipline. By the 1580s, smaller universities such as Parma, Pavia, and Siena had chairs in medicinal simples; in 1638, even the University of Messina could boast a position in this subject.[35] As Aldrovandi remarked in his 1573 survey of the University of Bologna for his patron Gabriele Palcotti, natural history was a subject "as necessary as any other of the *studio:* I say necessary to Physicians, useful to Philosophers and delightful for every sort of scholar."[36] The institutional genealogy of natural history, as a discipline first established by humanistically trained physicians who turned their attention to *materia medica* and popularized by subsequent generations of students, testifies to Aldrovandi's acuity in understanding the nature of the curricular reforms in which he participated.

Aldrovandi's own career clearly demonstrates the transformations then occurring in the faculties of medicine. Having first studied law and then philosophy, logic, and mathematics at the universities of Bologna and Padua in the 1540s, Aldrovandi took his degree in medicine and philosophy from the University of Bologna in 1553. In the fall of the following year, he was appointed to teach logic; at this point, his interest in natural history had already begun to crystallize.[37] Within a year, he began to teach philosophy,

---

[33]For Bologna, see Sabbatani, "La cattedra dei semplici fondata a Bologna da Luca Ghini," *Studi e memorie per la storia dell'Università di Bologna*, ser. 1, 9 (1926): 13–53. The first lecturer in simples at Perugia was Francesco Colombo (called "Platone"); Giuseppe Ermini, *Storia dell'Università di Perugia* (Bologna, 1947), pp. 207, 506.

[34]Benedetto Varchi, *Questione sull'alchimia,* Domenico Moreni, ed. (Florence, 1827), p. 34.

[35]Alessandro d'Alessandro, "Materiali per la storia dello *studium* di Parma (1545–1622)," in Gian Paolo Brizzi, Alessandro d'Alessandro, and Alessandro del Fante, *Università, Principe, Gesuiti. La politica farnesiana dell'istruzioni a Parma e Piacenza* (1545–1622) (Rome, 1980), pp. 61, 63; Gioelli, "Gaspare Gabrieli," p. 8; Danilo Marrara, *Lo studio di Siena nelle riforme del Granduca Ferdinando I (1589 e 1591)* (Milan, 1970), p. 20; Cascio Pratilli, *Università e Principe*, p. 80; Arturo Nannizzi, "I lettori dei semplici nello Studio senese," *Bullettino senese di storia patria* 16 (1909): 42–50; Dollo, *Modelli scientifici e filosofici nella Sicilia spagnola*, p. 150.

[36]BCAB, s. XIX. B. 3803, c. 5r (Ulisse Aldrovandi, *Informazione del rotulo del Studio di Bologna,* 27 September 1573).

[37]As early as 1551, Aldrovandi befriended Ghini, then teaching at Pisa but often in Bologna in the summers. Like his contact with the French naturalist Guillaume Rondelet during a stay in Rome (1549–1550), Aldrovandi's relationship with Ghini had a great impact on the direction of his intellectual interests.

reading lectures on most of Aristotle's major works. "I already have known of your reputation as a teacher of philosophy for many months," wrote Maranta, "but I was not aware of [your knowledge] of simples."[38] In 1556, Aldrovandi began to lecture on Dioscorides's *Materials of Medicine*; by 1560 he had initiated a series of lectures on Theophrastus's *On the Causes of Plants,* "placing all of natural philosophy in practice" as he described it.[39] Aldrovandi was the first to lecture on Theophrastus's botanical writings in a university course. He was similarly quick to incorporate Galen's writings on pharmacy into the curriculum. Despite the fact that the botanist Cesare Odone still held the *lectura de simplicibus,* a position inherited from Ghini in 1543, the popularity of Aldrovandi's lectures in natural history convinced the Senate of Bologna to create a separate post covering all of natural history, in the Plinian and Dioscoridian sense (*lectura philosophiae naturalis ordinaria de fossilibus, plantis et animalibus*) to which they appointed Aldrovandi on 22 February 1561. That same year, he read Galen's *On Theriac to Piso* to his medical students to inaugurate a new series of lectures in *materia medica* in his capacity as the first natural philosopher to teach natural history at the University.[40] By 1570, Fioravanti called him "the true professor of all three parts of medicine."[41]

In the space of ten years Aldrovandi moved from the teaching of logic and philosophy to the teaching of *materia medica,* producing numerous commentaries on Aristotle, Theophrastus, Galen, Dioscorides, and Pliny. "I teach what plants one should truly choose for medicinal uses to whoever makes use of medicines," he wrote to the Grand Duke Ferdinando I in 1588.[42] The pattern of Aldrovandi's teaching career ran parallel to that of Gabrieli at Ferrara, who also taught natural philosophy before taking a chair in botany.[43] In contrast, Falloppia at Padua saw the teaching of *materia medica* as secondary to his more important duties in the faculty of medicine. In a famous letter to Aldrovandi, written shortly after the Bolognese naturalist's acceptance of the

[38]Vallieri, "Le 22 lettere di Bartolomeo Maranta all'Aldrovandi," p. 743 (Naples, 6 March 1558).

[39]Aldrovandi, *Vita,* p. 13. Ghini, in fact, asked Aldrovandi for an index of the lectures he prepared on Theophrastus; De Toni, *Cinque lettere di Luca Ghini ad Ulisse Aldrovandi* (Padua, 1905), p. 11 (Pisa, 16 October 1553). Schmitt, "Science in the Italian Universities in the Sixteenth and Early Seventeenth Centuries," in *The Emergence of Science in Western Europe,* Maurice Crosland, ed. (New York, 1976), p. 43 (see BUB, *Aldrovandi,* ms. 78(1)). For a general discussion of the reappearance of Theophrastus's writings, see Schmitt, "Theophrastus in the Middle Ages," in his *The Aristotelian Tradition and Renaissance Universities* (London, 1984), III, pp. 251–270.

[40]Aldrovandi himself provides the best summary of the progress of his career in his *Vita,* pp. 10–14; see also BUB, *Aldrovandi,* ms. 25, c. 300r; Tugnoli Pattaro, *Metodo e sistema delle scienze nel pensiero di Ulisse Aldrovandi,* pp. 148–151; Mattirolo, p. 373.

[41]Fioravanti, *Tesoro della vita umana.*

[42]From a description of his research projects, dealing particularly with his *Commentaria in libros quinque Dioscoridis de Materia Medica;* Mattirolo, p. 782.

[43]Gioelli, "Gaspare Gabrieli," p. 11.

new chair in natural history, Falloppia chastized him for leaving medicine further and further behind. In his opinion, natural history was a supplement to the more important teaching of medicine and therefore not worthy of a philosopher with Aldrovandi's evident talent. "I do not wish to imply that you will be simply a herbalist," wrote Falloppia,

> However, it displeases me that you have made this transition—not because I dislike the profession, which you know that I still perform unworthily, but because I liked the first one better. It seems to me to be the more worthy one in every respect, and I will embrace you as a true and faithful friend if you return to [medicine] at the first opportunity that you can do so with honor, leaving the other to whoever wishes it. Thus I am able to leave my [duties in *materia medica*] and those in anatomy to attend only to medicine, as I would and will do voluntarily with the occasion arises.[44]

While *materia medica* had become an important and integral part of the medical curriculum by the middle of the sixteenth century, it still had neither the status nor the salary that the teaching of theoretical and practical medicine brought. Naturalists such as Ghini and Aldrovandi were relatively well paid compared to many contemporaries, yet their stipends were significantly lower than those of Cardano and Mercuriale, both of whom taught theoretical medicine in the faculty of medicine in Bologna while Aldrovandi was there.[45]

Clearly Aldrovandi's decision to teach natural history was motivated by more than professional success since he surely would have commanded a better salary had he followed a more traditional path. For him, like many collectors of nature, the universities provided the surest method of establishing natural history as a legitimate field of inquiry and of gaining common acceptance for the observational and textual practices developed by naturalists.[46] While admittance to court gave naturalists high visibility, their presence in the university allowed them to impact directly the education of aspiring physicians and natural philosophers. The institutional sanctioning of *materia medica* provided naturalists with the opportunity to expand their

---

[44]Di Pietro, "Epistolario di Gabriele Falloppia," p. 56 (Padua, 23 January 1561).

[45]Cardano received 521 scudi to teach *theoria* (1562–1570) in comparison with Aldrovandi's salary of 400 *scudi;* Alessandro Simili, *Gerolamo Cardano lettore e medico a Bologna, Nota II* (Bologna, 1969), pp. 9–10. Mercuriale, one of the most famous physicians in sixteenth-century Italy, received the extraordinary sum of 5400 *lire* to teach *theoria* (1587–1593) in one of the special posts reserved for professors who were not citizens of Bologna; Simili, *Gerolamo Mercuriale lettore e medico a Bologna, Nota II* (Bologna, 1966), p. 6.

[46]The emergence of natural history as a discipline follows a trajectory similar to fields such as astronomy, mathematics, and anatomy. See Mario Biagioli, "The Social Status of Italian Mathematicians, 1450–1600," *History of Science* 27 (1989): 41–95; Robert Westman, "The Astronomer's Role in the Sixteenth Century: A Preliminary Study," *History of Science* 18 (1980): 105–147. There has been no comparably broad study for the field of anatomy, although Daniel Brownstein is currently at work on this topic.

collecting practices from the museums and botanical gardens of private individuals into the official space of learning. Institutional recognition gave an added impetus to the activities already practiced by Renaissance naturalists by providing funding for the botanical gardens that became the centerpiece of the success of natural history in the medical curriculum.

LIVING MUSEUMS

While the observation of nature occurred in the course of travel and in the museums, it also took place in the university botanical gardens that gathered together plants from all corners of the earth for the perusal of medical students taking the new courses in *materia medica*. In conjunction with the establishment of a chair in medicinal simples at the University of Pisa in 1543, provision was made for the formation of a botanical garden, the first to appear in Europe. Naturalists praised the Grand Duke Cosimo I for being "the first to have a public garden of simples constructed, so that the Most Excellent Ghini could show the simples to the scholars after the public lessons."[47] Ghini was the ideal candidate to head the botanical garden in Pisa; as early as 1539, he lectured on Galen by showing simples.[48] As the first hereditary ruler of Tuscany, Cosimo I perceived the University of Pisa to be a public symbol of the well-being of the state. Prominent academicians and innovative teaching facilities such as the anatomy theater and botanical garden added luster to Tuscany by increasing the reputation of its university.[49] Cosimo I's activities challenged other rulers to match his innovations to maintain the reputation of their medical faculties. Within a short time, botanical gardens appeared at the universities of Padua (1545), Bologna (1568), and Messina (1638), as well as in Florence (1550) and Rome (1563).[50] Using the materials planted in the botanical gardens, naturalists reinforced the importance of demonstration in studying nature. "The Physician should know the instruments with which he induces health in the sick," observed Aldrovandi, citing Galen.[51] Collecting plants was an important step in the fulfillment of this goal.

[47]Aldrovandi, *Vita*, p. 27.

[48]Sabbatani, "La cattedra dei semplici," p. 20.

[49]The reorganization of the University of Pisa is discussed by Nancy Siraisi, "Giovanni Argenterio and Sixteenth-Century Medical Innovation," *Osiris*, ser. 2, 6 (1990): 163–165.

[50]By the late seventeenth century, there was also a botanical garden in Palermo, founded by Nicolò Gervasi. Much of this material, albeit with some inaccuracies, is summarized in P. A. Saccardo, "Contribuzioni alla storia della botanica italiana," *Malpighia* 8 (1894): 476–517, and Chiarugi, "Le date di fondazioni dei primi orti botanici del mondo," pp. 785–839. See also Azzi Visentini, *L'orto botanico a Padova;* Tongiorgi Tomasi, "Il giardino dei semplici dello Studio pisano," in *Livorno e Pisa;* Baldacci, "Ulisse Aldrovandi e l'orto botanico di Bologna," in *Intorno alla vita e alle opere di Ulisse Aldrovandi*, pp. 161–172; Dollo, *Modelli scientifici e filosofici nella Sicilia spagnola*, pp. 149–151; Cesare de Seta, Gianluigi Degli Esposti, and Cristoforo Masino, *Per una storia della farmacia e del farmacista in Italia. Sicilia* (Bologna, 1983), pp. 38–39, 42.

[51]BUB, *Aldrovandi*, ms. 25, c. 304r.

Ghini's garden soon became a congregating point for scholars from other universities, who visited it as they visited each other's museums. In 1553, shortly after taking his degree at Bologna, Aldrovandi traveled to Pisa where he

> selected and described all the rare herbs located in that garden of the most excellent Messer Luca Ghini, prefect of the garden of simples, constructed first for him by the commission of Grand Duke Cosimo, to whom all scholars of these beautiful things are indebted.[52]

Like the museum and the pharmacy, the botanical garden facilitated exchanges among scholars by providing a common location in which they could meet. While medical students studied plants there, patricians enjoyed the botanical garden as one of the "conversable spaces" that organized their leisure. In 1609, Lorenzo Pignoria informed the Vicenzan collector Paolo Gualdo that Prospero Alpino had invited him to "the first lesson that will open the botanical garden" in Padua.[53] In this fashion, professors used their botanical lessons to gain an audience beyond the confines of the university.

The medical significance of the activities within the botanical garden furthered its importance to the city as a whole. In an age obsessed with the purity of ingredients, the presence of a botanical garden assured each municipality of a steady supply of medicinal simples, particularly in times of plague when the Health Boards forbade the traffic of medicines between different regions. Many items highlighted in the botanical gardens were ingredients for theriac. "It is about forty years since I saw true *Amomo*," reminisced Aldrovandi in a letter to Evangelista Quattrami, "shown to me in the garden at Padua by Signor Aloigio Anguillara."[54] Collecting nature in the botanical garden materially improved the state of medicine. In this setting, scholars made decisions about the identification of simples essential to the making of authentic Galenic antidotes. Praising the Venetian Senate for underwriting the cost of the botanical garden at Padua, Mattioli commented that "all students of medicine and likewise physicians who come here, easily will be able to make themselves skilled and learned in the knowledge of simples quickly, without traveling through various parts of the world for many years."[55] Botanical gardens were not only constructed out of the materials naturalists brought back from their voyages but they became a replacement for travel itself—a laboratory of nature that allowed the observer to absorb the collective medical and botanical knowledge of the age. "These public and private gardens, with the lectures, are the reason that natural things are

[52]Aldrovandi, *Vita*, p. 27.

[53]Ongaro, "Contributi alla biografia di Prospero Alpini," p. 110.

[54]BUB, *Aldrovandi*, ms. 70, cc. 13r, 41r. Anguillara was the first prefect of the *orto botanico* at Padua 1546–1561.

[55]"Il Matthioli a gli studiosi lettori," in *I discorsi di M. Pietro Andrea Mattioli ne i sei libri della materia medicinale di Pedacio Dioscoride Anazarbeo* (Venice, 1557 ed.), n.p.

elucidated," remarked Aldrovandi, "joined to the New World that we are still discovering."[56]

Modeling himself after Ghini, Aldrovandi encouraged the Senate of Bologna to establish a botanical garden as early as 1554. In 1568, formal approval was finally granted for the formation of a garden, which Aldrovandi co-directed with Cesare Odone until 1571 and then directed by himself until his death.[57] "I believe that Your Most Illustrious Monsignor easily can understand if my lesson, and likewise the Garden annexed to this lesson like its limb, is most useful and necessary to the university," he wrote to Paleotti in 1573, "as the principal foundation of the *Protomedicato* and of all Pharmacy, like those that were introduced in Padua and Pisa with great expense many years ago."[58] Aldrovandi connected the presence of the botanical garden directly to the regulatory apparatus of medicine and pharmacy that oversaw the licensing and inspection of all medical practitioners; it exemplified the increased desire to incorporate training in practical problems into the medical curriculum.

Drawing upon his wide range of contacts with physicians, apothecaries, and other directors of university botanical gardens, Aldrovandi collected more than 1500 simples within a decade, creating one of the richest repositories of medicinal plants in Europe.[59] "I was so pleased to hear that you have created a public garden of simples in Bologna," wrote Maranta in 1570, "and that it is enriched with such a great number of plants and seeds in such a short time."[60] Visitors to the garden included other prefects of botanical gardens such as Casabona, who saw Aldrovandi's personal garden as well as the university's during a visit in 1583.[61] Similarly, Aldrovandi made a point of stopping in Padua to compare specimens with Wieland, upon returning from a visit to Verona in 1571. When his students traveled, Aldrovandi instructed them to collect samples from other botanical gardens and report on what they had seen.[62] Through an endless cycle of visits and exchanges, the botanical gardens grew in size and importance.

[56]BUB, *Aldrovandi*, ms. 70, c. 62r.

[57]Antonio Baldacci, "Ulisse Aldrovandi e l'orto botanico di Bologna," in *Intorno alla vita e alle opere di Ulisse Aldrovandi*, p. 4.

[58]BCAB, s. XIX, B. 3803, c. 6v.

[59]BUB, *Aldrovandi*, ms. 70, c. 13r. According to Aldrovandi, no more than 2000 plants, ancient and modern, had been catalogued by mankind until he personally surpassed this number through his collecting, indicating how quickly the inception of botanical gardens challenged such limits; *Discorso naturale*, pp. 182–183.

[60]Vallieri, "Le 22 lettere di Bartolomeo Maranta all'Aldrovandi," p. 770 (Molfetta, 9 April 1570).

[61]De Rosa, "Ulisse Aldrovandi e la Toscana," p. 213 (Bologna, 1 September 1583). Casabona was the first prefect of the *Galleria pisana*, the museum that adjoined the botanical garden; Tongiorgi Tomasi, "Inventario della galleria e attività iconografica dell'orto dei semplici dello Studio pisano tra Cinque e Seicento," *Annali dell'Instituto e Museo di Storia della Scienze di Firenze* 4 (1979): 22.

The formation of university botanical gardens depended upon the prefect's ability to collect specimens and seeds. Bartolomeo Ambrosini, custodian of the botanical garden and *Studio Aldrovandi* in Bologna, was praised in 1629 for "housing foreign gardeners who bring new plants into his home a few times every year."[63] The lists of plants that passed between Aldrovandi and his fellow naturalists indicate the extent to which this process of exchange permeated every aspect of their teaching and research. Collecting was not confined only to summer botanical excursions but continued throughout the year. "Remember that Signor Camillo Paleotti spoke to Bardello and Cesalpino, and that [the Grand Duke] has authorized Bardello to send me whatever I want from the garden in Pisa," scribbled Aldrovandi in a note to himself.[64] Such exchanges promoted the image of the botanical garden as a living laboratory, constantly permutating with the ebb and flow of simples among naturalists, who lectured and learned in such settings.

The presence of natural history museums as an adjunct to the university botanical gardens underscored the new emphasis on demonstration and observation in the medical curriculum. Just as dissection was limited to the winter months when bodies did not decompose as readily, the use of living specimens was, by definition, seasonal. While Pirro Maria Gabbrielli compiled a list of 130 plants "to show to scholars," the time allotted for his lectures at the University of Siena allowed him to demonstrate only 40 in 1684–1685.[65] But the new orientation of the medical curriculum demanded that observation be an on-going part of a physician's education. The skeletons, stones, minerals, herbaria (Aldrovandi claimed to have perfected the technique of drying and preserving plants[66]), and illustrations preserved in museums supplemented the anatomical and botanical demonstrations that were performed seasonally. The botanical gardens in both Pisa and Padua had collections attached to them by the end of the century. "The Grand Duke of Tuscany, Ferdinando, plans to have a room for subterranean things, as well as those above ground, installed in his garden in Pisa," observed Aldrovandi.[67] The Paduan garden was surrounded by "rooms and apartments that will serve for the various and diverse operations pertaining to the materials of medicine, for example, as foundries, distilleries, and other such

[62]Aldrovandi, *Vita*, p. 19. Aldrovandi's student Everard van Vorsten visited Carolus Clusius in Leiden in 1596 to deliver a letter from Aldrovandi; BUB, *Aldrovandi*, ms. 136, Vol. XXV, c. 133r.

[63]ASB, *Assunteria di studio. Requisiti dei lettori*, vol. I, n. 27.

[64]BUB, *Aldrovandi*, ms. 143, Vol. III, c. 220v. Cesalpino succeeded Ghini as professor of simples at Pisa.

[65]Nannizzi, "I lettori dei semplici nello Studio Senese," p. 49.

[66]Aldrovandi, *Vita*, p. 26.

[67]BUB, *Aldrovandi*, ms. 143, Vol. X, c. 2v; Tongiorgi Tomasi, "Inventario della galleria," pp. 21–27; idem, "Il giardino dei semplici dello studio pisano," in *Livorno e Pisa*, pp. 514–526.

works."[68] In Pisa and Padua, the botanical garden preceded the museum. In Bologna, Aldrovandi's *studio* preceded the formation of the botanical garden and served as a substitute for the construction of a university museum.

Aldrovandi's collection, which he used for demonstration purposes during his tenure as a university professor, essentially became the university museum in the seventeenth century. In 1606, the Flemish naturalist Johann Cornelius Uterwer, Aldrovandi's successor to the chair in natural history and the first custodian of his *studio*, petitioned the governing board of the University of Bologna to let him borrow specimens from the museum to use in his lectures:

> The most excellent Signor Doctor Ulisse Aldrovandi, of recent memory, left his library together with his *studio* of natural things to Your Most Illustrious Signors so that it should be conserved for the benefit of scholars. And he appointed me custodian of these things for the long service that I had shown him, conceding to me that I could use all of his things when the occasion warranted. Therefore, needing several things this term to show manually to scholars after each lecture, according to its content, I ask that you be so kind as to lend them to me.[69]

By the seventeenth century, as this petition demonstrates, museums and botanical gardens had been thoroughly integrated into the teaching of *materia medica* in the Italian universities. The culture of demonstration and experimentation begun in the private museums of humanists found its place in the medical curriculum.

As a sign of the success of *materia medica* in the curriculum, in 1564 the University of Padua split the chair into two positions: the lectureship in simples and a separate position demonstrating plants (*ostensor simplicium*), often held by the custodian of the botanical garden. The University of Perugia also chose to separate the teaching of medicinal simples into theory and practice.[70] The appearance of this division reflected the successful integration of natural history into the universities. Despite Aldrovandi's attempts to combine all these functions in one person—a move akin to Versalius's dramatic proclamations that he could both lecture and perform human anatomies—the traditional structure of the medical curriculum resisted the effacement of such divisions. In many universities, the professor lectured in the garden while the *ostensor* demonstrated the plants he described, an act that undermined the rhetoric of observation prevalent in natural history by

---

[68]Girolamo Porro, *L'Horto de i semplici di Padova* (Venice, 1572), sig. +4v–5r.

[69]ASB, *Assunteria di Studio. Diversorum.*, tome 10, n. 6 (*Carte relative allo Studio Aldrovandi*, Bologna, 21 October 1606).

[70]Chiarugi, "Le date di fondazione dei primi orti botanici del mondo," p. 826; Ermini, *Storia dell'Università di Perugia*, pp. 210, 288. We might use the 1790 description of the duties of the lecturers in *theoria* and *practica* to understand how the material was divided: "il lettore di teorica dei semplici parlerà nell'orto botanico delle piante pertinenti alla medicina, mentro il pratico avrà cura di preparare le piante da illustrare a lezione."

reinforcing the distance between the naturalist, as philosopher, and the objects he strove to understand. While naturalists made observation and demonstration integral to the study of medicine, reinforcing the lessons taught by anatomists during the same period, they also came to terms with the ambivalent meaning of success. Institutional recognition meant conformity. And the norms of the medical curriculum placed empirical study on the lowest rung of the academic hierarchy. The efforts of naturalists such as Aldrovandi, and particularly Cesalpino, to make natural history more philosophical reflected the difficulties of introducing a new form of learning into an essentially conservative academic culture.

The new medical curriculum of the Renaissance universities produced a generation of physicians trained to understand better the relationship among living specimens, textual descriptions, and artistic representations. "I saw the plant that Mattioli describes as *androsaces*," wrote Ambrosio Mariano to Aldrovandi, "but I am certain I did not see anything conforming to Mattioli's illustration among your plants when I was with you." Aldrovandi's mentor Francesco Petrollini described a visit to the *studio* of a physician who "on every herb discoursed with Dioscorides and Pliny constantly in hand."[71] While collectors initiated these practices through their interactions with like-minded scholars, the reforms effected within the medical curriculum gave official sanction to this new scientific culture. And the efforts of the first professors of natural history and the early custodians of museum and botanical gardens opened up the study of *materia medica* to a wider audience than it previously embraced.

POLICING THE PROFESSION

The introduction of *materia medica* into the medical curriculum brought a realm of knowledge traditionally monopolized by apothecaries under the domain of physicians. This had important professional ramifications. Mastery of nature, as many physicians argued, intellectually justified their subordination of less educated practitioners. "It is not my place to dispute this material now," wrote Pietro Fumagalli to Aldrovandi in response to his plans for reforming medical practice through the teaching of natural history, "but I say, for the foolish the knowledge of plants pertains to apothecaries and gardeners, that of fish to fishermen . . . and birds to falconers, so that each may defend his own ignorance."[72] The encyclopedic programs of naturalists such as Aldrovandi called into question these traditional divisions of knowledge by presenting the physician as someone capable of eroding these barriers. The physician's task, as many collectors envisioned it, was to combat the ignorance of other medical practitioners by instructing them in the

[71] BUB, *Aldrovandi*, ms. 38(2), Vol. I, c. 260 (Macerata, 25 May 1555); De Toni, *Spigolature aldrovandiane VII*, pp. 511–512 (15 November 1553).

[72] BUB, *Aldrovandi*, ms. 38(2), Vol. IV, c. 19r (Rome, 27 April 1565).

new techniques of natural history and by more closely regulating their activities. Apothecaries should "learn from physicians, and not physicians
from them," argued Aldrovandi. Evangelista Quattrami, botanist to the
Duke Alfonso II d'Este, concurred with this assessment, recommending the
study of *materia medica* to every physician so they would know "how terrible
it is to let every sort of quack apothecary (*Spetialetto*) compose such antidotes
without the knowledge of true simples."[73] The introduction of new mechanisms of control within the medical profession made this common practice
by the end of the sixteenth century.

While the desire to legislate medical practice from the top had long been
a goal of physicians, the collection and study of nature became new tools
with which to regulate the medical profession. By the sixteenth century, surgeons, apothecaries, midwives, and unlicensed practitioners all found themselves under intense scrutiny by the learned medical world.[74] Citing Galen,
physicians invoked their philosophical credentials as justification for the
new requirements and restrictions they imposed on other practitioners.
"Thus we conclude that the physician ought to be above the apothecaries
like the architect above the bricklayers, as Galen testifies in many places, so
that errors are not committed."[75] Increasingly, physicians believed that
greater intervention would improve the quality of medicines distributed to
their patients. Rather than leaving pharmacy to apothecaries, they used
their new knowledge of *materia medica* to officiate the activities surrounding
the preparation and dispersal of medicines. Museums and botanical gardens became the basis for physicians' claims to know nature better than any
apothecary. In these settings, they combined the *experientia* of an apothecary
with the *scientia* of a physician. Collecting, in effect, became a mechanism
for regulating pharmacy, a competitive strategy for countering the practices
of less sophisticated medical practitioners.

Since the Middle Ages, the regulation of medical practice had primarily
been the responsibility of the guilds and universities. During the sixteenth
century, two supervisory agencies that were instrumental in reorganizing
the medical profession achieved greater prominence: the College of Physicians and the *Protomedicato*. The increased importance of both organizations
exemplified the new forms of professional legitimation that arose from po-

---

[73]BUB, *Aldrovandi*, ms. 70, c. 14v; ASMo., *Archivio per le materie. Botanica. Naturalisti e Semplicisti*, f. 2 (Ferrara, 12 September 1595).

[74]For comparable studies of this subject dealing with other regions, see Alison Klairmont
Lingo, *The Rise of Medical Practitioners in Sixteenth-Century France: The Case of Lyons and Montpel-
lier* (Ph.D. diss., University of California, Berkeley, 1980), and Charles Webster, ed., *Health,
Medicine and Mortality in the Sixteenth Century* (Cambridge, 1979). For a discussion of medieval
licensing procedures, see Vern L. Bullough, "Training of Nonuniversity-Educated Medical
Practitioners in the Later Middle Ages," *Bulletin of the History of Medicine* 27 (1959): 446–458,
and Pearl Kibre, "The Faculty of Medicine at Paris, Charlatanism, and Unlicensed Medical
Practices in the Later Middle Ages," *Bulletin of the History of Medicine* 27 (1953): 1–20.

[75]Aldrovandi, *Discorso naturale*, p. 203.

litical centralization and the progressive aristocratization of society.[76] In Florence, the Grand Duke Cosimo I personally appointed the twelve men who composed the initial membership of the College of Physicians in 1560. Cities inhabited by hereditary rulers normally tied the creation of this new medical body to the increased status of physicians at court, linking the medical hierarchy to prevailing social structures.[77] The formation of colleges of physicians in cities such as Bologna, Verona, Milan, and Naples reflected their subsidiary status in the various territorial states of the Italian peninsula. Appointments did not come directly from the ruler but were channeled through local political officials who represented the interests of the territorial ruler *in absentia*. Regardless of the means of appointment, the College of Physicians quickly established its authority, officiating degrees and licenses, advising the health boards about the prevention of epidemics, and usurping the regulatory responsibilities that traditionally belonged to the guilds.[78]

In a parallel move, many apothecaries' guilds transformed themselves into colleges of apothecaries. The criteria for selection operated according to the same principles. Men with high social standing in their profession and close allegiances with the local government were chosen to supervise the activities of apothecaries. In cities such as Florence and Rome, court apothecaries figured prominently in the College of Apothecaries. Generally, the apothecaries chosen to head these associations were men with strong ties to the local medical establishment since they would be required to collaborate with the College of Physicians and *Protomedicato* in inspecting the work of their colleagues. This last factor made apothecaries with museums likely candidates for the office. As naturalists who collected to improve the state of medicine, they shared the same goal as physicians whose museums arose out of a similar desire to ennoble their profession. Both Calzolari and Imperato, for example, enjoyed a high degree of professional visibility. In Verona, Calzolari and his nephew both held the offices of prior and "visitor" on different occasions. In Naples, Imperato was a member of the council governing the Guild of Apothecaries, a consequence of his close ties with the Spanish Viceroy as well as his reputation as a naturalist.[79] Possessing na-

[76]I am indebted to Anthony Grafton for this point. For a more detailed overview of the development of these two offices, see Richard Palmer, "Physicians and the State in Post-Medieval Italy," in Andrew W. Russell, ed., *The Town and State Physician in Europe from the Middle Ages to the Enlightenment* (Wolfenbüttel, 1981), pp. 47–61.

[77]My discussion of the College of Physicians is drawn from Carlo Cipolla, *Public Health and the Medical Profession in the Renaissance* (Cambridge, U.K., 1976), pp. 72–75; see also Richard Palmer, "Medicine at the Papal Court in the Sixteenth Century," in Vivian Nutton, ed., *Medicine at the Courts of Europe 1500–1837* (London, 1990), pp. 49–78.

[78]For an overview of medical practice in Italy before the sixteenth century, see Katharine Park, *Doctors and Medicine in Early Renaissance Florence* (Princeton, 1985).

[79]ASVer., *Arte Speziale*, vol. 26, c. 19v; vol. 31, c. 94r and passim; Belloni Speciali, "La ricerca botanica dei Lincei a Napoli," p. 75.

ture helped to articulate not only divisions among medical practitioners but also differences *within* professional groups. Whether physicians or apothecaries, naturalists who collected rose to the top of the medical hierarchy.

The creation of the *Protomedicato,* which first appeared in Italy in the Kingdom of Sicily in 1397, further accentuated the divisions between physicians and other medical practitioners by creating a category of physician specifically empowered to oversee all aspects of the medical profession. It also linked the leading members of the College of Physicians and the College of Apothecaries together by placing the apothecaries most interested in reform under the supervision of physicians who enjoyed a certain autonomy from the jurisdiction of the local college. They formed the new elite within the medical community. The *Protomedicato* epitomized the concept of "expertise" developed by naturalists such as Aldrovandi. Consisting of a few prominent physicians, often appointed personally by the ruler, it allegedly was divorced from special interest groups such as the guilds and colleges of physicians and apothecaries and therefore was able to evaluate their activities impartially. The *protomedico* was often a collector—such as Aldrovandi in Bologna or Lodovico Settala who became *protofisico* for the Duchy of Milan in 1628[80]—whose museum gave him the authority to arbitrate the disputes that arose over correct procedures in the fabrication of medicines.

In Spain, Charles V established a *Protomedicato* in 1523 that soon established branches in the Kingdom of Naples and the Duchy of Milan, both under Spanish rule.[81] In the papal state, it first appeared in Rome in the late fifteenth century, as an office held by the prior of the College of Physicians. Leo X inaugurated the first *Protomedicato* in Bologna in 1517, where it grew out of the older *assumpti contra empyricos.* Until 1553, the year Aldrovandi entered the College of Physicians, the *Protomedicato* in Bologna remained under the direct supervision of the one in Rome and therefore had little power to effect wholesale reforms.[82] During the 1560s, it was removed from papal

[80]Rota Ghibaudi, *Ricerche su Lodovico Settala,* p. 40. Settala was also a member of the directorate of the Health Board of Milan in 1630, a position that gave him the authority to regulate yet another aspect of medical practice; Cipolla, *Public Health,* p. 37. For Gesner, see Mario Maragi, "Corrispondenze mediche di Ulisse Aldrovandi coi paesi germanici," *Pagine di storia della medicina* 13 (1969): 106.

[81]Goodman, *Power and Penury,* pp. 222–229; John Tate Lanning, *The Royal Protomedicato: The Regulation of the Medical Profession in the Spanish Empire,* John Jay Tepaske, ed. (Durham, NC, 1985); and L. De Rosa, "The 'Protomedicato' in Southern Italy: XVI–XIX Centuries," *Annales cisalpines d'histoire sociale,* ser. 1, 4 (1973): 103–117.

[82]For a general discussion of the *Protomedicato,* see Palmer, "Physicians and the state in postmedieval Italy," pp. 57–59. For Bologna, see Giuseppe Olmi, "Farmacopea antica e medicina moderna: la disputa sulla Teriaca nel Cinquecento bolognese," *Physis* 19 (1977): 277; Edoardo Rosa, "La teriaca panacea dell'antiquita approda all'Archiginnasio," in *L'Archiginnasio: Il Palazzo, l'Universita, la Biblioteca,* Giancarlo Roversi, ed. (Bologna, 1987), Vol. I, pp. 327–328. For Tuscany, see A. Garosi, "Medici, speziali, cerusici e medicastri nei libri del protomedicato senese," *Bullettino senese di storia patria,* n.s. VI, 42 (1935): 1–27; idem, "I protomedici del Collegio di Siena dal 1562 al 1808," *Bullettino senese di storia patria,* n.s. IX, 45 (1938): 173–181.

supervision and affiliated with the College of Physicians to better align its supervisory duties with the College's own interests in reforming medical practice within the city. The changes occurred largely due to Aldrovandi's efforts. "It is a great shame that a university town such as Bologna does not have its own *Protomedici,* who should be present at the composition of medicines, and an Antidotarium for the apothecaries to follow as a guidebook, like many other cities have," reflected Aldrovandi, somewhat rhetorically since he already held the office of *protomedico* when he wrote these words.[83] Like many of his contemporaries who defined the powers of the *Protomedicato,* Aldrovandi believed that greater autonomy and closer coordination among the different supervisory agencies would yield more effective medical reforms.

Physicians appointed to the *Protomedicato,* particularly in a city such as Bologna that perceived itself to be a center of medical innovation, enjoyed relatively unrestricted powers. "Who can correct the errors in the pharmacies if not the physicians and *protomedici* who are provided by the College for this purpose?" queried Aldrovandi.[84] Interest in regulating the profession provided an impetus to establishing controls over every stage of the apothecary's career. The licensing of apothecaries gradually became the responsibility of the College of Physicians and the *Protomedicato,* who no longer allowed the Guild of Apothecaries to regulate itself. According to the 1568 statutes of the Guild of Apothecaries in Verona, an apothecary needed to be examined by two physicians to open a shop.[85] Such exams tested one's knowledge of simples (including etymology and classification), basic literacy, and the ability to prepare medicines from written recipes.[86] Most importantly, the licensing of apothecaries by physicians clearly established their hierarchical relationship. "I will treat Physicians with due reverence," read the fourth statute of the Guild of Apothecaries in Modena.[87]

The academic medical world devised numerous methods to assure themselves that the apothecaries would not act independently. In his *Dialogue on the Deceptions of Certain Wicked Apothecaries* (1572), the physician Giovan Antonio Lodetto recommended to the Consul of Brescia that "no apothecary may make any sort of composition without the presence of a physician

[83]Aldrovandi, *Vita,* p. 16.

[84]BUB, *Aldrovandi,* ms. 70, c. 14v.

[85]ASVer., *Arte Speziali,* n. 26, c. 12r.

[86]Francois Prevet, *Les statuts et règlements des apothicaires* (Paris, 1950), Vol. V, p. 1142. It is interesting to note just how closely changes in licensing and medical curriculum in France and Italy paralleled each other. As Alison Lingo describes, licensing for apothecaries in Montpellier began in 1572, within a decade of the creation of courses in botany at the medical faculty. Under Laurent Joubert, chancellor of the Faculty of Medicine at the University of Montpellier, we see a reorientation of medical teaching and practice that mirrors the changes occurring in Bologna under Aldrovandi's tenure at the University; Lingo, *The Rise of the Medical Practitioners in Sixteenth-Century France,* pp. 68, 132–134, 145.

[87]BEst., Mss. Campori, y.Y. 5.50 (APP.1694), c. 9v.

and another apothecary," as was already standard practice in Milan, Florence, and Ferrara.[88] In addition to supervising the licensing of apothecaries, the *protomedico* made biannual or trimestral visits to all the pharmacies in the city to check the quality of ingredients in the antidotes and ensure that such compounds were made in accordance with the recipes standardized by the College of Physicians. In Siena, for example, the 1560 statutes of the apothecaries's guild recommended the election of a committee, consisting of the *protomedico* and two apothecaries, to inspect medicinal compounds and their ingredients. By the mid-sixteenth century, apothecaries could not make "any medicinal composition without first notifying the *Protomedico*," according to the 1568 statutes the Duchy of Savoy.[89] Fines were assessed for flagrant abuses of standardized procedures: 40 soldi in Siena in 1560, 25 ducati in Verona in 1568, 50 scudi in Bologna in 1573, and so on. The varied level of fines in different cities undoubtedly reflected the difficulties of translating ideals into practice, since apothecaries, for economic reasons as well as convenience, continued to transgress the guidelines established for them by physicians. Nonetheless, the physicians across Italy who were instrumental in effecting these reforms persevered in their inspection of medicinal simples and compounds, creating a system of regulation that exhibited a remarkable degree of continuity throughout the peninsula. "Thus it will be necessary, as we do in Bologna, that no medicine can be composed by an apothecary, if it is not first seen by the *protomedici*, so that they may inspect the legitimate ingredients," advised Aldrovandi in a letter to Quattrami, then undertaking a similar reform in Ferrara.[90]

The establishment of a specific office to regulate medical practice coincided with the creation of antidotaria, which were published catalogues of medicines and ingredients approved by the College of Physicians. The *Bologna Antidotarium* (1574), published by the College of Physicians largely because of Aldrovandi's efforts, was one of the first of its kind in Europe.[91] As the statutes of the Guild of Apothecaries in Verona reveals, antidotaria played a central role in the regulation of pharmacy by standardizing the recipes for medicines. "Everyone who will be approved to create medicines

[88]Giovan Antonio Lodetto, *Dialogo de gl'inganni d'alcuni malvagi speciali* (Venice, 1572), p. 19. The copy I consulted in the BUB was Aldrovandi's personal copy, demonstrating the extent of his interest in developing practical applications for the newly trained physicians that the medical program in Bologna and elsewhere produced.

[89]Siena, *Biblioteca Comunale degli Intronati*, ms. 116, A IV, 7, cc. 32v–35v; Prevet, *Les statuts et reglements des apothicaires*, Vol. VII, p. 1643 (20 October 1568).

[90]Siena, *Biblioteca Comunale degli Intronati*, ms. 116, A IV, 7 c. 33r; ASVer., *Arte Speziale*, n. 26, c. 13r; ASB, *Archivio dello studio bolognese. Collegi di medicina e d'arti. Nucleo attuario*, ms. 248, tome V (*Atto delli speziali dell'Arte*, 15 April 1573); BUB, *Aldrovandi*, ms. 70, c. 52v.

[91]Olmi, "Farmacopea antica e medicina moderna," p. 224. An antidotarium was not published in Sicily until 1637, and one did not appear in Milan until 1668; Dollo, *Modelli Scientifici e Filosofici nella Sicilia spagnola*, pp. 152–155; Luigi Belloni, "La medicina a Milano," in *Storia di Milano* (Milan, 1958), Vol. XI, p. 693.

is now obliged to avail himself, in good faith and all sincerity, to compose good simples and whatever other ingredients are ordained in the compositions from the antidotarium of the most excellent College of Physicians."[92] Particularly important compounds such as theriac, mithridatum, and acqua vitae were checked biannually by the *protomedico,* who often affixed seals to the jars to attest to their conformity to the standardized recipe. "And because many make acqua vitae with diverse sorts of ingredients such as coriander, cinnamon, cloves, sugar, and other similar things, and then dare to prescribe them as if they were physicians, for colic, uterine fevers, and stomachaches," declared the 1618 statutes for the Guild of Apothecaries in the Duchy of Savoy,

> we order that they do not dare to prescribe the said water without the approved order and license of the physicians. They must keep the seals above the vases that declare the ingredients with which the acqua vitae is made, under penalty of law. In order to ensure our people from abuses, we command our *Protomedico* to make visits to the said vases and said waters every six months, so that the above-mentioned orders may be inviolably executed.

During such visits, the *protomedico* was also advised to check the freshness of ingredients in the pharmacies and to supervise the disposal of simples that had been in the shop for over a year.[93] In all instances, the elaboration of the regulatory bureaucracy distanced authorized from unauthorized practice, articulating social differences that were becoming increasingly important among medical practitioners whose success depended upon their status.

One of the major concerns of the naturalists who attempted to regulate pharmacy regarded the quality of the ingredients used by apothecaries. "If I were to count all of the simples that have been substituted, either for fraud or from ignorance, the longest day in the year would not be enough time," wrote Lodetto.[94] Increased travel and the exploration of lesser known parts of the world produced an explosion of ingredients. Everywhere one looked, where there had been one choice before, there were now hundreds of simples. This situation produced two different results. On the one hand, it made available ingredients for classical antidotes that had been obscured since the time of Galen. Chastizing contemporaries who used false ingredients believing the originals to be extinct, Quattrami observed, "if they had read the various authors who write [on theriac], they would have discovered that Your Highness can have all the principal simples believed to be lost with

[92] ASVer., *Arte Speziale,* n. 26, c. 13r.

[93] Prevet, *Les statuts et reglements des apothicaires,* pp. 1659–1660 (1 April 1618). Fioravanti suggested that the stale ingredients used in medicaments were often the cause of their inefficacy: "Perche sono infiniti semplici, che stanno nelle botteghe, mutano qualità"; Fioravanti, *De' capricci medicinali* (Venice, 1573), p. 41r.

[94] Lodetto, *Dialogo,* p. 21r.

the greatest of ease from the different parts of the world—the very same ones that grew in the time of those who invented so glorious a compound."[95] The recovery of ancient simples, like the recovery of ancient texts, produced a sense of optimism among physicians who saw their efforts to reform *materia medica* yield tangible fruit. Collectors highlighted the possession of authentic simples as a sign of their skillfulness in constructing a material reality that replicated the medical programs of the ancients. Gherardo Cibo was so excited by his discovery of a two-headed viper conforming to Galen's description that he wrote immediately to Aldrovandi, who confirmed his classification.[96] Physicians praised apothecaries such as Calzolari and Imperato who specialized in increasing the number of "true" ingredients used in medicines.[97] Collecting and comparing simples was a necessary step in the recovery of ancient, hence "authentic," pharmacy.

On the other hand, travel and collecting introduced new simples into the pharmacopoeia. Like the guaiac wood that became a standard treatment for syphilis,[98] they represented a new material reality with which the medical profession had to come to terms. Initial reaction to these novelties was fairly conservative. Criticizing physicians and apothecaries who succumbed to the fashionability of *exotica*, Jan Eck remarked, "They wish to put Africa, Asia, Europe, and the New World together into their recipes, continuously concocting new things for their sick patients [that are] both dangerous and untested—something I have never tried."[99] Such admonitions reflected not only an intellectual conservatism on the part of scholars trained to revere the past but also a growing concern about the number of unorthodox medicines, old and new, on the market. Contemporaries were horrified at the shape complicated antidotes such as the theriac took, when the substitutions produced a drug that bore little resemblance to the ancient medicine. "Truly I am amazed upon seeing the composition of Theriac," exclaimed Fioravanti, "considering how so many things are put into it, one contrary to the other. . . . Similarly there are things put into Mithridatum that would kill people instantly, if [the ingredients] were given to them alone."[100]

---

[95]ASMo., *Archivio per le materie. Botanica. Naturalisti e semplicisti* (Dalle stanza della castellina [Ferrara], 12 September 1595).

[96]De Toni, *Spigolature aldrovandiane III*, p. 11.

[97]See Maranta's praise of Imperato's "good Myrrh" in his *Della theriaca et del mithridato* (Venice, 1572), p. 92. Karl Dannenfeldt discusses a classic example of the confusion in identifying simples in his "Egyptian Mumia: The Sixteenth Century Experience and Debate," *The Sixteenth Century Journal* 16 (1985): 163–180.

[98]Anna Foa, "The New and the Old: The Spread of Syphilis (1494–1530)," in *Sex and Gender in Historical Perspective*, Edward Muir and Guido Ruggiero, eds. (Baltimore, 1991), pp. 26–45; and Robert S. Munger, "Guaiacum, the Holy Wood from the New World," *Journal of the History of Medicine and Allied Sciences* 4 (1949): 196–229. For a similar attempt to replace ancient with modern medicines, see Karl H. Dannenfeldt, "The Introduction of a New Sixteenth-Century Drug: *Terra Silesiaca*," *Medical History* 28 (1984): 174–188.

[99]BANL, *Archivio Linceo*, ms. 18, c. 10r (San Candriglia, 5 February 1603).

[100]Fioravanti, *Dello specchio di scientia universale* (Venice, 1564), pp. 35v–36r.

More critical inspection of the ancient pharmacopoeia and the increased availability of a broader spectrum of ingredients led physicians and apothecaries to compose new medicines that claimed to succeed where traditional antidotes failed. Despairing that a perfect theriac could ever be concocted in his own age, Mattioli combined the ingredients of theriac and mithridatum to produce the most famous version of a compound known as "scorpion's oil." Antonio Bertioli in Mantua described it as "truly a liquid Theriac," reputed to have miraculous effects against the plague epidemic of 1575. Containing 123 ingredients (almost twice as many as theriac), Mattioli's scorpion's oil was prepared by his close friend Calzolari who sold it alongside his famous theriac in his pharmacy at the *Campana d'oro* in Verona.[101]

While Mattioli's medicine exemplified the salutatory aspects of new knowledge and was readily approved by physicians because of his status in the profession and its basis in ancient authority (Mattioli had combined two ancient medicines to produce a modern one), it also provided a dangerous precedent. Less learned practitioners created new drugs unsanctioned by the College of Physicians, the College of Apothecaries, and the *Protomedicato* such as "angelic oil," which a 1656 broadsheet advertised as having "all of Theriac's virtues, but greater."[102] While Mattioli invented his scorpion's oil as an alternative to the imperfectly made theriac and mithridatum, which failed to meet the standards set by the ancients, unlicensed practitioners created compounds that competed directly with ancient medicines, undermining their authority. By observing and standardizing the fabrication of pharmaceuticals, the medical profession hoped to eliminate the success of unorthodox practitioners such as Fioravanti, whose elixir vitae and electuaries drew business away from legitimate physicians and apothecaries.

The bewildering array of medicines available by the mid-sixteenth century led *protomedici* such as Alfonso Pancio to advise their rulers to appoint only physicians knowledgeable about *materia medica* to the *Protomedicato*: "It is very necessary that those who have the duty of visiting the pharmacies, besides sincerity, have a good knowledge of simples."[103] Collecting and observing nature became a precondition to the successful regulation of pharmacy. "Do we not read in Galen that the doctors in Rome, in his day,

[101]Antonio Bertioli, *Delle consideratione di Antonio Berthioli spora l'olio di scorpioni dell'eccellentissimo Matthioli* (Mantua, 1585), p. 8; Tergolina-Gislanzoni-Brasco, "Francesco Calzolari," p. 9. Don Antonio de' Medici recorded a recipe for *Olio contro a veleni del mattiolo* in his notebooks from the Casino at San Marco; BNF, *Magl.* XVI, 63, vol. I, ff. 105–108. The physician Girolamo Donzellino recommended it for external use in treating pestilential fevers, in contrast to the internally administered theriac; Linda Redmond, *Girolamo Donzellino, Medical Science and Protestantism in the Veneto* (Ph.D. diss., Stanford University, 1984), pp. 94–95. For a discussion of Mattioli's pessimism about the restoration of ancient medicines, see Thorndike, *A History of Magic and Experimental Science*, Vol. VI, p. 225; and Palmer, "Pharmacy in the Republic of Venice," p. 100ff.

[102]The broadsheet is reproduced in Marco Ferrari, "I secreti medicinali," in *Cultura popolare nell'Emilia Romagna. Medicina, erbe e magia*, p. 87.

[103]ASMo., *Archivio per materie. Storia naturale. Naturalisti e semplicisti* (1579).

conserved medicinal simples, as the apothecaries do today in their shops?" wrote Aldrovandi.[104] Galen surely would have approved of the fact that many specimens in Renaissance museums were ingredients for his antidotes. "Imperato most willingly would accept a little of that white cinnamon and true *costo* sent to you by Father Quattrami, whose letter on theriac ingredients has been in my possession for about three days," reported Pinelli to Carolus Clusius in Leiden.[105] Aldrovandi sent the Grand Duke Francesco I an Indian plant that he identified as *pseudocalamus,* an approved modern substitute for one of the classic theriac ingredients. When he visited the museum of the apothecary Stefano Rosselli during his second trip to Florence, he noted the presence of Peruvian balsam and true *costo,* the first a modern substitute for the ancient balsam described by Galen and Dioscorides and the second an example of an ancient simple recovered by naturalists during the sixteenth century.[106] Both were essential ingredients in theriac.

Naturalists relied heavily upon the contents of their museums to provide them with counterexamples to the simples they saw in pharmacies. The purpose of collecting, as most naturalists agreed, was to confront old specimens with new, true with false, and determine the safest and most effective combinations of ingredients for medicinal uses. Possessing medically potent objects established one's credentials as an authority in *materia medica.* Collectors coveted specimens of unicorn's horn, which reputedly drew forth poisons without even touching the body. From Prague in 1565, Mattioli sent Calzolari a sample of a "true and legitimate horn" personally presented to him by Maximilian I "so that it could be used as the most potent antidote against poisons."[107] Bezoar stones, the mineral calcifications found in the stomachs of various animals, also appeared in museums. As in the case of the unicorn's horn, great stock was placed in the possession of a true example since so many fakes were in circulation. "I would not advise anyone to buy [the stone] without first experimenting upon it," counseled Aldrovandi. "I have a piece of [the stone] in my museum that experience proves to be most perfect. It was donated to me many years ago by a Flemish man who had experimented with the same stone."[108]

Of all the simples prized by collectors, balsam—one of the last ingredients for theriac to be recovered—proved to be most elusive. Not until the 1580s did naturalists claim to possess it. Alpino, the naturalist credited with its recovery, made his reputation on this particular simple, arguing that the

[104]Aldrovandi, *Discorso naturale,* p. 203.

[105]In Neviani, "Ferrante Imperato," p. 75 (Venice, 16 February 1598).

[106]BUB, *Aldrovandi,* ms. 6, Vol. I, cc. 5v–6r; ms. 136, Vol. XI, cc. 66v–67r.

[107]In Bruno Accordi, "The Musaeum Calceolarium of Verona Illustrated in 1622 by Ceruti and Chiocco," *Geologica Romana* 16 (1977): 39; see also Andrea Bacci, *L'Alicorno.*

[108]Simili, "La pietra 'bezoar' in una relazione inedita dell'Aldrovandi e del Fonseca," *Atti del XVI Congresso Nazionale della Società Italiana di Storia della Medicina* (Bologna-Ravenna, 1959), pp. 401–402. See also Francesco Pona, *L'Amalthea overo della pietra bezoar orientale* (Venice, 1626).

Egyptian opobalsam collected during his tenure as physician to the Venetian ambassador in Cairo (1581–1584) was the original aromatic. When the apothecaries Giovan Paolo and Antonio Bertioli made *balsamo orientale* in Mantua in 1594, they asked Alpino to certify it. He did so by comparing it against his own "authentic" specimens in the botanical garden at Padua.[109] Others suggested that the balsam available in New Spain could easily replace the Egyptian brand, which was difficult to acquire.[110] Certainly Stefano Rosselli's display of Peruvian balsam to Aldrovandi in 1586 bears out this assertion. Regardless of which balsam one preferred to use, both the old and the new could be found in museums throughout Italy. When two apothecaries sent Giovanni Nardi samples of true opobalsam in 1639, he compared it against the balsam in the grandducal fondery. Even then, the foundry still contained samples sent by Aldrovandi and Alpino to Francesco I.[111]

The culture of demonstration that made the museum a site of medical knowledge was formalized in the statutes of the College of Physicians and the College of Apothecaries and in the regulations governing the *Protomedicato*. The 1569 reforms of the Guild of Apothecaries in Palermo demanded that all apothecaries "should make and place [their preparations] above their windows so that anyone may see them." Display became the primary means of authentication. In Messina in 1637, the learned Roman physician Pietro Castelli, founder of the botanical garden at the University of Messina, presided over the making of a sixty-two-ingredient theriac with the help of the College of Physicians: "seen, smelled, tasted, touched and with every simple most diligently examined; finally everything was approved with great applause."[112] Medicine was part of the theatrical culture of science that made the spectacle a legitimate form of inquiry. Rather than concocting medicines in secret like alchemists did, physicians and apothecaries insisted on their public display as the final stage in the process of authentication. In Verona, the apothecary Giovanni Pona submitted his ingredients for theriac to the *Protomedicato*. In Naples, the Lincean Stelliola invited the *protomedico* Giovan Antonio Pisano to examine the composition of the antidotes he described in his *Pamphlet on Theriac and Mithridatum* (1577); he and Della Porta, in turn, authenticated Fra Donato d'Eremita's elixer vitae in a display of Lincean fraternity.[113] Through such activities, the role of the spectator was further defined as someone capable of discerning truth. Social standing, the ability to engage in humanistic pursuits, and the possession of a high degree

[109]Giovan Paolo and Antonio Bertioli, *Breve avviso del vero balsamo, theriaca et mithridato* (Mantua, 1596), pp. 5, 9.

[110]Palmer, "Pharmacy in the Republic of Venice," pp. 109–110; Nicolo Monardes, *Delle cose che vengono portate dall'Indie Occidentali*, Italian trans. (Venice, 1584), p. 15.

[111]Giovanni Nardi, *Due lettere sopra il balsamo* (n.p., 1639), sig. A2r–A3v.

[112]Cesare de Seta, Gianluigi Degli Esposti, and Cristoforo Masino, *Per una storia della farmacia e del farmacista in Italia. Sicilia*, pp. 55, 57.

[113]Stelliola, *Theriace et mithridatia Nicolai Stelliolae Nolani libellus* (Naples, 1577); Belloni Speciali, "La ricerca botanica dei Lincei a Napoli," in Lomonaco and Torrini, *Galileo e Napoli*,

of civility all completed the image of the spectator, a category that defined the duties of the physicians and apothecaries assigned to the *Protomedicato*.

The three-day demonstrations of medicines mandated by colleges of physicians throughout Italy epitomized the relationship between professional privilege and public display. As the 1568 statutes of the Guild of Apothecaries in Verona prescribed,

> no one may presume to compose Theriac, Mithridatum or any other medicine . . . if he has not first displayed all of the ingredients for three days, so that they may be seen to the benefit of everyone who wishes to see them. After three days have passed, he may not compose the said antidote until the ingredients have been approved by the College of most excellent Physicians, under penalty of ten ducati.[114]

Descriptions of the composition of true balsam in 1594 by the Bertioli brothers in Mantua sheds further light on this process. As the ducal printer Francesco Osanno reported,

> For three whole days and nights . . . all the Ingredients were made in a large and most beautiful display in Piazza del Duomo before the pharmacy of [the Duke of Mantua]. Afterward the Physicians of the College and foreigners in the City [conducted] a most exact examination of everything, praising and admitting them as good and exquisite.

At the end of those three days, the balsam itself was "made with such fortune, ceremony, and honor as never before" (*con tanto fausto, festa & honore, che nulla piu*).[115] As Osanno suggested, the quality of the ceremony as well as the quality of the ingredients determined the success of the Mantuan balsam. Displayed in front of the leading physicians of the city and visitors such as Alpino, all scholars turned spectators, it gave testimony to the public culture of science that transformed medicine and pharmacy in the sixteenth century.

COLLECTORS TEST THEIR AUTHORITY

The relationship among collecting, new attitudes toward natural history, and the regulation of pharmacy is most clearly delineated in a series of disputes that arose in the 1560s and 1570s over the making of theriac. More than any particular professional group, it was the collectors of nature who emerged as the most learned and authoritative arbiters in the theriac disputes of the late sixteenth century. The obsession with the purity of ingredients and the desire to achieve and then surpass the level of knowledge set by the ancients created a social hierarchy within the medical profession that

---

pp. 76–77; and De Toni, "Il carteggio degli italiani col botanico Carolo Clusio," pp. 161–162 (Verona, 8 August 1606).

[114]ASVer., *Arte Speziali*, n. 26, c. 13v.

[115]Bertioli, *Breve avviso*, p. 5.

clearly reflected its intellectual order. The debates over theriac offer an example of how language informed social practice, for it was through the identification and articulation of a new descriptive model of science that collectors mapped the social contours of the medical world.

Discussing his plans "to write a treatise on the composition of Theriac and Mithridatum" in 1595, published two years later as the *Very Useful and Necessary Treatise on Theriac and Mithridatum,* Evangelista Quattrami reflected on the turbulent history of these antidotes: "Reading diverse treatises by many Colleges [of Physicians] in various Republics on this subject, I find many controversies—long and useless disputes."[116] To what specifically did Quattrami refer? Besides the numerous differences of opinion over the identification of ingredients—Quattrami, for example, took issue with Aldrovandi that Italian vipers corresponded to those described by Galen[117]—the botanist was undoubtedly aware of the controversies that had erupted surrounding Calzolari's and Aldrovandi's theriac in the previous decades. While these two naturalists were not the only ones to find themselves embroiled in acrimonious disputes over the composition of medicines, their behavior in each situation exemplifies the new relationship among hierarchy, knowledge, and authority that has been discussed in previous sections of this chapter.

In 1566, Calzolari published a *Letter on Certain Lies and Calumnies Made Against His Theriac,* responding to the criticisms of Ercolano Scalcina, an apprentice at the *spetieria dall'Angelo* in Verona. Scalcina boldly attacked not only the manner in which Calzolari made his antidote but also the inclusion of "lost" simples such as *apio, orobio,* and *scilla* that naturalists had recently recovered. Five years earlier, Scalcina had publicly criticized the inclusion of these simples in theriac, alleging that they were not really authentic, but his objections were overruled by the College of Physicians. After all, Calzolari had consulted the learned Aldrovandi, who gave the new ingredients his blessing. "This Easter I will make the theriac according to the process which you prescribe," wrote Calzolari to Aldrovandi on 3 March 1561.[118] What better proof was needed of the authenticity and potency of Calzolari's theriac than the approval of Italy's most famous naturalist? In 1566, several months after a second version of Calzolari's theriac appeared, earning the approval of the College of Physicians in Verona and well-known naturalists such as Gesner and Mattioli, Scalcina challenged the definitive nature of the apothecary's recipe with a theriac of his own. "Seeing that I had already composed my first Theriac, the desire came to him to make that [Theriac] for his employer with which he seeks to challenge mine and all others."[119]

[116]ASMo., *Archivio per le materie. Botanica. Naturalisti e semplicisti* (Dalle stanze della Castellina, 12 September 1595), f. 1.

[117]BUB, *Aldrovandi,* ms. 70, c. 18v.

[118]Cermenati, "Francesco Calzolari," pp. 114–115 (9 February and 3 March 1561).

At the time when Scalcina chose to attack Calzolari, the Veronese apothecary had just begun to enjoy an international reputation as a maker of authentic Galenic medicines. His *Voyage to Monte Baldo*, written to inform naturalists how to "directly and easily collect herbs that are necessary in the pharmacies," added to his prestige as a collector of medicinal simples; physicians throughout Europe came to Verona to visit his museum and consult with him on pharmaceutical questions. "But what can I say of the most accurate man in this profession, Messer Francesco Calzolari, apothecary at the *Campana d'oro?*" exclaimed Fioravanti. "[He] possesses almost all types of medicinal compositions so that the world marvels at him."[120] Contemporaries praised the skill of Calzolari as an apothecary, "whose theriac one can compare to any other made since the time of Galen." The fame of Calzolari's 1561 and 1566 theriac was due to its remarkable proximity to the Galenic compound. While the first theriac contained six substitutes, fewer than any made previously, the second reduced the number to three. By collecting and comparing specimens, Calzolari uncovered samples of such elusive ingredients as *balsamo, amomo, costo, folio, aspalatho, terra lemnia, marmo,* and *calamo aromatico*—the very materials whose whereabouts had confounded the medical community two decades earlier when Mattioli wondered if a true theriac could ever be made. To publicize his success in restoring ancient medicine, Calzolari devoted an entire room of his museum to theriac ingredients to underscore the purity of the compound visitors could buy in the shop below (see Museum Plan 1).[121] Thus, when Scalcina called into doubt the veracity of Calzolari's findings, he questioned the very foundations of the new social order emerging within the medical profession. "Having consumed my entire life in the search for true simples with the help of great men and my friends, and seeming to have found almost all the theriac ingredients, I began to make it, not secretly (as Scalcina would have you believe) but with the greatest approval that one may have in this subject, within and outside of my city," replied Calzolari.[122] In his response, the apothecary underscored his close connections to the world of official and orthodox medicine that helped him create a museum of medicine and that formally sanctioned the compounds he produced.

[119] *Lettera di M. Francesco Calceolari Spetiale al segno della Campana d'Oro, in Verona. Intorno ad alcune menzogne & colonnie date alla sua Theriaca da certo Scalcina Perugino* (Cremona, 1566), sig. Br.; Tergolina-Gislanzoni-Brasco, "Francesco Calzolari," p. 8.

[120] Calzolari, *Il viaggio di Monte Baldo*, p. 4; Fioravanti, *De' capricci medicinali*, p. 259.

[121] Prospero Borgarucci, *Della fabrica de li speziali* (Venice, 1566), p. 400; Cermenati, "Francesco Calzolari," p. 118 (Verona, 18 January 1568); Palmer, "Pharmacy in the Republic of Venice," p. 109. Documentation on the 1561 and 1566 theriac can be found in the ASVer., *Antico Archivio del Comune*, Reg. 610 (Minutes of the College of Physicians 1469–1569), ff. 197v, 221r–v. On Calzolari's room of *theriacali*, see Tergolina-Gislanzoni-Brasco, "Francesco Calzolari," p. 12.

[122] Calzolari, *Lettera*, sig. B.2r.

The success of Calzolari's quest for botanical truth, and its translation into medical practice, was attested to in numerous testimonials (*fedi*) by the College of Physicians and the College of Apothecaries in Verona and by prominent naturalists. Like the witnesses who legitimated the experiments of the Royal Society, sixteenth-century physicians and apothecaries also developed mechanisms to confirm the privileged status of their knowledge.[123] Experiments placed their knowledge in front of an audience capable of discernment. Mattioli, for example, described several tests he performed successfully with Calzolari's theriac as "miraculous proofs" of and "testimonials" to its efficacy.[124] While published texts formalized the process of authentication, the letters that circulated among naturalists provided more immediate verification. Apothecaries and physicians used these written statements as a form of proof; the more prestigious the authors and the more numerous the testimonials, the more efficacious the antidote. Calzolari collected testimonials about his 1566 theriac to confirm the honor and the legitimacy of his position. When the apprentice attacked the regulatory mechanisms that legitimized Calzolari's theriac, he further highlighted his role as an "outsider" in this affair. Scalcina specifically criticized the system of public testimonials as a means of perpetuating the social hierarchy within the medical profession.[125] For him, the success of Calzolari's theriac symbolized the close connections between social privilege and orthodox medicine. Calzolari's museum, his botanical expeditions, and his contact with the academic world of medicine defined the realm from which he, an ignoble apprentice, was excluded.

The medical profession, as Scalcina learned to his chagrin, was a self-perpetuating system that allowed little room for independent judgment. "I do not marvel that the theriac of the apothecary at the *Campana d'oro* works wonders," remarked Mattioli, "knowing that it was made with fewer substitutes than any other made in our time."[126] Mattioli's appraisal of Calzolari's theriac was noteworthy primarily for its predictability. Since he had inspected and certified samples of the *amomo, costo,* and *aspalatho* that Calzolari inserted in his theriac as "true" ingredients, his own authority as a naturalist depended on the success of this antidote. Like every physician who authenticated Calzolari's theriac, Mattioli used the debates surrounding it as an opportunity to publicize his own skills in identifying Galenic simples. Sanctioned by the local medical authorities and lauded by every naturalist of note, Calzolari's theriac was a testimony to the role of collecting in the material improvement of medicine. Its success reflected the ascendancy of

[123]Ibid., sig. Bv, B.3v–B.4r. The testimonials are published at the end of the *Lettera.* For a more detailed treatment of Shapin and Schaffer's notion of "virtual witnessing," see chapter 5.

[124]Mattioli, *Discorsi,* pp. 7, 12.

[125]Calzolari, *Lettera,* sig. Dv–D.2r.

[126]Mattioli, *Discorsi* (Venice, 1565 ed.), VI. 40, in Tergolina-Gislanzoni-Brasco, "Francesco Calzolari," p. 8.

humanist medicine and the techniques naturalists employed to reform medical practice.

More than anything, the controversy over theriac in Verona confirmed the perception that "truth," as the medical profession and its patrician clients defined it, lay within the legitimate medical community. Practitioners who did not participate in this culture merited the label of "charlatan" or worse. Calzolari underscored Scalcina's affiliation with this group by calling him an "empiric," casting him among the makers of "recipes for smelly fats, pomades, . . . fragile balsams, and quintessences that deafen the piazzas daily in the mouths of charlatans."[127] Scalcina, in other words, was someone who would *not* have been invited to sign Aldrovandi's catalogue of visitors. His words, unlike the legitimate statements made by practitioners upholding the norms of the profession, were of no significance for they lacked social credibility. His displays, in contrast to the approved ceremonies surrounding the production of Calzolari's theriac, were empty spectacles. Perhaps Mattioli was even thinking of Scalcina when he derided the charlatans who created "spectacles of themselves for the populace on the embankments . . . praising their false theriac with a sack of lies as the best in the world."[128] Scalcina, unable to secure a permanent position in any pharmacy and forced to wander from city to city, embodied the sort of ambulatory practitioner that respectable and honorable members of the medical profession most despised. If he did not already belong to this category when he contested Calzolari's theriac, his attempts to discredit one of the most famous apothecaries in Italy sealed his fate.

Scalcina was a man who desperately wished to belong to the community of naturalists that supported Calzolari, but found himself incapable of acquiring the skills to gain entry into this privileged circle. Before attacking Calzolari, he attempted to gain his friendship on several occasions. In the 1550s, Scalcina tried to join one of Calzolari's famed botanical expeditions to Monte Baldo to argue about plant classification, but he was ejected forcibly from the group. Since Calzolari only took his assistants and men of reputation on these expeditions, Scalcina hardly merited inclusion. His exclusion on this occasion foreshadowed future disappointments. Returning to Verona in 1561 after a short stay in Perugia, Scalcina sought employment with Calzolari at the *Campana d'oro*, but again was rebuffed. As Scalcina came to realize, he did not possess the appropriate credentials to join the apothecary's sodality. In his response to Scalcina, Calzolari underscored the social and professional distance between himself and the contentious apprentice.

[127]Fabrizio Cortesi, "Alcune lettere inedite di Giovanni Pona," *Annali di botanica* 6 (1908): 424 (Verona, 19 March 1625). Although these are not Calzolari's own words, the sentiment, from the mouth of another Verona apothecary–collector, is appropriate.

[128]Mattioli, *Discorsi*, p. 684. For more on charlatans, see Lingo, "Empirics and Charlatans in Early Modern France," *Journal of Social History* 20 (1986): 588, and Camporesi, "Speziali e ciarlatani," in *Cultura popolare nell'Emilia Romagna. Medicina, erbe e magia*, pp. 137–159.

"He would censure the Theriac composed by me," exclaimed Calzolari, "evaluating himself and his works, as if he were not the vile and mercenary apprentice that he is, but Master of Masters."[129] The latter category was one Calzolari reserved for himself. As a collector and the most prominent member of the local Guild of Apothecaries, he felt he had earned his place in the medical profession.

Scalcina's fate, after the fury of his exchange with Calzolari had subsided, underscored the inevitability of the different outcomes for these two makers of theriac. While Calzolari reaped honors and rewards from the encounter, enhancing his visibility within the profession, Scalcina was "buried alive with great shame" (*sepulto vivo cum gran sua vergogna*) by the medical establishment.[130] Rather than discrediting Calzolari, he foreclosed any further avenues of professional advancement for himself. Scalcina found himself wandering from city to city, unsuccessfully trying to practice in Padua, Venice, and Ferrara, places where Calzolari had many supporters. Finally, he returned to Perugia, where he took orders and was heard from no more. Even as Scalcina slid into obscurity, Calzolari enjoyed the benefits of notoriety. His honor had not only been restored during this episode but enhanced. Several years after the incident, he remarked to Aldrovandi that the controversy had increased the sale of his theriac: "I give away so much theriac that it is a marvel, and serve many more testimonials of the miraculous effects that it has caused and causes." Customers lined up at the *Campana d'oro* hoping to obtain any amount of Calzolari's precious antidotes. "Two days ago a gentleman from Milan came to me to procure theriac for a Senator," wrote Calzolari to Aldrovandi, "and he paid sixteen *scudi d'oro* for it and the scorpion's oil."[131] By the 1580s, when Joachim Camerarius wrote to Calzolari from Nuremberg requesting some of his theriac, the apothecary could respond, "It is a rare thing that is expensive, as my honored College, magnificent City, and our Senate testify, because it is made most competently with every ingredient as Galen did for those emperors." Expanding further on this point, he boasted, "I believe it goes all over the world. These past few days, I sent some of it to Lyon and Antwerp."[132] Calzolari, the apothecary who had put the world in his medicines, now enjoyed the satisfaction of selling his medicines to the world.

VIPERS, ANTIDOTES, AND COLLECTORS

Throughout the theriac debates of the 1560s, Aldrovandi had been a sympathetic ally for Calzolari, offering his advice and support. In the following decade, he called upon Calzolari to return the favor. The unity of the med-

[129]Calzolari, *Lettera*, sig. A.4v.
[130]Cermenati, "Francesco Calzolari," p. 117 (Verona, 18 January 1568).
[131]Ibid., p. 121 (Verona, 20 November 1571).
[132]In Olmi, "Molti amici in varii luoghi," p. 15.

ical profession against those who trespassed into privileged territory was equally apparent in the theriac controversy that occurred in Bologna in 1574–1577. In this instance, however, the challenge came from above rather than from below, for it was Aldrovandi himself who disputed the right of the College of Physicians to contravene his judgment as *protomedico* on the quality of the municipal theriac. His success in this affair, in the face of decided opposition from local physicians and apothecaries, reinforces the moral of the Calzolari tale: collectors were the men on top.

In 1574, Ulisse Aldrovandi and Antonio Maria Alberghini[133] were elected *protomedici* in the city of Bologna. As already discussed, the *Protomedicato* not only defined the nature of licensed medical practice but also mediated among the competing spheres of authority in urban medical culture, in this instance, the local government, the papal bureaucracy, the College of Physicians, and the College of Apothecaries. This was not the first time that Aldrovandi had held the position. In the 1560s, the papal legate specifically called upon Aldrovandi to accept the office of *protomedico* and oversee the publication of the local antidotarium, "seeming to him that Aldrovandi was well suited to this *Protomedicato* due to the public lectures that he gives on this material."[134] No doubt the legate recalled Aldrovandi's success in bringing the apothecary's guild under the direct supervision of the College of Physicians during his tenure as Prior of the College in 1554. Despite Aldrovandi's obvious qualifications, the College of Physicians strongly contested his appointment, indicating that they would prefer a foreigner if a nonpracticing physician were to hold the post. Their preference was to elect senior members of the College, as they proceeded to do with the second position, initially held by the oldest member of the College, Antonio Fava, and subsequently by Alberghini. Aldrovandi himself was sensitive to his status as an outsider because he neither practiced nor taught medicine.[135] Despite some misgivings, he accepted the office of *protomedico* and initiated a program of medical reform.

Like Calzolari, Aldrovandi enjoyed a high degree of visibility in the community of naturalists. By the time of his election as *protomedico*, numerous physicians and apothecaries consulted him on the preparation of antidotes, requesting copies of his recipes. "Otherwise, send me the means by which mithridatum is made in Bologna and what substitutes it contains," wrote Maranta to Aldrovandi in 1558, "because we do not wish to make it here, and have written to all the famous cities with the same request, so that we can decide which one seems the best to us." Calzolari valued Aldrovandi's

[133]Alberghini was professor of medicine at the University of Bologna, teaching philosophy (1533–1538), *theoria* (1539–1543, 1545–1562), and *practica* (1574–1587). See Dallari, *I rotuli . . . dello studio bolognese*, Vol. II.

[134]Aldrovandi, *Vita*, p. 16; Rosa, "La teriaca panacea dell'antichità approda all'Archiginnasio," pp. 328–330.

[135]BUB, *Aldrovandi*, ms. 70, cc. 17v–18r.

opinion so highly that he felt his own theriac would not be successful with-
out the Bolognese naturalist's approval. "I expectantly await . . . some news
of the said theriac, which I cannot possibly make without your help," he told
Aldrovandi in 1558, after requesting a copy of the Bologna antidote. "And
if you can think of any good advice regarding theriac," he reminded
Aldrovandi, three years later, "please tell me it for friendship's sake."
Imperato similarly depended on Aldrovandi. "From my dear Signor
Aldrovandi I would like something rare—among others, some Theriac ingredients," he wrote in 1573. The Neapolitan apothecary also consulted
with Aldrovandi about the identification of vipers suitable for inclusion in a
proper Galenic theriac. After dissecting vipers killed at three different times
of the year, Imperato found himself unable to discern the difference between a barren and a pregnant viper and turned to Aldrovandi for help.[136]

Like other collectors, Aldrovandi publicized the extent of his collection
of theriac ingredients. "I have in my museum a great stick of real cinnamon
from Portugal. Twice already I have used it in Theriac as true cinnamon."[137]
Though Aldrovandi did not create a special exhibit of *theriacali,* as Calzolari
did, the effect was virtually the same. His museum became an authoritative
site in which to compose Galenic medicines. While etiquette normally dictated that the *Protomedicato* meet at the house of the senior member (in this
instance, Fava or Alberghini), instead its members chose to congregate in
Aldrovandi's *studio.* "But everyone unanimously agreed that it should meet
at Aldrovandi's house to be able to see, from meeting to meeting, all the sensory things about which they harbored doubts, having congregated in his
Museum with great diligence."[138] During the 1560s and early 1570s, the
committee met frequently there to prepare the materials for the *Bologna
Antidotarium.*

Aldrovandi's appointment to the *Protomedicato* only accelerated his interest in collecting medicinal simples. With the help of patrons and friends
throughout Europe, he hunted down as many specimens for pure Galenic
theriac as he could find. As we have already seen in chapter 5, Aldrovandi did
not simply observe simples but actively examined them. Responding to a question about the location of the fetus in a viper's egg by Quattrami, Aldrovandi
recalled that the last time he had been asked this question, "I was dissecting
many vipers that I had at home, and saw everything with Signor Tagliacozzi
in the presence of many scholars." In a manner that foreshadowed Redi's experiments with poisons in Baroque Florence, Aldrovandi also tested theriac
recipes by experimenting upon a rooster, forced to ingest numerous poisons

[136]Vallieri, "Le 22 lettere di Bartolomeo Maranta all'Aldrovandi," p. 752 (Naples, 23 January
1558); Cermenati, "Francesco Calzolari," p. 107 (Verona, 10 September 1558), p. 114 (Verona,
6 February 1561); BUB, *Aldrovandi,* ms. 38(2), Vol. I, c. 253 (Naples, 10 July 1573); Neviani,
"Ferrante Imperato," p. 66 (Naples, 10 July 1573).

[137]BUB, *Aldrovandi,* ms. 70, c. 32v.

[138]Aldrovandi, *Vita,* p. 17.

to test the efficacy of the fabled panacea.[139] These were the "sensory things" Aldrovandi displayed before the members of the *Protomedicato*.

While Calzolari contented himself with the replication of the original Galenic theriac, Aldrovandi more boldly claimed that modern recipes— those composed by himself and his friends—surpassed the ancient formula. "I am of the opinion, as I have said at other times, that today one can make a more perfect theriac than was made in the time of Galen."[140] Collecting and experimenting made him the inheritor of rather than the commentator on ancient traditions. This perception of his own success made him a particularly vigilant inspector of the pharmacies that he visited every three months as *protomedico*. Aldrovandi paid special attention to the quality of theriac sold in the shops, checking it against the recipe approved by the College of Physicians. Apothecaries were forced to throw corrupt simples and mismade compounds into the public latrines.[141] Comparing himself to Tridentine reformers such as his friend Paleotti, Aldrovandi conceived of himself as a crusader against medical heresy whose goal was to restore medicine to its original state. Physicians and apothecaries who opposed him were identified as "heretics" who needed to be either converted to the new medical orthodoxy or expunged from the profession. In Tridentine Italy, "reform," an idea first introduced by humanists to describe their restoration of ancient learning, took on new and more directed meanings. In a move that anticipated attempts to coordinate the health boards in different regions, Aldrovandi proposed an expansion of the *Protomedicato* into a global system of controls spanning the length and breath of the Catholic world.[142] Thus, the creation of a new and more perfect theriac grew out of attempts to form a new medical and religious order.

By the time of his reappointment as *protomedico* in 1574, Aldrovandi had gained as many enemies as supporters. Illuminating the inconsistencies in pharmaceutical practice and the inability of the medical profession to prevent these errors, Aldrovandi presented himself as an outsider whose field of research gave him a privileged position in the medical community. "You [doctors] deceive yourselves in thinking that the deception [of the apothecaries] is occult, but I tell you that whoever wishes to open his eyes will understand it easily," observed Lodetto in his *Dialogue on the Deceptions of Certain Wicked Apothecaries*.[143] Aldrovandi's attitude toward the blindness and ineptitude of the College of Physicians reflected a similar impatience with

[139]BUB, *Aldrovandi*, ms. 70, cc. 22r, 25v. On Tagliacozzi's role in the theriac debates, see Teach Gnudi and Webster, *The Life and Times of Gaspare Tagliacozzi*, pp. 67–75.

[140]BUB, *Aldrovandi*, ms. 70, c. 26.

[141]Aldrovandi, *Vita*, p. 18; idem, *Discorso naturale*, pp. 201–202, 218; Olmi, "Farmacopea antica e medican moderna," p. 228.

[142]Olmi, "Farmacopea antica," p. 221. For a discussion of attempts to link the Health Boards, see Cipolla, *Fighting the Plague in Seventeenth-Century Italy*, pp. 19–50.

[143]Lodetto, *Dialogo de gl'inganni d'alcuni malvagi speciali*, p. 11v.

their lack of interest in the study of nature, for this was the knowledge that qualified them to inspect pharmacies. He begrudged their involvement in the making of the antidotarium and the implementation of reforms; they resented him as an interloper. Thus, the tensions between Aldrovandi and other members of the College had surfaced well before the discord over the 1574 theriac:

> In 1574, being *Protomedico* of the College of Physicians, he made the theriac in the pharmacy of San Salvatore with the greatest diligence possible, using fewer substitutes than had ever been done before, having found true *costo* and *amomo*. The display of this theriac was made public for four or six days so that everyone could see it, and it was visited and approved by the *Protomedici* and the entire College.[144]

The theriac standardized by the 1574 antidotarium contained sixty-one of the Galenic ingredients and only two substitutions—one less than the famous theriac composed by Calzolari in 1566. Aldrovandi's decision to recompose the Bologna theriac was based upon the success of these additions to theriac made in Venice, Verona, Padua, Naples, and Ferrara by other collectors of simples.[145] Once again, the larger community of naturalists dictated the shape of local medical practice.

After initially accepting the theriac proposed by Aldrovandi, the College of Physicians and the Guild of Apothecaries, in an aboutface, rejected the additions because "the *amomo* and *costo* were placed in the theriac without the consensus of the College."[146] To counter such criticisms, Aldrovandi presented proofs that the new ingredients were "true and legitimate simples." Distinguished physicians such as Mercati reported its success in the plague epidemic spreading at an alarmingly quick rate throughout Italy. Indeed, the resurgence of plague in 1575–1577 was one of the reasons the debates over Aldrovandi's theriac became so heated. In the 1560s, a period of relatively few outbreaks, the composition of theriac had been predominantly a matter of professional competency. In the 1570s, it became a matter of life or death. "Our city lives in fear," wrote Aldrovandi to Camillo Paleotti in 1576, "of the great plague that continues."[147] Over the protests of his critics, Aldrovandi's theriac was put on sale in the local pharmacies. For the time being, he had gained the upper hand.

In 1575, three years after Gregory XIII's election to the papacy, the physicians and apothecaries of Bologna resolved to produce a Jubilee theriac in honor of the success of their local son that explicitly challenged Aldrovandi's claim to make the best theriac. On 11 June 1575, the Prior of

[144]Aldrovandi, *Vita*, p. 20; see also BUB, *Aldrovandi*, ms. 21, Vol. III, c. 133v.

[145]Rosa, "La teriaca panacea dell'antichità," p. 334; BUB, *Aldrovandi*, ms. 21, Vol. III, c. 135.

[146]Aldrovandi, *Vita*, p. 20.

[147]BUB, *Aldrovandi*, ms. 97, c. 372v (Bologna, 19 September 1576).

the College of Physicians and *protomedici* congregated at the *speciaria del melone* to view the troches, the cakes made by mixing viper's flesh with breadcrumbs and other ingredients, that had recently been prepared. Inspecting them with a practiced eye, Aldrovandi noted that the vipers in question were freshly killed, in specific contravention of the Galenic formula which recommended late April (*quando il sole è in tauro*) as the optimum time for the slaughter.[148] His anatomical experience—or was it simply his conviction that apothecaries would err without his guidance?—told him that the female vipers were pregnant. Several even appeared to be male. Even worse, the vipers came from Ravenna: salty, maritime specimens that would inevitably produce a theriac that made people thirsty.[149] Needless to say, Aldrovandi refused to approve the troches as legitimate ingredients for theriac, a decision with which the other *protomedico*, Alberghini, concurred. As Aldrovandi triumphantly noted, the Jubilee theriac had not surpassed his own incomparable antidote.

The response of the College of Physicians was swift and decisive. Nine of the twelve members sided with the apothecaries who had made the theriac, voting to expel Aldrovandi and Alberghini, for five and two years, respectively—a punishment that undoubtedly assessed the relative burden of responsibility that each had incurred. From their perspective, Aldrovandi's actions further delayed the sale of a valuable antidote during a horrendous epidemic and undermined the credibility of the College of Physicians. "And Aldrovandi marveled that they should prefer votes to reason."[150] In retaliation, the papal legate and the Senate intervened, prohibited the sale of the Jubilee theriac. Unlike the College of Physicians, they could not afford to anger a naturalist who was not only a blood relative of Gregory XIII but also his consultant in natural portents. In contrast to Calzolari's situation in Verona, which pitted all forms of legitimate medicine and pharmacy against unlicensed practice, the situation in which the Bolognese naturalist now found himself was infinitely more complex. More than anything, it clearly demonstrated the conflicting competencies of the numerous overlapping authorities—medical, civic, and ecclesiastic—who claimed some role in the decision-making process. Unlike Calzolari, whose full membership in the corporate medical structure of Verona and in the broader community of collectors assured him unified support against Scalcina, Aldrovandi found himself with little support from the local medical establishment. Instead, he

---

[148]Aldrovandi does not make clear the significance of the astrological date ("when the sun is in Taurus"), but certainly it is a reminder of the highly charged symbolism surrounding the production of theriac.

[149]Aldrovandi, *Vita*, pp. 21–22; BUB, *Aldrovandi*, ms. 21, Vol. III, cc. 133–183; ASB, *Archivio dello Studio Bolognese. Collegi di medicina e d'arti. Nucleo Antico*, ms. 197. *Misc. Protomedicatus* 1559–1600 (1 July 1575).

[150]ASB, *Archivio dello Studio Bolognese. Collegi di medicina e d'arti. Nucleo Antico*, ms. 218. *Libro segreto dall'anno 1575 al 1594*, c. 3r (9 July 1575); Aldrovandi, *Vita*, p. 22.

depended entirely on his standing within the community of naturalists and his political ties to the local government and the papal bureaucracy to rectify the situation.

"Indeed I quickly was oppressed by many malicious people, and only for having upheld the truth, as you know," wrote Aldrovandi to the papal *nuncio* in Florence, Monsignor Bolognetti, in 1576,

> because of those vipers that were killed to put in the Theriac, and prepared out of season. And my opinion was supported by the foremost Colleges in Europe as the truth. . . . And this business has so tormented me until recently, due to this annoying lawsuit, that it has distracted my soul from the study of nature, having applied myself to the defense of my honor.[151]

Like Calzolari, Aldrovandi established his credibility by collecting testimonials. Bacci, author of *On the Dignity of Theriac* (1583), Cardano and Imperato all supplied personal affidavits. Those from the Colleges of Physicians in Florence, Ferrara, Mantua, Rome, and Naples were signed by prominent physicians such as Donati, Mercuriale, and Pisano, all *protomedici*.[152] "You will hear from Ferrante Imperato, from whom I received one of your most learned letters, that I have procured the opinions of our College so that it would be more authoritative, confirming the correctness of your judgment of the time to collect the Vipers and that those Troches were badly made," wrote Giovan Andrea Pisano, the *protomedico* in Naples. "I wrote it myself, showed it to everyone and had it confirmed by the Prior of our College."[153] Through his network of correspondents, Aldrovandi received widespread support from physicians and apothecaries who perceived the quarrel as a dispute between collectors, who as a group advocated greater regulation and standardization of pharmacy, and those medical practitioners who saw the problem in terms of the more traditional divisions between theory and practice.

Discussing the circulation of mistaken information about vipers within the medical community, Aldrovandi speculated that "they had learned it from the most vile people—charlatans, fortune tellers, and gossips (*ceretani, chioromatori e circonforani*)."[154] Like Calzolari, he equated his own position with medical orthodoxy. While Aldrovandi could not afford to label his detractors "charlatans"—after all, they belonged to the College of Physicians to which he wished to be reinstated—he impugned their judgment by insinuating that they acquired their knowledge of simples from disreputable practitioners, the Scalcinas who walked the streets of Bologna. In contrast,

[151]BUB, Cod. 596-EE, n. 1, c. 2v (Bologna, 9 April 1576).

[152]Aldrovandi, *Vita*, pp. 23–24; for copies of the testimonials, see BUB, *Aldrovandi*, ms. 21, Vol. IV, cc. 348–355. Cardano's opinion is reproduced in Simili, *Gerolamo Cardano lettore e medico a Bologna*, p. 124.

[153]BUB, *Aldrovandi*, ms. 21, Vol. IV, c. 348r (Naples, 10 December 1575).

[154]BUB, *Aldrovandi*, ms. 70, c. 20v. For the piazza as the domain of the charlatan, see Camporesi, *La miniera del mondo*, p. 274.

Aldrovandi's knowledge came not from the piazza, a site of untrustworthy and uncanonical information, but from his museum. Shortly after his expulsion from the College of Physicians, he invited its members, the *gonfaloniere* and archbishop of Bologna, to observe the proper state of a viper. Witnesses recorded going "to the house of Signor Aldrovandi to see the new state of a viper cut by Signor Gaspar Tagliacozzi in the presence of Doctor Alberghini and many other scholars. After the dissection, Signor Doctor Aldrovandi stated publicly that the viper troches made by the commission of the doctors from the College for theriac are not good because the vipers were not taken at the right time."[155] On this occasion, the public spectacle of theriac and the experimental culture of the museum converged.

While the incident in Verona was resolved in a matter of months, Aldrovandi's problems with the College of Physicians in Bologna dragged on for three years. Unable to resolve the dispute locally, the papal legate brought the case to Rome. As Aldrovandi wrote to his brother in March 1577, "You can believe that my coming to Rome will be the last remedy."[156] He reflected further on these events in his autobiography:

> Armed with the truth and with the opinions of Cardinal Paleotti and Signor Giovanni Aldrovandi, he went to Rome on 2 March 1577, where he presented himself at the feet of Gregory XIII and narrated the entire story truthfully, with all the judgments and confirmations of the other Colleges. Finally the Pope gave him his judgment, and asked if he wanted justice, concluding by his personal decision that Aldrovandi be reinstated *ad omnes honores et dignitates.*[157]

The reinstatement of Aldrovandi and Alberghini into the College of Physicians testified to the status of the Bolognese naturalist within the Italian medical community. Aldrovandi's personal ties to Gregory XIII and the prominence of his family in the local political culture of Bologna influenced the success of his petition against the judgment of the College of Physicians, but his fame as a teacher and collector of *materia medica* ensured the inevitability of the outcome. The College of Physicians was discredited; his honor was restored.

The 1574–1577 dispute between Aldrovandi and the physicians and apothecaries of Bologna further elevated the fame of his theriac. "The best Theriac that one finds today is made in Bologna," wrote Baldassare Pisanelli in his *Discourse on the Plague* (1577). "I esteem this Theriac, and affirm that it is better than any of the others, first, because it was made with great diligence

---

[155]ASB, *Archivio dello Studio Bolognese*, ms. 197. *Misc. Protomedicatus 1559–1600*, n.p. (1 July 1575).

[156]BUB, *Aldrovandi*, ms. 97, c. 353v (Bologna, 9 March 1577).

[157]Aldrovandi, *Vita*, p. 24. For the specifics of the papal bull of 24 May 1577, see Aldo Andreoli, "Ulisse Aldrovandi e Gregorio XIII," *Strenna storica bolognese* 11 (1961): 11–19, and idem, "Un inedito Breve di Gregorio XIII a Ulisse Aldrovandi," *Atti e memorie dell'Accademia Nazionale di scienze, lettere e arti. Modena*, ser. 6, 4 (1962): 133–149.

and because Signor Ulisse, most skilled in simples, uses only true and unadulterated things."[158] As the successful plaintiff in the most publicized theriac controversy of the late sixteenth century, Aldrovandi became an arbiter of subsequent debates. "I am now busy with one of our apothecaries because of theriac," wrote the naturalist Pompilio Taliaferri from Parma in the 1590s, "over which I have resorted to disputing many things with our physicians. And I still have many doubts about which I first would like your opinion."[159] In the decades that followed Aldrovandi's altercation with the College of Physicians, the number of scholars who wrote to him, soliciting his opinion about problematic simples and seeking samples of them from his museum, increased. Likewise, the purity of his theriac, replacing Calzolari's 1566 composition, became a benchmark that future generations could challenge.

Throughout the 1570s, naturalists heatedly debated the composition of antidotes and used the materials of medicine to argue their case. In 1570–1572, the College of Physicians in Brescia expelled the physicians Girolamo Donzellino and Giuseppe Valdagno for their advocacy of theriac as an antidote for all pestilential fevers.[160] In 1577 Stelliola defended Maranta against the criticisms of Wieland and Marco Oddo of Padua who disapproved of his substitution of *acoro* for *amomo*. Twenty years later, Quattrami was still embroiled in this debate.[161] In Milan, the College of Physicians continued to query throughout the 1590s whether or not theriac could be prepared as Galen had done for the Roman emperors.[162] All of these controversies took place around the testing of ingredients that occurred in the museums and pharmacies, where ancient knowledge confronted modern materials. At the end of the sixteenth century, collectors such as Aldrovandi, Calzolari, Imperato, and Mattioli could indeed be satisfied that they had wrenched natural history out of its "dark ages" by revitalizing the study of nature through their collecting, observations, and experiments. They had challenged the notion that medical knowledge could never reach the level it had achieved during the time of Galen and Dioscorides and could never

---

[158]Pisanelli, *Discorso sopra la peste* (Rome, 1577), in Olmi, "Farmacopea antica e medicina moderna," p. 206 (see footnote 40). Pisanelli taught *theoria* at the University of Bologna (1559–1562).

[159]BUB, *Aldrovandi*, ms. 136, Vol. XIX (1592–1593), c. 152v (Parma, n.d.).

[160]Redmond, *Girolamo Donzellino*, pp. 59–123; see also Giuseppe Valdagno, *De theriaca usu in febribus pestilentibus* (Brescia, 1570); Vincenzo Calzaveglia, *De theriaca abusu in febribus pestilentibus* (Brescia, 1570); Girolamo Donzellino, *De natura, causis, et legitime curatione febris pestilentis . . . in qua etiam de theriacae natura ac viribus latius disputatur* (Venice, 1570); idem, *Libri de natura, causis, et legitima curatione de febris pestilentibus* (Venice, 1571); idem, *Eudoxi Philalethis adversus calumnias et sophismatat cuiusdam personati, qui se eu androphilacten nominavit. Apologia* (Verona, 1573).

[161]Stelliola, *Theriace et mithridatia libellus*, esp. p. 15v; Quattrami, *Tractatus perutilis atque necessarius ad theriacum, mitridaticamque* (Ferrara, 1597). Andrea Bacci addressed his *De dignitate theriacae* (Padua, 1583) to Oddo.

[162]Thorndike, *A History of Magic and Experimental Science*, Vol. V, pp. 470–471.

surpass the Authorities of Nature. The revival and reconstitution of theriac in the sixteenth and seventeenth centuries was considered the most prominent measure of the success of the naturalistic collecting enterprise.

Collecting, as we have seen throughout this chapter, was about competing spheres of knowledge or, more precisely, the different ends to which the possession of such information could be put. While the apothecaries of early modern Italy collected to preserve their status as inventors and distributors of medicines to the elite, physicians and university professors envisioned the museum as a repository that opened up the world of natural history to educated practitioners. In doing so, it enveloped the empirical world of medicine and pharmacy within the academic world created by Ghini, Aldrovandi, and other professors of *materia medica*. It would be an exaggeration to assume that the absorption of empirical practice into academic discourse signaled the irreversible decline of the unlicensed medical practitioner. However, the increased bureaucracy created by the College of Physicians and the *Protomedicato* attempted to curtail the activities of the "charlatans"—a category whose pejorative meaning crystallized only as the idea of legitimacy took hold—whose presence undermined their own authority by accentuating the gulf between orthodox and unorthodox practice.

By the mid-seventeenth century, theriac ingredients ceased to be objects of note in museums of natural history. With the disappearance of plague from Italy, collectors no longer had reason to concern themselves with the quality of antidotes as they had done in the preceding century. Medical objects did not vanish from museums; they were given different frameworks of meaning. At the Roman College, Kircher conducted extensive tests on the serpent's stone, comparing its powers to theriac. In 1663 he treated spectators to an experiment in which a dog was bitten by a poisonous snake and subsequently cured with the serpent's stone.[163] The serpent's stone fascinated Kircher not because it was medically useful but because it was a dramatic proof of his theory of universal magnetism.

At the same time in Florence, Redi conducted many of his experiments on animals such as the viper that formed the basis of the antidotes the grand-ducal pharmacy dispensed. But Redi was not particularly interested in the curative powers of the viper per se. He studied vipers to understand the physiology of venom more than their medicinal uses. Dissecting enormous quantities of them, he undermined the entire premise of Aldrovandi's authority as a Galenic anatomist when he declared that the idea of searching for barren vipers was completely ludicrous. Didn't they all have eggs? Certainly they had them in January as well as May?[164] Continuing his critique of Renaissance

---

[163]Kircher, *Magneticum naturae regnum* (Amsterdam, 1667), p. 61. This event is also discussed in idem, *China Illustrata*, and Petrucci, *Prodomo apologetico*.

[164]*Francesco Redi on Vipers*, p. 27. For more on these events, see Tribby, "Cooking (with) Clio and Cleo."

medicine, Redi dismissed the idea of testing antidotes as a valid form of knowledge. "The most difficult and fallacious experiments are those regarding medical things because a great and general uncertainty accompanies all medicines, for the most part," he wrote in his *Experiments on Diverse Natural Things*.[165] Redi's skepticism about the sort of experiences advocated by naturalists from Aldrovandi to Kircher foreshadowed the reconfiguration of the natural history museum at the end of the seventeenth century, when classification rather than therapy became the principal issue of contention.[166] The objects remained the same; only the meaning had changed.

[165]Redi, *Experienze intorno a diverse cose naturali*, p. 153.
[166]Pomian, *Collectors and Curiosities*, p. 101.

PART THREE

Economies of Exchange

While the appearance of museums significantly reoriented the study of nature, they also redefined the identity of the naturalist. "Natural history" had not been a distinctive enterprise before the mid-sixteenth century, in part because the image of the "naturalist" had not yet crystallized. The revival of natural history and its growing popularity in the courts, academies, and universities, and more generally in the patrician culture of early modern Italy, made the men who studied nature more self-conscious about the uniqueness and potential appeal of their work. They responded by paying greater attention to their presentation of self. Autobiographies and portraits allow us to chart this progression. Concerns about image did not begin in print, however. They were initiated in action. Naturalists, like many socially mobile groups, viewed their successful negotiation of the patronage networks that bound together the different sectors of elite society as an important measure of their social ascendency. Patronage, established through letters, gifts, and brokers, was the instrument that determined success or failure. The princes, popes, and nobles who witnessed experiments and lent their imprimatur to the medicines produced in the museums also furthered the entry of the most successful naturalists into the courts and academies. Through their patronage, they propelled the continuous exchange of objects and arrival of important visitors that made the museum a central feature of the patrician world.

Part III considers these different activities as they relate to the process of exchange. "Exchange," in all its different forms, engendered the collector and the museum. It was the primary social mechanism that defined elite society and perhaps early modern society as a whole. Chapter 7 explores the metaphoric uses of exchange as it related to the identity of the collector. Imitation, a revered form of humanist expression, was central to the identity of the collector; the articulation of organic links between past and present allowed naturalists to feel that their "novelty" was preordained. In search of models and classic *exempla* to give meaning to their lives and work, they perceived themselves to be the living embodiment of key master narratives prized by men of letters. Drawing upon the language of encyclopedism, the images of the famous men who had preceded them in their endeavors, and the metaphors of discovery, they "invented" themselves. This was achieved through a calculated manipulation of literary, philosophical, and religious images and through an ingenious synthesis of ancient and contemporary *exempla*. Indeed, the ability to exchange identities continuously became the hallmark of the collector.

The symbolic economy forged out of the humanist invention of self found its parallel in the social interactions of patrons, brokers, and collectors. Chapter 8 outlines the network of human relationships behind the collecting of nature. Collecting offers an extraordinary opportunity to inspect the patronage system of early modern Italy. As the product of numerous collaborations among men of diverse social standing, museums reflected the hierarchical structure of society and the process of communication that facilitated the interaction of different sectors of the elite world. While patrons bestowed great gifts, privileges, and occasionally money on naturalists, brokers made this munificence possible. Few naturalists had risen high enough in the social world to communicate directly with powerful princes. They depended on court brokers to provide access to the top. Outside of the court hierarchies, brokers introduced naturalists and collectors in different regions to one another, mediated debates between individuals, and occasionally intervened to resolve disputes. With the exception of high-ranking individuals, who could only be patrons, many collectors had the opportunity to be patrons, brokers, and clients. Multiplying their roles, they ensured the success of their museums and secured their position within the community of naturalists. Natural history had begun in the humanist's *studio* and the pharmacy, neither a setting with great social prestige at the beginning of the sixteenth century. By the end of the seventeenth century, the naturalist could rightfully claim his place in the courts, academies, and salons, in short, all the most socially desirable settings. Having begun his journey as a humanist in search of courtly patronage, he ended it as a courtier ready to dispense favors.

SEVEN

Inventing the Collector

Self-knowledge must precede all other knowledge.
—TORQUATO TASSO

In his 1595 will, the Venetian senator, mathematician, and instrument collector Jacomo Contarini included a bequest "for the perpetual conservation of his study and books" that began with the following statement: "One of the most dear things I have had, and have, is my *studio,* from which have come all the honors and esteem for my person."[1] Contarini's special affection for his museum as the font of his good reputation reflected a sentiment common to most collectors. The detailed instructions of Aldrovandi's will of 1603 and of Alfonso Donnino's bequest to the Roman College in 1651, for example, reveal a similar preoccupation with collecting as a way of constituting and preserving one's image. What is most interesting about Contarini's statement is the clarity with which he articulated the relationship between collecting and social virtues. "Honor" and "esteem" were not words to be mentioned lightly. Both defined the nature of civil society to which collectors belonged. When Contarini described his museum as the *exclusive* source of these traits, he underscored its importance to the formation of patrician identity. The museum was not simply a product of his desires and interests; in its most essential sense, it was his self. In his *studio* in Venice the core of his very being was on display.

Like many members of the patrician elite, collectors were virtually obsessed with the question of identity. They explored this problem in their writings and in the presentation and interpretation of objects in their museums. Their patrons, visitors, and readers contributed to the discursive formulation of the identity of the collector by confirming the appropriateness of the images with which collectors represented themselves and by adding their own embellishments to the canon of exemplary figures. "Men are not

[1]Paul Lawrence Rose, "Jacomo Contarini (1536–1595), a Venetian Patron and Collector of Mathematical Instruments and Books," *Physis* 18 (1976): 120.

born but fashioned," wrote Erasmus.[2] Collectors epitomized this moral. Accumulating knowledge not only led to a better understanding of nature but also defined the identity of the people who engaged in this activity. In the courtly and patrician world of early modern Italy, collecting was a way of aestheticizing the self.[3] The museum provided the materials out of which to construct an identity and the means to publicize it. Self-knowledge, in other words, was social knowledge. Even Montaigne, who separated himself from society to write about himself, understood self-knowledge as a form of display. His goal, as he stated in the second book of his *Essays,* was "to appear on the Theater of this world."[4] While men such as Erasmus and Montaigne displayed themselves through their writings, collectors publicized their persona through their museums and the written and visual works that emerged from them, bringing the humanist notion of "self-fashioning" to its complex and utterly spell-binding conclusion.

The sixteenth century, as Frank Whigham, Stephen Greenblatt, and many others have noted, ushered in a "new notion of personal identity."[5] While scholars continued to celebrate the Augustinian model of introspection as a path to inner knowledge, they increasingly found reason to follow different itineraries in their search for the self. The world around them did not exist in opposition to the inner self, tempting man away from more spiritual endeavors. Instead, it provided the means of illumination. Travel, discovery, and collection all served to deepen one's sense of identity.[6] Only by going out into the world and bringing the world into the home could one achieve the sort of knowledge that constituted "identity," as most early modern patricians understood it. The courtesy literature that emerged during this period—such works as Castiglione's *Book of the Courtier* and Della Casa's *Galateo*—reinforced the message that each man played a role in molding his own image. There were choices to be made, models to be emulated, images to be displayed. The essential self needed to be articulated and embellished. Humanism was the primary locus for these new forms of introspection and extraversion. The actions of individuals, past and present, guided collectors in their quest for knowledge and self-knowledge. "The rediscovery of the

[2]In Thomas Greene, "The Flexibility of the Self in Renaissance Literature," in *The Disciplines of Criticism: Essays in Literature, Theory, Interpretation and History* (New Haven, CT, 1968), p. 248. See also Stephen Greenblatt, *Renaissance Self-Fashioning from More to Shakespeare* (Chicago, 1980).

[3]For an interesting discussion of this process in Tudor and Stuart England, see Patricia Fulmerton, *Cultural Aesthetics: Renaissance Literature and the Practice of Social Ornament* (Chicago, 1991).

[4]Montaigne, *Essais,* II, in Paul Delany, *British Autobiography in the Seventeenth Century* (London, 1969), p. 12.

[5]Frank Whigham, *Ambition and Privilege: The Social Tropes of Elizabethan Courtesy Theory* (Berkeley, 1984), p. 186. My arguments about where the novelty lay differ from those of Whigham, who sees it as a shift from "ascribed" to "acquired" characteristics.

[6]For a brief overview of this process, see Defert, "The Collection of the World," pp. 11–20.

classical tradition in the Renaissance was as much an act of imagination as of criticism, as much an invention as a rediscovery," observes Anthony Grafton.[7] The uses of collecting to shape identity provides us with a particularly rich example of the creativity of humanist discourse. Combining the best imagery that antiquity offered with the best objects that nature and art provided, the collector invented himself.

The ability to create one's place in society was primarily the domain of exceptional individuals with extraordinary means at their disposal. While all members of early modern society (and probably any society) had the capacity for self-transformation, very few took full advantage of the possibilities they were offered. Even fewer commanded the resources to make their transformation complete. We more readily think of individuals such as Cellini, Michelangelo, and Galileo as examples of men who successfully shaped their identities than we do of the peasant and imposter Arnaud du Tilh.[8] Like the collectors of nature, all three were relatively well-educated men whose ambitions brought them within the orbit of various Italian courts. In this competitive social environment, their desire to aggrandize and mythologize their selves was magnified. Italy, perhaps more than any other part of Europe, was a region particularly conducive to these sorts of activities; political complexity led to cultural competitiveness and ultimately to the social elevation of individuals whose activities served the needs of various rulers. During the sixteenth century, the status of humanists engaged in artistic, musical, literary, and scientific pursuits increased in direct proportion to their visibility at court. In 1506, Albrecht Dürer commented upon his favorable reception in Italy in comparison to his treatment in Germany, remarking, "Here I am a gentleman, at home a sponger."[9] Such observations were borne out by the self-conscious and self-confident pronouncements of men such as Michelangelo and Cellini about their status as "artists." The growing popularity of other cultural activities at court engendered an equivalent rise in status of those humanists who embellished the life of the court, among them naturalists, philosophers, and collectors. By the sixteenth century, displaying knowledge was a prized courtly virtue. Both inside and outside the courts— and certainly patrician society as a whole mirrored the values of court society—collectors, like artists, found themselves placed centerstage. Their

[7] Grafton, *Defenders of the Text*, p. 103. Jay Tribby also discusses humanism as an inventive discourse in his essays, "Body/Building," "Cooking (with) Clio and Cleo," and "Of Conversational Dispositions and the *Saggi*'s Proem," in Elizabeth Cropper, Giovanna Perini, and Francelco Solinas, eds., *Documentary Culture: Florence and Rome from Grand-Duke Ferdinand I to Pope Alexander VII* (Bologna, 1992), Villa Spellman Colloquia, vol. 3, pp. 379–390.

[8] Biagioli, "The Social Status of Italian Mathematicians, 1450–1600," pp. 48–49. For a nonelite discussion of early modern conceptions of identity and the details of Arnaud du Tilh's imposture, see Natalie Zemon Davis, *The Return of Martin Guerre* (Cambridge, MA, 1983).

[9] In Peter Burke, *The Italian Renaissance: Culture and Society in Italy* (Princeton, 1986), p. 76 (13 October 1506).

encyclopedic curiosity epitomized the sort of creative and public inventiveness that Italian patricians most admired.

Constituting an identity presented a particularly interesting challenge to men who had absorbed the lessons of a humanist education. In contradiction to the Burckhardtian image of identity as a product of Renaissance individualism, they defined themselves in relation to a series of communities.[10] Family, professional, social, and religious groups and the world of learning all provided the humanist with the tools to form his identity. Collectively, they defined the possible avenues of inquiry that an educated man, who knew how to squeeze every ounce of meaning out of the events and symbols that shaped his life, could explore. In many respects, collectors were ideally suited to the task of constituting the self as an object of display. The epic tales and moral treatises they first encountered in the classroom were filled with stories of exemplary figures whose virtues were considered worthy of imitation. Seeing and understanding the world through this literary canon gave them a common framework of understanding that they shared with others who had benefited from the same humanist pedagogies. It also separated them from the rest of society, those who were denied access to museums, because, among other things, they did not participate in the aestheticization of knowledge. "For those fictions and illusions are what hold the gaze of spectators," wrote Erasmus, underscoring the necessity of invented images in the reconstruction of a life.[11] Collectors, among the most visible participants in humanist culture, successfully deployed all their social and intellectual talents to create a dazzling bricolage.[12] Like their museums, they were mosaics, composed of the fragments of the culture they had inherited. They were at once profoundly unique and profoundly derivative, determined to achieve individual recognition for their incomparable activities yet forever defined and contained by the communities they inhabited.

Humanist vocabulary offered collectors three important models through which to achieve self-definition: *imitatio, exempla,* and *inventio.*[13] These formal categories of knowledge presented the past as a flexible entity whose renarration shed light on the present. *Imitatio* described the general practice of "imitation" prevalent in Renaissance discourse and was highly valued by humanists. No journey of self-discovery could ever be more memorable than one that led the collector ever backward to the original source of his expe-

[10]Here I am in agreement with Natalie Zemon Davis, "Boundaries and the Sense of Self in Sixteenth-Century France," in *Reconstructing Individualism: Autonomy, Individuality, and the Self in Western Thought,* Thomas C. Heller, Morton Sosna, and David E. Wellbery, eds. (Stanford, 1986), p. 53.

[11]Erasmus, *In Praise of Folly* (1511), in David Quint, *Origin and Originality in Renaissance Literature: Versions of the Source* (New Haven, CT, 1983), p. 12.

[12]The use of bricolage in establishing an identity is also discussed in Biagioli, *Galileo Courtier,* and Shapin, "A Scholar and a Gentleman."

[13]There are other related categories such as *aemulatio,* but I simplify the discussion of formal genres for the sake of clarity.

rience, whatever that source might be. Imitation, as Thomas Greene observes, offered two "rival possibilities of humanist wholeness."[14] Was it better to imitate one person or an entire culture? Scholarly convention validated both models, and collectors accordingly made use of them, with a slight preference for the latter (after all, they were collecting identities). Through imitation, collectors positioned themselves within familiar narratives that their educated visitors and readers immediately recognized. Yet they remade these narratives into personal itineraries. While participating in the universal enterprise of revivifying ancient culture, they also were drawn to its intellectual and ethical norms due to their unshaken belief that they *personally* contributed to the perpetuation and transformation of this discourse. Without their participation, it might cease to exist altogether.

Collectors particularized the global enterprise of imitation through their choice and combination of models. Like many of their contemporaries, they valued the "rhetoric of exemplarity" as a framework that gave meaning to their lives. *Exempla* defined the "examples" worthy of imitation, the famous men whose words and deeds constituted the moral canon of western culture. "Nothing moves me like the examples of famous men," wrote Petrarch. Identifying with and reenacting the lives of these individuals brought one closer to the essential truths of the society. As Timothy Hampton writes in his discussion of exemplarity, "For the Renaissance reader, to 'practice those great souls of the best ages,' as Montaigne phrased it, was to define the self in relation to ideal images from the past."[15] Like the objects assembled in museums, *exempla* provided clues to the identity of the individual capable of possessing knowledge. They reinforced the historically contingent nature of the humanist enterprise by underscoring the past as a source of ideas that molded the present.

Inventio signaled the process by which humanists bridged the temporal distance between their own culture and the classical civilizations they valued. In contemporary language, "invention" had two principal meanings. First, it identified unique moments of human artistry and ingenuity, as outlined in Polydore Vergil's *On the Inventors of Things* (1499). In imitation of Pliny, Polydore Vergil defined inventors as the "first begetters of things."[16] Inventors were individuals whose contributions added to the accumulation of knowledge that defined "civilization." They were the new Prometheuses. Under the category of "inventor," John Evelyn included "the diligent and

[14]Thomas M. Greene, *The Light in Troy: Imitation and Discovery in Renaissance Poetry* (New Haven, CT, 1982), p. 151.

[15]Timothy Hampton, *Writing from History: The Rhetoric of Exemplarity in Renaissance Literature* (Ithaca, NY, 1990), pp. 3–4. See also John D. Lyons, *Exemplum: The Rhetoric of Example in Early Modern France and Italy* (Princeton, NJ, 1989).

[16]Copenhaver, "The Historiography of Discovery in the Renaissance: The Sources and Composition of Polydore Vergil's *De inventoribus Rerum*, I–III," *Journal of the Warburg and Courtauld Institutes* 41 (1978): 192–214; Denys Hay, *Polydore Vergil: Renaissance Historian and Man of Letters* (Oxford, 1952), pp. 52–78.

curious collectors of both artificial and natural curiosities."[17] The invention of knowledge included both the process of discovery and the notion of intellectual property that possessing objects denoted. Aldrovandi, for example, described himself as superceding "ancient inventors" when he accumulated more medicinally useful objects than they had known. He also perceived the acknowledgment in print of individuals who had shown him particularly noteworthy objects as a way of "giving honor to the inventor."[18] From his perspective, "inventors" were the possessors of knowledge, an image probably culled from reading one of the many editions of Polydore Vergil in circulation. While other collectors—for example, Della Porta, Settala, and Kircher—tied *inventio* specifically to technological innovation, Aldrovandi used the broadest humanist definition of this term.

Inventio also described the narratives through which collectors displayed their talents. Encompassing both the idea of *imitatio* and the use of *exempla*, it defined "the selection and interpretation of subject" for an iconographic program and the visual narrative that resulted from this imaginative undertaking.[19] As the Florentine court humanist Vincenzo Borghini explained, invention was a mental exercise (*una fatica tutta mentale*), a most excellent form of learning "common to poets, historians, philosophers, etc., painters, sculptors, and architects."[20] *Invenzioni* decorated the *studiolo* of the Grand Duke Francesco I in the Palazzo Vecchio in Florence and also the walls of Aldrovandi's country villa outside of Bologna. In both instances, they drew upon classical imagery to articulate the experience and authority of the collector. Francesco I was the new Prometheus just as Aldrovandi was the new Ulysses. Through the display of objects and images, the collector invented himself, over and over again.

DISPLAYING THE SELF

While all objects in a museum shaped the image of the collector, certain artifacts highlighted the self-referential aspects of collecting; positioned within the humanist narratives about the shape of knowledge, they emblematized the difficulties of gaining the transcendental wisdom that all collectors sought. Two items in particular made visible the dialectic between knowledge and self-knowledge: the chameleon and the mirror. Their

[17]Evelyn, *Numismatica* (London, 1697), p. 282, in Murray, *Museums: Their History and Their Use*, Vol. I, p. 171. This also confirms Adalgisa Lugli's image of the collector as "author," in her "Inquiry as Collection," p. 69.

[18]BUB, *Aldrovandi*, ms. 70, c. 26v; ms. 136, Vol. XXV, c. 157.

[19]Zirka Zaremba Filipczak, *Picturing Art in Antwerp 1550–1700*, p. 31. See also Bolzoni's "Parole e immagini per il ritratto di un nuovo Ulisse: l'"invenzione' dell'Aldrovandi per la sua villa di campagna," in Cropper, *Documentary Culture*, pp. 326–330. My discussion of *invenzioni* throughout this chapter is highly indebted to Lina Bolzoni's excellent work on this subject.

[20]In Bolzoni, "L'"invenzione' dello Stanzino di Francesco I," p. 261.

presence placed the museum at the intersection of two competing images of man, Proteus and Narcissus. In his *Metamorphoses,* Ovid described Proteus as a demi-god "who has no settled shape" and Narcissus as a beautiful young man who, gazing into a pool of water, was "excited by the very illusion that deceived his eyes."[21] While Proteus symbolized man's and nature's capacity for transformation—what Thomas Greene has called the "metamorphosis of the self"[22]—the fable of Narcissus was a cautionary tale about the dangers of mistaking self-adulation for self-knowledge, the exterior for the interior. Narcissus, as Ovid put it, had been "spellbound by his own self" and finally vanished as the result of looking and loving too much. Collectors played with these different images by displaying objects whose inherent properties activated classic mythologies in the minds of the visitors who saw them.

Chameleons were omnipresent in museums. As creatures with the unique capacity to take on the color of "that which is nearest," they exemplified the sort of natural paradox in which collectors delighted.[23] Giganti listed a chameleon and an illustration of one in his museum inventory, while Settala had both male and female specimens. Naturalists such as Kircher conducted experiments on these creatures to determine the origin and extent of their transformative powers.[24] Della Porta included a chameleon in his *On the Physiognomy of Man* (1586) that "we preserve alive at home," juxtaposed to one of the ancient statues that he used to illustrate the art of reading faces[25] (fig. 21). Both appeared in the text as examples of the plasticity and malleability of identity. Undoubtedly, Della Porta kept a chameleon in his museum to remind him of the protean features of nature that he made a special part of his research. "With what knot can I hold this Proteus whose face is ever changing?" wrote Horace.[26] The chameleon and the statues in Della Porta's museum provided the "knot" that bound the identity of man to the Renaissance image of nature. They allowed collectors to study and, in a sense, possess the shifting features of nature and memorialize the equally protean features of humans.

Chameleons were objects replete with contemporary as well as classic allusions. No educated visitor could see a chameleon without thinking of Pico della Mirandola's *Oration on the Dignity of Man* (ca. 1486). In this classic humanist description of humans' place in the world, Pico described his species

[21]Ovid, *Metamorphoses,* Mary M. Innes, trans. (Harmondsworth, Middlesex, U.K., 1955), pp. 50, 80. For more on Narcissus in Renaissance iconography, see Cristelle Baskins, "Echoing Narcissus in Alberti's *Della Pittura,*" *Oxford Art Journal* (1993, in press).

[22]Greene, "The Flexibility of the Self," p. 260.

[23]Della Porta, *Della fisonomia dell'uomo,* Mario Cicognani, ed. (Parma, 1988), p. 62.

[24]Fragnito, *In museo e in villa,* pp. 196, 198; Ambr., Cod. Z.388 sup, f. 79; Petrucci, *Prodomo apologetico,* p. 79.

[25]Della Porta, *Della fisonomia dell'uono,* p. 62.

[26]Horace, *Epistles* I. 1, 90, in Greene, "The Flexibility of the Self," p. 247 (n. 9).

Figure 21. Della Porta's chameleon. From Giovan Battista della Porta, *Della fisonomia dell'uomo*, Mario Cicognani, ed. (Parma: Ugo Guanda, 1988).

as a "creature of indeterminate nature." Humans were metamorphic, capable of absorbing all the images of nature into themselves due to their semidivine status. "Who would not admire this chameleon?" exalted Pico.[27] In light of this passage, the necessity of the chameleon in the museum becomes clearer. It was the object, among all others, that most closely resembled man in the humanist interpretation of the world. As Edgar Wind writes in his analysis of Pico's imagery, "In his adventurous pursuit of self-transformation, man explores the world as if he were exploring himself."[28] The museum became the primary locus in which to reflect and comment on the intersections of these two essentially indistinguishable activities.

In his treatise *On Chameleons*, the seventeenth-century Dutch naturalist Isaac Schookius compared the chameleon's transformation to a form of theatrical representation, "just as an actor puts on a mask."[29] Della Porta, author of several plays as well as numerous scientific treatises, surely would have agreed with this analogy. Had Schookius lived in Italy rather than the Dutch republic, he probably would have compared the chameleon to a

[27]Pico della Mirandola, *Oration on the Dignity of Man*, in Kristeller, Cassirer, and Randall, *The Renaissance Philosophy of Man*, pp. 224–225.

[28]Edgar Wind, *Pagan Mysteries in the Renaissance* (New Haven, CT, 1958), p. 158. Pico's *Oration* also is analyzed in Greene, "The Flexibility of Self," pp. 242–243.

[29]In Edmund Goldsmid, ed. and trans., *Un-Natural History, or Myths of Ancient Science* (Edinburgh, 1886), Vol. III, p. 11

courtier; blending in with its surroundings, it made the social virtue of dissimulation a natural trait. Chameleons, in other words, exemplified the preferred behavior of collectors and their visitors who constantly altered their appearance to suit their environment, masking and at times risking the loss of the inner self in the process. Collectors, who often found themselves being everything to everyone, imagined their chameleons as an ironic commentary on their own place in society. From a philosophical perspective, embodied by Pico's neo-Platonic pronouncements, the chameleon recalled the human ability to be a microcosm and maker of his or her own destiny. From a social vantage point, it evoked a world that had transformed the humanist image of mankind into a courtly ethos in which display had become an end in itself.

Pico's *Oration* also made the image of Proteus central to the humanist conception of the self. "It is man who Asclepius of Athens, arguing from his mutability of character and from his self-transforming nature, on just grounds say was symbolized by Proteus in the mysteries."[30] In his interpretation, the metamorphosis of the self was something higher than the metamorphosis of nature, but nonetheless derived from it. Proteus, representing both man and nature, symbolized these connections. As Jean-Claude Brunon argues, despite Pico's admiration of Proteus, he did not appear in the emblematic literature until the mid-sixteenth century. Even then, he portrayed vices rather than virtues—inconstancy, heresy, and diabolic magic. Not until the mid-seventeenth century did Proteus regain his more positive attributes, as the god of metamorphoses. While Renaissance society allowed Proteus to reign over nature, epitomized by the chameleon and other physically indeterminate creatures, the Baroque world remade him into an inventor who manipulated nature at will.[31] Proteus increasingly represented art rather than nature, the collector rather than the objects collected. Transforming himself, he remade the world in the process.

Kircher, a great admirer of Pico whom he considered to be a true Catholic philosopher, made the image of Proteus one of the central themes of his museum. There one could enter the *Proteus Catoptricus* a room of moving images that gave the spectator the illusion that his head was transformed into various animal shapes, or see a *Proteus Metallorum*, one of Kircher's numerous experiments in which he transmuted metals (though not in the impious alchemical fashion, as he constantly reminded his readers).[32] The protean images scattered throughout Baroque museums reflected a growing sense that "man the chameleon" at last was making his presence felt in the material world. While Kircher's museum continued to celebrate "changing nature," it placed even greater emphasis on the idea of a "nature that one

[30]Pico, *Oration*, in Kristeller, Cassirer, and Randall, *The Renaissance Philosophy of Man*, p. 225.

[31]Jean-Claude Brunon, "Protée et Physis," *Baroque* 12 (1987): 15–22.

[32]De Sepi, pp. 38, 46.

can change."[33] Exploring the image of Proteus through the selection and display of objects—from the chameleon to the shifting mirrors that undermined Della Porta's view of physiognomy as the art of reading essential truths—collectors enhanced their image as interpreters and ultimately masters of nature. They were the epitome of the metamorphic self.

Mirrors were equally crucial to the reworking of the tale of Narcissus into a moral object lesson. Few if any Renaissance museums had actual mirrors in them, but then the mirror (in its improved, silver-backed form) did not appear in Europe until the early sixteenth century.[34] Despite the virtual absence of the mirror as a museum artifact until the seventeenth century, it existed as a metaphor for collectors to explore. On the wall of the second room in Aldrovandi's villa, essentially his private museum, was an image of a man looking at himself in the mirror. Underneath it one could read the following motto: *Nosce te ipsum* ("Know yourself"). Below, a harpy and a chimera, representing false paths and obstacles on the road to self-knowledge, framed the image of Narcissus. In his commentary on this emblem, Aldrovandi wrote: "Admiring one's own face is most easy, [but] knowing one's internal self always has been reputed to be difficult."[35] For Aldrovandi, such knowledge was achieved through the contemplation of the artifacts in his collection that surrounded the Narcissus emblem in his villa in their allegorical form.

As Lina Bolzoni observes in her study of the iconographic program of Aldrovandi's villa, the thirty-seven *imprese* decorating three rooms primarily featured objects in the naturalist's collection. Transforming his artifacts into emblems, Aldrovandi clarified their meaning by fitting them within the formal categories of interpretation validated by humanist culture. At the center of this "moralized natural history museum" lay the image of Narcissus whose mirror led him to lose all sense of identity.[36] In contrast, Aldrovandi's mirror—the world and the museum rather than himself—would not lead him astray. Like Cardano, who also chose "know yourself" as his guide in the writing of his famous autobiography, Aldrovandi imagined his desire to look outward rather than inward as a means of avoiding the fate of Narcissus.[37] The emblems adorning his villa constantly reminded him of this goal.

By the seventeenth century, the tale of Narcissus had become a moral for spectators as well as collectors to contemplate. New technologies in the mu-

[33]Brunon, "Protée et Physis," p. 21.

[34]Georges Gusdorf, "Conditions and Limits of Autobiography," in James Olney, ed., *Autobiography: Essays Theoretical and Critical* (Princeton, NJ, 1980), p. 32.

[35]BUB, *Aldrovandi*, ms. 99, c. 40v. This also is discussed by Mario Fanti, "La villeggiatura di Ulisse Aldrovandi," *Strenna storica bolognese* 8 (1958): 30, and Bolzoni, "Parole e immagini," p. 341.

[36]Bolzoni, "Parole e immagini," p. 337.

[37]Cardano, *The Book of My Life*, p. 49. I have altered the translation that reads "know thyself."

seum rather than emblems contributed to its revival. Della Porta described a silver drinking vase in which he had written *forma ne capiaris tua* ("do not be seduced by your own image"). A visitor who drank out of it was warned not to "fall in love with oneself and die like Narcissus."[38] The abundance of mirrors and other shimmering surfaces in Baroque museums placed Narcissus in competion with Proteus, the One and the Infinite. Both images flirted with the danger of losing one's identity, either through excessive contemplation or excessive imitation. In his museum in Milan, for example, Settala strategically positioned mirrors to catch visitors by surprise and even sought to recreate the conditions that led Narcissus to see himself so clearly by displaying "purged or crystalline waters in which whoever admires himself sees his own image reflected."[39] Cospi placed a mirror directly overhead so that visitors could picture themselves in the museum, joined to the objects through its reflection.[40] Baroque mirrors ironicized and subverted the image of Narcissus. Distorting, transforming, and multiplying the image of the viewer, they suggested that seeing was hardly a transparent activity. Their mirrors, unlike the one into which Narcissus gazed, contained the possibility of revealing more than the exterior. At the very least, they suggested that seeing had more than one aesthetic. Visitors to Baroque museums, more clearly than their Renaissance counterparts, understood themselves to be objects of display. Rather than simply *looking at objects,* they were constantly *looking through objects* and often looking at themselves to discern the meaning embedded in them.

Like the chameleon, the mirror symbolized both the collector and the museum. As Lina Bolzoni observes in her study of Francesco I's collection, "Thus entering [the *studiolo*], the prince enters also into himself."[41] Each object reflected a fragmentary image of the collector whose identity was only revealed in the encyclopedic totality of the museum. As Aldrovandi suggested when he chose Narcissus as his emblem, collecting was the true path to self-knowledge. The identity of man lay hidden in nature, just as Pico proclaimed when he labeled him a chameleon. In a society that placed a high value on physiognomy, the collector was the preeminent reader of signs, the person most capable of understanding these hidden affinities. When Juan Caramuel described Kircher as the "mirror of wisdom," a common encyclopedic image, he alluded to the Jesuit's ability to absorb and reflect all the knowledge of the age.[42] The collector was not the mythological Narcissus, who saw only himself, but a new Narcissus who discovered himself by looking at the world.

[38]Gabrieli, *Contributi,* Vol. I, p. 753.
[39]Scarabelli, *Museo o galleria adunata dal sapere,* p. 1.
[40]Legati, *Museo Cospiano,* p. 213.
[41]Bolzoni, "L''invenzione' dello Stanzino di Francesco I," p. 282.
[42]PUG, *Kircher,* ms. 564 (X), f. 181v (n.p., 4 August 1663).

ULISSE/ULYSSES

Self-knowledge, writes Georges Van Den Abbeele, is "an Odyssean enter-prise."[43] We have already seen in Chapter 4 how travel shaped the experi-ence of Renaissance naturalists. In the case of Aldrovandi, the voyage of self-discovery that all humanists claimed as their common patrimony took on a more personal meaning, for his name was Ulisse. *Nomina sunt omina:* names are omens.[44] "Who was Ulysses?" Aldrovandi asked himself repeatedly. A hero of the Trojan war, a "man of wide-ranging spirit . . . who wandered af-terward long and far." For Castiglione, Ulysses was the model of "suffering and enduring," a survivor who learned from his adventures.[45] For mid-sixteenth century humanists such as Ludovico Dolce he was an eminently civil man who exhibited the moral virtues prized by patricians, someone who did not allow *fortuna* to triumph over *virtù*. In all his different forms, Ulysses was one of the most important *exempla* cultivated by Renaissance hu-manists. He represented triumph over adversity, the knowledge that came from experience, and the wisdom acquired through careful guidance. He was a man who traveled only so he could return home—the perfect model for a collector to emulate.

Homer's writings inspired many other humanists besides Aldrovandi. After journeying through Europe, Duke August of Brunswick-Lüneburg, one of Kircher's main patrons and a collector in his own right, referred to his hometown of Hitzacker as "Ithaca."[46] Retracing the paths that the ancients took, humanists appropriated different aspects of their identity. Collectors such as Giganti prized marble busts of Homer that canonized him as one of the *viri illustres* worthy of being emulated and displayed. Aldrovandi's contemporary, Joseph Scaliger, recalled Homer with great fondness since he allegedly taught himself Greek in twenty-one days by reading and memorizing all of Homer's works.[47] While Scaliger made the challenge of reading Homer's original words the centerpiece of his autobi-ography, in which he presented himself as leading an exemplary life of scholarship, Aldrovandi read *The Odyssey* not as an initiation into the mys-teries of ancient languages but as the key to self-understanding. Like *The*

[43]Van Den Abbeele, *Travel as Metaphor*, p. xv.

[44]Bolzoni, "Parole e immagini," p. 317.

[45]Homer, *The Odyssey*, Walter Shewring, trans. (Oxford, 1980), p. 1; Castiglione, *The Book of the Courtier*, p. 331. In essence, Aldrovandi's quest was defined by Virgil's query, "Sic notus Ulysses?"

[46]Fletcher, "Kircher and Duke August of Wolfenbüttel," in Casciato, p. 283. For a general study of the impact of Homer on early modern German culture, see Thomas Bleicher, *Homer in der deutchen Literatur (1450–1740)* (Stuttgart, 1972).

[47]Fragnito, *In museo e in villa*, p. 197; George W. Robinson, ed. and trans., *Autobiography of Joseph Scaliger* (Cambridge, MA, 1927), p. 31. For a discussion of Scaliger's molding of his iden-tity as a scholar in his autobiographical writings, see Anthony Grafton, "Close Encounters of the Learned Kind: Joseph Scaliger's Table Talk," *American Scholar* 57 (1988): 581–588.

Odyssey, his own tale was "the story of a hero," a life of contemplation that emerged from a life of action. Reinterpreting this epic narrative, Aldrovandi met his destiny.

Aldrovandi's life, molded by the hidden affinities that his name revealed, recapitulated the Homeric epic. The Aldrovandi family seemed to have had a special connection to Homer. Aldrovandi's father Teseo (Theseus), described by the naturalist as "a very intelligent man of humane letters and vernacular eloquence," named two of his three sons after the protagonists of *The Iliad* and *The Odyssey:* Achilles and Ulysses.[48] When Aldrovandi's younger brother Achille took orders, he renamed himself Teseo, in memory of their father. Unwilling to let the name Achille disappear from the family, Aldrovandi bequeathed it to his illegitimate and only son. Unfortunately, it proved to be prophetic. In July 1577, the teenage Achille was killed in a fall from the corridor of the family palace. Aldrovandi was inconsolable, writing grief-stricken letters worthy of the fallen victor of Troy.[49] Yet at the same time, he reminded himself that he could not escape the fate the Homeric epic decreed for him. His source of fame was also the root of his sorrow. He could not enjoy the former without accepting the latter. Within a decade of his son's death, Aldrovandi began to map out the connections between the Homeric Ulysses and the Bolognese Ulisse in two different media: the cycle of paintings decorating the main room of his villa in San Antonio di Savena (located just outside the walls of the city at Porta San Vitale) and his autobiography. They were the primary loci in which he exercised the humanist passion for invention, drawing upon Homeric wisdom, his own experiences, and the objects in his collection to narrate his life.

Both the beginning of Aldrovandi's autobiography and the opening of the Homeric cycle emphasized the Odyssean framework of his life. "This will be a narrative or representation of the most illustrious facts of Ulysses, following their actual historical order" (*che realmente segue l'ordine de tempi*), wrote Aldrovandi in his analysis of the Homeric cycle. As Lina Bolzoni notes, Aldrovandi consciously chose to deviate from the textual order of events to create a temporal framework better suited to the expression of his personal itinerary and his sense of history.[50] Thus, the cycle of thirteen panels in the *sala maggiore* began with Ulysses's attempt to avoid leaving Ithaca by feigning insanity and his recognition of Achilles among the daughters of Lykomedes, and ended with Minerva's restoration of peace to the world and the accidental slaying of Ulysses by Telegon, his son by Circe. Not coincidentally, given the depth of his grief about the loss of his son, the two images that most reminded Aldrovandi of Achille—Achilles and Telegon—

---

[48]Aldrovandi, *Vita,* p. 5.

[49]See his "consolatory letters" to his brother and his nephew; BUB, *Aldrovandi,* ms. 91, cc. 417–428r; ms. 97, cc. 339–345r; and ms. 150.

[50]BUB, *Aldrovandi,* ms. 99, c. 2r, in Fanti, "La villeggiatura di Ulisse Aldrovandi," p. 27; Bolzoni, "Parole e immagini," p. 328.

framed the Homeric cycle. In between, the cycle recounted the signal events of Ulysses's travels.

Reinforcing the message of the *invenzioni* painted on the walls of the villa, Aldrovandi's autobiography also established the image of Ulisse the traveler as a predetermined feature of his identity. As Aldrovandi recalled, "his desire to go out [into the world]" was one of his earliest memories. It led him to wander through Italy, Spain and parts of France, and fueled his taste for the exotic and the unknown. Like his namesake, Aldrovandi encountered various dangers along the way. Neptune, sworn enemy of the mythological Ulysses, crossed his path several times—an event portrayed in the tenth panel of the Homeric cycle in his villa and complemented by tales of near drownings during his travels. He endured hunger and attacks by bandits and pirates, yet survived. In imitation of the other Ulysses, he took sword in hand "to save his life and his shipmates."[51] Aldrovandi, in other words, enjoyed a youth filled with perilous and heroic exploits, or at least this was the way he recalled it when he sat down to compose his *vita* at the age of sixty-four. Like Ulysses who twice dressed as a beggar to avoid detection—events featured in the third and eleventh panel of the cycle—he at times found it expedient to travel in disguise. Ulisse the pilgrim and Ulysses the beggar both gained knowledge through their assumed humility. Ulysses had observed Sisyphus in the original *Odyssey*; Aldrovandi did not neglect to record the fate of a man whose task was never finished (*interminabilis hominis labor*), portraying Sisyphus in one of the emblems painted in his villa. With Minerva as their guide and their conscience, both the ancient and modern Ulysses undertook a great moral voyage. At the end lay peace as well as knowledge, the prophecy of a tranquil death at home after a turbulent and sorrowful life abroad. "And death will come to you far from sea, a gentle death that will end your days when the years of ease have left you frail and your people round you enjoy all happiness."[52] Meditating on these images in the mid-1580s, some years after the death of his son and the unpleasantness of the theriac debates, Aldrovandi saw the wisdom in Homer's epic. His name indeed had determined the signal events of his life.

Aldrovandi intended the Homeric cycle, and the three rooms decorated with *imprese* that adjoined it, as a continuation of the display that began in his museum in Bologna. Rather than choosing between the city and the countryside, the classic humanist dilemma, Aldrovandi enjoyed the benefits of both settings. The villa did not replicate the purpose of the museum—to display and examine all of nature—but instead refracted its contents. Removed from urbanity, Aldrovandi reflected on the personal rather than the professional and pedagogical meanings of his activities, presenting the results of his meditation in a series of elaborate iconographies. The Latin tags

---

[51]Aldrovandi, *Vita*, pp. 6–7.
[52]Homer, *Odyssey*, p. 131.

below each of the Odyssean scenes and all the *imprese,* created in imitation
of the emblem books popular during this period, provided visitors with the
necessary clues to understanding their message. Aldrovandi's written commentary on their iconography gave them further insight into his personal
interpretation of the villa; in effect, it was a moral autobiography that appropriately began with a consideration of the etymology of his name. Just as
visitors decoded nature by looking at the synoptic tables in Aldrovandi's museum, they gained privileged access to his "other self" by reading the words
that he inscribed around a series of carefully chosen images. The villa completed what the museum began, transforming the materials of the collection
into a self-conscious moral artifact.

The iconographic program of the villa also resolved various problems left
unfinished in Aldrovandi's autobiography, which ended with the successful
conclusion of the theriac debates in the 1570s. In the writing of a *vita,*
Aldrovandi could not look into the future; the genre did not permit this sort
of speculation. In fact, the material added at the end of the autobiography
clarified details of his early life rather than completing his history. Instead,
Aldrovandi accomplished this task through his renarration of the *Odyssey*
and choice of emblems. The first room of *imprese* in his villa, for example,
contained an image of fate as a young boy holding a book and a cornucopia,
sitting on top of the world. *Reliquorum vicissitudo* ("the vicissitude of the future"), commented Aldrovandi. The portrayal of fate as the possessor of the
two canonical forms of knowledge surely reflected the ambitions of a collector who saw books as the embodiment of authority and the cornucopia
as a symbol for the humanist encyclopedia. Placing them in the hands of
fate, Aldrovandi obliquely suggested that collecting was a means of gaining
mastery over the future. Only the possession of knowledge could prepare
one for the vicissitudes wrought by time. Subsequent *imprese* such as the image of a crane supporting itself by placing one foot on a stone—*Cura sapientia crescit* ("wisdom increases with care")—reinforced this message.

Decorating his villa about fifteen years after the Grand Duke Francesco I
had commissioned his own cycle of *invenzioni* for his *studiolo* in Florence,
Aldrovandi was certainly aware of the power of allegorical images in late Renaissance culture. The creation of a "moral museum," displayed to a select
group of patrons, placed Aldrovandi among the leading interpreters of humanist culture. Defining *imprese* as a form of knowledge that "made manifest the virtuous and magnanimous intentions of the noble soul,"
Aldrovandi presented the three rooms of emblems as revelations of his inner self.[53] They were the fruit of his voyage into nature. Just as his commentary on the Homeric cycle explained the new Odyssey to visitors, the *imprese*
gave testimony to his ability to achieve the wisdom he desired. Unfortunately, we have no specific sense of how visitors responded to the *invenzioni*

[53]BUB, *Aldrovandi,* ms. 99, in Fanti, "La villeggiatura di Ulisse Aldrovandi," p. 29.

decorating the villa. Certainly they must have been impressed that a private individual could create an aesthetic worthy of a prince. Among its many different objectives, Aldrovandi's villa reinforced his status as a man of learning capable of courtly knowledge. He was the equal of Borghini, the humanist who designed the *invenzioni* for Francesco I with the help of Vasari, yet also the equivalent of a Renaissance prince when he suggested, not very subtly, that his life was worthy of representation and interpretation.

Aldrovandi concluded his commentary on the Homeric cycle with the following words: "This is what we wished to elucidate . . . so that this new Odyssey, so to speak, or series of the most notable deeds and errors of Ulysses, will be rendered more easily to viewers who are unfamiliar with it."[54] While only privileged visitors saw the images in Aldrovandi's villa, they certainly were not private; we might describe them instead as select images for select patrons. Aldrovandi often retreated to his villa when he wished to avoid the *negotium* of the city, inviting intimates—visitors capable of appreciating the display of his inner self—to join him there. Important associates such as Gabriele Paleotti studied nature in this setting. The villa was not a site of solitude, in its most absolute sense, but of leisure. As the motto above one of the four doors opening into the *sala grande* stated, "It is fitting that worthy men collect the common fruit from the prosperity of civil company."[55] Aldrovandi's villa, a product of *otium*, emerged from the conversation of learned men. While transforming Aldrovandi's life into a vast moral emblem, it also memorialized the forms of learning that he and his associates most treasured.

The degree of premeditation in the selection and composition of the *invenzioni* decorating Aldrovandi's villa leaves us with no doubt that he imagined it as his ultimate self-portrait, more revealing in many ways than the autobiography that recounted his travels and career. But Aldrovandi did not stop there. In addition to the four decorated rooms, filled with images of the best artifacts in his collection—vipers, birds of paradise, ostriches, butterflies, crabs, baboons, unicorns, sunflowers, and so on—Aldrovandi recorded the presence of six portraits in the villa. They included images of himself and his second wife, Francesca Fontana, both painted by Lorenzo Benini, and portraits of Francesco I and Ferdinando I de' Medici, a "wild man" (*homo sylvester*) and his daughter the "hairy girl" (*puella hirsuta*).[56] In this restricted portrait gallery, Aldrovandi once again defined the parameters of his self. The two Medici princes recalled the importance of patronage to his activities, defining his social identity. "You, Francesco, have

[54]Ibid., p. 27.

[55]Ibid., p. 32. For a fascinating discussion of this theme as it regards the activities of other sixteenth-century humanists, see Fragnito, "Il ritorno in villa," *In museo e in villa*, pp. 65–108.

[56]All are discussed in Fanti, "La villeggiatura di Ulisse Aldrovandi," pp. 35–36, where the inscriptions under each portraits are reproduced. See also Bolzoni, "Parole e immagini," pp. 346–347.

demonstrated that great things suit great men through your virtues, appearance, name, and genius," wrote Aldrovandi, suggesting that his qualities as client mirrored those of his patron.

The images of two monstrous creatures underscored the concept of natural identity. Pedro Gonzales, Aldrovandi's *homo sylvester*, and his children visited Bologna during the 1580s as Aldrovandi was completing his villa. They attracted the attention not only of the city's greatest collector but also the artist Lavinia Fontana, who made a sketch of Gonzales's daughter.[57] Beneath the portrait of the wild man, Aldrovandi recounted the story of a peasant who mistook a similar creature for a monkey. Below the portrait of the hairy girl he wrote, "My portrait shows me hairy of face and hand but under my clothes I am all skin" (*at sub veste rigent caetera membra pilis*). Both images return us to the theme of man's indeterminate nature first raised by the presence of the chameleon in the museum. We are never quite what we seem to be, suggested Aldrovandi. Despite our attempts to elevate ourselves to the divine, nature always finds ways to remind us of our earthly, animal side. The four portraits surrounding the images of Aldrovandi and his wife reflected two different ends of the moral spectrum in which they were positioned. On one side lay dignity, virtue, and fame, the nobility of man as seen through the portraits of great princes and eminent patrons; on the other side lay two images worthy of any cabinet of curios whose physical indeterminancy marked the boundaries between the collector and the world he collected. Situating himself *in medias res*, Aldrovandi interpreted the role of the collector as a man who maintained equilibrium in the world.

Underneath his own portrait, Aldrovandi inscribed the following words: "This is not you, Aristotle, but an image of Ulysses: though the faces are dissimilar, nonetheless the genius is the same." In 1599 Aldrovandi reproduced this image in the first volume of his *Ornithology* (fig. 22). Like the other images in his villa, many of which appeared in his publications in their "natural" form, it testified to the strong connections between his public persona and private self. The 1599 portrait, perhaps more than any other created by Aldrovandi, underscored the relationship between intellectual achievement, moral virtue and the emulation of the past. Beneath the "image" (*imago*) of Ulysse lay the "genius" (*ingenium*) of Aristotle. *Ingenium* was an attribute of the inventor. Even as Aldrovandi completed his new Odyssey, he denied its ability to capture the completeness of his life. The itinerary that led visitors through the Homeric cycle and three rooms of emblems ended in front of this remarkable portrait, where Aldrovandi gave a new and deeper level of meaning to the autobiographic impulse that ordered the images in the villa.

The published version of the portrait—unfortunately the only one available, as the original has not survived—has an emblematic quality about it.

[57]Kenseth, *Age of the Marvelous*, pp. 329–330, 333. See Lavinia Fontana, *Portrait of the Daughter of Pedro Gonzales* (ca. 1583), in the Pierpont Morgan Library, New York.

Figure 22. Aldrovandi as Aristotle. From Ulisse Aldrovandi, *Ornithologiae hoc est de avibus historiae libri XII* (Bologna, 1599).

Surrounding the image of Aldrovandi as Aristotle are some of the creatures he studied. Despite the fact that the portrait appears in his history of birds, its content is not strictly bound to its context. In fact, the accompanying images—a lion, a butterfly, a peacock, an owl, an eagle, and so on—recall the *imprese* of the villa more than the subject of his ornithology. Their presence forms the basis of Aldrovandi's claim that he has inherited the intellect of Aristotle. Yet the form of inheritance has a peculiarly humanist flavor. *Ingenium* was a characteristic ascribed to talented artists who created images that seemed, at least in the eyes of contemporaries, to call into question the fixity of the divide separating appearance from essence. Aldrovandi's portrait shares that same quality of instability. The words that define its theme seem to will forth the *ingenium* of Aristotle from the *imago* of Ulysses, effecting an Ovidian metamorphosis. Like the wild man and hairy girl, Aldrovandi also represents the indeterminacy of human identity.

The words accompanying another portrait of Aldrovandi in the 1602 volume of his *Ornithology* further underscore this point. As the inscription below Giovanni Valesio's portrait of the octogenarian stated, no doubt in explicit reference to the 1599 portrait, "He can paint your image but not the miraculous endowment of the soul" (*Effigiem potuit pingere, non animi dotes mirificas*).[58] Both engraved portraits suggested that one could gain insight into Aldrovandi's identity as Aristotle only through his written words. *Animus* and *ingenium* were found in texts, not in images. Despite the profusion of visual materials surrounding Aldrovandi, he always returned to words as the source of meaning. No image in his books or in his villa could be understood without their mediation.[59]

One other portrait of Aldrovandi merits our attention in relation to his villa. It is an image created by Agostino Carracci for Cardinal Odoardo Farnese entitled *Hairy Harry, Mad Peter and Tiny Amon*, completed about 1598–1600 (fig. 23). While art historians remain undecided about whether the Bolognese painter Carracci actually intended to portray Aldrovandi, the resemblance of the figure in the upper right to that of Aldrovandi in other portraits is very strong. Even more noteworthy is the assemblage of creatures, all present at Cardinal Farnese's court in Rome. With the exception of the dwarf, they also appeared in the different images decorating Aldrovandi's villa.[60] Perhaps we might conclude that Carracci, an acquaintance of

---

[58] Aldrovandi, *Ornithologia* (Bologna, 1602), n.p. Both the 1599 and 1602 engravings are discussed briefly in Diane DeGrazia Bohlin, *Prints and Related Drawings by the Carracci Family* (Washington, DC, 1979), pp. 334–335. Both images bear the signature of Aldrovandi's assistant and successor, Johann Cornelius Uterwer, who evidently had a hand in their composition. Giovanni Valesio (ca. 1579–1623) was a Bolognese artist.

[59] This point is also raised by Bolzoni, "Parole e immagini," pp. 329–330.

[60] For discussions of this painting, see Fanti, "La villeggiatura di Ulisse Aldrovandi," pp. 36–38; and Giuseppe Olmi and Paolo Prodi, "Art, Science, and Nature in Bologna circa 1600," in *The Age of Correggio and Carracci: Emilian Painting of the Sixteenth and Seventeenth Centuries* (Washington, DC, 1986), pp. 213–236. The image is reproduced and identified on pp. 261–262.

Figure 23. Agostino Carracci, *Hairy Harry, Mad Peter and Tiny Amon* (ca. 1598–1600). From Museo e Gallerie di Capodimonte, Naples.

Aldrovandi's from his academy days in Bologna, where he and his two brothers dominated the artistic scene, intended the painting as a double portrait of the two worlds he inhabited—the courtly culture of Rome, where cardinals populated their households with living wonders, and the encyclopedic culture of Bologna, where Carracci's interest in portraying nature accorded well with the activities of Aldrovandi. We have no proof that Carracci actually saw the *imprese* or the portrait of the hairy man in the villa. Yet he was in contact with the local group of humanists who had access to this setting. Even if Carracci himself never set foot in the villa, he surely knew of its existence and probably had a good idea of its contents, given the number of artists who entered the villa to complete Aldrovandi's commission. At the very least, we know that in 1596 he engraved a portrait of the naturalist—the very one that found its way into the 1599 volume of the *Ornithology*—and therefore had some involvement in the publicizing of Aldrovandi's image.[61]

[61]Bohlin, *Prints and Related Drawings by the Carracci*, p. 334; Olmi and Prodi, "Art, Science, and Nature," p. 229. Carracci created the image of Aldrovandi but not the external borders,

Carracci's painting brings together many elements present in the villa. However, here their relationship is unmediated by Aldrovandi's words, which gave them formal meaning in their emblematic presentation. *Homo sylvester* and *homo sapiens* are engaged in mute conversation. The body of "Mad Peter" leans into the painting, while his eyes make contact with those of the wild man. Certainly it would have been a quintessentially humanist gesture to make the image of a madman resemble the face of one of the most learned men in Italy, especially since the first panel of the Homeric cycle depicts Ulysses's feigned insanity. The gesture of the wild man's left hand draws our attention to the rest of the scene, in which all the different parts of nature interact harmoniously. All the creatures in the composition define man through their proximity to him. Both the parrot and the monkey remind the viewer that many human traits can be found in the animal world, while the two dogs recall the human ability to domesticate nature. "Tiny Amon" completes the protean imagery that begins with the interchange between "Hairy Harry" and "Mad Peter." He represents yet another aspect of nature in error, the monstrosities whose appearance undermines the distinctness of the human species from nature. In his painting, Carracci manages to touch upon most of the themes raised in Aldrovandi's own self-portraits. As with the appearance of the naturalist's official portrait in his *Ornithology,* Carracci's painting reflects the continued interaction between the private contemplation of the self and the public display of the collector.

Aldrovandi's image making did not stop with the reinvention of Ulysses and Aristotle. At various points, the Bolognese naturalist also proclaimed himself a new Galen and a new Pliny, images that completed his identity as the reincarnation of an ancient naturalist. They were the famous men who "moved him" just as Petrarch was moved by his encounters with Cicero, Virgil, and above all, Augustine. Yet Aldrovandi did not confine his *exempla* to hallowed halls of ancient learning. He juxtaposed his image as a naturalist *redivivus* to his image as a discoverer, styling himself as a new Columbus. The choice of Columbus as a model represented continuity rather than disjuncture. Columbus was the new Ulysses, a wandering spirit whose voyages transformed the world; Aldrovandi too hoped to launch a similarly ambitious program of discovery. When Giacomo Antonio Buoni described Aldrovandi as "almost a new Ulysses" due to the numerous botanical expeditions the naturalist made, he linked the classical and contemporary images of the explorer.[62] Yet Aldrovandi's travels did not quite merit equivalency with those of Ulysses or Columbus, who traveled the world. During the 1570s, at the height of his career, Aldrovandi attempted to rectify this situation. He would complete the voyages of exploration begun by ancient travelers such as

which, according to Bohlin, were designed by Giovanni Valesio. An oil painting of Aldrovandi, now in the Accademia Carrara di Belle Arti in Bergamo, is also attributed to Agostino Carracci (ca. 1584–1586).

[69]Giacomo Antonio Buoni, *Del terremoto* (Modena, 1572), p. 45r.

Ulysses and Galen (in a medical capacity) and renewed by Columbus, journeying to the New World to complete his study of nature.

Like many Renaissance naturalists, Aldrovandi avidly read all the literature on the voyages of exploration. He owned two copies of Columbus's letters and the Italian translations of Oviedo's and Acosta's natural histories of the Indies. Around 1562, he first began to articulate his vision of the place he would occupy in the new narratives of discovery then being written. "It is already ten years since I conceived of this fantasy of going to the newly discovered Indies," he wrote in his *Natural Discourse* in 1572–1573. "And disregarding every difficulty, I would find a way to make this voyage in the guise of Christopher Columbus." Aldrovandi saw himself as uniquely equipped to undertake such a voyage. Like Columbus, he was an avid reader of Pliny. Unlike the Genoese sailor, his humanist education and philosophical training made him better prepared to comprehend the full significance of what he observed. "And knowing what the Arabs, Greeks, Latins, and other writers have written on these matters, I would bring great profit to the World, if I were to go in these places. If any man in Europe is fit to do this, I believe (it may be said without boasting) I am."[63] In his *Catalogue of Scholarly Men Who . . . Have Enriched the Theater of Nature,* Aldrovandi included Columbus whom he praised for his "diligence, effort, and vigilance."[64] Columbus exemplified the modern man who created novelties without rejecting the virtues of antiquity and therefore was worthy of emulation.

Throughout the 1570s and 1580s, Aldrovandi attempted to interest various patrons in funding an expedition to the New World. He addressed his *Natural Discourse* to the son of Gregory XIII in an effort to interest the pope in his projects. When Gregory XIII did not respond, Aldrovandi turned to other prospects. By the mid-1570s, he approached representatives of the Spanish king, Philip II, in Bologna. "In this affair my intention is similar to that of the diligent and ingenious admiral Christopher Columbus, well remembered, who was able to find new countries with great utility and honor to his King and himself with the help of the Catholic King," explained Aldrovandi to the Cardinal Protector of the Spanish College.[65] He encouraged him to pass on the message to the king. By reminding Philip II of a previous instance of Spanish patronage of an Italian discoverer, Aldrovandi hoped that his own identification with Columbus would spur the current ruler of Spain to fulfill his historically preordained role as the patron of explorers, thus completing the image Aldrovandi had chosen for himself. Unfortunately, Philip II decided to shower one of his own naturalists with his

[63]Aldrovandi, *Discorso naturale*, pp. 208, 210. Aldrovandi's interest in the New World is discussed in greater detail in Mario Cermenati, "Ulisse Aldrovandi e l'America," *Annali di botanica* 8 (1906): 3–56.

[64]BUB, *Aldrovandi*, ms. 21, Vol. II, c. 580. Also discussed by Tugnoli Pattaro, *Metodo e sistema delle scienze nel pensiero di Ulisse Aldrovandi*, p. 91.

[65]BUB, *Aldrovandi*, ms. 66, c. 361r (Bologna, 12 November 1576).

munificence. From 1570 to 1577, he sent the Spanish court physician Francisco Hernandez to Mexico to catalogue the flora and fauna and collect information about local medical customs.[66]

While wooing the Spanish king, Aldrovandi did not neglect the patrons he had immortalized in the portrait gallery of his villa. Shortly after Ferdinando I had become Grand Duke of Tuscany upon the death of his brother in 1587, Aldrovandi wrote to him to request support for his New World voyage. "And the generous piety of the Catholic rulers Ferdinand and Isabella in helping the judicious plan of Columbus to discover the New World has affected me as well," he confessed to Ferdinando I.[67] By this point, Aldrovandi was sixty-five years old, no longer capable of enduring the ardors of a lengthy trip. Yet he nonetheless persisted in indulging his fantasy, urging the Medici princes not to let other rulers claim all the glory of discovery; he would be the Medicean Columbus. Once again, Aldrovandi watched another complete the voyage he intended to take. In the 1590s, he stayed at home, tending his failing eyesight and bruised ego, while the more youthful Casabona, botanist to the Grand Duke, went to Corsica and Crete. Instead, Aldrovandi contented himself with directing these expeditions from his museum.[68] In this instance, the given identity as Ulysses proved to be more powerful than the chosen identity of Columbus.

Despite Aldrovandi's proclamation that "modern geniuses are not inferior to ancient ones but different," his words and actions at every stage of his life made evident the strength of the latter.[69] And it was his relationship to the past that his early biographers celebrated. Lorenzo Crasso opened his 1666 eulogy of Aldrovandi with the following words: "Our age does not envy antiquity its Aristotle . . . while we have Ulisse Aldrovandi." Reversing the order in which Aldrovandi developed his different images, Crasso began with Aristotle and ended with Ulysses. "Worn out by his mature Age but even more by incessant Studies, worthy of Ulysses, having crossed the spacious Ocean of sciences, he passed on to a better life on 4 May 1605."[70] Aldrovandi's new odyssey was effected not by retracing the path of Columbus but by mastering the various forms of knowledge pertinent to knowing nature. Whether in the guise of Aristotle, Galen, or Pliny, or through his attempts to become the next Columbus, Aldrovandi led an exemplary life worthy of the name that determined the course he would take. Collecting nature, he had made himself an object worthy of collection. Appearing in Crasso's *Elogies of Men of Letters* and lauded in poetry by no less a patron than Maffeo Barberini, the future Urban VIII, Aldrovandi entered the gallery of

[66]For more on this episode, see Goodman, *Power and Penury,* pp. 234–238.

[67]BUB, *Aldrovandi,* ms. 21, Vol. II, c. 12 (Bologna, n.d.).

[68]This already has been discussed in chapter 4. See BUB, *Aldrovandi,* ms. 21, Vol. IV, c. 170r (Bologna, 29 August 1595); Galluzzi, "Il mecenatismo mediceo e le scienze," p. 199.

[69]BUB, *Aldrovandi,* ms. 21, Vol. IV, c. 49r.

[70]Lorenzo Crasso, *Elogi d'huomini letterati* (Venice, 1666), Vol. I, pp. 135, 137.

famous men that he had helped to shape, reinforcing its function as an un-interrupted dialogue with the past.[71]

MUSAEUM SUI IPSIUS: DELLA PORTA

Aldrovandi was not the only collector who explored nature to know himself. He represented a supremely articulated version of an essential humanist topos. Self-awareness was the most difficult and therefore the most desirable form of knowledge to attain, and nature was the object whose contemplation brought into relief the identity of the naturalist. Through the use of emblems, mythologies, and literary allusions, collectors spun elaborate allegories, clothing themselves with the images of antiquity remade in the mirror of their own culture. Like the objects in their museums, they had the capacity to produce mutiple forms of knowledge.[72] Publicizing this skill was essential to their success in maintaining and expanding the encyclopedic tradition.

No naturalist better reflected this philosophy than Della Porta. In contrast to Aldrovandi, Della Porta left few autobiographical fragments behind. He wrote no *vita* nor did he commission a series of *invenzioni* for his villa in Vico Equense, a seaside resort for the nobility south of Naples. On the whole, Della Porta led a considerably more "private" life than Aldrovandi, perhaps a reflection of his desire for secrecy. While Aldrovandi made the events of his own life as visible as the objects in his museum, Della Porta preferred to enhance the mystery surrounding the activities at his villa and in his family palace in Naples. He reflected the continued strength of a somewhat different philosophical tradition that cast the naturalist in the role of a magus. For Della Porta, it was essential that some of the details of his life remain obscure; the cloaking of his identity completed his image as a man who controlled the secrets of nature. Despite these important differences, Della Porta made use of essentially the same tools to fashion his self. He presented his villa and museum as settings of complementary action whose dialectical relationship completed his identity, and he drew upon the emblematic conventions of the day to develop key symbols that expressed both his *imago* and *ingenium*.

Like many of his contemporaries, Della Porta conceived of his villa as a place of *otium*. "When he tired of the continuous visits [to his museum], he found himself often withdrawing to one of his villas, a short distance from Naples."[73] As a setting in which to recover one's sense of self, the Villa delle

---

[71] Crasso reproduces Barberini's three poems to Aldrovandi; ibid., pp. 137–138. They also appear in his *Poemata* (Rome, 1631), pp. 186, 215–216. Lina Bolzoni describes Barberini as the "new Homer" for the "new Ulysses" in her "Parole e immagini," p. 318.

[72] This agrees well with Krzysztof Pomian's image of the museum as a "semiophore."

[73] BANL, *Archivio Linceo*, ms. 4, in Gabrieli, *Contributi*, Vol. I, p. 678. This image of the villa accords well with the one presented in Fragnito, *In museo e in villa*.

Pradelle was essential to Della Porta's image as a magus; only by retreating from society (*ritirarsi*) could he discover his own identity (*ritrovarsi*). In this setting, he conducted agricultural experiments, some of which appeared in his *Natural Magic* and later were collected together in his *Villa* (1592).[74] According to the Lincean biography of Della Porta, Vico Equense also was the setting in which he wrote his comedies as a means of relaxing from the "more serious studies" he conducted in the city. There he invited Cesi and other gentlemen of note to partake in the pleasures of "philosophical contemplation," removed from the ardors of civil society.[75] While Aldrovandi used his own pastoral enclave to recreate his museum in a different form and for a different purpose, Della Porta celebrated his villa as a site of leisure and productive usefulness. It completed his identity as a natural philosopher interested in operative magic and enhanced his image as a master of the civil arts of learning.

While we have no evidence of any *invenzioni* decorating Della Porta's villa, we do know that he was well versed in emblematics, a reader of such works as Camerarius's *Collection of Symbols and Emblems* (1593-1604) that made all of nature a moral museum. Emblems were a form of knowledge better created in a villa than in a museum; as a genre that brought together the different components of humanist learning, they were visual displays of rhetorical wit, the supreme products of scholarly leisure.[76] They enhanced a scholar's standing among the princes and nobles whom he wished to cultivate as patrons and whose identities also were bound to this form of display.[77] With Aldrovandi, we can trace his use of emblems between the villa and the museum, from his *invenzioni* to his publications. With Della Porta, the trail is not as evident, although the trajectory was probably similar.

Emblems found their way into his publications and formed the core of intellectual exchanges with his associates. The emblem on the frontispiece of the *Phytognomonica* (1588), for example, inspired Cesi to make the lynx a symbol of his academy. But it first stood for Della Porta, who collected nature in order to read her signs. As Della Porta wrote to Cesi, "Having discovered the 'Physiognomy of plants,' until now considered by no one, it seemed to me

[74]Prior to publishing his *Villae . . . libri XII* (Frankfurt, 1592), Della Porta published two sections of this agricultural encyclopedia: *Suae villae pomarium* (Naples, 1583) and *Suae villae olivetum* (Naples, 1584).

[75]Gabrieli, *Contributi*, Vol. I, pp. 678, 680; see also Louise George Clubb, *Giambattista Della Porta Dramatist* (Princeton, 1965), pp. 4–7, 19–20.

[76]There is a large and growing literature on emblematics. Most relevant to this study is Ashworth, "Natural History and the Emblematic World View," and Wolfgang Harms, "On Natural History and Emblematics in the Sixteenth Century," in Allan Ellenius, ed., *The Natural Sciences and the Arts*, Acta Universitatis Upsaliensis, Figura Nova, vol. 22 (Uppsala, 1985), pp. 67–83.

[77]For an interesting example of how natural philosophers used their skill in emblematics to develop allegorical relationships with their patrons, see Mario Biagioli, "Galileo the Emblem Maker," *Isis* 81 (1990): 230–258.

that I had penetrated the secret of plants with my thought. I made the lynx, whose sight passes through a mountain according to all writers, the *impresa* for this book and under it wrote the motto *Aspicit et inspicit* (He observes the exterior and the interior)."[78] As Della Porta proudly proclaimed through this emblem, he was a natural philosopher whose sharpened instincts allowed him to penetrate the secrets of nature by extracting nature's *ingenium* from her *imago*. The lynx became a metaphor for the process of reading nature that Della Porta popularized in his treatises on celestial, human, and plant physiognomy and invited his contemporaries to practice on himself. Who could decode the master decoder? This was a challenge designed to appeal to a humanist audience that thrived on such conundra.

During his tenure as a Lincean, Della Porta sent Cesi a list of twenty-five emblems he had created, the textual equivalent of Aldrovandi's three rooms of *imprese*.[79] While many emblems were gifts for friends, at least six were personal symbols. Most prominent was the lynx. Della Porta also portrayed himself as a compass, an oak tree, a grapevine, a candle, and a fish with eyes on top of its head. He was a compass "not because I proudly wished to demonstrate my profession as mathematician but in order to be measured in my desires with the motto *Metior, ne metiar* (I measure so that I would not judge)." Thus, he exemplified the virtue of prudence, modest in his abilities to encircle the world. The oak tree symbolized Virgilian humility: *Descendo ut ascendam* ("I descend so that I may ascend"). Other emblems depicted his selflessness. Della Porta was a grapevine, bowed by the weight of its fruit, with the motto *Perit ubertate sua* ("It perishes from its fertility") because he "had given himself away in order to help many of his friends."[80] Likewise a candle—*Morior mihi, ut vivam aliis* ("I die for myself so that I may live for others")—signified the fate of a man who "consumed himself to give pleasure to his friends." All of these different symbols reinforced the image of Della Porta as a man who made his knowledge available to others, often at the expense of his own peace of mind. His visibility at times threatened to dissolve his sense of self, something he attempted to recapture through the creation of meaningful objects that expressed his *ingenium*. They were the topoi of a scholar who, while expressing a decided preference for the quietude of his villa over the activity of his museum, nonetheless felt obliged to participate in civil life and in fact derived great satisfaction from the scholarly exchanges held in his museum.

[78]Gabrieli, *Contributi*, Vol. I, p. 754; Vol. II, pp. 1664, 1670. Gabrieli notes that Aldrovandi used this same motto in his discussion of the lynx and surmises that the common origin was Joachim Camerarius the Younger's *Emblemata animalia*. In Della Porta's case, the image certainly could not have come from Camerarius's published work since the first volume did not appear until 1593.

[79]They are found in BANL, *Archivio Linceo*, ms. 9, ff. 21–22, and reproduced by Gabrieli; *Contributi*, Vol. I, pp. 752–754.

[80]This was an image that Della Porta liked enough to use twice.

Della Porta's final emblem, the fish who gazed upward at the heavens, was meant to reinforce his contemplative side. Yet, as Della Porta recounts, some enemies maliciously reinterpreted it as a sign of his intellectual disorientation and continued disfavor with the Catholic authorities on account of his ambiguous orthodoxy. Rather than symbolizing his piety, as he intended, it indicated his monstrosity.[81] Emblematic knowledge, as this episode of competing interpretations suggests, was constantly knowledge in play. It represented the transformative possibilities of humanist culture in which the concept of identity was as protean as the metamorphic objects in the museum. Della Porta evidently circulated these emblems among the Neapolitan *letterati* and the Linceans. They formed the basis of a series of discussions about who and what he was that took place in the academies of Naples and Rome. Like Aldrovandi, Della Porta was a scholar molded by the opinions of the republic of letters. It provided him with the tools to create his intellectual mosaic and confirmed the appropriateness of the images that recalled his memory.

In response to Della Porta's emblematics, Cesi produced a medal of the Neapolitan magus in 1613 that solidified the official Lincean view of his identity. The academy biography of Della Porta described the medal as "alluding to his surname and arms, who truly was a Door through which nature exited from her most secret and deepest hiding places to show herself to philosophers."[82] On the front of the medal was an image of Della Porta. On the back, beneath the words *Natura reclusa* ("Nature disclosed"), was an image of a nude woman. With her hair in flames, holding a globe in her right hand and a cornucopia in her left hand, she stood in front of a cliff in which there was an open door. Above the door, a group of philosophers, undoubtedly his fellow Linceans, was vaguely discernible. Observing Nature's entry into the world, they were mute witnesses to Della Porta's pivotal role in unlocking the mysteries of nature for human eyes. While Aldrovandi's name preordained his fate as a traveler, Della Porta's patronym positioned him as the portal between hidden meaning and revealed wisdom. He was the door and his mind the key that unlocked it, releasing Nature from her long confinement in the prison of ignorance.

Such images also appeared in later editions of Della Porta's works—for example in the poem underneath his portrait in the 1616 Italian edition of his *On Celestial Physiognomy*—and in eulogies written after his death. Lorenzo Crasso, a late seventeenth-century Neapolitan scholar who possessed many of Della Porta's papers, published a poem by Marino that described Della

---

[81] Gabrieli, *Contributi*, Vol. I, p. 754: "Ha portato l'autore per impresa l'uranoscopio, pesce che ha gli occhi sopra la testa, che mira semplre il cielo, col motto: *Ex animo,* quasi volesse dire che aspira al cielo con tutto il core, ma i suoi inimici l'interpretano altramente: *Recedo lussatus, non sanatus.*"

[82] Ibid., Vol. I, p. 679. The use of the medal as a means of getting Della Porta to donate his museum to the Accademia dei Lincei is discussed in CI., Vol. II, pp. 347–348.

Porta as the "DOOR that closes the immortal treasure that no other wealth on earth equals."[83] The image of the door was a particularly compelling one in the case of Della Porta because it suggested the potential for openness as well as closure. Was Della Porta the philosopher who let nature emerge fully into the light of day, as the Linceans portrayed him, or the guardian who protected her most intimate secrets, allowing an occasional glimmer of her wealth to slip through the cracks? "DOOR of fine and incorruptible gold from which light escapes that dazzles every light," wrote Marino. Della Porta's janus-faced image reflected the ambiguity of his position in a world of competing natural philosophies in which he alternately represented old and new forms of inquiry. Yet, both interpretations acknowledged Della Porta's role as the gatekeeper to the wisdom of the ages. Like Aldrovandi, he was entrusted with the contents of the cornucopia that defined nature's encyclopedic image.

While Cesi and Della Porta exchanged emblems that refined the symbolic meaning of their identities, Della Porta also used his publications to complete his identity as an encyclopedist and Renaissance magus. Two portraits of Della Porta, the first done in his late forties and the second at age sixty-four, comment on his relationship to the study of nature. The first portrait appeared in the 1589 edition of his *Natural Magic* and was reprinted in subsequent editions of his treatises on physiognomy[84] (fig. 24). In this image, Della Porta became the object of physiognomic analysis; the eyes of the lynx turned inward. Just as Della Porta prefaced his *On Human Physiognomy* with a physiognomy of his most important patron, Cardinal Luigi d'Este— "the ornament, dignity, and majesty of your face is so noteworthy"[85]—he now provided viewers with the materials to analyze the face of the physiognomer. Like his patron, Della Porta fit the classical ideal of a noble face. His head was "a little larger than small," a size that indicated prudence and intelligence. His nose bespoke his social status: "The man who has a long nose reaching down to the mouth is a gentleman (*uomo da bene*) and audacious, as Aristotle writes to Alexander." Finally, the gravity of Della Porta's expression placed him in the company of luminaries such as the Duke of Ferrara, Alfonso II d'Este, and Cardinal Bessarion, men of "excellent genius" and

[83]Crasso, *Elogii d'huomini letterati*, Vol. I, p. 174. In full, Marino's poem reads as follows:

Ecco la PORTA, ove con bel lavoro
   Virtù suoi fregi in saldo cedro intaglia
PORTA che chiude l'immortal tesoro,
   Cui null'altra richezza in terra agguaglia.
PORTA di fino, e incorrutibil'oro
   Ond'esce luce che ogni luce abboglia.
Si che può ben del Ciel dirsi la PORTA,
   Poscia ch'al mondo un si ben SOLE apporta.

[84]For the publishing history of both portraits, see Gabrieli, *Contributi*, Vol. I, p. 752. A third portrait of Della Porta was done by El Greco.

[85]Della Porta, *Della fisonomia dell'huomo*, p. 7.

Figure 24. Della Porta's physiognomy. From Giovan Battista della Porta, *Della fisono-mia dell'uomo* (Vicenza, 1615 ed.)

"goodness of spirit."[86] Della Porta, in sum, embodied all the natural attributes and social virtues that men of quality and learning desired. Situated at the intersection of the collecting of antiquities and nature, both of which provided him with the materials to complete his physiognomy, he was the mirror (or as his contemporaries would have said, the door) that made visible the connections between these two different learned practices.[87]

Della Porta's 1589 portrait was merely a prelude to the image that appeared in his *On Distillation* (1608) (fig. 25). As Michele Rak notes, it is a *musaeum sui ipsius*—a museum in itself. Like the 1599 portrait of Aldrovandi, the image surrounds Della Porta with the objects of his study. In the upper left we see the plants, animals, and human faces that form the subject of his *On Human Physiognomy* and *Phytognomonica*. Directly above Della Porta's head are three animals closely associated with humanity: the monkey, the elephant, and the parrot. Below, in the lower left, lies the cornucopia emerging from his investigations of agriculture, collected together in his *Villa*. While the left side of his portrait alludes to his knowledge of nature, the right side displays the variety of human artifice. Astrology, geometry, optics, military technology, secret ciphers, and the alembics used in alchemical and chemical experiments encircle this half of Della Porta's image. They evoke the contents of works such as his *Natural Magic*, *On Ciphers* (1563), *On Refraction* (1593), *Curvilinear Elements* (1601), and *On Fortification* (1608). As Michele Rak writes, "in this image a total representation of the work of the author is realized, a micromuseum of his activities and . . . the best discoveries of a lifetime of research."[88] It was the visual culmination of Della Porta's career. While the 1589 portrait depicted Della Porta as the master of one type of learning, physiognomy, the 1608 portrait established his control over all forms of knowledge.

The objects surrounding Della Porta presented him as the centerpiece of a sustained philosophical inquiry into the natural world. However, their mnemonic function was quite different than those appearing in Aldrovandi's portrait. The image of Della Porta reflects the changing meaning of *inventio* in humanist discourse. While Aldrovandi's portrait recalled a series of moral virtues, the artifacts framing Della Porta publicized his discoveries. In imitation of Cardano, he saw himself as a "man of inventions" (*vir inventionum*).[89] The intellectual distance between Aldrovandi and Della Porta is bridged in many ways by Cardano. Like Aldrovandi, Cardano gave great weight to the activities of Columbus and other explorers in making his life unique. "Among

[86]Ibid., pp. 119–120, 161, 179.

[87]This point also is underscored in Lina Bolzoni, "Teatro, pittura e fisiognomica," esp. pp. 492–495. See also Giovan Battista Masculi's poem in Crasso, *Elogii d'huomini illustri*, Vol. I, pp. 173–174, which purports to be a physiognomy of Della Porta.

[88]Michele Rak, "L'immagine stampata e la diffusione del sapere scientifico a Napoli tra Cinquecento e Seicento," in Lomonaco and Torrini, *Galileo e Napoli*, p. 320.

[89]Cardano claimed in his autobiography that Andrea Alciati had given him this title; Cardano, *De propria vita liber*, in his *Opera Omnia* (Lyon, 1663), Vol. I, pp. 40, 47. See also *The Book of My Life*, pp. 219, 254.

Figure 25. Della Porta the encyclopedist. From Giovan Battista della Porta, *De distillatione* (Rome, 1608).

the extraordinary though quite natural circumstances of my life, the first and most unusual is that I was born in this century in which the whole world became known," he wrote in his 1575 autobiography.[90] For Cardano, discovery described not only the age in which he lived but also the method with which he investigated nature. His continuous testing of knowledge—he claimed to have solved some 240,000 problems in his lifetime—earned him the title of *vir inventionum*. In his own eyes and in the view of many contemporaries, Cardano was an inventor because he repeatedly established first knowledge on a variety of different subjects. This was the model that Della Porta strove to emulate when he surrounded himself with the various technologies that completed his identity as a magus.

As William Eamon details in his study of technology and magic, the revival of natural magic traditions in the late sixteenth and seventeenth centuries enhanced the image of the magus as inventor.[91] Della Porta epitomized this trend. While he took great pride in his physiognomies of nature, he assiduously promoted his ability to surpass the artifice of earlier natural magicians. In a discussion of burning lenses in his *Natural Magic*, for example, Della Porta placed himself in direct contest with the ancients. "I shall describe what is found out by Euclide, Ptolemy, and Archimedes; and I shall add our own inventions that the Readers may judge how far new inventions exceed the old."[92] The centerpiece of these investigations was his recapitulation of Archimedes's famous experiments with burning mirrors that reputedly kept the Roman ships from entering the port at Syracuse by engulfing them in flames. While questioning Archimedes's ability to perform this experiment successfully, Della Porta advertised himself as a magus capable of exceeding anything done by the ancients:

> I will shew you a far more excellent way than the rest, and that no man as ever I knew writ of, and it exceeds the invention of all the Ancients, and of our Age also; and I think the wit of man cannot go beyond it. This Glass does not burn for ten, twenty, a hundred, or a thousand paces, or to a set distance, but at infinite distance.[93]

These mirrors, prominently displayed in his 1608 portrait, emblematized his ability to harness nature's powers. They also established his credentials as an inventor in the Plinian sense: a philosopher who created new knowledge from his engagement with nature.

Two years after his portrait appeared, Della Porta regretted the fact that he had not included an image of a telescope, now popularized by Galileo. "The invention of the eyepiece in that tube was mine," he wrote to Cesi,

[90]Cardano, *The Book of My Life*, p. 189.

[91]Eamon, "Technology as Magic in the Late Middle Ages and the Renaissance," *Janus* 70 (1983): 171–212, esp. pp. 197–198.

[92]Della Porta, *Natural Magic*, p. 371.

[93]Ibid., p. 375.

"and Galileo, lecturer in Padua, has adapted it."[94] While Della Porta saw nature as an infinite source of emblematic knowledge, and therefore as the groundwork for shaping his identity as a naturalist, he increasingly saw artificial magic as the means through which to establish his claims to know nature. In doing so, he set the stage for Baroque collectors such as Settala who derived their identities entirely from their ability to create objects that mediated the relationship between humans and nature. Della Porta, as his biographer Crasso wrote, was a philosopher who, "shrewd in the inventions of scientific things," surpassed the "limits of ordinary knowledge."[95] Collecting provided him with the tools to create new forms of knowledge, and the museum became the setting in which to display his extraordinary powers.

SETTALA THE INVENTOR

More than any other collector, Settala cast himself in the role of the inventor. The portrait of Settala done by Daniello Crespo, probably executed shortly after Manfredo's travels in Europe and the Levant in 1622–1629, shows the young cleric holding one of his inventions, a delicately turned ivory tower emerging from a base out of which a series of interlocking, hollow balls were carved (see fig. 6). Settala's expression is as studied as it is serious. In contrast to the portraits of Aldrovandi and Della Porta, the image of Settala emphasizes his hand rather than his face. A courtly hand, it is raised in a gesture reminiscent of one of Bronzino's paintings of young nobles. Gracefully accommodating the invention, the "hand of Signor Manfredo," as contemporaries frequently referred to it, transformed technology into courtly *sprezzatura*. It is the same hand that, in a later portrait, reaches outward in an expansive gesture designed to accommodate all the marvels displayed on the table at which Settala is seated (fig. 26). Here a cluster of artifacts visually summarized the breadth of his ingenuity. The bagpipes with bellows (called a *sordellina* by Scarabelli), the elegantly turned figurine, and the armillary sphere were all objects of his own devising. Brought together in this image, a microcosm of Settala's museum, they alluded to the diversity of arts and sciences that Settala had mastered by acquiring the skills necessary to perform each craft. "He produced things that did not seem generated but worked," wrote his biographer Pastorini.[96] Settala reflected the ennoblement of the mechanical arts that had begun in the late sixteenth

[94]Gabrieli, *Contributi*, Vol. I, p. 644. In a letter to Faber, Della Porta wrote, "Scribis te magnopere admirari Anglos, Belgas, Francos, Italos et Germanos sibi Telescopii inventum arrogari; me solum, qui inventor extiterim, inter tantos rumores conticescere." Ibid., p. 648. For a more detailed discussion of this controversy, see Albert van Helden, "The Invention of the Telescope," *Transactions of the American Philosophical Society* 67, pt. 4 (Philadelphia, 1977).

[95]Crasso, *Elogii d'huomini illustri*, pp. 170–171.

[96]Giovan Battista Pastorini, *Orazion funebre per la morte dell'illustriss[imo] Sig[nor] Can[onico] Manfredo Settala* (Milan, 1680), p. 19.

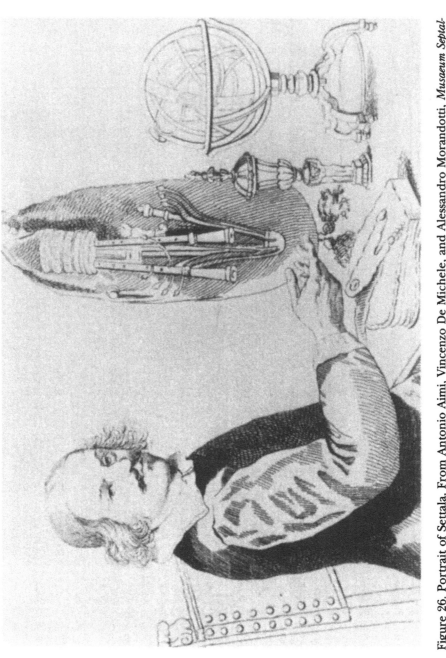

Figure 26. Portrait of Settala. From Antonio Aimi, Vincenzo De Michele, and Alessandro Morandotti, *Musaeum Septalianum: una collezione scientifica nella Milano del Seicento* (Florence: Giunti Marzocco, 1984).

century through the efforts of mathematicians such as Galileo and natural magicians such as Cardano and Della Porta.[97] The success of his image as an inventor gave testimony to the new status of machines and machine-produced knowledge in the seventeenth century.

Settala's career as an inventor began in his student days at Pisa and Siena, where he attracted the attention of powerful patrons such as the Medici. His museum contained an armillary sphere encased within a glass globe "made by my hand when I was in Pisa in 1621 at the Studium." Lest anyone mistake his role in the creation of this object, Settala added, "And I was the inventor of these bagatelles."[98] Perhaps through his connections, he was able to see the workshops of princely inventors such as Don Antonio de' Medici, or the foundry used by the then deceased Francesco I and Ferdinando I. Certainly they must have been the inspiration for his own laboratory. After returning to Italy in 1629 from his own version of a new odyssey, he settled in Milan where he was appointed the canon of San Nazaro, a church near the family palace on Via Pantano, in 1630. With the death of his father, the famous physician Lodovico Settala, in 1633, Manfredo inherited the task of maintaining the family collections. Working between the museum and the cloister of San Nazaro, where he set up his workshop, Settala soon established himself as a collector and inventor of note. Visitors such as Walthar von Tschirnhaus, one of the pioneers of porcelain manufacture in Europe, traveled to Milan specifically to meet Settala and discuss his inventions with him. By the time of his death in 1680, he had earned the title of the "Milanese Archimedes."[99]

While naturalists lauded Aristotle, Galen, and Pliny as scholars worthy of emulation, mathematicians and engineers accorded Archimedes a similar role.[100] The fifteenth-century architect Filippo Brunelleschi was given the title "second Archimedes" because of the skillfullness of his building designs, and numerous sixteenth- and seventeenth-century natural philosophers interested in the mechanical arts took Archimedes as their role model. Galileo called him "the most divine Archimedes," praise he pointedly did not lavish on Aristotle or any of the other ancients. The Florentine court mathematician and philosopher applied this label to contemporaries such as the Jesuit mathematician Bonaventura Cavalieri, whom he called the "new Archimedes." Cavalieri may well have fueled Settala's own interest

[97]For more on this subject, see Biagioli, "The Social Status of Italian Mathematicians"; W. R. Laird, "The Scope of Renaissance Mechanics," *Osiris*, ser. 2, 2 (1986): 43–68; Paul Lawrence Rose, *The Italian Renaissance of Mathematics* (Geneva, 1975). A classic point of departure for this subject is Paolo Rossi, *Philosophy, Technology and the Arts in the Early Modern Era*, Salvator Attanasio, trans. (New York, 1970).
[98]Ambr., Cod. Z.387 sup., f. 18.
[99]Rota Ghibaudi, *Ricerche su Lodovico Settala*, p. 46; Aimi, De Michele, and Morandotti, *Musaeum Septalianum*, pp. 29–30.
[100]The material for this section is drawn from W. R. Laird, "Archimedes Among the Humanists," *Isis* 82 (1991): 628–638.

in mechanics since he also was a native of Milan and participated in the
debates about burning mirrors. Settala read his treatise on *The Burning Mirror* (1650) and used it as the basis for a number of his own experiments with
parabolic mirrors.[101] Archimedes exemplified the ability to combine philosophical learning with practical knowledge. He became an emblem of
the elevation of the mechanical arts in early modern Italy, and his experiments became a literary topos whose reconstruction was a form of emulation. Followers such as Settala could earn no higher praise than to be compared to him.

In his capacity as an artisan–scholar, Settala carefully cultivated his image
as someone who invented for philosophical rather than material gain. Like
Francis Bacon, he perceived technology as an example of utilitarian knowledge whose mastery would advance the cause of natural philosophy (and
while we have no concrete evidence that Settala read Bacon's works, his correspondence with the Royal Society suggests a certain familiarity with the
goals of the Baconian program). Despite the high value placed upon his inventions, particularly his creations made from luxury goods such as gems,
ivory, and porcelain, Settala never sold anything. As Scarabelli assured his
patrician readers, "He was far from the sordidness of money, a distinct attribute of plebian souls."[102]

While clearly distinguishing himself from the lowly craftsman, who earned
a living with his hands, Settala also separated himself from collectors such as
Cospi and Moscardo, whose museums were—like Aldrovandi's and Della
Porta's villas—places of leisure. Cospi, who began collecting "as a noble pastime in his adolescence at court," and Moscardo, whose museum was the
product of a "leisurely life," typified the audience who came to Settala's
gallery.[103] But they did not exemplify his own image of the museum as a site
of scholarly labor where "philosophical workers" congregated. "Inventing,
toiling, and collecting"—the very words Pastorini used to describe Settala—
were complementary activities that defined the experimental life of the museum.[104] When Settala died in 1680, the rectors of the Jesuit College of Brera
in Milan erected four columns in the temporary mausoleum they created in
his memory, each representing the different forces that had guided the deceased collector: work, wisdom, virtue, and art. On the first column, they emblazoned the following motto: *Potius mori quam otiari* ("It is better to die than

---

[101]Gino Fogolari, "Il Museo Settala," *Archivio storico lombardo*, ser. 3, 14 (1900): 91; Tavernari,
"Il Museo Settala: Presupposti e storia," pp. 27, 42 (see footnote 59). See *Lo specchio ustorio di
P. F. Bonaventura Cavalieri* (Bologna, 1650).

[102]Scarabelli, *Museo o galeria adunata dal sapere*, p. 199.

[103]Legati, *Museo Cospiano*, n.p.; *Note overo del museo di Lodovico Moscardo*, n.p. By the time his
second catalogue appeared, Moscardo had become a count, perhaps as a result of the fame
he gained from his museum; Olmi, "Science-Honour-Metaphor," in *Origins of Museums*,
pp. 13–14.

[104]Scarabelli, *Museo o galeria adunata dal sapere*, p. 35; Pastorini, *Orazion funebre*, in Tavernari,
"Il Museo Settala," p. 26.

to be at leisure").[105] In choosing Archimedes as his alter ego, Settala cast himself as a natural philosopher who gave labor intellectual legitimacy. The hands that greeted princes and turned the pages of learned tomes also worked to perfect human knowledge and the beauty of its creations.

Settala stood at the crossroads of two different traditions that gave Archimedes new prominence among humanist philosophers and mathematicians. On the one hand, he greatly admired Galileo and corresponded regularly with members of the Accademia del Cimento. His museum recapitulated the Galilean program of mechanical invention, participating in its popularization within the Italian patriciate. Filled with long, elegant telescopes made of ebony and "other foreign woods," microscopes, armillary spheres, astrolabes, and examples of "Galileo's compass made by my own hand," it testified to the material impact of the new philosophy on the culture of collecting.[106] All of these objects highlighted the relationship between human ingenuity and scientific knowledge, presenting Settala as a model of the new synthesis between *ars* and *scientia*. On the other hand, Settala also embodied the older image of Archimedes as a technological wizard. Like Roger Bacon and Albertus Magnus, Archimedes became the focal point of learned mythologies about the powers of human inventiveness that developed within the natural magic tradition. Galileo was not the only natural philosopher to admire Archimedes; occult philosophers such as Settala's contemporary Robert Fludd portrayed him as "the archetype of the perfect natural magician."[107] In the eyes of natural philosophers such as Della Porta, Campanella, and Fludd, Archimedes's divinity took on a more potent and literal meaning. He was an engineer who made the impossible happen. Settala also cultivated this image. As Scarabelli commented in a description of the automata in the museum, "while he revived the eye with the spirit of his motions, he disheartened the soul in tracking down the true cause of them."[108]

The story of Archimedes's single-handed defeat of the Roman fleet in Syracuse with his burning mirrors (ca. 214–212 B.C.) formed the core of his reputation as a natural magician. The heroic circumstances of this event and Archimedes's noble death in a later battle made it a topos equal to the image of Ulysses the traveler and therefore worthy of imitation. By the mid-seventeenth century, popular works of mathematical magic transformed this tale into a fiery, superhuman spectacle in which Archimedes, metamorphosizing into Jupiter, hurled thunderbolts down on the invading fleet.[109] Following in the wake of Cardano, Della Porta, and Campanella, Set-

[105]I. A. Alifer, *Manfredo Septalio Academia Funebris publice habita* (Milan, 1680), in Fogolàri, "Il Museo Settala," p. 121.

[106]Tavernari, "Il Museo Settala," p. 23; Ambr., Cod. Z.387 sup., f. 11.

[107]Eamon, "Technology and Magic," p. 200.

[108]Scarabelli, *Museo o galeria adunata dal sapere*, p. 38.

[109]See Heinrich van Etten, *Mathematicall Recreations* (London, 1653), pp. 129–130, in Eamon, "Technology and Magic," p. 201.

tala made this event the *locus classicus* for his own experiments with parabolic mirrors. As Scarabelli wrote at the beginning of his catalogue of the Settala museum, the legend of Archimedes was a "historical tale" (*racconto historico*) and therefore an appropriate point of departure for his own description of Settala's endeavors.[110] It situated Settala's museum of inventions within the grand humanist narrative that made collecting part of the continuous dialogue with the past. While Settala bore the *imago* of Manfredo, he carried the *ingenium* of Archimedes, continuing a pattern established by predecessors such as Aldrovandi.

The "great genius of Archimedes" was matched by the inventiveness of Settala, in matters as mundane as the construction of an armillary sphere and as extraordinary as the restaging of the events at Syracuse.[111] In effecting the latter, Settala enlisted the help of Kircher. During his trip to Sicily, Kircher traveled to Syracuse to determine the feasibility of Archimedes's alleged feat. After measuring the harbor, he concluded that it would have been possible with the right sort of mirror. Returning to Rome, he eagerly sought out the best craftsmen in Italy to construct a mirror capable of achieving the desired effect. Through these efforts, he encountered Settala, at work on the same problem; they exchanged reports on the varied success of their experiments from the 1640s through the 1670s. By the time Kircher published his *Great Art of Light and Shadow* in 1646, Settala had succeeded in restaging the Archimedean experiment with a mirror that burned at a distance of fifteen paces.[112] Kircher even reproduced letters from other clerics who testified to the ability of Settala's mirrors to ignite objects from a distance that equaled half the calculated span between Archimedes's mirror and the ships in the harbor. In July 1668, Settala reported success with a mirror that liquified silver and lead from a distance of sixteen *bracchia*. The following month, he informed Kircher of the "construction of a larger mirror . . . and I hope to attain the desired effect from it."[113] In steady increments, Settala slowly worked his way up to achieving the effect wrought by Archimedes during the siege of Syracuse, some 2000 years before.

For Baroque natural philosophers, the successful recapitulation of the Archimedean experiment discredited earlier efforts by Cardano and Della Porta to surpass Archimedes with their infinite burning mirrors and deepened the ties between the reformed natural magic of the Jesuits and the revival of ancient mechanics. Settala, a careful reader of Kircher's and Schott's

[110]Scarabelli, *Museo o galeria adunata dal sapere*, p. 4.

[111]Ibid., p. 20.

[112]Kircher, *Ars magna lucis et umbrae*, pp. 764–765. An engraving depicting Archimedes's original feat is reproduced on p. 764. According to the two testimonials, Settala performed his experiments in 1644.

[113]PUG, *Kircher*, ms. 564 (X), ff. 87r, 99r (Milan, 22 August and 11 July 1668). Kircher's correspondence provides a representative sample of the interest in Archimedes's experiment and other forms of mathematical magic; see PUG, *Kircher*, ms. 557b (IIIb), ff. 237–238.

treatises on artificial magic, embodied the orthodox position on natural magic that did not place these effects beyong the limits of human ingenuity. "Real artificial magic produces real effects, as when Archytas fabricated a flying dove out of wood.... Archimedes set fire to a distant fleet with a lens turned toward it, Daedalus made statues that moved by the action of weights or by quicksilver," wrote Campanella, listing the technological *mirabilia* that preoccupied collectors such as Kircher and Settala. "Indeed this art does not produce any miracles."[114] While claiming the powers of Archimedes, Settala did not invest them with supernatural significance. The mystery of viewing his machines resulted from human artfulness rather than divine or diabolical intervention. And his museum, as he repeatedly emphasized in its portrayal, was a theater of inventions, not miracles.

Upon Settala's death in 1680, his relatives, the congregation of San Nazaro, and the Jesuit academicians at Brera College organized a public "academy" in honor of the dead inventor. Orations were written, emblems manufactured, and an elaborate mausoleum was constructed in the Jesuit college. All the emblems published in Giovan Maria Visconti's *Exequies in the Church of San Nazaro for Manfredo Settala* (1680) used objects in Settala's collection to interpret the meaning of his life as well as his death. A compass tracing the outline of a circle bore the motto *Satis est implevimus orbem* ("It is enough that we enriched the world"), alluding to Settala's contributions to the perfection of navigational instruments. Like Della Porta, Settala used the compass as an emblem for his ability to circumscribe the world. His bagpipes, now deflated rather than standing upright, wittily symbolized the end: *Deficit spiritus meus* ("My breath runs out"). In other images, Death replaced Settala as the inventor, minting a coin in honor of the Milanese cleric with the inscription *Maius ab exequis nomen* ("The name is greater from the exequies") and experimenting with some of the most noteworthy objects in the museum. In Visconti's emblems, all the objects in the museum became a form of *memento mori*. Once animated by Settala, they became memorabilia that symbolized the passing of time, the fleeting virtues of fame and the inevitability of death.

Two emblems in particular exemplify the reappropriation of Settala's inventions as a form of visual commentary. In the first emblem, the hand of death reaches out to prevent the ball in a perpetual motion machine from dropping. *Nihil perpetuum* ("Nothing is forever") (fig. 27). Settala's perpetual motion machine was one of the most celebrated objects in his collection. It appeared in the illustration of his museum, prominently displayed in the entry to the left gallery and merited an entry in the watercolor illustrations done of various museum artifacts in the 1640s to 1660s. Scarabelli described this invention in his catalogue: "In the perpetuity of these motions, the hand of Signor Manfredo always proved itself equal to [his] ingenuity. He made

---

[114]Campanella, *Magia e grazia*, in Eamon, "Technology and Magic," p. 198.

Figure 27. "Nothing is forever." From Giovanni Maria Visconti, *Exequiae in tempio S. Nazarii Manfredo Septalio Patritio Mediolanensi* (Milan, 1680).

[a perpetual motion machine] consisting of twelve brass circles, turned on a lathe, carved out, engraved, and soldered with singular artifice."[115] During Settala's lifetime, the perpetual motion machine stood at the entryway of his museum as a monument to his inventiveness, which continuously created new artifacts and improved on old ones. Now the hand of death had stopped "the hand of Signor Manfredo" and with it the perpetual motion machine,

[115]Ambr., Cod. Z.387 sup., f. 1r; Scarabelli, *Museo o galeria adunata dal sapere,* p. 37.

whose presence in the funeral ceremonies reminded mourners of the transitory and finite nature of all things. As a natural philsopher whose religious affiliation committed him to some form of Aristotelian physics, Settala undoubtedly would have enjoyed this moral.

Visconti's emblematics did not neglect Settala's parabolic mirrors. The Archimedean experiment became the subject of another emblem in which Death, standing in front of mirror, held aloft his scythe which burst into flames. *Sic splendor collectus abit* ("Thus the splendor collected vanishes"). Rather than being the manipulator of Archimedes's burning mirror, Settala and his museum have fallen victim to its powers. This final emblem recalled the cautionary moral about gazing at oneself in mirrors that we already have encountered in the tale of Narcissus; it also underscored the ephemeral quality of Settala's collection, and perhaps all forms of human production, whose existence was no more permanent than that of its creator. As Pastorini lamented in his *Funeral Oration* (1680), "Those same marvels, those same works that I used to see with such great delight through the gentility of Signor Manfredo, I now stare at with equal torment, for nothing seems worthy of admiration after him."[116] A museum without a collector was no museum at all.

Settala's friends and family did not simply stop with this pronouncement; they acted upon it. All the objects in his museum were carried in procession by the rectors and students of the Jesuit college from his home to Brera College where they participated in an elaborate funeral ceremony. Personified by the members of the college, his inventions recited Latin epigrams in competition with the various muses under which Settala had worked—physics, optics, and music. The moral pageant commemorating Milan's greatest collector opened not with the traditional blast of a trumpet but with a resounding bellow by "fame" from Settala's speaking tube. Settala's exequies were a form of Jesuit theater in which his objects, family, and friends all participated in the collective rehearsal of his life. In the niches of the four columns supporting the canopy, statues of his most famous ancestors appeared. Three scenes from his life decorated his tomb: a perilous encounter with the Turkish fleet during his trip to Egypt, the ease with which he extricated himself from the Roman catacombs, and his success in teaching the art of polishing to Don Juan of Austria (illegitimate son of Philip IV), Vincenzo Gonzaga, and the marchese of Caracena.[117] While the first two panels illustrated formative moments in Settala's youth—travel and the conquering of the Christian labyrinth—the third panel depicted an exemplary scene from the life of the collector. Scarabelli, in fact, recalled the visit of the three princes, each of whom "honored this museum in person" and apprenticed

[116]Pastorini, *Orazion funebre*, in Aimi, De Michele, and Morandotti, *Musaeum Septalianum*, p. 29.

[117]The funeral academy is discussed by Alifer, *Academia funebris*, and summarized in Fogolari, "Il Museo Settala," pp. 119–121.

himself to the canon.[118] Lifting episodes from the pages of Settala's museum catalogues and transporting his possessions into the mausoleum, the rhetoricians at the Jesuit college relived the life of the Milanese Archimedes and symbolically buried his museum with him. The objects continued to exist. But what were inventions without the inventor?

AENEAS INTO HERMES

Absorbing all the different techniques of invention employed by his predecessors and contemporaries, Kircher took the process of image making to new heights. Expanding the canon of *exempla* to include figures from the mythical Egyptian past, as Kircher and other hermeticists understood it, he reached backward in time in search of an identity that adequately depicted his quest for original knowledge. While Aldrovandi surpassed Aristotle to become Columbus, and Settala enjoyed his role as a new Archimedes because it allowed him to emulate Galileo, Kircher began as Aeneas and ended as Hermes. All used antiquity as their point of departure; but while others perceived it as a means of advancing to the present, Kircher resisted this trend. His goal was not to become the first of the "new men" but the most ancient man alive. While others shed the garments of antiquity to emulate contemporary heros, Kircher traded one form of antiquity for another. Only by exploring the whole of antiquity, he argued, could he complete his archaeology of the self.

From the start, it seemed preordained that Kircher would enter orders. His doubly Christian name—"Athanasius the Churchman"—singled him out at birth as someone intended to fulfill a divine mission. Born on the feast day (2 May) of Saint Athanasius, Bishop of Alexandria in the turbulent fourth century and one of the founders of the monastic movement, his father named him Athanasius as a thanksgiving to the saint for the birth of his son and the return of his patron, the Archbishop of Fulda, exiled from Hesse-Darmstadt for twenty-five years during the religious upheavals of the Reformation. Certainly Johann Kircher found a parallel in the "outstanding constancy and imperturbable fortitude" of Saint Athanasius in his defense of Christian orthodoxy during the Arian controversies, for which he suffered exile and persecution, and the stalwart behavior of the local archbiship, also a defender of the Church against new forms of heterodoxy.[119] By naming his son Athanasius, Johann Kircher hoped to impart to him the moral virtues of his patron saint and revered archbishop. When Kircher entered the Society of Jesus, he fulfilled one aspect of his inheritance by com-

---

[118]Scarabelli, *Museo o galeria adunata dal sapere*, p. 214.

[119]Kircher, *Vita*, pp. 1, 4; Reilly, *Athanasius Kircher*, pp. 24–25. On Athanasius of Alexandria (A.D. 295–373), see W. H. C. Frend, *The Early Church* (Philadelphia, 1965), pp. 146–157; and Henry Chadwick, *The Early Church* (Harmondsworth, Middlesex, U.K., 1967), pp. 137–144.

mitting himself to a life of service in the Catholic Church in one of the orders noted for its defense of the faith.

In imitation of his namesake Athanasius, Kircher endured exile and religious persecution at the hands of Protestant soldiers. Yet he was unable to return to his own "Alexandria," fleeing Germany during the turmoil of the Thirty Years' War. Arriving in Rome in 1634, Kircher identified not only with Saint Athanasius, who had spent several years in Rome, but also with a classical hero who had been forced to depart from his homeland: Aeneas. "Pious Aeneas," as Virgil called him, was a wanderer frustrated in his attempts at heroic adventure and weighed down by his responsibilities; he was no Ulysses. "In sum, it is not so easy to find the Homeric Ulysses in our days," wrote Petrucci in his synthesis of Kircher's work, perhaps recalling the political and religious turmoils of the seventeenth century that frequently uprooted scholars.[120] While Aldrovandi and Settala portrayed their travels as new Odysseys, armed with the foreknowledge of their safe return, Kircher imagined his own voyage as a new Aeneid. He was the traveler for whom there was no return. Fate decreed his exile from Germany and led him to Italy. As Juno said of Aeneas, "He came to Italy at the fates' command."[121] Kircher's arrival in Italy also seemed foretold. After two unsuccessful attempts to continue his trip from Marseilles to Vienna, where he was to become court mathematician, Kircher unexpectedly found himself in Rome, where his superiors informed him that, due to the intervention of Peiresc and Francesco Barberini, he instead would succeed Christopher Scheiner as professor of mathematics at the Roman College. "I could not wonder enough at Divine Providence," wrote Kircher in his autobiography.[122]

For Kircher, Virgil's writings became the *locus classicus* through which he interpreted his first encounters with Italy. Educated on the *Aeneid* as a schoolboy, he probably used treatises such as his fellow Jesuit Jacobus Pontanus's exhaustive commentary on Virgil to guide his humanist fantasies.[123] Like Aeneas, Kircher witnessed the destruction of his culture and participated in the birth of a new civilization in another land. Both wandered from country to country in search of a new homeland. Aeneas's description of himself in the first book of the *Aeneid* might well have come from the

---

[120]Petrucci, *Prodomo apologetico*, p. 159. For a succinct analysis of the differences between the two epic heroes, see R. D. Williams, *The Aeneid* (London, 1987), p. 80ff.

[121]*The Aeneid of Virgil*, Brian Wilkie, ed., and Rolfe Humphries, trans. (New York, 1987), p. 234 (X, 67). See also Aeneas's response to Dido that Italy was not his chosen destination (IV, 361). The editors of this volume have renumbered the verses. I include the standard citations for readers using another edition.

[122]Kircher, *Vita*, p. 52.

[123]Jacobus Pontanus, *Symbolarum libri XVII quivus P. Virgilii Maronis Bucolica, Georgica, Aeneis ex probatissimis auctoribus declarantur, comparantur, illustrantur* (Augsburg, 1599). For a discussion of Virgil's impact on early humanistic culture, see Craig Kallendorf, *In Praise of Aeneas: Virgil and Epideictic Rhetoric in the Early Italian Renaissance* (Hanover, NH, 1989).

mouth of Kircher: "I am Aeneas, a good, devoted man; I carry with me my household gods, saved from the Greeks; I am known in heaven; it is Italy I seek, a homeland for me there, and a race descended from lofty Jove."[124]

Kircher's autobiography indicates the great impact of the Thirty Years' War on him. On several occasions, he barely escaped with his life—he had been forced to flee the Jesuit college at Würzburg, leaving all his books and notes behind—and equated the Protestant victory over various Catholic strongholds with the fall of Troy (and later the fall of Egypt). Kircher perceived himself as a uniquely situated participant in the vast human narrative of the rise and fall of civilizations that repeated itself at regular intevals. Exiled from the ruins of Germany, he consoled himself with the thought that his mission lay elsewhere and that ultimately his efforts in the papal city would contribute to the cultural rebirth of the Habsburg lands. His task coincided with the one Virgil credited to Rome: "to build civilization upon a foundation of peace."[125] Contributing to the restoration of Catholic imperium through his role in the renovation of the obelisks and use of his museum as a center in which to display the new learning of the post-Tridentine church, he became the new Aeneas. Appropriately, one of Kircher's last publications was a study entitled *Latium* (1671). The civilization Aeneas founded, Kircher excavated.

Kircher's perception of himself as another Aeneas gained further credence during his trip to Naples and Sicily in 1637–1638. I have already discussed his invocation of Virgil in relation to the volcanic eruptions of Etna and Vesuvius. Certainly Aeneas's voyage into the underworld was a logical point of departure for Kircher's explorations of the subterranean world, which he also populated with rivers "running fire."[126] The frontispiece of the first volume of the *Subterranean World* directly alluded to the Virgilian context of his adventure. Depicting the twelve winds shaping the earth with their breath, as described in the sixth book of the *Aeneid*, it bore the following quotation: *Spiritus intus alit totamque per artus infusa, Mens agitat molem* ("The Spirit supports from within: infused through its every member, Mind sets mass in motion").[127] In his rereading of the *Aeneid*, Kircher transformed Anchises's speech to his son Aeneas in the underworld into a poetic confirmation of his own natural philosophy, predicated on a similarly animistic conception of the world. This passage, in fact, became a leit motif that reappeared constantly in Kircher's work. Virgil's cosmological pronouncements

[124]Virgil, *Aeneid*, p. 13 (I, 378–380).

[125]Ibid., (VI, 852). I have taken this translation instead from Williams, *The Aeneid*, p. 39.

[126]Ibid., p. 141 (VI, 550). For more on this episode, see chapter 4.

[127]Virgil, *Aeneid* (VI, 726–727). I base my discussion of this frontispiece on Tongiorgi Tomasi, "Il simbolismo delle immagini: i frontespizi delle opere di Kircher," in Casciato, p. 173. The translation in Godwin, *Athanasius Kircher*, p. 86, is preferable to others I have consulted. For a discussion of the classical context of this passage, see Philip R. Hardie, *Virgil's Aeneid: Cosmos and Imperium* (Oxford, 1986), pp. 51–83.

circumscribed Kircher's depiction of Egyptian cosmology in the *Egyptian Oedipus* and decorated the magnetic demonstrations in his museum. On one of his "zodiac spheres," Kircher attached a bird which, flying between two columns through the movement of a hidden magnet, marked the passage of time. The other half of Virgil's famous epigram encircled it: *Mens agitat molem, et magno se corpore miscet* ("Mind sets the mass in motion and mingles itself with the great body").[128] In the *Aeneid*, Kircher found succinct expression of the occult affinities that infused and ordered the world. Meditating on Virgil, he learned more about himself and nature. His museum became the site in which to display this knowledge.

Kircher's explorations of his identity did not end with a declaration of allegiance to his patron saint, Athanasius, and his chosen muse, Virgil. These early attempts to define his Christian and classicized self only encouraged Kircher to pursue other forms of exemplarity. By the 1660s, he had assembled a formidable array of identities, some of his own making and others awarded to him by admirers. Kircher was a "resuscitated Pythagoras" due to his deft manipulation of numbers, an image confirmed by the presence of the Greek mathematician on the frontispiece of his *Arithmology* (1665).[129] Georg Philip Hartdörff felt that Kircher demonstrated the *ingenium* of Daedalus and Archimedes beneath the *imago* of Athanasius.[130] He was a new Archimedes because of the passion he shared with Settala for burning mirrors, perpetual motion, and other demonstrations of classical mechanics. His fascination with automata, flying machines, and the image of the labyrinth made him a new Daedalus. The labyrinth Daedalus created was Kircher's to unravel, as the possessor of Ariadne's thread.[131]

Kircher's demonstration of the flying dove of Archytas, one of his many magnetic experiments, further confirmed his image as Daedalus. The success of this museum exhibit, straight out of the pages of natural magic textbooks, led to a rumor that Kircher himself knew how to fly but was prohibited from doing so by the pope.[132] While denying this, Kircher nonetheless cultivated his image as a man who possessed the powers of both Daedalus and his son Icarus. Unlike Icarus, however, Kircher would not fall when he approached the sun. As the Jesuit Nicolaus Mohr wrote in the preface to Kircher's *Ecstatic Voyage*, "It is a fable that Icarus traversed the Heavens in flight . . . but it is not a fable that Athanasius Kircher encircled the Heavens with his genius. . . . Moreover, how much more happily did Kircher fly than

---

[128]Kircher, *Oedipus Aegyptiacus*, II, i, p. 418, in Godwin, *Athanasius Kircher*, p. 60; de Sepi, p. 21. Late Renaissance commentators on Virgil also recognized the neo-Platonic and hermetic possibilities of these words; Pontanus, *Symbolarum libri XVII Virgilii*, pp. 1503–1505.

[129]PUG, *Kircher*, ms. 568 (XIV), f. 52 (Naples, 7 December 1652).

[130]PUG, *Kircher*, ms. 557b (IIIb), f. 251r: "Daedalus ingenio monstras imitabile Coelum, atque Archimedis machina quaq[ue] tua est."

[131]For more on this image, see chapter 2.

[132]Gaspar Schott, *Magiae universalis naturae et artis* (Bamberg, 1672 ed.), p. 253.

Icarus."[133] As Kircher constantly suggested through the association of his *ingenium* with key religious, philosophical, and mythological figures, there was no identity worthy of imitation that he could not absorb. "Jesus, what universality! What profundity of knowledge!" exclaimed Giacomo Scafili in a 1652 letter to the Jesuit.[134]

With the publication of his *Egyptian Oedipus* in 1652–1654, Kircher revealed a new set of images to his readers, proclaiming himself the master of the most ancient form of wisdom known to the humanist world. His given name already presaged his turn toward Egypt. As Athansius, Kircher imagined himself to be the reincarnation of a bishop who had presided over Egyptian Christianity in an attempt to hold East and West together; his new philosophy would bring harmony to a troubled world just as his renovation of the obelisks restored the splendor of Egypt to Rome. Hermeticism provided the locus for this supreme synthesis. While the original Athanasius surely condemned hermeticism as a dilution of the orthodox Christianity he practiced, this did not deter his namesake from harmonizing these different forms of divine wisdom. Nurtured in the speculative culture of late Renaissance and Baroque humanism that privileged Egypt as the site of the most impenetrable secrets, Kircher appropriated hermetic imagery to publicize his ability to possess *prisca sapientia.*

Kircher's mastery of language, demonstrated repeatedly in his publications and museum, became the basis for his image as a scholar who had unlocked the mysteries of Egypt. Based on his success as an interpreter of the hieroglyphs, Kircher initially identified himself as the new Oedipus. As his works on Egypt poured forth from the presses, contemporaries challenged Kircher with new riddles, in the form of ancient inscriptions, modern ciphers, and other hidden languages, to determine his right to bear this title. "You, most ingenious Oedipus, who have the Sphinx, easily elicit the interpretation," wrote Gottfried Aloysius Kinner in a letter than included a cryptogram for Kircher to decode.[135] In an age obsessed with the problem of communication, Kircher's skill with languages gave him quasi-divine status.

Identifying with Oedipus was simply a prelude to Kircher's revelation as Hermes. As Oedipus, he unlocked the mysteries of hermetic language; as Hermes he became an initiate into their secrets, a scholar capable of producing this form of knowledge as well as interpreting it. In the preface to his *Egyptian Oedipus*, Kircher included the following poem from a British admirer:

> To thee belongs the fame of Trismegist
> A righter Hermes, th'hast outgone the list
> Of's triple grandure.[136]

---

[133]I have excerpted these passages from a dedicatory poem in Kircher, *Iter exstaticum* (Würzburg, 1660 ed.), n.p.

[134]PUG, *Kircher*, ms. 568 (XIV), f. 143r (Trapani, 15 June 1652).

[135]PUG, *Kircher*, ms. 557 (XIII), f. 248 (Reichenbach, 4 January 1656).

[136]Kircher, *Oedipus Aegyptiacus* (Rome, 1652), n.p. Also in Yates, *Giordano Bruno*, p. 416.

The dedication announced what the content of the book confirmed: Kircher had surpassed the wisdom and breadth of learning possessed by Hermes, earning the title of his successor. Hermes, as Lactantius wrote, "was most fully imbued with every kind of learning, so that he acquired the name of Trismegistus through the knowledge of many subjects and arts." Clement of Alexandria gave Kircher a more specific goal to attain when he wrote that Hermes had completed a total of forty-two works, thirty-six of which dealt with Egyptian philosophy.[137] Publishing thirty-eight books, Kircher nearly equaled Hermes's total. If we count unpublished manuscripts, which Kircher surely did, he surpassed it, just as Aldrovandi overtook Aristotle and Pliny. While claiming to resuscitate Egyptian learning, Kircher nonetheless remade its hero in his own image, a new Hermes for a new Egypt.

Kircher's identification with Hermes was most apparent in the frontispieces of his books. Images of the scribe and the messenger—two dominant hermetic topoi—are omnipresent. In the frontispiece to the second volume of the *Subterranean World* (see fig. 13), Kircher, depicted as a female muse and seated beneath a statue of Isis, draws the hieroglyphs.[138] It is unclear from his pose whether he is in the process of deciphering or creating them, but he nonetheless meditates upon the obelisks on which Hermes had inscribed his "hidden writings."[139] Kircher is interrupted by Mercury whose intervention, predictably reminiscent of his guidance of Aeneas, directs the Jesuit to explore the underworld. Mercury also guides the mythologized Kircher, again with quill in hand, his arm resting on a stack of books labeled "Chaldean Astrology, Greek Philology, Pythagorean Mathematics and Egyptian Wisdom," on the frontispiece of the *Pamphilian Obelisk* (1650). As the Latinate Hermes, he is the alter ego of the German Hermes, both of whom reincarnate the Egyptian Hermes in Kircher's historical narrative.

The portrayal of Kircher as scribe underscores his comparison with Hermes, scribe of Osiris. "Hermes saw all things and understood what he saw and had power to explain to others what he understood . . . for what he discovered he inscribed on tablets, and hid securely what he had inscribed, leaving the larger part untold, that all later ages of the world might seek it."[140] In the frontispiece to Petrucci's *Apologetic Forerunner*, for example, Kircher sits atop a crocodile, symbol of the difficulty of his studies, and below a solar lion, symbol of Osiris and the fecundity of nature[141] (fig. 28). In front of him lie the weighty volumes he has written and read to achieve his wisdom. Kircher, however, writes on a scroll rather than in a book. Unfurled

---

[137] *The Divine Pymander and Other Writings of Hermes Trismegistus,* John D. Chambers, trans. (New York, 1975, 1882), pp. viii–ix, p. 141. I have modified the translation of Lactantius.

[138] Alternately Kircher might be Orpheus, given the following inscription.

[139] Walter Scott, ed. and trans., *Hermetica* (Boston, 1985, 1924), Vol. I, p. 493.

[140] *Stobaeus,* Excerpt XXIII, in ibid., p. 459.

[141] For more detailed discussion of the iconography of Kircher's illustrations, see the analysis accompanying the plates in Rivosecchi, *Esotismo in Roma barocca.*

Figure 28. Kircher the encyclopedist. From Gioseffo Petrucci, *Prodomo apologetico alli studi chircheriani* (Amsterdam, 1677).

across the frontispiece, it displays the totality of knowledge he has mastered—geometry, numerology, music, astronomy, alchemy, and so forth. While Della Porta encircled his own image with essentially the same encyclopedia, Kircher chose to reveal his wisdom furtively in keeping with his image as a hermetic sage. He is the scribe who completes the recording of knowledge begun by Hermes and continued by philosophers ever since. Portraying himself as both the interpreter and inventor of hermetic learning, Kircher claimed as his own the inherited wisdom of the ages.

Kircher's image as Hermes was completed in his *Ecstatic Voyage*, his only treatise on astronomy, written in the form of a dialogue between Theodidactus and his angel Cosmiel, a Christian Mercury. In the frontispiece to this treatise, Kircher finishes the portrait of himself begun in the *Subterranean World* (fig. 29). While he descends with Virgil, he ascends with Hermes. Kircher's hermetic identity allows him to avoid the fate of Icarus, who had no divine intervention in his own perilous flight. Guided by the words of the *Corpus Hermeticum*, he roamed the universe, "but man ascends even to heaven, and measures it; and what is more than all beside, he mounts to heaven without quitting the earth; to so vast a distance he can put forth his power." Schott confirmed the possibility of undertaking such an immaterial voyage in the introduction to Kircher's dialogue, where he recounted observing his master in a trancelike state during which Kircher's soul wandered the heavens. As the Mind explained to Hermes in the eleventh book of the *Corpus Hermeticum*, "Bid [your soul] fly up to heaven, and it will have no need of wings."[142] Just as his earlier works announced Kircher's successful imitation of Oedipus, the *Ecstatic Voyage* publicized his attainment of a new level of spiritual and philosophical fulfillment equaled only by Hermes. In this dream, the ascent of the soul, advancement of knowledge, and rebirth of Egypt came together.

Not coincidentally, Kircher published his *Ecstatic Voyage* only two years after the *Egyptian Oedipus* appeared. The 1650s were the height of Kircher's image making. At work on various obelisks, he witnessed the election of Fabio Chigi as Alexander VII in 1655. That same year Christina of Sweden arrived in Rome and took the name of "Christina Alexandra."[143] The fateful conjuncture of these three individuals in Rome—Kircher, Chigi, and Christina—cemented the image of Rome as meeting ground for a Catholic hermetic revival and Kircher's role as the Christian Hermes. Finding himself in the new Alexandria, patronized by a Christian Osiris and Isis, he was confident that he at last understood what Fate, who had sent him out of Egypt some twenty years before, had in store for him. Witness to the decline of one empire, he gave testimony to its rebirth in another land. In this cycle of events, Kircher's

[142]Scott, *Hermetica*, Vol. I, pp. 205, 211. For an interesting discussion of these passages, see Garth Fowden, *The Egyptian Hermes: A Historical Approach to the Late Pagan Mind* (Cambridge, U.K., 1986), p. 109.

[143]See Åkerman, *Queen Christina of Sweden and Her Circle*, pp. 259–261.

Figure 29. Kircher's *Ecstatic Voyage*. From Athanasius Kircher, *Iter exstaticum coeleste* (Würzburg, 1671 ed.).

brought together the identities of Athanasius, Aeneas, and Hermes, collapsing them into the humanist whole he presented in his autobiography.

When Philipp von Zesen described Kircher in 1662 as "easily the Phoenix among the learned men of this century," he referred explicitly to Kircher's ability to constantly transform and regenerate himself.[144] Kircher made the phoenix a prominent symbol in his work. It appeared on the frontispiece of Petrucci's *Apologetic Forerunner,* a hieroglyph of nature's ability to renew itself, and echoed throughout his museum. The "phoenix reborn" was one of the famous hermetic experiments Kircher demonstrated to visitors.[145] Kircher's identity emerged from a syncretic amalgam of classical, patristic, and hermetic imagery; the objects in his museum continually reaffirmed the intersections of these different forms of authority. Rebirth lay at the heart of Virgilian and hermetic natural philosophy and was the motive force behind his museum.[146] The complementarity of these two discourses reinforced the complementarity of the two main identities Kircher chose for himself. Like the objects in his museum, the Jesuit collector embodied multiple truths, possessing a series of identities that traversed the vast chronological expanse between the beginning of civilization and the events of his own times.

Nurtured on texts such as Augustine's *Confessions* and Loyola's *Spiritual Exercises,* Kircher was trained by vocation to look inward in the search for true knowledge. Yet while Augustine retreated into the "cloisters" of his memory, Kircher instead labeled those same internal edifices "galleries." In them he deposited the images of the world. Unlike Aldrovandi and Della Porta, however, Kircher had no villa that offered him the repose necessary for self-contemplation. Toward the end of his life, Kircher found a means of retreating from the worldly cares of the city. In 1661, he discovered the ruins of an ancient church marking the spot where Saint Eustace had seen a vision of Christ in a stag's horns. Kircher restored the church in Mentorella and made it into a Marian shrine. He also used it as his villa to which he retreated at regular intervals. As he wrote to one correspondent in 1674,

> I have read your very welcome letter, not in Rome, but in the vast solitude of Mount Eustachianum or Vulturelli, where it is my custom to spend each autumn, away from all worldly noise. I put aside all my usual employments and engage entirely in those of God . . . alas, however, even here I am followed by the request of visiting scholars, so that you could say of me. "He who tries to avoid Charybdis falls into Scylla."[147]

[144]In Fletcher, "Astronomy in the Life and Correspondence of Athanasius Kircher," *Isis* 61 (1970): 52.

[145]De Sepi, p. 45.

[146]See Fowden, *The Egyptian Hermes,* p. 108, on *palingenesia.* Virgil's image of renewal centers around the passage from the sixth book of the *Aeneid,* discussed earlier in this chapter.

[147]Kircher, *Historia Eustachio-Mariana* (Rome, 1665), in Reilly, *Athanasius Kircher,* p. 176. The latter part of Kircher's autobiography is concerned with the discovery and use of this

Like the other collectors whom we have considered, Kircher found it impossible to separate fully the *otium* of his retreat from the *negotium* of the city. While he left behind the museum, he brought the cognitive space of his "galleries" with him.

THE SCIENTIFIC SELF

Aldrovandi, Della Porta, Settala, and Kircher represent several different trajectories in articulating the identity of the collector. Aldrovandi and Settala relied on key classical topoi to develop their images, dividing up the different forms of scientific exemplarity between them as representatives of the revival of natural history and mechanics. Della Porta characteristically denied the ability of any one ancient topos to contain him, preferring instead to draw upon classical materials to create emblems of his own devising. Kircher instead progressed through a series of images to achieve the status of a new Hermes, his supreme vision of himself. Creating and exchanging these identities, all four collectors depended upon the resources of learned culture to complete their personal mosaics. They testified to the continued vitality of humanist thought during a period in which scientific culture was undergoing rapid transformations.

Even as they put the finishing touches on their images, a new intellectual culture denied the validity of the humanist exercise of invention, claiming to offer new techniques of analysis that led to new and more modern identities. Galileo embodied this other culture. While reaping the benefits of humanist learning, he nonetheless rejected its claims to scientific authority. Describing the type of natural philosopher whom he opposed, in a letter to Kepler, he wrote, "This kind of man thinks that philosophy is a sort of book like the *Aeneid* and *Odyssey* and that truth is to be found not in the world or in nature but in the collection of texts (I use their terminology). I wish I could spend a good long time laughing with you."[148] Certainly he must have been thinking of naturalists such as Aldrovandi and Kircher, inheritors of the tradition of learning that reached back to the epic narratives of the Greeks and Romans, when he wrote these acerbic lines. What Galileo refused to acknowledge, unlike the recipient of his letter, Kepler, was the complementarity of these different forms of inquiry. Participating in the culture of the Accademia dei Lincei, which made Della Porta and his emblematics a central feature of its activities, Galileo belonged to a group of

---

shrine, an indication of how important it became to him. In contrast, Kircher says very little about the museum in his autobiography, perhaps because he had said so much elsewhere.

[148]Kepler, *Gesammelte Werke*, Vol. XVI, p. 329, in Grafton, *Defenders of the Text*, p. 2. My argument here about the relationship between humanism and science runs along similar lines to the argument in Grafton's work. For a discussion of the narratives shaping Galileo's own identity, at least in the eyes of his followers, see Michael Segre, *In the Wake of Galileo* (New Brunswick, NJ, 1991), pp. 107–126.

individuals who denied the validity of the cultural tools that they constantly put into play. We might take Agostino Scilla who portrayed himself as "nude of learning" as the prototype of the new image of the natural philosopher.[149] Rejecting the humanist bricolage of the collector, they invented the "scientific" self.

[149]Scilla, *Vana speculazione,* in Morello, *La nascita della paleontologia,* p. 158. On Della Porta's own indebtedness to the classical tradition, see Andrea Garaffi, *La filosofia del Manierismo. La scena mitologica della scrittura in Della Porta, Bruno e Campanella* (Naples, 1984).

EIGHT

Patrons, Brokers, and Strategies

This century of ours does not lack supporters of Virtuosi in diverse parts of Europe.
—ULISSE ALDROVANDI

The museum was the quintessential product of the patronage culture of early modern Europe. Its appearance in Renaissance Italy coincided with the rise of court culture and attendant forms of behavior; its entrenchment in Baroque Italy testified to the importance of such cultural institutions in the absolutist states.[1] As humanist learning increasingly found its home in the courts and the academies, the museum became a visible expression of the new ties between the worlds of learning and politics. The prince, as Machiavelli wrote in his famous treatise on this subject, was a "lover of ability" who valued all forms of excellence as a mirror of his own virtue.[2] By patronizing collectors, princes publicized their role as learned and munificent benefactors. Acquiring clients with the same passion as they acquired objects, princes demonstrated their command of the natural and human resources within their grasp. They made themselves the most important audience for the culture of display that formed the museum.

Naturalists equally benefited from their cultivation of princely patronage. They collected patrons with a similar passion, displaying them in the portraits and medals that hung in their museums and in encyclopedic lists such as Aldrovandi's *Catalogue of Men Who Have Helped Our Studies.*[3] The risks

---

[1] The relationship between absolutism and scientific culture is discussed in Biagioli, "Scientific Revolution, Social Bricolage and Etiquette"; Evans, *The Making of the Habsburg Empire,* ch. 9 and 12; David S. Lux, *Patronage and Royal Science in Seventeenth-Century France: The Académie de Physique in Caen* (Ithaca, NY, 1989); and Smith, *The Business of Alchemy.* For a discussion of the relationship between collecting and political culture, see Olmi, "Dal 'Teatro del Mondo' ai mondi inventariati," p. 251, and idem, "Ordine e fama," pp. 247–249, 256–257.

[2] Machiavelli, *The Prince,* in Eamon, "Court, Academy and Printing House," in Moran, *Patronage and Institutions,* p. 32.

were high, the path long and arduous, but the rewards were great. In the short term lay the promise of fame and either a position at court or public association with one of Italy's leading families. In the long term, lay the possibility of immortality. Publishing their works and donating their museums, they attempted to ward off the ravages of time. Collectors and their patrons harbored the fear that, despite all their efforts to bring themselves to the forefront of society, they nonetheless would recede into obscurity, ultimately sharing the fate of the less fortunate, talented, and well-positioned members of society. Participating in a competitive system of cultural exchange fostered by the growing strength of the court as a political institution, collectors elevated the museum and the discipline of natural history as a means of improving their own social status. Patrons, responsive to the dictates of humanist culture that showered accolades upon a cultured ruler, similarly benefited from these relationships. They achieved recognition as "prince–practitioners" and "prince–savants," rulers capable of understanding the connections between learning and power.[4]

By now, early modern patronage is a well-studied subject. Recently, it has begun to receive its share of attention from historians of science as well as historians of art, literature, music, and politics.[5] Patronage has been invoked as an explanation for the spectacular success of individual practitioners, the growing status of natural philosophers as a group, and the intellectual legitimation of various scientific disciplines. As Mario Biagioli argues in his work on Galileo, patronage cannot be regarded simply as a means of economic survival for cash-starved philosophers; in many instances, it was their *raison d'être*. While scientific practitioners tailored the structures of patronage to fit their particular goals, all acknowledged the necessity of understanding these mechanisms to succeed in the world. To achieve the fame and status that lent credibility to all forms of expression, particularly "novelties," one had to have patrons. Only through their efforts to publicize

[3]This is the third and fourth part of BUB, *Aldrovandi*, ms. 110, which has already been discussed in chapter 3.

[4]On this subject, see R. J. W. Evans, *Rudolf II and His World: A Study in Intellectual History 1576–1612* (Oxford, 1973); Bruce Moran, *The Alchemical World of the German Court: Occult Philosophy and Chemical Medicine in the Circle of Moritz of Hessen (1572–1632)*, Sudhoffs Archiv: Zeitschrift für Wissenschaftsgeschichte, 29 (Stuttgart, 1991); and idem, "German Prince–Practitioners: Aspects of Development of Courtly Science, Technology and Procedures in the Renaissance," *Technology and Culture* 22 (1981): 253–274.

[5]Two general collections on this subject are *Patronage in the Renaissance*, Guy Fitch Lytle and Stephen Orgel, ed. (Princeton, 1981); and Kent and Simons, *Patronage, Art and Society in Renaissance Italy*. On scientific patronage, see Mario Biagioli, "Galileo's System of Patronage" and his *Galileo Courtier*, which parallel many remarks made here. Moran, *Patronage and Institutions* contains a representative sample of the most recent work in this area. Scholars such as Nicholas Clulee, Lisa Sarasohn, Richard Westfall, and Robert Westman have also incorporated patronage into their work on various aspects of early modern science.

the work of favored clients could one belong to the new, court-based cultural elite that dominated patrician society.

In their ability to bring together different sectors of elite society, museums allow us to study this process in detail. The exchange of identities and objects that occurred constantly in this setting reflects its significance for the patronage system. Collectors offered patrons multiple ways to express their devotion to them as clients; gifts, visits, and publication subsidies all contributed to the splendor of a museum and its creator. In return, they showered princes with numerous signs of their devotion. Countergifts, dedications, and the strategic positioning of objects referring to a particular patron all contributed to the process of image making.[6] They learned their lessons from men like the imagined courtier in Tasso's *Malpiglio*, who declared, "The courtier's end is the reputation and honor of the prince, from which his own reputation and honor flows as a stream from a spring."[7] As the inventors of a learned space in which most of elite society entered, collectors enabled a prince to extend and magnify his image outside of his own court. They were important publicists for the new political culture of Italy, as republican forms of expression gave way to more aristocratic behavior.[8]

As an example of this process, let us consider the case of Calzolari. In 1590, the apothecary had himself painted holding a medal with a gold chain containing a portrait of Vincenzo I Gonzaga. The Duke of Mantua presented it to him during or shortly after a visit to his museum that year. Apparently this was the duke's preferred form of acknowledgment, for in 1604 we find him sending the same gift to Galileo in response to his military compass. Like many early modern rulers, Vincenzo I could imagine no greater gift to a client than a portrait of himself made of precious metals. Frequently clients sold the chain and kept the medal. Situated in a city too far removed from Florence and Rome to make it possible for him to attract other princely patrons, and limited in his possibilities for social advancement due to his profession

⁶For more on the importance of gift giving among collectors and their patrons, see Findlen, "The Economy of Scientific Exchange in Early Modern Italy," in Moran, *Patronage and Institutions*, pp. 5–24. For related discussions of the importance of gift giving in early modern culture, see Biagioli, "Galileo's System of Patronage," *History of Science* 28 (1990): 18–25, 38–41; Natalie Zemon Davis, "Beyond the Market: Books as Gifts in Sixteenth-Century France," *Transactions of the Royal Historical Society*, ser. V, 33 (1983): 69–88; Marcello Fantoni, "Feticci di prestigio: il dono all corte medicea," in *Rituale, ceremoniale, etichetta*, Sergio Bertelli and Giuliano Crifo, eds. (Milan, 1985), pp. 141–161; Fragnito, *In museo e in villa*, p. 167; Paolo Galluzzi, "Il mecenatismo mediceo e le scienze," in *Idea, istituzioni, scienze ed arti nella Firenze dei Medici* (Florence, 1980), p. 207; Sharon Kettering, "Gift-giving and Patronage in Early Modern France," *French History* 2 (1988): 133–151; and Olmi, "Molti amici in varii luoghi."

⁷Tasso, *Dialogues*, p. 171. Also discussed in Eamon, "Court, Academy and Printing House," in Moran, *Patronage and Institutions*, p. 30.

⁸For a general overview of this subject, see Lauro Martines, *Power and Imagination: City-States in Renaissance Italy* (New York, 1979), pp. 218–337; and Cochrane, *Italy 1530–1630*, pp. 33–49.

of apothecary, Calzolari treasured his association with the Gonzaga. In his 1602 will, he left his nephew the entire contents of his museum save for "the ring with a ruby valued at 100 ducats." This the apothecary bequeathed to his wife not simply for its financial worth but because it was "a gift of the Duke of Mantua" and therefore a precious reminder of his service to the Gonzaga.[9] The presence of the ducal medal, proudly displayed by Calzolari in his portrait, and the ducal ring in his museum, amidst the jars of theriac, dried plants, and stuffed animals, ennobled the practice of collecting and, by extension, the collector. Calzolari's status as a Gonzaga client may in fact have been more important to his identity than the purity of his theriac, for in many circles it was the former that secured the reputation of the latter.

Chapter 8 broadly sketches the contours of museum patronage and its implications for the discipline of natural history and the careers of individual collectors. It explores both the continuities and discontinuities in the patronage system. Structural features such as the use of brokers, the language of dedication, and the patronage of publications united collectors from Aldrovandi to Kircher. They also shared certain patrons through the overlap of successive generations of naturalists and the continued political vitality of particular ruling families. Within Italy, only Medicean and papal patronage spanned the entire period under study. While the Medici initially cultivated clients who could culturally legitimate their new political powers as hereditary rulers, by the mid-seventeenth century, their political visibility depended on their reputation as great patrons of arts and learning. As Edward Goldberg perceptively argues in his studies of art patronage in grandducal Florence, with the waning of the Tuscan state as a military and economic force, patronage became the primary means for the Medici to maintain their position among the leading rulers of Europe.[10] Unable to compete in size and strength with emerging absolutist states such as France and Savoy, the Grand Duchy of Tuscany nonetheless continued to equal if not surpass them in its role as a center of cultural production. Thus, the Medici attracted and encouraged the most well-known collectors in Italy as their clients and agents.

While the papacy long had made patronage an important part of its political structure, the growth of the papal court in the sixteenth and seventeenth centuries accelerated this preexisting tendency. Like the Medici grand dukes, early modern popes successfully made the transition from Renaissance to absolutist politics. By the time of Urban VIII (1623–1644), the

[9]Tergolina-Gislanzoni-Brasco, "Francesco Calzolari," p. 15; Franchini et al., *La scienza a corte*, p. 124; Biagioli, "Galileo's System of Patronage," pp. 19–20; Fantoni, "Feticci di prestigio," p. 143. The portrait of Calzolari is reproduced in Franchini.

[10]Goldberg, *Patterns in Late Medici Art Patronage;* idem, *After Vasari: History, Art and Patronage in Late Medici Florence* (Princeton, 1988). For a general overview, see Cochrane, *Florence in the Forgotten Centuries,* and R. Burr Litchfield, *Emergence of a Bureaucracy: The Florentine Patricians 1530–1790* (Princeton, 1986).

papal court was the largest court in Europe and the pope the most luminous prince.[11] Despite the exceptional nature of papal power, Renaissance
and Baroque popes shared many common features with secular rulers.
Chosen from among the leading families in Italy, who were often involved
in their local political culture, it is hardly surprising that they would trans-
port secular practices into ecclesiastical governance. The papacy, in essence, had become the premiere office to which Italy's leading families aspired and through which they extended their claims to power. Not least of
these was a desire for fame. Patronage provided a means of achieving that
goal by offering one pope after another a series of images that reinforced
his role as ruler of the holy city and Christendom.[12] Sensitive to their weakened reputation in the wake of the Reformation, early modern popes transformed the face of urban Rome and supported clients who strengthened
the papacy's authority as a religious institution while enhancing the visibility of the pope as a prince. Collectors such as Mercati and Kircher offered
both services to their patrons. Museums captured in miniature the imperial
visions of the early modern papacy, representing the possession of nature
and knowledge as a Catholic prerogative and glorifying the activities of individual popes. They were ready-made instruments for the renovation of
Catholic culture. As in the case of the Medici, the waning fortunes of the papal territorial state in the mid-seventeenth century only accelerated programs of patronage. Splendor did not mend a troubled economy or revive
a faltering militia, but it allowed the papacy to maintain its honor and reputation in the face of uncertainty.

The prominence of other Italian ruling families as patrons ebbed and
flowed with their political fortunes. In the sixteenth century, the Dukes of
Ferrara, Mantua, Parma, and Urbino offered the Medici and the pope stiff
competition for the title of most beneficent prince. By the mid-seventeenth
century, the Dukes of Savoy were the only territorial rulers indigenous to the
Italian peninsula who commanded the status of the Medici or leading families such as the Barberini and Chigi who occupied the papal throne and monopolized the cultural life of Rome. Outside of Italy, both the Spanish and
Austrian Habsburgs continued to attract collectors as clients. The Spanish
who occupied large portions of the peninsula—the Duchy of Milan and King-
doms of Naples and Sicily—had never been particularly inclined to invest in
the cultural pursuits of their Italian subjects. While appointed officials such
as the Spanish Viceroys patronized the museums of Della Porta, Imperato,
and Settala, the Spanish kings expressed no comparable degree of interest
in the scientific activities of these collectors, unless a client was willing to

[11]Particularly illuminating on this subject are Paolo Prodi, *The Papal Prince: One Body and Two
Souls, The Papal Monarchy in Early Modern Europe,* Susan Haskins, trans. (Cambridge, U.K.,
1987), and Laurie Nussdorfer, *Civic Politics in the Rome of Urban VIII* (Princeton, 1992).

[12]On the process of image making, see Charles L. Stinger, *The Renaissance in Rome* (Bloomington, 1985).

travel all the way to the Spanish court as Della Porta did in his youth. As Aldrovandi learned to his chagrin, the Spanish Habsburg princes naturally preferred to support the efforts of local rather than imported talent. In the wake of their own declining fortunes as a political force in the seventeenth century, this attitude only became more pronounced.[13]

In contrast, the Austrian Habsburgs found Italy an increasingly attractive place to seek out promising clients. As the Holy Roman Emperors, they felt a special link to Rome, the ancient source of legitimation for the crown that they precariously wore. The growing strength of the Society of Jesus in the late sixteenth and seventeenth centuries provided a vehicle to link the learned culture of these two Catholic empires. Through the movement of scholars between Italy and Central Europe, embodied by Kircher, they came to see Rome as a setting in which to refine their political mythology as the Catholic princes *par excellence.* The relocation of prominent Jesuit intellectuals out of war-torn regions to Rome and the influx of scholars at the German College ensured the Habsburgs of a steady supply of clients in the papal city. Their patronage of collectors such as Kircher provided a model for other German princes, Catholic and Protestant, to imitate. By the mid-seventeenth century, fewer German-speaking students attended the Italian universities in the way that they had done previously; the new and expanded universities of Central Europe offered them equivalent educational opportunities close to home.[14] But many more German princes, nobles, and scholars traveled to Rome to see its attractions, foremost among them Kircher's museum. He was a client who spread the glory of the Holy Roman Empire as far as Rome, linking imperial and papal ambitions through his museum.

The social and political structures that underlay the practice of collecting testify to the complex relationships among artifacts, individuals, and institutions. The ability of key individuals to enter the world of the courts virtually ensured the success of collecting as a noble pastime, which in turn solidified the position of the naturalist at court. The search for new ways to symbolically represent the power of the prince engendered new opportunities for the establishment of fruitful patronage relationships. The ease with which naturalists moved among universities, academies, pharmacies, and courts illuminates the intersections of these diverse institutions of culture and reinforces the importance of the museum as a setting that facilitated certain forms of social interaction. The museum captured and preserved a system of social understanding that was central to early modern culture. Patricians created museums not only to possess knowledge but also

[13]For a discussion of Spain's political and cultural role in the sixteenth and seventeenth centuries, see J. H. Elliot, *Spain and Its World 1500–1700* (New Haven, CT, 1989), and Anthony Pagden, *Spanish Imperialism and the Political Imagination* (New Haven, CT, 1990).

[14]Evans, "German Universities After the Thirty Years' War," *History of Universities* 1 (1981): 169–190.

to replicate the dynamics of court life within the home. Reconstructing the exchanges that breathed life into a collection, we gain a more precise understanding of the relationship between the new status of museums, collectors, and natural history and the growth of patrician culture.

ARISTOTLE IN SEARCH OF ALEXANDER

The classic *exempla* that shaped the identities of naturalists also informed their perceptions of patron–client relations. Foremost among them was the historical image of Aristotle as a client of Alexander. Naturalists as diverse as Aldrovandi, Mercati, and Bonanni and, outside of Italy, the English natural philosopher John Dee invoked this image as an analogy for the relationships they hoped to cultivate with their own patrons.[15] Alexander the Great embodied the virtues of leadership admired by princes: learned, militarily astute, and commander of a vast empire, he typified the successful ruler who combined *sapientia* with *potentia*.[16] Collectors suggested that their patrons "follow in the footsteps of Alexander the Great" or, more generically, "in imitation of ancient kings and princes"[17] because it established a code of conduct for their patrons and themselves that allowed them to reconcile intellectual programs with social expectations. In this spirit, the botanist Evangelista Quattrami attempted to interest his patron, Alfonso II d'Este, in subsidizing his research on theriac. "And with this opportunity, Your Most Serene Highness already will receive praise among the Princes of the world, for being the first Prince who had two such important compounds made in the way that the Ancient inventors, Emperors, and Kings composed them."[18] After some thirty years of dedicated service to the d'Este, he perceived himself to be the new Andromachus to the new Mithridates, the new Galen to a new Emperor, the new Aristotle to a new Alexander.

As tutor and philosopher to one of the most celebrated leaders of the ancient world, Aristotle embodied the virtues of a successful client. He was a naturalist who had secured the favor of a powerful prince willing to subsidize his work; in return, Aristotle offered Alexander lasting fame through his immortal words. Writers of etiquette manuals were quick to perceive the

[15]I have taken the title for this section from Nicholas H. Clulee, *John Dee's Natural Philosophy: Between Science and Religion* (London, 1988), pp. 189–199, where he discusses Dee's idea of himself as a "Christian Aristotle" in the service of a "Christian Alexander," variously identified as Elizabeth I, Rudolf II, or any other ruler who expressed an interest in his work. I discuss Aldrovandi's and Mercati's use of this imagery in this chapter. For Bonanni, see the preface to his *Ricreatione*.

[16]This combination is discussed in Eamon, "Court, Academy and Printing House," in Moran, *Patronage and Institutions*, p. 33, though not in reference to Alexander.

[17]In Ludovico Frati, "Le edizioni delle opere di Ulisse Aldrovandi," *Riviste delle biblioteche e degli archivi* 9 (1898): 164; Aldrovandi, *Discorso naturale*, p. 231.

[18]ASMo, *Archivio per materie. Storia naturale*, b. 1 (Dalle stanze della Castellina, 12 September 1595).

analogies between ancient and Renaissance patronage systems. "And Aristotle was the author of the deeds of Alexander, employing the methods of a good Courtier," wrote Castiglione in his *Book of the Courtier*.[19] Early modern naturalists viewed Aristotle as an appropriate model not only because of his intellect but also because of the methods by which he secured the resources to complete his research and record it in writing. He was a Renaissance courtier *avant la lettre*.

As naturalists constantly reminded their patrons, the might of Alexander survived primarily through the words of Aristotle. "Clearly Alexander the Great King of Macedonia acquired so much praise and glory only for having commissioned his tutor Aristotle to write the history of animals," observed Aldrovandi. Writing to different princes, he did not hesitate to invoke the fragility of Alexander's empire, whose physical monuments, like those of Ozymandias, had crumbled with time. In contrast, the words of Aristotle had survived some 1900 years. "If Alexander were alive today, he would marvel [at the fact] that, among the multitude of famous works that he did, those books of animals acquired greater glory than any other." The deeds of princes soon lost their luster, but knowledge was eternal. Just as Aldrovandi surpassed his alter ego, his patrons also inherited the capacity to better Alexander through great acts of patronage.

After meeting the Grand Duke Francesco I in 1577, for example, Aldrovandi did not neglect to praise him for his "incomparable fame of long duration" and "greater glory and honor" in comparison to Alexander.[20] While the Macedonian king patronized only the study of animals, the Medici prince made himself the patron of all nature. Aldrovandi's friend Mercati also invoked this image in the dedication of his *Metallotheca* to Clement VIII. "For if Aristotle bestowed great devotion and power upon Alexander the Great in composing the History of Animals," he wrote, "how much greater should this work of mine be anticipated by Your Holiness whose empire is spread throughout the entire Christian World?"[21] While Francesco I exceeded Alexander in his intellectual capacity, Clement VIII surpassed him in the breadth of his rule. Through the manipulation of this image, naturalists reminded princes of the danger of not taking patronage seriously.

On the frontispiece of the first volume of his *Ornithology* (1599), dedicated to Clement VIII, three images were prominently displayed. On the left, Aristotle offered his works to Alexander; on the right, Pliny gave his

[19]Castiglione, *The Book of the Courtier*, p. 332.

[20]BUB, *Aldrovandi*, ms. 6, Vol. I, cc. 38r–40r (Bologna, 15 September 1577). Many of the letters I cite in manuscript form in this chapter are also reproduced in Tosi, *Ulisse Aldrovandi e la Toscana*. For other examples of Alexander imagery, see BAV, *Vat. lat.* 6192, Vol. II, ff. 656v–657r (Bologna, 23 July 1577); BMV, *Arch. Mor.*, Vol. 103 (= *Marc.* 12609), ff. 27–28; BUB, *Aldrovandi*, ms. 6, Vol. I, cc. 38, 40r; ms. 66, cc. 355–356; Aldrovandi, *Discorso naturale*, p. 180.

[21]Mercati, *Metallotheca*, p. lii.

Natural History to the Emperor Vespasian. At the bottom, Aldrovandi himself presented his book to the pope[22] (fig. 30). While the unreliability of real patrons made the naturalist's status as a client tenuous, the firm association with an abstract patron ensured the continuance of the social model within which he defined himself. Assigning his patron an iconographically rendered identity that reinforced his own image as a successful client, in the best tradition of the Greeks and Romans, Aldrovandi publicized his arrival in the world of the courts. After forty years of sustained efforts, he finally had accumulated enough patrons to finance the publication of the first volume of his natural history. Like many early modern intellectuals, Aldrovandi perceived printing to be the primary vehicle through which he could rival the achievement of the ancients. "And it's incredible but true that one man alone can print more in one day than even the quickest writer could scribble in two years."[23] Despite his optimism about his chosen method of disseminating his ideas, Aldrovandi found it more difficult to achieve this goal than he had hoped. Publication, as many encyclopedists discovered, was a costly and time-consuming enterprise. For this reason, numerous scholars continued to perpetuate the manuscript tradition even if they preferred print culture. Such compromises only sweetened the victory of publication as a moment when a scholar indeed could proclaim that he had surpassed Aristotle not only in word and deed but also in the medium through which he made his patron and himself immortal.

For Aldrovandi, the image of Aristotle as a client of Alexander served as a leit motif for his own attempts to gain patronage. Establishing himself as the favored client of a prince would ensure the continued vitality of his museum, during and beyond his own lifetime, and provide him with the financial means to see his work into print. By the 1560s, Aldrovandi had exhausted the resources available to him in Bologna, the second city of the papal state. He enjoyed a position of high visibility in the university and could harbor little hope for further advancement in that setting. He was an intimate of the archbishop Paleotti, the papal legate, and the senators whom Cardano described as "all patricians, extraordinarily courteous, cultivated, experienced, and brilliant."[24] But Bologna lacked the court culture necessary for his stated objectives. During this period, he initiated a campaign to secure patronage from a prince worthy of his endeavors.

[22]Aldo Adversi, "Ulisse Aldrovandi bibliofilo, bibliografo e bibliologo del Cinquecento," *Annali della scuola speciale per archivisti e bibliotecari dell'Università di Roma* 8, n. 1–2 (1968): 180.

[23]I cannot find the reference, but it is located in BUB, *Aldrovandi*.

[24]Cardano, *The Book of My Life*, p. 63. As a member of one of the noble families with a hereditary right to a seat in the Senate, Aldrovandi himself took an active role in the political life of the city, serving numerous times in the *Sindici di Gabella* and even being elected to the *Gonfaloniere di Giustizia*, the executive body of the Senate, in 1569. Thus, the naturalist could count on them to support his periodic pleas for publishing funds, a higher stipend, and additional money for the botanical garden. BUB, Cod. 559 (770), Vol. XV, pp. 167, 226, 376, 682, 701, 979; AI, *Fondo Paleotti* 59 (F 30) 29/7, c. 1r (Bologna, 11 December 1585).

Figure 30. Aldrovandi giving his *Ornithology* to Clement VIII. From Ulisse Aldrovandi, *Ornithologiae hoc est de avibus historiae libri XII* (Bologna, 1599).

Mattioli's appointment as imperial physician to the Holy Roman Emperor Maximilian II provided Aldrovandi with his first opportunity to secure princely patronage. Shortly after Mattioli had arrived in Prague, in 1558 Aldrovandi wrote to inquire whether the Habsburg ruler would be interested in underwriting a scientific expedition to the islands surrounding the Italian peninsula. "I have considered what you wrote me concerning the voyage you would like to make to Crete, Cyprus, and other places in Greece, Sicily, and Corsica, if you had a King or Prince who would sustain the expense," replied Mattioli cautiously, "but whether or not my King or my Prince absolutely would do it I cannot permit myself to promise nor to affirm it so spontaneously for him."[25] Rebuffed on the imperial front, Aldrovandi continued to try his luck at home. Two years later, his name appeared in the correspondence of the prelate Ludovico Beccadelli, the employer of Giganti. Stopping briefly in Bologna, Beccadelli found himself swamped with requests from various clients. "This Bologna is a nightmare of recommendations and many apply to me as the servant of the most Reverend Monsignor Morone," he wrote to friends in Rome. Among the supplicants was Aldrovandi, whom Beccadelli described as a close friend. "He wishes to be recommended to the Most Illustrious Monsignor, understanding that [the Pope] wishes to do some good for this studium."[26] As a lecturer at the most famous university within the papal state, Aldrovandi hoped to be the beneficiary of whatever favors the pope chose to shower upon his institution. Friends such as Beccadelli, Paleotti, and the papal legates who came to Bologna did not neglect to bring him to the attention of the papal court.

Slow but steady success with patrons in Rome and Florence during the 1560s emboldened Aldrovandi to make another attempt at interesting a foreign monarch in his work. Philip II was a logical candidate; as a ruler with a territorial interest in Italy, whose citizens attended its universities, the Spanish king was well informed about activities in this part of the world. The presence of his representatives in Bologna gave Aldrovandi the opportunity to approach Philip II through trusted intermediaries. In 1567, he composed a letter to the Cardinal Protector of the Spanish College outlining his qualifications as a client worthy of a Habsburg prince. He began the letter by articulating the historical nature of the relationship he hoped to establish: "In my studies I need a patron like Alexander the Great, King of Macedonia, was to his tutor Aristotle." Describing the expedition he wished to undertake as a new Columbus, he argued that a comprehensive study of New World nature could only be financed by a descendent of Ferdinand and Isabella. Acquiring the patronage of a Spanish king was vital to the fulfillment of the ancient and modern identities that he chose for himself. In return, Aldrovandi promised to "dedicate all of my works and labors to His

[25]Raimondi, "Lettere di P. A. Mattioli," p. 38 (Prague, 29 January 1558).
[26]BPP, ms. Pal. 1010, c. 373v (Bologna, 17 August 1560).

Majesty, deservedly as my Patron, like Aristotle did for Alexander."[27] He and Philip II mutually would fulfill each other's destiny. Aldrovandi would affirm Philip II's role as the new Alexander by describing the flora and fauna of his Renaissance Macedonia, New Spain. Disappointed to learn that Philip II instead preferred to offer this position to one of his court physicians, Hernandez, Aldrovandi nonetheless continued to inform the Spanish monarch of his endeavors. When friends traveled to Madrid, they made sure to praise Aldrovandi to the Habsburg king.[28]

By the 1570s, Aldrovandi imagined that his time to secure a worthy patron had finally arrived. Paleotti had been appointed archbishop, his cousin Gregory XIII wore the papal tiara, and the inevitable succession of rulers had produced younger, more energetic princes who had not yet completed their lists of favorites. Aldrovandi began first with his cousin the pope. As part of the activities surrounding his anatomy of the papal dragon, he lost no time in suggesting how Gregory XIII could reciprocate. "I need a patron," pleaded Aldrovandi to Giacomo Boncompagni, natural son of Gregory XIII, shortly after he had become Castellan of Sant'Angelo in 1572.[29] Despite Aldrovandi's attentiveness to the wishes of the pope, Gregory XIII expressed little interest in the scientific work of his cousin. By 1576 Aldrovandi openly expressed his disgruntlement over their relations. "For many reasons, I had hoped to be able to place many of the great works done by me in your hands under this most happy Papacy," wrote Aldrovandi to the papal *nuncio* in Florence, "but I have had very little luck."[30] The following year, when Aldrovandi traveled to Rome to beseech the pope to intervene in the theriac dispute, he perceived this unfortunate event as a unique opportunity to renew his requests in person. Writing to his brother Teseo shortly before his departure, he suggested, "Perhaps I could obtain some subsidy for printing my works from His Holiness." Gregory XIII probably felt that his intervention in a physicians' dispute was all the patronage he cared to bestow on his troublesome cousin. Some months later, Aldrovandi encouraged his brother to assist him in gaining "the help of Princes to have the figures engraved" for his natural histories.[31] Whatever promises he elicited from the pope, they were not enough to assure him that he had found his Alexander.

Gregory XIII's recalcitrance led Aldrovandi to renew his efforts at gaining a foothold in the Medici court. In 1560 he had declined an offer of a teaching position at the University of Pisa from Cosimo I but maintained cordial relations with the Grand Duke. Cosimo I's death in 1574 offered Aldrovandi an opportunity to establish himself with his successor,

[27]BUB, *Aldrovandi*, ms. 66, cc. 355v, 367r (Bologna, 12 November 1567).

[28]BUB, *Aldrovandi*, ms. 38(2), Vol. IV, c. 55r.

[29]Aldrovandi, *Discorso naturale*, p. 180.

[30]BUB, Cod. 596-EE, n. 1, f. 2v (Bologna, 9 April 1576).

[31]BUB, *Aldrovandi*, ms. 97, cc. 354r, 319r (Bologna, 9 March and 14 December 1577).

Francesco I. His decision to stop in Florence on the return from Rome in 1577 only cemented the impression that Aldrovandi was using his increased visibility, in the wake of the theriac controversies, to acquire a new patron. After visiting the grandducal collections and renewing his friendships with various intimates at the Medici court, Aldrovandi returned to Bologna convinced that Francesco I was indeed his Alexander. Rushing to assure the Grand Duke that he had Aldrovandi's exclusive devotion as a client, a hyperbole he invoked in initiating any patronage relationship, Aldrovandi dashed off the following encomium from his *studio:* "I truly know of no other Prince in Europe who can more easily augment these natural sciences than Your Highness," he wrote in 1577.[32] Through letters, gifts, and his promised publications, Aldrovandi hoped to become the Medici court naturalist.

For a time, Aldrovandi enjoyed this role as his exclusive prerogative. Francesco I bestowed numerous favors upon him, among them the use of his court painter Jacopo Ligozzi, who not only illustrated objects in the Medici collections but also came to Bologna to transform the objects in Aldrovandi's *studio* into beautiful watercolors.[33] In his autobiography, Aldrovandi described his relationship with Francesco I as consummated by great gifts on both sides:

> Besides the other favors that His Highness did for Doctor Aldrovandi, he gave him many rare and beautiful things, promising him in the future that he would give him part of anything that befell him from foreign hands, and every time that he had two he would give him one, as he has done since, having sent plants, seeds, metals, birds depicted live and other things. One can see this in the Museum of Doctor Aldrovandi and likewise in his histories, where he remembers the Grand Duke, as he does for all the others who have liberally enriched his Theater of Nature.

As Aldrovandi wrote to Francesco I in 1586, "Words cannot express how dear are the things Your Most Serene Highness sent me. . . . Part of them I reserve in my Museum in perpetual memory of you, as I always do with all the things you send me, in which, as in a mirror, I gaze at and revere Your Highness from afar."[34] As his role as a Medici client became established, Aldrovandi began to advertise his museum as an extension of the Medici collections. Duplicating many of the objects found in Florence, it was the material version of the popular "mirrors of princes" that exalted the virtues of a ruler. The culmination of this form of representation was the portrait of Francesco I accompanying the portrait of Aldrovandi as Aristotle in his villa.

[32]BUB, *Aldrovandi*, ms. 6, Vol. I, c. 11 (Bologna, 19 September 1577).
[33]For example, BUB, *Aldrovandi*, ms. 70, c. 21r; on Ligozzi, see *Mostra di disegni di Jacopo Ligozzi*, M. Bacci and A. Forlani, eds. (Florence, 1961); Tongiorgi Tomasi, "L'immagine naturalistica a Firenze tra XVI e XVII secolo," in *L'immagine anatomico-naturalistica nelle collezioni degli Uffizi* (Florence, 1984). Many of Ligozzi's drawings still can be seen today in the prints and drawings room of the Uffizi and in the BUB, *Aldrovandi*, esp. *Tavole di animali*.
[34]Aldrovandi, *Vita*, p. 28; Mattirolo, pp. 375–376 (Bologna, 5 May 1586).

While the epigram beneath the Grand Duke's image did not explicitly name him Alexander, its allusions to the great deeds of great men made the analogy unmistakable.[35]

Francesco I was particularly receptive to this form of flattery. Like the Emperor Rudolf II, he was a prince who surrounded himself with men of talent to enhance his image as a learned ruler. Observing the relationship between the Grand Duke and important naturalists, his botanist Giuseppe Casabona wrote, "Ulisse Aldrovandi and Mercati write and converse with the Grand Duke as if they were brothers and sit at his own table."[36] Francesco I accorded naturalists a higher status than the court astronomers or mathematicians who were invited to dine at court but not at the table of the Grand Duke. His personal interest in natural history made Aldrovandi a Medici favorite, even though he hardly ever came to Florence. In exchange for his praise and loyalty, Francesco I wrote letters of supplication to the Senate of Bologna, as did Camillo Paleotti and Cardinal Alessandro Farnese, supporting Aldrovandi's demands for salary increases and special stipends to maintain the botanical garden. In 1585 when the death of an uncle threatened the Aldrovandi family's loss of an important title, which Gregory XIII chose to confer on his treasurer rather than on his kinsman, Francesco I interceded to have the "honor and dignity" of his client restored.[37] As Aldrovandi progressed in the completion of his natural history, he was reasonably confident that Francesco I would underwrite the costs of its publication as part of his continued affection toward the Bolognese naturalist.

The patronage market in early modern Italy fluctuated with the rise and fall of the various princes. Political disfavor or an untimely death could mean the end of numerous projects; when such changes occurred, clients regrouped around new patrons, made suddenly powerful, to offer them their fidelity. The death of Francesco I in 1587 was one such moment. Naturalists throughout Italy wondered whether Ferdinando I would display the same affection for their studies that his brother had demonstrated. In Florence, Casabona bemoaned the fact that he was "without a patron" and wondered if there was anything left for him at court. He urged friends who had enjoyed the patronage of Francesco I to write to the new Grand Duke "to confirm the initiation of your friendship." Within a short period, Casabona's fears were quickly dispelled. Ferdinando I instructed his botanist to correspond in his name with important naturalists and other prince–savants to publicize his interest in scientific matters. Soon afterward, Casabona found himself participating in the expansion of the botanical garden in Pisa, which included the construction of the gallery with a room for his own "*studiolo* or

[35]See chapter 7 for a discussion of the portrait.

[36]UBE, *Briefsammlung Trew, Casabona*, n. 17, in Olmi, "Molti amici in varii luoghi," p. 17. For more on Francesco I's activities, see Berti, *Il Principe dello studiolo*.

[37]Tosi, *Ulisse Aldrovandi e la Toscana*, pp. 284–288, 301–303; Ronchini, "Ulisse Aldrovandi e i Farnesi," p. 9 (Bologna, 23 September 1586).

natural museum."[38] Ferdinando I chose to enhance rather than dismantle the patronage of learning established under his brother.

In a letter to Ferdinando I, written shortly after his succession as Grand Duke, Aldrovandi rushed to assure the new ruler of his willingness to continue his relationship with the Medici:

> The continuous, affectionate protection with which Your Highness favors me
> . . . certainly aroused me to appeal immediately to [you] with all of my informed vigils and efforts, knowing that I was appealing to the protector and promoter of virtues and virtuosi, the Grand Duke Ferdinando de' Medici. But I retained my usual humble circumspection with reason, so that [my request] would not be among the great cares that recently made princes carry with them.

He reminded the new Grand Duke of their encounter in Rome, when Ferdinando I had been a cardinal, and hoped that the affection Ferdinando I held for his brother would translate into support of Francesco I's clients. For this reason, he glorified the "most Serene and Glorious House of Medici which . . . has always embraced, protected, helped, favored, and carried [virtuosi], claiming among them the generous Leos, Clements, Giovannis, Cosimos, Alessandros, and Francescos whom, by hereditary succession, Your Most Serene Highness Ferdinando succeeds."[39] Describing Ferdinando I as the inheritor of a long tradition of princely patronage, Aldrovandi argued that it was his destiny to become the new Alexander. Only by patronizing the new Aristotle, however, could he claim this patrimony.

While acknowledging Aldrovandi's clientage, Ferdinando I did not offer him the favored status that Francesco I had done. Aldrovandi was no longer the naturalist who sat at the table of the Grand Duke but one of many supplicants who courted his favors. "Certainly the Grand Duke demonstrates a great desire to do great things," his court physician Mercuriale wrote to Aldrovandi.[40] Despite these reassurances, Ferdinando I was remarkably unforthcoming in his support of Aldrovandi's publication projects. He was happy to have Casabona exchange *naturalia* with Aldrovandi—after all, this was a relatively inexpensive favor repaid in kind—but he exhibited little interest in financing the cost of publishing Aldrovandi's *Ornithology*. While Aldrovandi had successfully cast Francesco I as his Alexander, Ferdinando I continued to resist this label.[41] Aldrovandi's promise to dedicate all his works to the Grand Duke, the same gift he had offered Philip II some twenty years before, did little to change Ferdinando I's stance. He was happy to

[38]UBE, *Casabona*, n. 23, 25, in Olmi, "Molti amici in varii luoghi," pp. 19–21.

[39]BUB, *Aldrovandi*, ms. 21, Vol. II, cc. 11–12, n.d.

[40]BUB, *Aldrovandi*, ms. 136, Vol. XXV, c. 3v.

[41]See Mattirolo, "Le lettere di Ulisse Aldrovandi a Francesco I e Ferdinando I," p. 384 (Bologna, 27 November 1591). For more on the Medici sense of destiny in the sixteenth century, see Janet Cox-Rearick, *Dynasty and Destiny in Medici Art: Pontormo, Leo X, and the Two Cosimos* (Princeton, 1984).

continue the family association with the learned naturalist but not at any further expense to the overextended governmental purse. In 1593, Mercuriale advised him to drop the matter, citing Aldrovandi's age and the incompleteness of the work as two of the main reasons. "Today the minds of Princes are inclined more to earning than to spending," he counseled. When Mercati died suddenly that same year without publishing his *Metallotheca*—and this despite that he had been a favored client of at least four popes[42]—Aldrovandi began to fear that the same fate lay in store for him. "Today this is the luck that men of letters have with many Princes who apply [their] minds to everything else," he wrote bitterly.[43] Rather than concentrating exclusively on the Medici, Aldrovandi resolved to renew his contacts with other princes throughout Italy.

By the 1590s, Aldrovandi was no longer optimistic about the prospects of finding a new Alexander. Instead, he was willing to settle for a patron "not like the one Aristotle had but somewhat less significant."[44] Contacts at the courts of Mantua, Parma, Ferrara, and Urbino, not to mention Rome, offered him other possibilities. While Gregory XIII had cared little for Aldrovandi's natural history, later popes expressed greater interest. Though Clement VIII did not publish Mercati's *Metallotheca*, he did underwrite the costs to complete the first volume of the *Ornithology* after visiting Bologna in 1598, thus earning the right to be represented as Alexander on the title page (see fig. 30). Despite his decision to depict Clement VIII as the Alexander to his Aristotle, Aldrovandi understood his *Ornithology* to be the result of accumulated debts to numerous individuals. Many princes and scholars provided him with examples of the birds he described; others underwrote the cost of the woodcuts that copiously illustrated the treatise, while still others paid the printer's bills. As his friend Castelletti observed, the production of Aldrovandi's *Ornithology* "overcomes the force of every private person."[45] Rather than being the product of an exclusive relationship with one patron, his work owed its completion to multiple patrons and friends who had sustained him over the years.

Public acknowledgment was a delicate operation. The inclusion of too many patrons belittled their individual importance, while the omission of certain benefactors might morally offend them. A letter from Mattioli to Aldrovandi illustrates the complex maneuvering involved in bringing a work to press. Mattioli lamented the loss of a friend, who no longer wrote

[42]Mercati was a favorite of Pius V, Gregory XIII, Sixtus V, and Clement VIII as papal physician and naturalist. He may also have worked for the other three popes who reigned briefly during the late sixteenth century. Such continuity was remarkable, given the usual turnover in court appointments; see Palmer, "Medicine at the Papal Court in the Sixteenth Century," in Nutton, *Medicine at the Courts of Europe*, pp. 49–78.

[43]BUB, *Aldrovandi*, ms. 136, Vol. XIX, c. 154r (Pisa, 1593); ms. 21, Vol. IV, c. 169r (San Antonio di Savena, 29 August 1595).

[44]BUB, *Aldrovandi*, ms. 66, c. 356v.

[45]BUB, *Aldrovandi*, ms. 76 (Genoa, 26 November 1598).

to him because he was not acknowledged in the latest edition of Dioscorides. "If all of the important friends that I have in diverse parts of Europe, who write to me daily, were obliged to become enemies for my not having named them, I would like to swear to no longer write books nor anything good, since doing so makes me lose friends."[46] Aldrovandi also had to contend with these problems as he committed his words and his thanks to print. His manuscript dedications became a testing ground for the acknowledgments that appeared in his publications. Friends such as Calzolari, brokers such as the *nuncio* Bolognetti and patrons such as Gregory XIII all received treatises directed personally to them.[47]

While dedications were the most public form of acknowledgment, the presentation of books also secured ties between patrons and clients. As Natalie Zemon Davis points out, while historians have often lingered over dedications to unearth the mechanisms of patronage in early modern culture, they have paid little attention to the role that books played as actual gifts.[48] The dedication of a book cemented a relationship between patron and client by linking them on the printed page. However, the gift of a book also reaffirmed the role of patrons and friends in an author's intellectual production. It was a means of acknowledging past support and soliciting new patrons for future projects. By obliging princes for a gift as costly and lavish as a book, particularly a highly illustrated one such as a museum catalogue or a work of natural history, Aldrovandi hoped to widen his circle of patrons. Copies of his *Ornithology* sent to princes often included embellishments such as hand-colored illustrations and personalized epigrams written on the flyleaf. Aldrovandi carefully included instructions on how to read his epigrams as a gloss on the frontispiece in letters accompanying copies of his work.[49] Other books came with instructions on how and when to read them, such as the copy presented to Federico Borromeo "which you will deign to receive with your usual graciousness and courtesy, and then favor me by skimming it, when you are not busy."[50] Such modifications underscored the ambiguity of the "official" patronage message of the frontispiece.

---

[46]Raimondi, "Le lettere di P. A. Mattioli," p. 22 (Goritia, 20 May 1554).

[47]BUB, *Aldrovandi*, ms. 6, Vol. II, c. 57r; ms. 34, Vol. I.

[48]Davis, "Beyond the Market," pp. 69–70.

[49]As Aldrovandi wrote to the Florentine secretary of state, Belisario Vinta: "ho voluto mandarne uno a Vostra Signoria con l'epigrammo incluso quale desidero ch'ella faccia legare dirimpetto al frontispicio del libro per maggior segno dell'osservanza mia verso lei, et che l'uno e l'altro le serva per tener memoria di me"; Tosi, *Ulisse Aldrovandi e la Toscana*, p. 389 (Bologna, 6 April 1599). No letter exists with similar instructions for Ferdinando I, but the Biblioteca Universitaria in Pisa holds the Grand Duke's copy with Aldrovandi's epigram (also reproduced by Tosi): "Quamvis magna sciam magnos Dux Magne decere/Hunc tibi mitto tamen, munera parva, librum./Et te quem veneror semper, venerabor et unum,/Oro velis aequo suscipere illum animo:/Susceptumque leges: neque enim tunc tempseris, ut qui/Principibus cellis plurima digna tenet."

[50]Ambr., G. 186 inf. (118) n.p. (Bologna, 11 October 1600).

Aldrovandi canonized Clement VIII as an exemplary patron in print, but he undermined the pope's exclusive hold upon the title of Alexander in the manner in which he presented his work. The identity of Alexander, Aldrovandi suggested, was a mutable as his own image in the role of a client.

With the publication of the first volume of the *Ornithology* in 1599, almost forty years after the appearance of his treatise on the antiquities of Rome, Aldrovandi thanked his patrons by presenting them with copies of the book. During the mid-1590s, as the *Ornithology* neared completion, he scribbled several lists of patrons and friends in his notebooks. These individuals would receive a copy from the author upon publication. In the initial list of 1595, Aldrovandi included Paleotti and Gaetano at the top of the list, just behind the Grand Duke, the Pope, and the current legate Montalto. By 1598, a year before the book was published, the list had shifted slightly. Untimely deaths, the continuous circulation of appointed officials within the papal state, and the providential appearance of new patrons altered its composition. Clement VIII now headed the list, followed by his cardinal nephews, Montalto, the Habsburg Emperor Rudolf II, and the heads of all the Northern Italian territorial states, including the Medici, Farnese, and Della Rovere princes, the Republic of Venice, the Senate of Bologna, and the president of the Senate of Milan. Aldrovandi did not neglect to note the type of edition to present. All heads of state received bound folio editions. Cardinals such as Gaetano and Borromeo, as well as the archbishop and the vice-legate of Bologna, were presented with smaller versions, while senators, relatives, and friends such as Mercuriale, Camerarius, Alpino, and his assistant Uterwer received loose, unpaginated copies.[51] The physical appearance of the book denoted the status of the recipient. Patrons received the largest, most luxurious editions, essentially works of art, while associates contented themselves with mere printed words.

In writing up the initial agreement with his printer in 1594, Aldrovandi had allowed a special provision for "one run of twenty-five books for each work that will be printed. They will be made only so that the abovementioned signor Aldrovandi can give them to Princes, Cardinals, and others to whom he is most obliged, for being favored by them in the acquisition of various and diverse things for the Museum."[52] The limited run was even printed on special paper to further set it apart from the ordinary appearance of the book. Aldrovandi presented these special editions to important patrons such as Alfonso II d'Este, Duke of Ferrara and Modena, "particularly for the singular devotion that I have held and hold toward the Most Serene House of Este."[53] Aldrovandi echoed almost the same words in the

---

[51]BUB, *Aldrovandi*, ms. 136, XXV, cc. 84v–85r; ms. 136, Vol. XXVII, cc. 64v–65r, 214v–215r. See also the list for donations of the second volume, ms. 136, Vol. XXX, c. 304.

[52]In Sorbelli, "Contributo alla bibliografia delle opere di Ulisse Aldrovandi," in *Intorno alla vita e alle opere d'Ulisse Aldrovandi*, pp. 73–74.

[53]BEst., *Est. It.* 833 (Alpha G, I, 15) (Bologna, 2 May 1599).

letter accompanying his gift to the Duke of Parma and Piacenza: "Great and
signal are the favors I and others in my family always have received from the
Most Serene House of Your Highness," he assured the Farnese duke in 1599:

> I humbly beg that Your Highness deign to appreciate this devoted demon-
> stration of my feelings, as weak as it may be, as your brother the Most Illustri-
> ous Cardinal has already done. Having presented the same book to him when
> he passed through Bologna, he was kind enough to want to see my Museum,
> treating me as the Most Serene House of Farnese does with its servants on
> every occasion, that is, gratifying them.[54]

Thus, the presentation of a book provided an opportunity to reaffirm the
commitments between patron and client. The delivery of a book was a sign
of gratitude on the part of the naturalist, and its acceptance a token of the
continued goodwill of his patrons. Even the recalcitrant Ferdinando I felt
obliged to respond favorably to the gift of a book. "At certain hours, when
I can steal the time, I slip away to read some part, as a pastime, and therefore I most dearly welcome the gift that you sent me of it," he wrote to
Aldrovandi.[55]

While continuing to pursue the most visible patrons, Aldrovandi did not
neglect the local officials who facilitated his contact with the pope and many
other members of the papal court. He dedicated the second and third volumes of the *Ornithology* to Alessandro Peretti, cardinal Montalto, who had
been a papal legate in Bologna.[56] Despite his effusive letters to the d'Este
and Farnese dukes, Aldrovandi chose to dedicate his next publication, *On
Animal Insects* (1602), to Francesco Maria II della Rovere, Duke of Urbino:

> I send Your Most Serene Highness my History of Insects, dedicated by me to
> your most glorious name, thus to satisfy some part of the great obligations that
> I owe you and the desire that I have had for many years to leave some sign of
> my infinite devotion toward you behind me.[57]

The last volume that Aldrovandi prepared before his death in 1605 (further
edited by his assistant Uterwer) appeared posthumously with a dedication
written by his wife, Francesca Fontana.[58] Appropriately, she dedicated *On the*

[54]Ronchini, "Ulisse Aldrovandi e i Farnese," p. 14 (Bologna, 4 May 1599).

[55]BUB, *Aldrovandi*, ms. 136, Vol. XXXI, c. 238r (Villa Ferdinanda, 16 October 1603). See also
Tosi, *Ulisse Aldrovandi e la Toscana*, pp. 394, 398.

[56]Other contributors to defraying of the cost of Aldrovandi's publications included Francesco
and Ferdinando de' Medici, Francesco Maria II della Rovere, Gabriele Paleotti, Giovan Battista
Campeggi, Giovan Vincenzo Pinelli, and Aldrovandi's brother Teseo; Adversi, "Ulisse Aldrovandi bibliofilo, bibliografo e bibliologo del Cinquecento," pp. 96–97. Campeggi, the Archbishop of Majorica, contributed a single gift of 1000 *scudi* to the publishing costs; Frati, "Le edizioni delle opere di Ulisse Aldrovandi," p. 162.

[57]Aldrovandi, *De animalibus insectis libri septum* (Bologna, 1602); Mattirolo, p. 392 (Bologna, 4
September 1602).

[58]Aldrovandi, *De reliquis animalibus exanguibus libri quatuor, post mortem eius editi nempe de mollibus, crustaceis, testaceis, et zoophytis* (Bologna, 1606).

Remains of Bloodless Animals (1606) to the Senate of Bologna, which by then had committed itself to publishing the remainder of his treatises as a condition of the donation of his museum to the city. "Many Signors can attest to the utility that these Histories will bring to the world," wrote Camillo Paleotti in a letter supporting Aldrovandi's petition to the Senate to underwrite the cost of his publications, "particularly some of the most Illustrious and Reverend men who have seen his Museum, such as Paleotti, Gaetano, Sforza, Valieri, Borromeo, Ascolano, Sfondiato, Sega, and many others."[59] Aldrovandi had begun his search for a patron by going outside of his native city, but ultimately it was in Bologna that he found his most lasting support.

POWERFUL CONNECTIONS

To achieve his position as the preeminent naturalist in late Renaissance Italy, Aldrovandi depended on a wide range of contacts scattered throughout the courts and republics. While patrons sustained clients with their favors, brokers facilitated the interactions among different parties. Courtly and patrician culture operated on the principle of indirect access; scientific culture mirrored these strategies. If a prince made himself too readily available to his subjects, he diminished his magnificence. Distance as well as intimacy defined one's power, a lesson Aldrovandi absorbed in establishing his own identity.[60] The inaccessibility of a ruler only heightened the sense of accomplishment that aspiring clients felt when they established direct contact, for this signaled their entry into a privileged inner circle and allowed unmediated access to the center of power. Such intimacy, as we have seen in the case of Aldrovandi, was achieved only after careful negotiation. To attract the attention of a prince, one had to befriend his brokers. They provided the gift of access. Brokers occupied a central place in patronage negotiations, guiding clients in their ascent of the social and political ladder.[61] They represented both the patron and the client *in absentia*. Quite often, their intervention determined the outcome of an exchange. As the presence of brokers reminds us, no single act of patronage concluded a relationship; the cultivation of relationships over weeks, months, and years determined the level of a client's success with his patron.

The brokerage system of early modern Italy was a curious mixture of frequent avowals of friendship and a calculated measurement of the precise social distance between patrons, brokers, and clients. Mediated relations

---

[59]AI, *Fondo Paleotti*, ms. 59. F30 (30/2), c. 2r (*Memoriale del Dottore Aldrovandi*, n.d., ca. 1596).

[60]Elias, *The Court Society*.

[61]To date, there have been few studies that discuss the role of the broker historically. See Biagioli, "Galileo's System of Patronage"; Findlen, "The Limits of Civility and the Ends of Science" (unpub. paper); and Kettering, *Patrons, Brokers and Clients in Seventeenth-Century France* (Oxford, 1986).

were intrinsically preferable because they involved an allegedly disinterested party, someone known and trusted by the people on both ends of the exchange whose presence clarified the context of the encounter. The position of the broker was not confined simply to one stratum of the social scale; while elite culture formalized and articulated most clearly the process of mediation, it was central to all aspects of early modern society. Highly placed patricians within the Italian courts acted unofficially in this capacity, while papal legates and ambassadors negotiated exchanges as part of their professional duties. For example, Ghini utilized the Grand Duke's diplomatic courier and the Florentine ambassador in Venice to acquire plants from Anguillara in Padua, thereby reminding the latter that this was an affair of state and not simply an exchange between friends.[62] At court, individuals with unique proximity to the ruler, particularly court physicians and secretaries of state, enjoyed the mutual confidence of patrons and clients. Their knowledge of a prince's medical and political "secrets" made them powerful in the eyes of contemporaries, yet they were educated men who spoke the same humanist language used by naturalists who attempted to gain princely patronage.

Papal physicians such as Bacci, Cesalpino, and Mercati and court physicians such as Pancio in Ferrara and Mercuriale and Redi in Florence enjoyed high status within the community of naturalists. Their interest in natural history contributed directly to its arrival in the Italian courts, and they frequently used their positions to elevate friends such as Aldrovandi who worked in a city without a court. Men such as the Medici secretary of state, Belisario Vinta, listened to the pleas of scholars and passed on their requests to the Grand Duke when it suited him.[63] The ability of Aldrovandi to establish contact with such highly placed courtiers was a measure of his own visibility in this status-conscious environment. Not only did he offer a service attractive to early modern rulers—the glorification of their reign through the mastery of nature—but he was careful to utilize the appropriate channels when he made his requests, demonstrating his own mastery of court protocol.

Even outside of the court, patrician scholars preferred to initiate contact with one another via a known intermediary. Scholars such as Ghini, Aldrovandi, Pinelli, Peiresc, and Kircher, who devoted themselves to creating and maintaining the humanist intellectual community, rose to great prominence. They connected scholars with common interests, introduced promising students to men of great reputation, and mediated disputes over

---

[62]A. Sabbatani, "Il Ghini e l'Anguillara negli orti di Pisa e di Padova," *Rivista di storia della scienze mediche e naturali*, n. 11–12, ser. 3 (1923): 308.

[63]Vinta is discussed in Biagioli, "Galileo's System of Patronage," passim, and his correspondence with Aldrovandi appears in Tosi, *Ulisse Aldrovandi e la Toscana*, pp. 274–275, 368–379, 383, 388–389, 399–402.

intellectual property. Through the cycle of introductions, negotiations, and intrigues that surrounded the work of collecting nature, brokers enhanced their own visibility by presenting themselves as the men who bound the different parts of elite society together. Most importantly, they forged a sense of continuity between one generation of scholars and the next. When Pinelli died in 1601, Peiresc, still in Italy, wrote immediately to his correspondents, promising "to be a servant to all his friends."[64] This impulse followed the same guidelines as that which guided Aldrovandi in his assumption of Ghini's role upon the death of his mentor in 1556. Through the natural succession of brokers, the early modern scholarly community periodically regenerated itself without destroying the cultural inheritance of the previous generation. The durability of this framework and its appropriateness for the political culture of the age kept the patronage networks of the learned world virtually intact for almost 200 years.

Aldrovandi's use of brokers and his assumption of this role on the behalf of other scholars illuminates the mechanisms of patronage. As a place of conversation and exchange, museums were ideally suited to this form of interaction. Brokers provided all the tools necessary to form a successful museum—gifts, visitors, and information. Through their networks of communication, they expanded the reach of the collector beyond his local environment, increasing his presence in the republic of letters and the courts by promoting his name. Frequently, Aldrovandi asked his most promising protégés to participate in this form of advertisement. Traveling throughout Europe, they carried his letters and represented his interests in person. When Evert van Vorsten returned to the Dutch Republic in the 1590s, for example, he promptly delivered a letter to Clusius in Leiden and inspected Paludanus's museum in Enkhuizen to report on anything there that might interest his mentor.[65] Similarly, Camerarius acted as a courier between different naturalists during his travels in Italy. "I received your letter from the hand of messer Joachim Camerarius," wrote Maranta to Aldrovandi, "who certainly is a rather learned, decent young man and truly demonstrates that he had been your disciple."[66] While the use of younger scholars such as Vorsten and Camerarius to deliver letters certainly did not enhance Aldrovandi's social standing, it projected his intellectual presence to every corner of the learned world. Aldrovandi frequently patronized younger scholars to secure future favors from them once they took their place in the academic community or at court; in the meantime, they facilitated his

[64]In Rizza, *Peiresc e l'Italia*, p. 19.

[65]BUB, *Aldrovandi*, ms. 136, vol. XXV, c. 133 (Dordrecht, 19 July 1596).

[66]Vallieri, "Le 22 lettere di Bartolomeo Maranta all'Aldrovandi," p. 767 (Naples, 4 March 1562). Giacomo Antonio Cortuseo, prefect of the botanical garden in Padua, and Lodovico Maietano in Venice also received letters via Camerarius; BUB, *Aldrovandi*, ms. 382, Vol. I, c. 238r (Padua, 14 August 1562), c. 267 (Venice, 23 September 1562).

continued dialogue with the community of naturalists and gathered information to sate his encyclopedic curiosity.

The news that poured into Aldrovandi's *studio* via his emissaries alerted him of the appearance of new collectors worthy of his acquaintance. He approached them with all the caution of a man conditioned by court rituals and gentlemanly protocols. "I hear that there is a doctor in the city of Rotterdam in Holland who amuses himself infinitely with natural things and has collected a forest of them," wrote Guglielmo Mascarelli, "and who has the most rare and precious things from the Indies brought to him daily. I would like Your Most Excellent Signor, by means of Signor Vorsten or some other Dutchman, to try to befriend him (*fare amicizia*) in order to obtain something rare from him to augment your *studio*."[67] Aldrovandi's use of his student as a broker mirrored the process by which princes established that his museum was worthy enough to command their attention. For example, the Duke of Savoy did not immediately visit Aldrovandi's museum in person. Instead, he sent an envoy, Pompeo Viziano, who duly posted a complete description of the *studio* as it appeared in 1604 to the Duke, "who wished to be informed of the quality of this Museum."[68] This visit was probably one of the last of numerous ceremonial entries that courtiers or princely "agents" arranged to Aldrovandi's museum. In their use of intermediaries, both Aldrovandi and the Duke of Savoy shared a common concern about honor. Were they to initiate a relationship with someone whose reputation was unworthy of their own, they would diminish themselves in the process.

Early in his career, Aldrovandi succeeded in gaining the confidence of Ghini and Mattioli, the most prominent naturalists in mid-sixteenth century Italy. Almost immediately this established his reputation as a broker who wielded considerable influence. Friends such as Calzolari paid their respect to Ghini—"your patron and mine"—through Aldrovandi. By the mid-1550s, many scholars considered Aldrovandi to be the most important broker within the community of naturalists. "I do not know nor could I ever know another friend and great patron who brings me more affection and love than you," wrote Wieland in 1555.[69] When Calzolari wished to meet Mattioli, he begged Aldrovandi to arrange the introduction. "This will be one of the greatest favors that you can do for me," he explained, "because I then will have the means to write to him often and learn a few things." So anxious was Calzolari about the whole affair that he repeated his request a few days later. "I believe that I have written you in another letter how dear it would be, through your efforts, to make the slightest acquaintance with the magnificent Messer Pietro Andrea Mattioli."[70]

[67]BUB, *Aldrovandi*, ms. 136, Vol. XXV, c. 109r (Cologne, 3 November 1596).

[68]BCAB, B. 164, c. 302r.

[69]Fantuzzi, p. 100; De Toni, *Spigolature aldrovandiane XI*, p. 12 (Padua, 4 January 1555).

[70]Cermenati, "Francesco Calzolari," p. 95 (20 September 1554); p. 100 (23 September 1554).

With the death of Ghini in 1556, Aldrovandi formally inherited the role of scientific broker. Almost immediately he found himself mediating an acrimonious dispute between Maranta and Mattioli regarding the inheritance of Ghini's papers and herbarium. Understandably reluctant to involve himself, Aldrovandi nonetheless was persuaded by friends that it was his duty to intervene "as a neutral party, friend of the two and the most knowledgeable in the matter."[71] While Aldrovandi continued to mediate disputes throughout his career, he also found himself attracting clients, drawn to him for his learning as well as his fame. The nephews of both Calzolari and Mattioli appeared on his doorstep, hoping their family name would gain them some favor. Paolo Calzolari even went so far as to compose a poem expressing his devotion to Aldrovandi:

> Of Aldrovandi I too
> Hope to be the servant, as my Uncle was,
> And if to serve him now I am deprived
> I will not hesitate to love him as long as I'm alive.[72]

No doubt Aldrovandi must have appreciated the sentiment even if he found the poetry abominable.

Engaged in the process of helping younger scholars and less well-placed associates, Aldrovandi did not neglect to advance his own interests. His success as a collector lay in his ability to secure the services of important brokers who generated new forms of support for his projects. One of Aldrovandi's most profitable encounters was with the humanist Pinelli. Originally from Naples, Pinelli spent most of his time between there and Padua, where he held informal academies in his *studio,* attended by luminaries such as Aldrovandi, Galileo, and Peiresc. Prior to settling in Padua, he had created a botanical garden in Naples that attracted naturalists such as Maranta and Imperato.[73] A tireless collector, he traveled throughout Europe in search of books, manuscripts, *naturalia,* and antiquities, bearing letters to and from scholars along the way. Aldrovandi first met Pinelli in person in 1558 when he stopped in Bologna to see the museum.[74] He had corresponded with him for at least two years prior to the encounter. In October, 1556, Luigi dal Leone introduced himself to Aldrovandi as a protégé of Ghini and Pinelli, "my signors and patrons."[75]

Due to his frequent travels, Pinelli became the axis through which many naturalists came to know one another. Imperato aptly called him the "patron of scholars."[76] When the Neapolitan apothecary wished to meet

---

[71]Fantuzzi, pp. 180–187, 227–228; *Spigolature aldrovandiane XVIII,* pp. 304–305 (Padua, 18 November 1561).

[72]BUB, *Aldrovandi,* ms. 136, Vol. XXVII, c. 8r. The poem was written in 1599.

[73]De Toni, *Spigolature aldrovandiane XVIII,* pp. 297–299.

[74]BUB, *Aldrovandi,* ms. 38(2), Vol. I, c. 91 (Naples, 18 September 1558).

[75]Ibid., c. 220 (Naples, 3 October 1556).

Aldrovandi, he asked Pinelli to negotiate the introduction. Writing to Aldrovandi in 1572, Pinelli described Imperato as "your ardent [admirer] and in whose name I donated several little things, together with the book *On Theriac,* here to you." A year later Pinelli reminded Aldrovandi of Imperato's desire to initiate correspondence:

> Now this Messer Ferrante is so enamored of your virtues that he desires nothing other than to serve you; and you may be assured that you cannot give him a greater present than to command him. . . . I beg you to write him a letter conforming with his appetite in which, besides praising the great knowledge that he has of *materia medica* and the great collection that he has made of singular things, you invite him to follow up, offering this warmly to him, because I promise you that you will acquire a good friend.

Aldrovandi's continued silence led the exasperated Pinelli to remind the naturalist of the obligations he owed to Pinelli for his successful mediation of Aldrovandi's own affairs. "If you ever desire to express your gratitude to me (*farmi cosa grata*), send me your response to Imperato, in which I remind you to praise him a little, for no other reason than to make him continue the enterprise of simples more boldly."[77] Without Pinelli's help, Aldrovandi and most naturalists in Italy would have found it difficult to communicate with their Northern European colleagues, for example, Jacob Zwinger in Basel, Clusius in Leiden, and Camerarius in Nuremberg, let alone with one another.[78] Needless to say, Aldrovandi soon initiated a correspondence with Imperato.

While Pinelli's support strengthened Aldrovandi's position in the republic of letters, as Peiresc did for Kircher in the following century, other brokers increased his contact with various princes. Their advocacy of Aldrovandi reinforced the fashionability of natural history as a courtly pastime for the political elite. In Tuscany, the friendship of Mercuriale was crucial to his success with the Medici. We have already encountered the court physician's cautionary advice about the limits of the Grand Duke's generosity. Despite these caveats, Mercuriale presented Aldrovandi's requests to Ferdinando I. In a letter written to Ferdinando I in June 1591, Aldrovandi requested that Mercuriale express his devotion (*far riverenzia*) to the Grand Duke as a prelude to asking for some items for his museum. He continued,

> And now I have entered into the hope that, since Signor Mercuriale was the reason that I knew of these beautiful and foreign animals, he might also be the cause of my receiving a copy of each of them, for the infinite graciousness of Your Highness and for *the value of the mediator.*

[76]Imperato, *Dell'historia naturale,* n.p.

[77]De Toni, *Spigolature aldrovandiane XVIII,* p. 308 (Padua, 25 August 1572); p. 311 (Padua, 21 May 1573); p. 312 (Padua, 6 August 1573).

[78]Migliorato-Garavani, "Appunti di storia della scienza nel Seicento," pp. 36–37; De Toni, "Il Carteggio degli Italiani col botanico Carlo Clusio," pp. 154, 170.

Several months later, Aldrovandi thanked the Grand Duke for the seven plants sent "by means of the excellent Signor Mercuriale."[79] While Mercuriale could not wrench any significant subsidies out of Ferdinando I for Aldrovandi's *Ornithology*, his status at court made it possible for him to extract precious objects from the Medici collections. When Aldrovandi wished to present copies of his publications to the Grand Duke, he asked Mercuriale to carry out this favor. The court physician not only delivered the books but reported on Ferdinando I's favorable reaction.[80]

Other Florentine court brokers confirmed Mercuriale's reports on the vicissitudes of princely patronage in Florence. Brokers such as Aldrovandi's nephew Giuliano Griffoni, Bolognese ambassador to Rome, and the archbishop Paleotti, delivered Aldrovandi's greetings to the Grand Duke, returning with whatever court gossip they had overheard.[81] During the reign of Francesco I, friends such as the papal *nuncio* Bolognetti, Casabona, and the court humanist Lorenzo Giacomini kept him abreast of events in Tuscany. Ever alert for new opportunities to insinuate himself with his brokers, Aldrovandi rushed to congratulate Bolognetti upon his appointment as *nuncio* in 1576:

> And I certainly cannot show you any gratitude at the moment other than to thank you always for the good will that you hold toward me. Without my requesting it, you wish to bring me to the attention of the most Serene Grand Duke, to whom I find myself infinitely obliged, knowing that mention has been made of me by other signors to His Highness, among whom I know that Signor Pietro Antonio Bardo remembered me to His Most Serene Signor after having seen my Museum.

A year later, Bolognetti repaid the naturalist's extravagant rhetoric by securing a personal introduction to Francesco I during a visit to Florence.[82] Patrons, clients, and brokers did not interact out of simple obligation; rather it was a complex interweaving of calculated risk, genuine affection, and social repayment that determined their decision to act.

As the court botanist, Casabona was also an important figure for Aldrovandi to cultivate. When he visited Bologna in 1583, Aldrovandi personally gave him a tour of his collections. "I showed him my garden and the public one," he wrote to Casabona's employer Francesco I, "and *made him the patron* of everything that you find here."[83] Patronage did not simply describe a relationship of mutual interest and benefaction, but entailed a symbolic possession of another person's cultural goods. When Aldrovandi made Casabona the patron of his museum, he acknowledged the botanist's status

[79]Mattirolo, p. 386 (19 June and 27 November 1591).

[80]Tosi, *Ulisse Aldrovandi e la Toscana*, p. 427 (Pisa, 22 April 1599). Casabona performed the same favor for Camerarius; Olmi, "Molti amici in varii luoghi," p. 17.

[81]Mattirolo, pp. 364, 368, 370, 377, 384.

[82]BUB, Cod. 596-EE, n. 1 (Bologna, 9 April 1576), f. 2r; Aldrovandi, *Vita*, p. 27.

[83]In De Rosa, "Ulisse Aldrovandi e la Toscana," p. 213 (Bologna, 1 September 1583).

as the official representative of the Grand Duke. By showing Casabona great hospitality, treating him as if he were Francesco I, he conveyed his willingness to serve the Grand Duke whose favorites were all "patrons" of Aldrovandi. Brokers such as Casabona derived their status from the princes whom they served. Despite the fact that Casabona was technically Aldrovandi's inferior—a younger, less educated, and less famous naturalist—he nonetheless enjoyed superior social status as a courtier.

Similarly, the humanist scholars employed at court enjoyed greater prestige than Aldrovandi whose primary employer was the university. Planning his second trip to Florence, Aldrovandi consulted Lorenzo Giacomini to inform himself better about the movements of the Grand Duke:

> Still it would be dear to me if you were to advise me on whether or not His Highness will find himself in Florence and when that may be; because I am almost thinking of coming there, during the vacation at the end of May or shortly before, and kissing the most Serene hands, visiting friends and patrons together, and especially in order to see the infinite rare things that His Most Serene Highness writes to me that he has gathered, and to embellish my histories with the mention of them.

When Francesco I died in 1587, Aldrovandi wrote Giacomini—assigned the task of writing the funeral oration—to express his grief of the loss of their mutual patron and test out strategies for winning the ear of the new Grand Duke. Moreover, he asked the courtier if he would "deign to do me the favor of presenting the letter in the hands of the Your Most Serene Highness, as a kindness, and also accompany it with some words."[84] With each important transaction—the visits of 1577 and 1586 and the publication of his *Ornithology* in 1599—Aldrovandi utilized a different broker to plead his case before the Grand Duke. This rotation reflected the mutability of the court hierarchy itself, in which the list of favorites was always changing. For his message to be successful, Aldrovandi needed to inform himself about the fluctuating reputations of his various brokers. As his representatives, their status reflected upon him.

Aldrovandi's ability to read the political climate of the court was also tested in Rome. Mastering the complexities of papal patronage proved even more time consuming and frustrating than his dealings with the Medici. Relatives such as his brother Teseo, well respected by many cardinals, and his nephew Griffoni guided him through the maze of the papal court. Like

[84]Ricc., Cod. 2438, pt. I, lett. 89r (Bologna, 29 April 1586); lett. 94 (Bologna, 2 November 1587). Aldrovandi did take this trip to Florence between 13 and 22 June 1586, visiting the museum of Casabona, the *studio* of the court painter Jacopo Ligozzi, the Grand Duke's botanical gardens and grotto, and many private collections; BUB, *Aldrovandi*, ms. 136, Vol. XI, cc. 32–66. For a few details on Giacomini's role in the literary production at court, see Berti, *Il Principe dello studiolo*, pp. 44, 58, 204, 223 (n. 6). Giacomini also kept Giganti informed of the goings on at court.

many of the secular courts, the papacy also began to promote natural history during the mid-sixteenth century. The choice of Bacci, Cesalpino, and Mercati as papal physicians, all prominent naturalists, bears out this trajectory. Even before the appointment of these friends to the position of *archiatro,* correspondents excitedly informed Aldrovandi of the possibilities that awaited him in Rome. When Antonio Compagnoni visited Rome in 1563, he discovered that natural history had become the rage among courtiers, facilitated by Mattioli's translations of Dioscorides. "Today it is rumored in the Pope's palace by men that amuse themselves with minerals, plants, and every sort of animal . . . that the Mattioli is rare . . . as I discussed with you," he informed Aldrovandi. Compagnoni took advantage of this happy circumstance to promote his friend's cause with the Pope:

> I told the one in charge of creating a *loggia* where the entire cosmography will be, who wishes to decorate several panels above the cosmographies that you could give him large naturalistic illustrations of plants and birds when I told him about your *studio.* He begged me to get a plant and a bird from you naturally depicted, with the name of each, so that he could show them to Cardinal Amulio who will mention it to His Holiness if he likes them. Amulio is in charge of decorating this *loggia,* which will be the most beautiful in the world, and from this will arise the opportunity for you to come to the attention of His Holiness.[85]

We may never know whether Aldrovandi responded to this request with one of his famous illustrations. Yet this was the first of many opportunities that friends and patrons created to bring his work to the attention of the pope.

During the 1560–1570s, his brother Teseo, Griffoni, and friends such as Mercati worked to establish Aldrovandi as a naturalist of repute in Rome. Teseo introduced Aldrovandi to the Roman naturalist Ippolito Salviani and to Fulvio Orsini, the Farnese librarian and antiquarian. When Aldrovandi visited Rome in 1577, he had lunch in Orsini's garden, where they discussed interesting bits of nature and confirmed an agreement to make Orsini the godfather of Aldrovandi's children.[86] Teseo must also have introduced Aldrovandi to Guglielmo Sirleto, prefect of the Vatican library. As custodian of its immense collection of books and manuscripts, Sirleto had his finger directly on the pulse of the papacy's cultural activities. After returning to Bologna, Aldrovandi composed a lengthy letter to the cardinal librarian to enlist him as a papal broker:

> I need a great Prince and Patron to give me an honorary stipend, not for personal use but for universal benefit, so that I could have many writers, copyists,

[85]BUB, *Aldrovandi,* ms. 38(2), Vol. I, c. 229 (Rome, 17 April 1563).
[86]Ibid., vol. IV, c. 66 (Rome, 30 May 1573); vol. II, cc. 1–4 (Rome, 1557–1560); Ambr., ms. S.80 sup., f. 260r, n.d. Mercati also knew Teseo, and letters to and from Aldrovandi were occasionally directed through him.

sculptors, and painters to bring forth my many honored labors so much sooner. I cannot imagine what greater benefit my most Reverend Monsignor could do for the world than to help me print my works.

Aware of the magnitude of his request, Aldrovandi concluded his letter with a strong rhetorical flourish to remind Sirleto that he understood the power he wielded by virtue of his access to the Pope. "I beg Your Most Illustrious Signor to forgive me if I transgressed the bounds of good writing (*se ho passato il segno del scrivere*), because I was invited [to do so] by your most affectionate offer made to me when I was in Rome."[87] In Sirleto, he recognized a broker potentially as powerful as his patron Francesco I who also received a version of the same letter in 1577. By appealing to Sirleto's sensibilities as a man of learning who understood the difficulties of scholarly work, Aldrovandi hoped to acquire a broker powerful enough to persuade the pope to finance his publications.

Aldrovandi's nephew Griffoni expanded the ecclesiastic connections initiated by Teseo. He dutifully brought Aldrovandi's requests to Mercati and delivered his uncle's treatises to ecclesiastic patrons such as Cardinal San Sisto, Enrico Gaetano, and Federico Borromeo. "I received your letter with the memo of the things that you want from Rome," responded Griffoni in 1567. His primary mission was to report any favorable signs of ecclesiastic patronage to his uncle. "His Holiness Signor Cardinal San Sisto asked me about you," he reported in 1573.[88] Griffoni's diplomatic responsibilities made him a logical successor to Pinelli as a courier to many cities. While his primary base was Rome, he also traveled as far north as Milan where he delivered letters from his uncle to Federico Borromeo. "I understand from my nephew, Signor Giuliano Griffoni, how much Your Most Illustrious Signor has deigned to exert himself and assert his authority with [the Pope] in my affairs, out of your innate generosity," wrote Aldrovandi to Borromeo in 1601, "which truly would never turn out well without you." In the same letter, however, Aldrovandi also indicated some dissatisfaction with his nephew's representation of his interests. "And I know very well . . . that the demands my nephew writes to me that are about to be fulfilled concern his interests more than mine."[89] Instead, he implied that Borromeo, a great personage who had no need to scramble for patronage, would be a more impartial broker.

Despite his qualms about Griffoni, Aldrovandi nonetheless gave him the honor of presenting his *Ornithology* to Gaetano. Legates such as Gaetano and

[87]BAV, *Vat. lat.*, ms. 6192, vol. 2, ff. 656v–657r (Bologna, 23 July 1577). On Sirleto, see Irena Backus and Benoit Gain, "Le Cardinal Guglielmo Sirleto (1514–1585), sa bibliothèque e ses traductions de Saint Basile," *Mélanges de l'école française de Rome* 98 (1986).

[88]Mercati, "Lettere di scienziati dell'Archivio Segreto Vaticano," pp. 67–68 (Bologna, 31 March 1599); BUB, *Aldrovandi,* ms. 38(2), Vol. IV, c. 342r (Rome, 29 October 1567), c. 349 (Rome, 13 May 1573).

[89]Ambr., G.188 inf. (233) (Bologna, 17 February 1601).

his successor Montalto were among the highest ranking dignitaries in the papal bureaucracy. The information that they provided to Rome about the territories under their jurisdiction assuredly reached the ear of the Pope himself. As part of their reports on Bologna, both brought Aldrovandi's studies to the attention of various popes and were instrumental in securing the publication of his manuscripts. With the copy of the *Ornithology*, Griffoni presented the following letter:

> The accumulation of favors done for me at various times by Your Most Illustrious Signor has so assured me of your good will toward me, that I do not doubt that you are about to receive my *Ornithology* with a smiling face. . . .This book of mine is a fruit brought forth with the help that Your Most Illustrious Signor already obtained for me from Sixtus V of happy memory.[90]

Like Pinelli and Mercuriale, Gaetano was one of Aldrovandi's most powerful brokers. He provided a means of approaching the Pope that the naturalist could never hope to achieve on his own, not even when his cousin wore the papal tiara.

While brokers in Rome smoothed the way for Aldrovandi's pleas of poverty, using their influence to extract funds from the papal treasury, other friends encouraged him to use his museum as a tool of persuasion. Prior to Clement VIII's arrival in Bologna in 1598, Bernardo Castelletti counseled him on the most appropriate way to approach the pope with his request:

> But meanwhile I cannot neglect to tell you that giving a brief guide to the things in Bologna to His Holiness, where certainly many of these Prelates and Cardinals still come, they will visit your Museum; and consequently the Pope ought to be well informed of the beautiful collection of such a variety of natural things and [the fact] that you have written the History of all of them with the intention of bringing it to light.[91]

A year later, the *Ornithology* appeared. Aldrovandi publicly expressed his debt to the pope and the other patrons who had underwritten its expense, but he did not neglect to acknowledge the importance of the brokers who truly had orchestrated the event.

Certainly Aldrovandi's reliance on intermediaries was not unique. Other naturalists utilized the same social mechanisms to communicate their discoveries, broaden their circle of acquaintances, and obtain patronage. When Giovanni Pona wished to obtain plants from Tobias Aldini, custodian of the Farnese botanical garden in Rome, he wrote to Faber asking him to negotiate the exchange.[92] If anything, the brokerage system became further

---

[90]Mercati, "Lettere di scienziati dell'Archivio Segreto Vaticano," pp. 67–68 (Bologna, 31 March 1599).

[91]BUB, *Aldrovandi*, ms. 76 (Genoa, 26 November 1598).

[92]Cortesi, "Alcune lettere inedite di Giovanni Pona," p. 419 (30 December 1618). In 1626 Aldini became the custodian of Francesco Barberini's museum.

articulated in the seventeenth century in conjunction with the emergence
of more aristocratic forms of behavior among the upper classes. Greater emphasis on hierarchy increased the power and prestige of court physicians
and naturalists. For example, Redi wielded significantly more power at the
Baroque Florentine court as the confident of Ferdinando II and Cosimo III
than Mercuriale did during the reign of Ferdinando I. Similarly Cospi's relationship with the Medici, his distant relations, was significantly more formal than Aldrovandi's had been. After an adolescence at the Medici court,
Cospi returned to his native city to become the "agent of the Grand Duke
in Bologna." Besides negotiating artistic transactions and acquisitions for
cardinal Leopoldo and Cosimo III, he and his son-in-law Annibale Ranuzzi
mediated exchanges between Tuscan and Bolognese virtuosi. Montalbani
and his assistant Legati received copies of Redi's publications from the hands
of Ranuzzi, who subsequently delivered their responses to Florence.[93] As the
case of Cospi suggests, collectors were not only the recipients of patronage
but often extended the influence of their patrons beyond the confines of
the court.

In Baroque Rome, Francesco Barberini and his coterie appointed themselves arbiters of taste in the papal city. Save for a brief period of exile after
the death of Urban VIII, they enjoyed a virtual monopoly over the cultural
life of the city.[94] The members of the Accademia dei Lincei would not have
enjoyed such prominence without Barberini patronage, nor could Kircher
have risen so quickly in Roman society without their support. We have already seen how Peiresc, protégé of Pinelli and intimate of Barberini, gave
Kircher his entry into the humanist culture of the papal city. The "commerce of letters" that animated the mid-seventeenth century scholarly community perpetuated the system of communication developed by Renaissance humanists such as Aldrovandi. The dissemination of ideas and the
social elevation of their proponents went hand in hand. Brokers, like collectors, were men of wit and learning. But they were also adept politicians
who saw the growing prominence of collecting and natural history as com-
modities worthy of management. Through their ministrations, the museum
became one of the primary locations in which princes publicized their reputations as patrons of learning and the arts and collectors advanced their careers as courtiers.

OF DRAGONS, BEES, AND EAGLES

While persuading brokers to represent their interests, naturalists did not neglect to create gifts encoded with messages designed to please patrons. Ever

[93]Goldberg, *Patterns in Late Medici Art Patronage*, p. 36; Laur., *Redi* 222, c. 33r (Bologna, 17
December 1667); c. 293r (Bologna, 18 September 1668).

[94]For more on this subject, see Haskell, *Patrons and Painters*.

in search of opportunities to communicate with princes, collectors naturally gravitated toward the study of objects with emblematic value. Galileo's transformation of the satellites of Jupiter into a form of Medici mythology was not a unique occurrence, although it is one of the most spectacular examples of how scientific culture made its place in the political climate of the courts. The ink had hardly dried on the dedication of his *Sidereal Messenger* (1610) when other natural philosophers began to contemplate how to imitate his actions. Peiresc, for example, contemplated reworking the political mythology of the four "moons" to direct its message toward the French monarchy. Accordingly, he proposed renaming two of the satellites Catherine and Marie in honor of the two Medici who had married into the French royal family.[95] Only the discovery that Galileo intended to publish further on this subject (and undoubtedly enhance the dynastic imagery he had created) prevented Peiresc from completing this project. Yet the facility with which he understood the significance of Galileo's dedication suggests how attuned natural philosophers were to these emblematic messages.

Galileo surely owed his success in deploying symbolic imagery to his observation of the efforts of other aspiring courtiers such as Aldrovandi. In fact, his attempts to bring his work to the attention of the Medici bore a remarkable resemblance to Aldrovandi's efforts to gain princely patronage. We have already observed Aldrovandi's skill as an emblematist in the fabrication of his own identity. He also brought these talents to bear in his strategies to secure patronage. To enter the courts, Aldrovandi made himself a master of the natural imagery that surrounded each prince. This allowed him to direct specific portions of his natural history toward different princes. Studying the whole of nature made it possible for him, over time, to accumulate materials on almost every princely emblem used by the Italian ruling families, many of which highlighted natural objects as uniquely appropriate symbols. Armed with this knowledge, Aldrovandi presented his natural history as a gloss on political culture.

Aldrovandi's analysis of the 1572 dragon represented his most significant attempt at creating a scientific emblem. The appearance of the dragon on Bologna, he argued, presaged a fruitful relationship with his chosen patron, Gregory XIII. The publicity surrounding this portent and Aldrovandi's role as its interpreter led him to believe that the pope would favor him with gifts and preferments in gratitude for his neutralization of this potentially subversive object.[96] When this did not occur, he increased his efforts in other directions. The Gonzaga princes in Mantua had a crest of four eagles. Aldrovandi did not neglect to produce a complete history of the significance of these birds in his *Ornithology*, accompanied by a detailed discussion of the

---

[95]Biagioli, "Galileo the Emblem Maker"; Gassendi, *Mirrour of True Nobility and Gentility*, Vol. I, pp. 145–146.

[96]I discuss this episode in greater detail in chapter 1.

Gonzaga emblem. In his histories of quadrupeds, he also praised the legendary Gonzaga horses.[97] Scattering the natural imagery associated with Italy's most powerful families throughout his treatises, Aldrovandi hoped that at least one of these princes would perceive natural history as a novel medium through which to glorify his family name. Unlike Galileo, Aldrovandi never created a natural emblem that became a permanent feature of a prince's image; instead, he articulated preexisting aspects of political symbolism. Yet his attempts to communicate with patrons through similar mechanisms indicate the importance of this form of representation in scientific culture. Princes were not drawn to natural philosophy or natural history for its content alone. Instead, they perceived it as a form of knowledge that magnified their power through its symbolic content.

Other collectors were more successful in manipulating political imagery. In Rome, the Linceans directed their publications to the Barberini family. Stelluti dedicated his treatise on the "wood-fossil-mineral" and his translation of Persius to Francesco Barberini. As he wrote in the introduction to the former, "I wished to dedicate it to Your Eminence both for your usual good taste in contemplating the occult parts of nature and also because of the worthy and eminent testimony that you bring to the forefront [of this work], reassuring others that I do not propose false and fabulous things to them."[98] The association with the Barberini family lent prestige and therefore credibility to the activities of the academicians. With the election of Urban VIII in 1624 and the addition of his brother the cardinal to the academy in that same year, the Linceans increased their efforts to present themselves as clients of the most powerful family in Rome. One year later, after an audience with Urban VIII, Faber reported, "It pleased him that I dedicated it to Signor Cardinal Barberini."[99]

The centerpiece of their efforts was the publication of the *Apiarium*, also dedicated to the cardinal nephew, which was a study of the bees that decorated the Barberini crest. Describing it as an investigation of "urban bees," an image that doubly connected it to the papacy, Cesi and his academicians openly referred to it as a work designed exclusively to advance their position at court, possibly inspired by Galileo's own use of political emblems. "This was made to signify even more our devotion to Patrons and to exercise our specific study of natural observations," Cesi informed Galileo. After ascertaining that Cesi had already presented the treatise to the pope and his nephew, Colonna suggested that he terminate his own study of Neapolitan bees "since this edition was made for no other end than to please the taste of patrons and not to bring to light a particle of his labors."[100] We have no specific sense of the extent to which the Lincean's efforts to turn their

[97]Franchini et al., *La scienza a corte*, pp. 98–99, 106–107.
[98]Stelluti, *Trattato*, p. 3–4.
[99]CL, vol. III, p. 1003 (Rome, 27 July 1625).
[100]CL, vol. III, pp. 1066, 1100 (?, 1625 and Naples, 13 February 1626).

microscopic studies into princely messages resulted in their increased status at the papal court. Unfortunately, the premature death of Cesi in 1630 and Galileo's condemnation in 1633 prevent us from seeing how this relationship might have played out. Yet the Linceans's incorporation of political messages into their scientific work certainly stamped them as naturalists in the process of continuing the patronage strategies initiated by Aldrovandi. They also could not succeed without the support of princes and accordingly made their work attractive to papal court by making the Barberini the "natural" patrons of the new natural history. By associating bees with the novelty of the new experimental philosophy and its instruments, they hoped to secure papal patronage for their academy.

Of all the early modern natural philosophers, Kircher exhibited the most skill at transforming his works into a political hieroglyph. He was the beneficiary of at least a century of successful and partially successful attempts at attracting princely patronage. His Jesuit education gave him formal training in the use of emblems, a skill he demonstrated not only in his allegorical frontispieces and dedications but also in his *Political Archetype of the Christian Prince* (1672), a Habsburg emblem book intended to instruct young princes in the symbols glorifying Catholic rule. Thus, Kircher was better prepared than either Aldrovandi or the Linceans to integrate political language and imagery into the presentation of his work.

Kircher's most successful emblem was the Habsburg eagle. As William Ashworth observes, he and many other Habsburg clients used it repeatedly to signify their allegiance to the Imperial court.[101] Kircher prominently displayed the double-headed eagle on many of his frontispieces, reminding readers of the status of his patrons. For him, the eagle was an eminently malleable object that appeared repeatedly throughout his museum and in his publications. Numerous experiments conducted in the Roman College and placed on permanent display in his museum revolved around the Habsburg eagle. Magic lanterns made it appear mysteriously on the wall of the museum, parabolic mirrors distorted it in every imaginable direction, magnetic toys, sundials, and other technological inventions accommodated it.[102] While other collectors included portraits of princes in the museum, Kircher made the crest of his most important patron a leit motif of the display. His creative use of political emblems helps us to understand why he was one of the most successful clients in Baroque Europe.

Like most rulers, the Habsburgs had no intrinsic commitment to any particular approach to nature. They did not chose their clients primarily for their scientific skills, although one's scholarly reputation laid the groundwork for an introduction at court. Aldrovandi's limited success with Gregory

[101]Ashworth, "The Habsburg Circle," in Moran, *Patronage and Institutions*, pp. 137–168.
[102]See the frontispieces of Kircher, *Magnes sive de arte magnetica* and *Ars magna lucis et umbrae*. Also see the images scattered throughout the pages of the *Ars magna lucis et umbrae* (e.g., p. 364) and the *Oedipus Aegyptiacus* (e.g., vol. III, p. 257).

XIII, the Linceans's courtship of the Barberini, and Kircher's ability to attract the Habsburgs were a relative measure of their skill as emblem makers.[103] Nature offered an infinite array of symbols for her students to grasp, each a potential link between patrons and scholars. The extent to which naturalists could present natural history as a form of courtly discourse was predicated upon their understanding of this principle. Their success in acquiring patrons depended on their ability to make natural history a form of humanist learning that made symbolic forms of knowledge central to the deciphering of nature. The accentuation of these tendencies in the hands of Kircher only reinforces the importance of the political context of his work. Presenting the study of nature as a subject suitable only for princes and their clients, Kircher used his erudition to secure his place in the Habsburg patronage system.

RETURN TO EGYPT

Client of four popes, favorite of two emperors and many German and Italian princes, and broker for the learned world, Kircher may well have been the most successful natural philosopher in early modern Europe. His spectacular rise to prominence in the mid-seventeenth century was due in no small part to his sophisticated understanding of the patronage networks. Visitors from all over Europe came to Rome to see him. He and his museum were considered an integral part of the cultural life of the city. As Lucas Schroeck wrote in 1670, "If, in fact, I leave Rome without greeting or seeing Your Reverence, I would believe that I had seen nothing of Rome."[104] Kircher's numerous publications contributed greatly to his reputation. Read throughout the world, due to the farflung network of Jesuit colleges and missions and his own efforts to distribute them to scholars of note, they advertised the research program displayed in the Roman College museum.

Like other collectors, Kircher recognized the importance of being printed. Books were the most effective way to record the exchanges that occurred in the museum and in correspondence in such a way that patrons and other scholars could appreciate its value. As Kircher realized better than any of his predecessors, books commodified the experience of collecting; they were objects produced from the gathering of other objects, collectibles in their own right. Filled with costly illustrations, multiple dedications, and frequent references to the gifts and patrons that created his museum, Kircher's books were the extravagant effluvia of a patronage culture. They gave material testimony to his success in attracting patrons. While Aldrovandi managed to publish only four of his approximately two hundred treatises in his lifetime, Kircher published almost forty. And this number

---

[103]I borrow this term from Biagioli, "Galileo the Emblem Maker."
[104]PUG, *Kircher*, ms. 559 (V), f. 140 (Rome, 17 October 1670).

does not include the publication of his experiments by disciples such as De Sepi, Kestler, Lana Terzi, Petrucci, and Schott. The scientific community varied widely in its reaction to the content of his work, but its members could not help but admire Kircher's ability to speak the language of princes. However such men as Oldenburg, Boyle, and Huygens might deride his encyclopedic speculations, they could not afford to ignore him. His connections made him a powerful broker within the republic of letters, just as Aldrovandi had been for a previous generation.

The growth of Kircher's museum and the appearance of his publications were parallel manifestations of his patronage strategies. One of the first sights that greeted visitors to the Roman College museum was a tribute to the Emperor Leopold I, followed by the sculpture collection of Donnino and "effigies of various Kings, Princes, Benefactors, and Disciples." Portraits of Urban VIII, Innocent X, Alexander VII, and Clement IX; the Emperors Ferdinand II, Ferdinand III, and Leopold I; and rulers such as the Spanish king Philip IV, Louis XIV, the Grand Duke Ferdinando II, Duke August of Brunswick-Lüneberg, and his son Ferdinand Albrecht filled the halls of the gallery. While the prominence of Donnino's collection was obligatory—his donation initially made the museum possible—the placement of images of various rulers signaled their importance as Kircher's patrons. The image of Leopold I introduced the museum, underscoring the importance of Habsburg patronage. Effigies of Alexander VII, the pope who most favored Kircher as a client, appeared throughout, sometimes in the most unexpected places.[105] The portrait gallery at the entrance made the museum a collective representation of absolutist rule. Louis XIV, who did not patronize Kircher, appeared to enhance the message of power; Duke August, a relatively minor political figure and a Protestant, earned a berth in appreciation for his support of Kircher. The omnipresence of imperial and papal images reminded visitors that the Roman College museum was a political theater in which all the might of nations came together to sanction the universal knowledge it contained.

Kircher's optical exhibits contained not only portraits of Alexander VII but also "the effigies of various Patrons positioned upside down at a certain distance."[106] To visitors, they appeared to float in midair. The whimsical appearance of princely images in the form of technological puzzles, literary emblems, and aesthetic embellishments directly integrated Kircher's patrons into the activities of the museum. He legitimated their power by advertising it in Rome, the center of the Catholic world and birthplace of antiquity in the humanist conception of learning. They implicitly advocated his approach to nature through their financial support of his experimental program. Creating artifacts that made Habsburg eagles "demonstrate"

[105]De Sepi, pp. 2, 6. On the images of Alexander VII, see chapter 1.
[106]Ibid., p. 38.

precepts of Aristotelian natural philosophy, for example, Kircher reinforced the relationships among intellectual, religious, and political orders by making the goals of his patrons his own.

While Kircher's portrait gallery expanded to encompass a succession of popes and emperors, his publications reflected the expanse of his network of princely supporters. His first task upon arriving in Rome was to gain papal support. Kircher's *Coptic Forerunner* (1636), the first fruits of his forays into Egyptian language, was dedicated to Francesco Barberini. To hasten its publication, Peiresc also financed this Coptic vocabulary, offering Kircher his first stipend. However, Peiresc was careful to warn his protégé that Barberini, "who chiefly had promoted his work and whose honor brought you there," should perceive himself to be the sole patron of this greatly anticipated book.[107] Rather than jeopardize his own relationship with the cardinal nephew, whose portrait had a place of honor in his own museum, Peiresc preferred to be an anonymous benefactor. His role, as he indicated repeatedly in his correspondence with another Barberini client, Dal Pozzo, was to facilitate Kircher's relationship with Barberini. One day after he promised Kircher a stipend, Peiresc anxiously wrote to Dal Pozzo encouraging him to "take on the duty of obstetrician" (*farvi officio di obstetrice*) and bring forth Kircher's book.[108] The Jesuit was a conduit for a form of esoteric learning that would bear the Barberini imprimatur, "fathered" by Cardinal Francesco, patron of learning.

Kircher's success with the Barberini established his position in Rome. As a result of Peiresc's and Barberini's intervention, his teaching duties were substantially curtailed so that he could devote himself full-time to scholarly work.[109] His favorable situation in Rome provided him with a base from which to renew his acquaintance with the German scholarly community and bring his work to the attention of the Habsburg emperor, Ferdinand III. Kircher realized this goal by drawing upon his contacts throughout the Holy Roman Empire. By 1640, the Bohemian scholar Marcus Marci of Kronland gained entry to the Imperial court, where he became Ferdinand III's physician and negotiated a stipend for his friend in Rome. Kircher informed the Emperor of his work and sent Ferdinand III gifts such as an engraving of a heliotrope. Through Marci's brokerage, Kircher became a favorite of the Habsburg emperor. The Emperor availed himself of Kircher's linguistic skills when a series of cryptograms from the Swedish commander came into his possession in 1641. Copies were promptly forwarded to Rome for Kircher to decode. Despite the distance between Rome and Vienna,

[107]PUG, *Kircher*, ms. 568, vol. XIV, f. 374 (Aix, 6 September 1634). See Fletcher, "Claude Fabri de Peiresc and the Other French Correspondents of Athanasius Kircher (1602–1680)," *Australian Journal of French Studies* 9 (1972): 260.

[108]Peiresc, *Lettres à Cassiano dal Pozzo*, pp. 146–147 (7 September 1634).

[109]Ibid., p. 254 (31 October 1636).

Kircher maintained close relations with Ferdinand III, his son Leopold I, and many scholars at their courts. Even before the end of the Thirty Years' War, the Emperor attempted to persuade Kircher to travel to Vienna and was willing to allow Marci to accompany Kircher on a trip to Egypt that never materialized.[110] Book dedications played an important role in strengthening these ties.

While Kircher dedicated several works to Italian patrons, particularly Alexander VII, he reserved this honor primarily for the Habsburg family. "Since Father Kircher is greatly obliged to the Imperial Majesty for many titles, thus he does not neglect the opportunity to deliver those signs of devoted observance that are owed to the most munificent Patrons," observed Petrucci.[111] Kircher dedicated his *Loadstone* (1641) and *Egyptian Language Restored* (1643) to Ferdinand III. With the deaths of Peiresc and Urban VIII, papal interest in his studies waned, and he was forced to increase his efforts to attract patrons who would support the publication of his most definitive and costly work on Egypt, the *Egyptian Oedipus*. Throughout the 1640s, Kircher repeatedly found himself frustrated in his efforts to see this work into print. He renewed his attempts to gain Habsburg patronage by dedicating his *Great Art of Light and Shadow* (1646) to the Archduke Ferdinand, the Emperor's son, and his *Universal Musurgia* (1650) to Archduke Leopold Wilhelm, the Emperor's brother. Again with the help of Marci, Kircher persuaded Ferdinand III to bequeath the enormous sum of 3000 *scudi* to pay for the special fonts necessary to publish his *Egyptian Oedipus*. Ferdinand III was so pleased with the two-volume encyclopedia that he awarded Kircher an annuity of 100 *scudi*.[112] Like many authors, Kircher dedicated different sections of the *Egyptian Oedipus* to different patrons. While the Emperor was its principal patron, Leopold Wilhelm, Ferdinand IV, Leopold I, Marci, and most of the prominent scholars of the Holy Roman Empire merited recognition. In content as well as presentation, the work was a collective tribute to the Imperial court.

In the preface of his *Egyptian Language Restored,* Kircher addressed Ferdinand III as the "Trismegistan King." In the *Egyptian Oedipus* he elaborated on this image, praising Ferdinand III, "an admirer of the secret philosophy of Hermes Trismegistus," for his restoration of *prisca sapientia* through his patronage of Kircher. Lauding Ferdinand in twenty-four languages, Kircher called him the "Austrian Osiris."[113] Just as the search for Alexander

---

[110]Fletcher, "Johann Marcus Marci Writes to Athanasius Kircher," *Janus* 59 (1972): 98–101.

[111]Petrucci, *Prodomo apologetico,* p. 18.

[112]Evans, *The Making of the Habsburg Monarchy,* p. 434. The imagery of the *Ars magna lucis et umbrae* frontispiece is discussed in Ashworth, "The Habsburg Circle," in Moran, *Patronage and Institutions,* p. 142. Kircher also dedicated his *Arithmologia* (1665) to Leopold Wilhelm.

[113]Kircher, *Oedipus Aegyptiacus,* vol. I, sig. ++v and Elogium XXVII.

completed Aldrovandi's image as Aristotle, the discovery of a new Osiris secured Kircher's claim to be the new Hermes. As the emperor who ushered in the end of the Thirty Years' War, Ferdinand III merited the title of Osiris. According to Kircher, Osiris was yet another name for Noah's son Cham, who founded the Egyptian state after the flood. The resurgence of the Habsburg state in the wake of its virtual destruction paralleled the rise of Egypt after this biblical calamity. In the mythology that unfolded in the *Egyptian Oedipus,* the political culture of ancient Egypt mirrored that of the Habsburg monarchy.[114] Kircher had become Ferdinand III's client, not simply to promote his own work but to fulfill his destiny.

Kircher's continued success at the Habsburg court suggests that he understood his patrons well. When Leopold I became emperor in 1658, he continued the family's association with the Jesuit collector. In addition to the annuity, Kircher received 100 *scudi* in 1669 and 600 *scudi* in 1670.[115] Leopold I became the main patron of Kircher's studies of artificial and universal languages. Kircher dedicated his *New Polygraphy* (1663), *Great Art of Knowledge* (1669), and *Tower of Babel* (1679) to him. When Kircher was unable to find an Italian patron for his *New Phonurgy* (1673), a study of acoustics, he turned to the Emperor for help. As he wrote to Hieronymus Langenmantel in 1672, "I may not try [this] Patron again but am forced to return to a foreign one." Undoubtedly, Kircher had hoped that one of the Roman nobles in whose "Princely Palaces the plays of sound are exhibited" would interest himself in the project.[116] When this did not occur, however, he counted on the strength of his connections at the Imperial court to see the *New Phonurgy* into print. By the 1660s, Kircher had become one of the main publicists for the Habsburgs. As Ferdinand III and Leopold I both recognized, the circulation of his writings contributed significantly to their reputation as learned monarchs. With their political power increasingly diminished, they, like the Medici, understood the value of this image in maintaining the dignity of their rule.

Like Aldrovandi, Kircher used the rhetoric of exemplarity to define his relationship to more than one patron. He described Alexander VII, to whom Kircher dedicated the first volume of his *Subterranean World* and his *Egyptian Obelisks* (1666) as "Osiris reborn."[117] The Chigi pope, friend of Settala from their studies in Siena and patron of Kircher, also presented himself as a prince–savant who patronized learning to enhance the splendor of

[114]Evans, *The Making of the Habsburg Monarchy,* p. 436; Rivosecchi, *Esotismo in Roma barocca,* p. 59.

[115]Fletcher, "Athanasius Kircher and the Distribution of His Books," *The Library,* ser. 5, 23 (1968): 114.

[116]Langenmantel, *Fasciculus epistolarum* (Augsburg, 1684), p. 48 (Rome, 9 October 1672); Kircher, *Phonurgia nova,* p. 91.

[117]De Sepi, p. 12. The material in this section also is discussed in Rivosecchi, *Esotismo in Roma barocca,* pp. 136–143.

his reign. Under his papacy, the rebuilding of Rome and the restoration of ancient monuments continued apace.[118] As Alexander VII's main consultant in the renovation of the obelisks, particularly the one erected on top of Bernini's elephant in front of Santa Maria sopra Minerva, Kircher had ample opportunity to develop an iconographic program that would appeal to his patron's and his own sense of manifest destiny. In the dedication of the *Egyptian Obelisks,* he underscored his role as a publicist: "This obelisk of the learned ancients, erected to make the glory of your name shine forth, may go to the four parts of the world and speak to Europe, Asia, Africa, and America of Alexander, under whose auspices it has come back to life and for whose command it is revived."[119] Kircher presented the restoration of ancient monuments as the symbol of a new world order that similarly mended the "ruins" of Europe through the proclamation of political and religious harmony. Alexander VII, who became pope when Ferdinand III was already ailing and when the Holy Roman Empire seemed once again on the verge of collapse, was hailed as the new Osiris. Always sensitive to the political climate that surrounded him, Kircher sensed renewed strength in the papacy, weakened after the financial and military excesses of the Barberini and Pamphili papacies, at the very moment when the power of his imperial patrons seemed diminished. Never one to miss an opportunity, he turned his attention away from Vienna and toward Rome.

Throughout the 1650s and 1660s, Kircher worked to expand his circle of patrons. As a client of several popes and emperors, he chose his new patrons carefully so as not to diminish the honor of his principal benefactors through inadvertent association with ignoble princes. In keeping with his desire to maintain a foothold both in Italy and in Germany, Kircher pursued relationships with two of the most promising prince–savants of Baroque Europe: Leopoldo de' Medici and Duke August of Brunswick-Lüneburg. Kircher's first contact with both men occurred in 1650. In January of that year, he initiated a correspondence with Duke August, already known for his love of books; in April, Prince Leopoldo traveled to Rome. As the official court diary records, on 27 April 1650, "He stopped at the house of the Jesuit Father Kircher, mathematician and great virtuoso." After Leopoldo's return to Florence, Kircher maintained contact with him and his brother, the Grand Duke Ferdinando II, by presenting them with his publications.[120] Knowing the Medici prince's fascination with universal languages, Kircher did not neglect to send an example of his new form

[118]De Michele et al., *Il Museo di Manfredo Settala,* pp. 2–3; Krautheimer, *The Rome of Alexander VII.*

[119]Kircher, *Obelisci Aegyptiaci nuper inter Isaei Romani rudera effossi interpretatio Hierogyphlica* (Rome, 1666), n.p.

[120]ASF, *Mediceo Principato,* ms. 5396, f. 761 (27 April 1650), in Goldberg, *After Vasari,* p. 19; BNF, *Autografi Palatini,* II, 70 (Rome, 31 May 1655). In this second letter, Kircher presented Leopoldo with the second volume of his *Oedipus Aegyptiacus.*

of communication. "I finally come to present Your Highness with the promised artificial secret of languages, enclosed within a little chest and communicated to no one until now except His Majesty the Emperor, my most August Patron, and to the most Serene Archduke Leopold, likewise a great promoter of my studies."[121] By informing Leopoldo of the company that he kept as a patron of Kircher's studies, the Jesuit collector accorded him a stature equivalent to that of the Emperor and his brother.

In March 1668, Leopoldo was elected cardinal. He returned to Rome to begin his duties. By late April, the cardinal found himself sufficiently at leisure to engage in the scientific and cultural pursuits that most delighted him. As he wrote to his brother the Grand Duke, "In such hours as I have free from these occupations, I gratify myself by going to see medals, pictures, and statues in the most famous cabinets and shops."[122] By May, Leopoldo expressed a desire to return to the Roman College to see the gallery that had been constructed since his previous visit. Kircher's absence when the cardinal chose to alter the date of his visit precipitated a flurry of activity in Rome. "We hope that Cardinal Medici will honor the College [with his presence] earlier," scribbled Domenico Brunacci in haste to Kircher,

> and therefore we send you and a companion a mount, so that you may return immediately.... Don't delay at all costs because your presence on this occasion is most important. And try to arrive early to put the things in the Gallery in order.[123]

Unfortunately, Kircher's correspondence does not tell us what actually transpired during this second visit or whether Kircher arrived in time. As the patron of the Accademia del Cimento and numerous other scientific and cultural initiatives in Florence, Leopoldo was a prince who met Kircher's criteria of association. Adding his face to the portrait gallery in the Roman College museum only increased its splendor.

Kircher was equally eager to place an image of Duke August in his hall of fame. August promised to send the portrait in 1656, but the plague that devastated Rome forestalled its departure. Kircher anxiously reminded the duke's agents that he would display it publicly in his museum, placing August among the "other learned Princes, German as well as Italian."[124] When the engraved image finally arrived in July 1659, Kircher "immediately had it framed in gold and put up in my Gallery as a mirror of the magnanimity, wisdom, and generosity of the high-born prince." He continued, "My Gallery or Museum is visited by all the nations of the world and a prince

[121]PUG, *Kircher*, ms. 563 (IX), f. 99 (Rome, n.d.).

[122]In Goldberg, *After Vasari*, p. 23 (Rome, 27 April 1668).

[123]PUG, *Kircher*, ms. 564 (X), f. 165 (Rome, 12 May 1668).

[124]Jacob Burckhard, *Historia Bibliothecae Augustae quae Wolffenbutteli est, duobus libris comprehensa* (Wolfenbüttel, 1744), vol. II, p. 147 (Rome, 7 March 1659).

cannot become better known 'in this theater of the World' than to have his likeness here." The portrait of the duke served the dual purpose of advertising August's exemplary patronage of Kircher and adding luster to his image as a ruler. Always sensitive to the political import of his actions, Kircher recognized that the placement of the duke's image in his museum would have much greater meaning for him than it did for more powerful rulers, habituated to this sort of display. The following winter Kircher was still talking about the marvelous reception that the portrait, displayed "in the prime location of my Museum," continued to receive.[125]

August was one of several prospective patrons to receive a copy of his *Universal Musurgia* (1650). He, Duke Friedrich III of Schlewig-Holstein-Gottorf, and Queen Christina of Sweden all responded favorably.[126] After an unpromising start, August's interest in Kircher's research grew, particularly during the period in which the Jesuit was at work on a new artificial language. In 1660, Kircher presented August with a manuscript copy of his *New Invention of All Languages*, a prelude to his *New Polygraphy*. Underscoring the exclusive nature of this publication, Kircher described it as a work designated only for princes, and he flattered August, as he did with Leopoldo de' Medici, by informing him that Alexander VII and Leopold I had also received copies of it.[127] Unlike the other recipients of this gift, August claimed some expertise in this area; he had authored a cryptography in 1624 under the pseudonym of Gustavus Selenus. The duke tactfully reminded Kircher of his own foray into this subject by presenting him with a copy of his *Cryptography* in 1664, which Kircher had neglected to mention in his *New Polygraphy*. Despite this oversight, August awarded him a stipend of 200 *imperiales* in addition to the 100 *ducati di banco* sent in gratitude for Kircher's chaperoning his son, Ferdinand Albrecht, in Rome.

In Duke August, Kircher found a patron genuinely interested in the scholarly significance of his work and eager to facilitate "the further publication of his pleasing writings." As Kircher wrote to Johann Georg Anckel, the duke's agent in Augsburg, in June 1664, "the duke's acts of kindness toward me have resulted in my publishing many things which otherwise would never have been published."[128] Kircher, increasingly disillusioned with local patronage—"Such princes are not to be found in Italy," he wrote in disgust

[125]In Fletcher, "Kircher and Duke August of Wolfenbüttel," in Casciato, p. 285; Burckhard, *Historiae Bibliothecae Augustae*, vol. II, p. 130 (Rome, 3 January 1660). The material comes from the correspondence preserved in PUG and the Herzog August Bibliothek, Wolfenbüttel. The correspondence is further summarized and partially reproduced in Fletcher, "Athanasius Kircher and Duke August of Brunswick-Lüneburg," in Fletcher, *Athanasius Kircher*, pp. 99–138.

[126]In November 1650, Friedrich sent Kircher 300 *imperiales;* August did not respond until 1651. Fletcher, *Athanasius Kircher*, p. 101. Other patrons such as the Habsburgs and the Elector of Bavaria also received copies.

[127]Fletcher, "Kircher and Duke August of Wolfenbüttel," in Casciato, p. 286.

[128]Ibid., p. 288 (Wolfenbüttel, 5 February 1664; Rome, 24 June 1664).

to the duke[129]—saw August as the new light on the horizon. He prominently displayed the gift of an amber-encased lizard in his *Subterranean World* and thanked the duke profusely in the preface to the *Egyptian Obelisks*. Kircher also announced plans to dedicate the second edition of his *Great Art of Light and Shadow* to August as well as his *Etruscan Journey*, a lost manuscript originally intended for the Grand Duke of Tuscany. August expressed some hesitation about the propriety of Kircher's actions, given the evidently Italian nature of the latter subject, but he was nonetheless pleased enough by the Jesuit's attentions to send him 700 *imperiales* during the last two years of his life. By 1666, after August had bestowed numerous gifts and stipends upon Kircher, including other portraits in the form of silver and gold medals, the Jesuit called him "the prototype of royal munificence, the example for all princes to imitate."[130]

While Kircher could not legitimately present August as another Osiris— by no stretch of the imagination could Brunswick-Lüneburg be considered an empire equivalent to Egypt—he did not neglect to give this patron a suitable identity, culled from a reading of Herodotus's *Histories*. As Kircher informed August in 1651, the duke was another "Amasis King of Egypt" to his "Polycrates, a most learned man."[131] Amasis, whose reign "was the most prosperous time that Egypt ever saw," was a learned and inventive king and a giver of great gifts. Polycrates, "the first of mere human birth" to aspire to an empire, made himself ruler of the island of Samos before his Christ-like death at the hands of Oroetes. Amasis presented two wooden statues of himself that still stood before the Temple of Juno in Samos when Herodotus composed his history. Polycrates honored his "contract of friendship with Amasis by "sending him gifts, and receiving others in return."[132] Presenting himself as an upstart scholar in comparison to August, who already had stabilized the reign of learning through his collecting of books and patronage of scholarship, Kircher hoped that the historical example of Amasis's and Polycrates's great friendship would catalyze their own relationship. Encouraging the duke to patronize his work as Ferdinand III had done with the gift of 3000 *scudi*, Kircher integrated August into his Egyptian conception of the Holy Roman Empire.

The year 1655 was the date of Kircher's own "marvelous conjuncture." The election of Alexander VII and the arrival of Queen Christina of Sweden in Rome offered a wealth of alluring possibilities for his projects. Christina had already expressed an interest in Kircher's work. After sending her several of his books, a tactic Kircher often used when he introduced

[129]Ibid., p. 291 (Rome, 24 July 1666).
[130]Ibid. (Rome, 10 July 1666). August died on 17 September 1666.
[131]Zacharias Goeze, *Ad Augustam D. B. & L. Athanasii Kircheri S. J. Epistolae tres* (Osnabrück, 1717), sig. A.4v.
[132]*The Histories of Herodotus*, E. H. Blakeney, ed., George Rawlinson, trans. (New York, 1910), vol. I, pp. 206, 208 (II.177, 182); vol. II, pp. 230, 268 (III.39, 122).

himself to a prospective patron, Christina sent a letter to Kircher thanking him for the gift and inquiring whether he planned to dedicate a work to her. In response, Kircher requested a subsidy from the Swedish monarch and dedicated his *Ecstatic Voyage* (1656) to the Queen, thanking her for her "obstetrical hands."[133] Within a month of her ceremonial entry into Rome and official abjuration of Protestantism, Christina made two visits to the Roman College. Kircher prepared well in advance for her arrival. Shortly before the arrival of Queen Christina of Sweden, "who has the greatest curiosity about simples and natural history," Kircher wrote to the Vatican librarian Lucas Holstenius, asking for a stipend to finish the catalogue of his museum so that he could present her with a description of "some machines in my Gallery."[134]

For the first visit on 18 January, undoubtedly under Kircher's supervision, the Jesuits transformed the entire college into a theater filled with emblems, epigrams, and inscriptions, particularly ones commemorating "celebrated heroines." For the second visit on 31 January, Christina toured the sacresty, library, pharmacy, and museum.[135] Kircher personally demonstrated his choicest experiments to the royal visitor. Christina particularly admired the magnetic clocks and was treated to a display of the famous "vegetable palingenesis" that recapitulated the myth of the Phoenix. For this particular visitor, the latter experiment had added significance, since Christina herself was a sort of Phoenix, a Protestant monarch transformed into a Catholic. Having recently taken the name of Alexandra as part of her entry into the church, she indeed had renewed her identity.

At the end of the tour, Kircher presented Christina with two gifts. The first was an Arabic translation of the Psalms of David with an index to passages on Solomon's Temple and Moses's Tabernacle, both in reference to the house of wisdom that Christina had come to Rome to build. Kircher's identification of Santa Maria sopra Minerva as the site of the Temple of Isis added yet another dimension to his message. Like Alexander VII and Kircher, Christina came to Rome to fulfill the destiny portended in the historic ruins of the papal city. She was the new Isis, consort of Osiris, and patron of Hermes. Kircher's second gift made this message transparent. It was a miniature obelisk that bore the following inscription in thirty-three languages: "Great Christina, Isis Reborn, erects, delivers, and consecrates this Obelisk on which the secret marks of Ancient Egypt are inscribed,"[136] Previously Kircher had created an obelisk with a "Hieroglyphic eulogy" for

---

[133]PUG, *Kircher*, ms. 561 (VII), f.50 (Rome, 11 November 1651); ms. 556 (II), f. 174 (n.d.); Fletcher, "Athanasius Kircher and the Distribution of His Books," pp. 114–115.

[134]CL, vol. III, p. 1254 (Rome, 1649–1650); BAV, *Barb. Lat.*, ms. 6499, f. 120 (Rome, 15 October 1655).

[135]Villoslada, *Storia del Collegio Romano*, pp. 276–277.

[136]De Sepi, pp. 12, 38; Åkerman, *Queen Christina of Sweden and Her Circle*, pp. 226–227; Fletcher, "Astronomy in the Life and Correspondence of Athanasius Kircher," p. 57.

Ferdinand III, prominently displayed in his *Egyptian Oedipus*. Alexander VII had no need of an invented obelisk because his position made him the patron of the ancient obelisks of Rome. By offering the Habsburg emperor and the Swedish queen their own hieroglyphs, Kircher publicized their participation in the restoration of ancient learning that he saw as the key to political and religious reunification. In April 1656, Schott wrote to Kircher, "I greatly rejoice at the visit of the Queen of Sweden to your Museum."[137] With the arrival of Osiris and Isis in Rome during the same year, Kircher had at last returned to Egypt.

As we have seen with other collectors, Kircher's understanding of the protocols of patronage determined his success in obtaining benefactors for his museum. Like Aldrovandi, he perceived collecting and publishing to be complementary forms of display that enhanced his reputation within the republic of letters and gave him access to court culture. While other Baroque collectors directed their supplications toward one principal patron, hoping to develop an exclusive relationship with a prince, Kircher instead made the entire political culture his theater. If we briefly consider the activities of Settala and Cospi, for example, we can see the contrast. For Settala, contact with the Medici during his student days at Pisa gave him the opportunity to travel and maintain contact with the scientific community in Florence. Both Ferdinando II and Cosimo III continued to send gifts to Milan. After Cosimo III visited his museum in 1664, Settala had Terzago dedicate his catalogue to the young prince, undoubtedly with the hope that he might repeat Galileo's success with an earlier Cosimo.[138] On a considerably smaller scale than either Aldrovandi or Kircher, Settala strove to maintain a fruitful relationship with his most promising patron.

While Settala was the primary Medici client in Milan, Cospi mirrored this position in Bologna. Like Aldrovandi and Kircher, Cospi distributed the one publication associated with his museum as a gift, donating most of the 800 copies of Legati's catalogue to "many Princes of Italy, Cardinals, and Knights of Merit."[139] But most particularly, he saw it as a way of strengthening his bond with his most important patron, Cosimo III, Grand Duke since 1670. In the engraved image of the museum (see fig. 9), the Medici crest was prominently displayed in one quarter of Cospi's shield, his right by birth; Legati's catalogue also highlighted his youth at the Florentine court and subsequent gifts and visits of various Medici to Cospi's museum. Nei-

---

[137]PUG, *Kircher*, ms. 561 (VII), f. 40 (Würzburg, 1 April 1656).

[138]Tavernari, "Manfredo Settala," pp. 47–48; idem, "Il Museo Settala," p. 29. Regarding one of his mechanical marvels, Settala observed, "Il primo scherzo, che feci in Pisa allo studio, che con questo m'acquistai la gratia del ser[enissi]mo Gran Duca, mentre ero Giovinetto, come anche della ser[enissi]ma sua madre, qual era sorella dell'Imperatore." Ambr., Cod. Z.387 sup., f. 21.

[139]ASB, *Fondo Ranuzzi-Cospi, Vita del Sig. March.e Bali Ferdinando Cospi*, c. 37, in Olmi, "Ordine e fama," p. 257; Goldberg, *Patterns in Late Medici Art Patronage*, p. 38.

ther Settala nor Cospi (nor a collector such as Moscardo) chose to engage in a widescale program of research that produced learned tomes. For them, display was an end unto itself instead of a prelude to publication. Rather than seeking patrons to finance the printing of their works, they invested in the publication of catalogues to repay the princes and patricians who had honored their museums with gifts and visits. As a result, they had little need for the level of patronage that Aldrovandi, Della Porta, and Kircher demanded to complete their encyclopedias of nature.[140]

More often dependent on patronage than stable sources of income for survival, few museums survived the demise of their initial creator. With the exception of Mercati, none of the scientific collectors enjoyed a stable relationship with one particular court, and yet the Vatican *metallotheca* was one of the first museums to disappear after Mercati's untimely death in 1593.[141] The survival of Aldrovandi's museum seems to have been the exception rather than the rule. Not until 1702, for example, did the Catholic Church give the Roman College museum a permanent stipend, due to Bonanni's efforts to gain the patronage of Clement XI, leaving us to wonder at the effort Kircher expended to maintain it on gifts, goodwill, and his own financial resources.[142] For this reason, collectors devised elaborate schemes to acknowledge their numerous patrons without offending them by the implication that their relationship was not unique. They understood all too well how crucial patronage was for the survival of their museums.

While most collectors were largely unsuccessful in their attempts to secure a unique relationship with a particular prince, as Galileo was to do when he became court mathematician and philosopher to Grand Duke Cosimo II in 1610, it was not for lack of effort. Yet, as even Aldrovandi discovered, it was particularly difficult to present a museum as a unique offering to a patron; unlike the Medicean stars, collecting had become fully integrated into elite culture by the end of the sixteenth century and therefore was no novelty.[143] In Kircher's day, almost every patrician concerned with his reputation—and who was not?—had a collection tucked away in some corner of the family palace. The act of collecting alone could hardly interest a prince. The messages that Aldrovandi and Kircher offered their prospective patrons were crucial to their success. Completing the epic narratives that began as an exercise in self-definition, they presented themselves as clients whose destiny lay in the completion of an image they already knew to be

---

[140]I have chosen to omit Della Porta in this chapter simply because of constraints of space, but he exhibited behavior analogous to that of Aldrovandi and Kircher; see Clubb, *Giambattista della Porta Dramatist*, pp. 13–56; and Eamon, "Court, Academy and Printing House," in Moran, *Patronage and Institutions*, pp. 39–41.

[141]Olmi, "Science-Honour Metaphor," in *Origins of Museums*, p. 6.

[142]ARSI, *Rom.* 138. *Historia* (1704–1729) XVI, f. 180v.

[143]This point also is raised by Mario Biagioli in his "Scientific Revolution, Social Bricolage and Etiquette."

associated with a particular ruler. Humanists in the service of a nascent absolutism, they gathered the resources of the world to create monuments at
the feet of their patrons. In many respects, collectors were the group best
able to appreciate the anxieties of the princes whom they served and to anticipate their demands. For they also engaged in mortal battle with time, in
an attempt to avoid the fate of many other men of learning whose wisdom
had vanished in the desert sands, a loss that even the hieroglyphs did
not record.

Epilogue
The Old and the New

Despite the valiant attempts of Renaissance and Baroque naturalists to preserve their vision of the world by guaranteeing the immortality of their museums, their worst fears were confirmed (though not for the reasons they had imagined). Within a century of Kircher's death in 1680, a new generation of scholars invalidated the premise of the humanist encyclopedia of nature and, with it, the museums that contained this knowledge. The technological mirabilia that animated the experimental life of Kircher's and Settala's museums, replaced by better microscopes, bigger telescopes, and Leiden jars, were no longer the tools of a working collection but historical curiosities. Improvements in techniques of preservation made the crumbling herbaria and fragile animal corpses left behind by Aldrovandi, Calzolari, and Imperato virtually useless for eighteenth-century naturalists who enjoyed the benefits of wax injection, saline solutions, and other methods to capture a moment of nature in perpetuity. Even more important than these material advances, however, was the intellectual premise that informed the practices of Aldrovandi's heirs. For they finally had succeeded in doing what the Linceans had attempted to do in the early seventeenth century: create a new encyclopedia of nature.

The novelty of Enlightenment natural history lay partially in the approach to nature but primarily in the rhetoric surrounding its creation. Men such as Buffon and Linnaeus, and followers such as Lazzaro Spallanzani, proclaimed themselves to be the first of a new generation, the refounders of a discipline that had lain dormant for more than 2000 years.[1] They rested their arguments on a polemic against earlier naturalists who, from their

---

[1] Certainly the debates surrounding Linnaeus—last of the Aristotelians or first of the Moderns?—point to the complexity of this issue. See Sten Lindroth, "The Two Faces of Linnaeus," in Tore Frängsmyr, ed., *Linnaeus: The Man and His Work* (Berkeley, 1983), pp. 1–62.

vantage point, embodied the excesses and ignorance of a culture paralyzed by its reverence for the past. Numerous anecdotes about the fate of their predecessors served to illustrate how far natural history had come. In the fifth volume of the *Academic Collections* (1755–1779), for example, the editor of Jan Swammerdam's *Bible of Nature* described the Dutch anatomist's thwarted attempts to sell his collection in the 1670s. The Amsterdam collector wished to retire from a long career of teaching and public research to pursue his intellectual interests in private. When the Grand Duke Ferdinando II offered him 12,000 florins for his cabinet in 1667, on the condition that he come to Florence to supervise its installation and join the scientific virtuosi at the Medici court, he refused. Now, a few years later, Swammerdam asked for 15,000 florins, apparently an overinflated price since he had no immediate takers. The narrator tells us that the naturalist eventually sold his collection at a lower price in frustration. With a sigh, the eighteenth-century biographer concluded, "but this was not yet the century of Natural History."[2]

The Enlightenment view of Renaissance and Baroque natural history provides an appropriate conclusion to a discussion of early collecting practices. As this anecdote demonstrates, the philosophes did not treat their predecessors kindly. Looking back upon previous efforts at understanding nature with that touch of condescension that comes with self-avowed superiority, eighteenth-century naturalists rewrote the historiography of their discipline. For them, the science of natural history did not come into being until the eighteenth century, when cabinets of curiosity were replaced with museums of natural history whose purpose was highly Baconian and whose organization attended to the debates about classification and taxonomy. Diderot underscored this point in his *Encyclopédie* article on "cabinets of natural history," when he presented the success of natural history in his century as a product of new collecting habits.[3] The Enlightenment periodization of the relationship between natural history and collecting still persists today. For we, like Buffon, Linnaeus, and the anonymous editor of Swammerdam's papers, continue to believe that Renaissance natural history can be sharply differentiated from the Enlightenment history of nature.

By the eighteenth century, the idea of the museum had entered the public and institutional domain. The Ashmolean Museum at Oxford had been charging its visitors admission for some thirty years when Peter the Great established a public museum in Saint Petersburg upon the advice of Leibniz in 1714, the very year that the museum of the Institute for Sciences in Bologna was established.[4] In England, Hans Sloane's generous bequest laid

---

[2] *Collections académiques concernant l'histoire naturelle et la botanique, la physique experimentale et la chymie, la medicine et l'anatomie* (Paris, 1755–1779), Vol. V, pp. vii–xii.

[3] Diderot, "Cabinet d'histoire naturelle," p. 489.

[4] Oleg Neverov, "'His Majesty's Cabinet' and Peter I's *Kunstkammer*," in *Origins of Museums*, pp. 54–61.

the foundation for the British Museum, which opened its doors in 1753. When the Prince of Biscary presented Thomas Hollis with a studiolo of shells and other marine products, culled from the natural history museum that opened at the Academy of the Etnians (*Accademia degli Etnei*) in Catania in 1758, the English virtuoso was so taken with the gift that he had it installed in the British Museum "to greatly preserve the memory of the Donor and the Gift."[5] A century earlier, Hollis probably would have kept the bequest himself to enhance the quality of his private collection, but the appearance of museums that defined a nation's identity made it possible to imagine memory as a more institutionalized concept, the collective representation of a nation rather than the portrait of an individual. By the late eighteenth century, enlightened rulers such as the Empress Maria Theresa and the Archduke Peter Leopold made their collections of art and science accessible to their subjects through the opening of the Brera Museum in Milan in 1773 and through the public donation of the Cabinet of Physics and Natural History in Florence in 1775 and the Uffizi galleries in 1789.[6] With the growth of public museums, collections such as the ones instituted by Aldrovandi and Kircher were either absorbed into larger museums or left to gather dust. In the first instance, dismemberment of a collection destroyed the integrity of the creator's vision of the museum. In the second instance, the decline of patrons and visitors to a museum signaled its demise as a monument of culture.

The revolutionary political climate that swept across the continent at the end of the eighteenth century transformed museums as well as the rulers who maintained them. Most famous of all was the opening of the Louvre on 27 July 1793; as George Bataille observes, in this fashion the "origin of the modern museum is thus linked to the development of the guillotine."[7] While Diderot had already speculated that as many as 1200 to 1500 people daily visited the cabinet of natural history in the *Jardin du Roi* in Paris in the mid-eighteenth century, by the end of the century many more would come to see the desacralized sights of the revolutionary republic and later the First Empire.[8] In the Rousseauian climate of revolutionary Europe, collections were no longer the property of a private individual, the church—the abolition of the Jesuit order in 1773 precipitated the decline of their cultural institutions[9]—or the personal possession of a monarch. They belonged to a disembodied state, which placed the museum alongside other institutions

[5]Abate Domenico Sostini, *Descrizione del museo d'antiquaria e del gabinetto d'istoria naturale di sua eccellenza il sig[no]r Principe di Biscary Ignazio Paterno Castello* (Catania, 1776), pp. 47–48.

[6]Marco Cuaz provides a quick survey of the chronology of eighteenth-century museums in his anthology, *Intellettuali, potere e circolazioni delle idee nell'Italia moderna 1500–1700* (Turin, 1982), p. 35.

[7]George Bataille, "Museum," *October* 36 (1986): 25.

[8]Diderot, "Cabinet d'histoire naturelle," p. 490.

[9]One example of this cultural defoliation is the Brera Museum in Milan, built primarily out of Jesuit spoils; Cuaz, *Intellettuali, potere e circolazioni delle idee*, p. 33.

of culture that it regulated, maintained, and reshaped to fit its new image. In this context, the rhetoric of Felice Fontana's museum catalogue of the new Cabinet of Physics and Natural History in Florence can be situated: "Everything is so well ordered that a person can profit more in a few days than in many years in the other Cabinets, which seem made more to demonstrate the greatness of Sovereigns than for public utility."[10] A literary flourish on the part of Peter Leopold in 1775, rhetorically distancing the state museum from the person of the monarch, prophesied the fate of other courtly and patrician collections by the end of the century.

Certainly we cannot see the eclipse of the courtly and urbane Renaissance museum in favor of the appearance of the truly "public" museum purely as an example of the mechanisms of social control and political upheaval. Such an interpretation would neglect the many exceptions—the eclectic and personal museums of the *grands savants*, the wax museums that embodied the journalistic spirit of the nineteenth century, and the freakish sideshows that turned medical monsters into carnival displays.[11] Collecting always has the potential to be a highly personal affair. However, the history of the *particular* type of museum that we have considered increasingly became a history of state-sponsored institutions from the eighteenth century onward. The formation of scientific institutions with a stable source of income, a permanency that lasted beyond the lifetime of one patron, and an identifiably pedagogical purpose that fit the eighteenth- and nineteenth-century notions of the public utility of science crystallized the ambiguous notion of scientific community circumscribing the early modern museum.

Behind this transformation lay a new conception of the political culture that informed and guided the museum. Creators of eighteenth-century museums made explicit statements about the *national* intent behind such institutions. Even in Italy, a region unified neither by government nor by language, *illuministi*, as the Italian philosophes called themselves, presented museums as one of the primary instruments through which to effect a more national culture. While we see glimmerings of these images in the late seventeenth century—for example, in the suggestion that Redi write a "Natural History of Italy" in 1667[12]—the idea was not fully formulated until the early eighteenth century. The Veronese noble and antiquary Scipione Maffei embodied this transformation. In designing his museum of inscriptions in Verona, begun in 1716 and opened to the public in 1745, Maffei envisioned a *patria* that stretched beyond the traditional connotations of the word to

[10]Felice Fontana, *Saggio del real gabinetto di fisica e di storia naturale di Firenze* (Rome, 1775), p. 34. For more on the genesis of this collection, see Ugo Schriff, "Il museo di storia naturale e la facolta di scienze fisiche e naturale di Firenze," *Archeoin* 9 (1928): 88–95, 290–324.

[11]For a recent critique of the overapplication of social control theory to early modern institutions, see Brendan Dooley, "Social Control and Italian Universities: From Renaissance to Illuminismo," *Journal of Modern History* 61 (1989): 205–239.

[12]Laur., *Redi* 209, f. 319 (Genoa, 9 January 1667).

include all of Italy.[13] One of the main purposes of his "lapidary museum" was to ensure that the valuable antiquities that made Italy so unique would "neither perish nor go out of Italy anymore." As he wrote to Anton Francesco Marmi, antiquarian to the Grand Duke of Tuscany, in 1717, "certainly this museum will bring some honor to all of Italy."[14] Maffei envisioned antiquity as a form of cultural patrimony, an image that became the foundation for later preservation and restoration programs. Even natural history collections were not immune to this rhetoric. Nature was also a form of patrimony, worthy of preservation. In his *Forerunner in the Form of a Letter of the Natural History of the Euganean Mountains* (1780), Antonio Carlo Dondi described the collecting of nature as a means of "rendering a good service to your Fatherland." He continued,

> I only complain that no one yet has had the useful idea of collecting the productions of our Mountains . . . and forming a National Cabinet with them, arranged in good order and with method. Many learned Naturalists in various times collected some particular things but treated them only for the particularity of the thing itself or because they fit some Theory of theirs, and not ever to form a complete collection of them, which would be desirable.[15]

Like Maffei, Dondi envisioned a "public Museum of Natural History" that would showcase Italy's natural resources, just as the *museo lapidario* in Verona displayed its historical treasures.

The museum took its place alongside the new public libraries, reading clubs, and other institutions of culture that embraced a more heterodox audience. "It will be enough to understand how commendably and how generously such a select fund of books has been transferred from personal and private use to public and common access by the Most Eminent Corsini," wrote the abbot Querci in 1755 when the Biblioteca Corsiniana opened for four hours daily.[16] The Corsiniana had been preceded by libraries such as the Ambrosiana in Milan in 1609, the Casanatense in Rome in 1698, the Medicea in Florence under the direction of Antonio Magliabecchi in 1747, and the Marucelliana, also in Florence, in 1752. The new publicity of scien-

---

[13]Giordana Mariani Canova, "Il Museo Maffeiano nella storia della museologia," *Atti e memorie di agricoltura, scienze e lettere di Verona*, ser. 6, 27 (1975–1976): 177–190; Licisco Magagnato, Lanfranco Franzoni, Arrigo Rudi, and Sergio Marinelli, *Il Museo Maffeiano riaperto al pubblico* (Verona, 1982).

[14]Scipione Maffei, *Epistolario* (1700–1755), Celestino Garibotto, ed. (Milan, 1955), Vol. I, p. 222 (Verona, 2 September 1716); pp. 242–243 (Verona, April 1717). The relationships of Italian nationalism, renewed notions of papal empire, and collecting are discussed by Carolyn Springer, *The Marble Wilderness: Ruins and Representations in Italian Romanticism, 1775–1850* (Cambridge, 1987), esp. pp. 21–38. On early eighteenth-century Italy, see Brendan Dooley, *Science, Politics and Society in Eighteenth Century Italy: The Giornale de' Letterati d'Italia and Its World* (New York, 1991).

[15]Antonio Carlo Dondi Orologio Padovano, *Prodromo in forma di lettere dell'istoria naturale de' Monti Euganei* (Padua, 1780), pp. 10–11.

[16]Abate Querci, *Novelle letterarie*, in Cuaz, *Intellettuali, potere e circolazioni delle idee*, p. 83.

tific and cultural institutions signaled the decline of the collector who personally controlled access to his natural history museum, library, or art gallery, restricting it largely to participants in the same social and cultural world, in favor of a curator who admitted a paying public. Private collectors continued to play an important role in the eighteenth century; as Diderot observed, their numbers increased during this period. But they were no longer the axis through which the entire learned world came together. Increasingly, institutions rather than individuals took on this function.

The openness of eighteenth-century cultural institutions contributed significantly to the redefinition of natural history. As museums became more of a public phenomenon, learned practitioners took greater pains to differentiate themselves from the unlearned audience who exhibited only curiosity and not virtuosity. Two defining features of sixteenth- and seventeenth-century natural history and its collecting practices were revised. Curiosity was no longer a valued premise for intellectual inquiry but rather the mark of an "amateur." Natural history was no longer subsumed under the categories of medicine and philosophy but had achieved a certain autonomy. Ironically, the recognition of natural history as a separate field of inquiry coincided with its disintegration as an encyclopedic structure that incorporated the observation of mankind and its productions alongside the investigation of nature. Just when the fashionability of theaters of nature increased for the general public, their close ties to contemporary research in the field of natural history declined. Natural history, in effect, became an object of public consumption, as did the museums themselves. Describing amateur collectors of nature, Buffon remarked,

> Most of those who, without any prior study of natural history, wish to have collections of this sort are people of leisure with little to occupy their time otherwise, who are looking for amusement and regard being placed in the ranks of the curious as an achievement.[17]

Some years later, Lamarck differentiated the "cabinet of curiosity" from the "cabinet of natural history" to underscore their diverse purposes; the former was for amusement and the latter for the progress of the sciences. While earlier naturalists had conflated these different activities, their Enlightenment heirs assiduously separated them. Terms such as *philosophical naturalist* and *dilettante* replaced *virtuosi*, a heterodox category that established common cultural ground for natural philosophers, physicians, apothecaries, courtiers, and patricians. As museums became more public, the new social and cultural elite of the eighteenth century worked harder to differentiate their purpose and audience.

The numerous public auctions of natural history collections in the eighteenth century contributed to the growing distinction between naturalists

---

[17]Buffon, "Initial Discourse," in *From Natural History to the History of Nature,* John Lyon and Phillip R. Sloane, eds. (Notre Dame, IN, 1981), p. 107.

and amateurs. "The curious rather love these sorts of sales," wrote Gersaint in 1736, "they come here with pleasure and regard them as an amusement."[18] This was the same audience who eagerly bought up the volumes of Buffon's *Natural History*, which sold out in six weeks when it first appeared in 1749. The learned curiosity of Renaissance naturalists, so integral to their conception of the nature of scientific inquiry, had been demoted to a sentiment expressing an amateur interest in science that contrasted unfavorably with the measured and reasoned practices of Enlightenment naturalists. As Buffon aptly put it in the introduction to his natural history, "This particular failing has been completely eliminated in this century."[19]

Concomitantly, argued the eighteenth-century naturalists, the true study of nature began only with their efforts to reorganize natural history along more economical lines, bypassing the Renaissance cabinet of curiosities for an ostensibly more systematic and normative approach to collecting. The founder of the Institute for Sciences in Bologna, Luigi Ferdinando Marsigli, derided collections that "serve more to delight young men and provoke the admiration of women and the ignorant rather than to teach scholars about nature."[20] Similarly Spallanzani, a later member of the Institute, disparaged contemporaries such as the "young knight whom I knew during my studies in Bologna who created a *studio* to place butterflies elegantly between two small crystal pieces sealed with wax, promptly nominating each of the butterflies marvelous with names of Linnaeus . . . but in fact he did not know an ounce of the natural history of butterflies."[21] In the sixteenth century, naturalists argued that the museum itself distinguished the new natural history from the old. By the eighteenth century, it was the method of ordering objects rather than the act of possessing nature that created this same division. As part of this movement, collectors became more attentive to the written descriptions that accompanied objects in a museum (undoubtedly in reaction to the more emblematic portrayal of objects that characterized earlier collections). New catalogues and new labels gave new meanings to old objects. "If this immense collection of wax anatomical models were left without any explanation," wrote Felice Fontana in his 1775 catalogue of the new public science museum in Florence, "it would have

[18]Lamarck, *Mémoire*, in Yves Laissus, "Les cabinets d'histoire naturelle," in *Enseignement et diffusion des sciences en France au XVIIIe siècle*, René Taton, ed. (Paris, 1964), pp. 667, 669. Approximately 1600 natural history auctions occurred in Great Britain alone in the eighteenth century, and an equal level of interest can be found in France; J.M. Chalmers-Hunt, ed., *Natural History Auctions 1700–1972* (London, 1976).

[19]Buffon, "Initial Discourse," in Lyon and Sloane, *From Natural History to the History of Nature*, p. 110. For a discussion of the seventeenth- and eighteenth-century critique of curiosity, see Pomian, *Collectors and Curiosities*, pp. 45–64.

[20]In Olmi, "Science-Honour Metaphor," in *Origins of Museums*, p. 15.

[21]Biblioteca Municipale di Reggio Emilia, *Ms. Regg.* B 144, in Maria-Franca Spallanzani, "La collezione naturalistica di Lazzaro Spallanzani," in *Lazzaro Spallanzani e la biologia del Settecento*, Giuseppe Montalenti and Paolo Rossi, eds. (Florence, 1982), p. 597.

been a Museum like so many others, more for the admiration of foreigners than for public utility."[22]

The origins of much of the new rhetoric surrounding natural history and collecting came from the Baconian aspirations of the Royal Society and the transposition of these ideals to the continent. The mission of the Royal Society repository, for example, was to make a private enterprise public by unifying all scientific collections. As its first historian, Thomas Sprat, observed in 1667, "the *Royal Society* will be able by degrees to purchase such extraordinary inventions, which are now close lock'd up in *Cabinets*; and then to bring them into one common Stock, which shall be upon all occasions expos'd to all men's use."[23] Under the direction of Robert Hooke and Nehemiah Grew, the Royal Society envisioned its museum as a dramatic contrast to the continental collections of Aldrovandi, Kircher, and the like, who did not limit their audience exclusively to natural philosophers (here defined by their membership in this and like-minded societies). "The use of such a Collection is not for Divertisement, and Wonder, and Gazing, as 'tis for the most part thought and esteemed, and like Pictures for Children to admire and be pleased with," explained Robert Hooke, "but for the most serious and diligent study of the most able Proficient in Natural Philosophy."[24] In theory, the museum was bound to the particular conception of order and scientific method that the Royal Society wished to imprint upon the sciences. "This *Repository* he [Hooke] has begun to reduce under its several heads, according to the exact Method of the Ranks of all the *Species of Nature*, which has been compos'd by Doctor *Wilkins*, and will shortly be publish'd in his *Universal Language*."[25] Thus, the collection was designed to showcase the "experimental philosophy" of the Royal Society, blending Baconian utility with revisionist taxonomy.

The emphasis on order and classification in the rhetoric of Restoration natural history reappeared with greater force in the practices of Enlightenment collectors. By the late seventeenth century, Baconianism had captured the imagination of some of Italy's leading naturalists. Marsili, founder of the Institute for Sciences in Bologna, belonged to this generation of scholars.[26] For him, like Malpighi, the outmoded organization of the *Studio Aldrovandi* must have been a daily reminder of Italy's "backwardness" as a center of sci-

[22]Fontana, *Saggio del real gabinetto*, p. 32.

[23]Thomas Sprat, *History of the Royal Society* (1667), Jackson I. Cope and Harold Whitmore Jones, eds. (St. Louis, MO, 1958), p. 75.

[24]Robert Hooke, *Posthumous Works* (1705), in Hugh Torrens, "Early Collecting in the Field of Geology," in *Origins of Museums*, p. 211.

[25]Sprat, *History of the Royal Society*, p. 251; on this and other scientific classification systems of the period, see Slaughter, *Universal Languages and Scientific Taxonomy*. In practice, however, the Royal Society museum hardly looked different from its predecessors since it, too, was filled with the natural and artificial marvels of the world.

[26]Marta Cavazza, *Settecento inquieto. Alle origini dell'Istituto delle Scienze di Bologna* (Bologna, 1990), esp. pp. 119–148.

entific learning. Little wonder that one of his first acts upon creating the Institute for Sciences was to form a new museum that eventually absorbed the wealth of the Aldrovandi and Cospi collections, transforming it into Baconian rather than Aristotelian knowledge. The organization of a collection provided an important method for distinguishing the *studiosi di natura* from scientific *dilettanti.*

Collectors such as Marsili and Spallanzani, creator of two different natural history museums in Pavia and Scandiano, searched for a "methodic order" that would highlight the difference between the "chimerical interpretations and relations" of Renaissance naturalists and Enlightenment amateurs and "true natural history." Marsili had the rooms designated for the Institute for Sciences in the Palazzo Poggi divided to reflect the new disciplinary divisions of the sciences, including the emerging subdivisions of botany, zoology, and paleontology. For this reason, the Abbé Coyer described it as "the Atlantis of Chancellor Bacon executed" after a visit in 1776.[27] Other eighteenth-century scientific collections such as the university museum of Turin also partitioned artifacts according to new guidelines, including separate rooms for botany, electrical instruments, and "ossified curiosities."[28] By physically separating the different subcategories that defined the whole of natural history, eighteenth-century collectors hoped to avoid the "false connections" made by Renaissance naturalists, who had placed all artifacts together.[29]

The collections of Aldrovandi and Cospi, incorporated in the Institute museum since 1742, were redistributed to reflect the eighteenth-century taxonomy of nature. By destroying the integrity of these earlier collections, the Institute members hoped to build a museum in keeping with its "modern" principles. "We need to render [it] less confused," wrote Marsili in a letter to the last custodian of the *Studio Aldrovandi* and the first president of the Institute for Sciences, Lelio Trionfetti, regarding the organization of the natural history section of the Institute.[30] Some forty years later, after a generous bequest to the Institute, Benedict XIV wrote to his broker in Bologna,

[27]In Spallanzani, "La collezione naturalistica di Lazzaro Spallanzani," p. 595; idem, "Le 'Camere di storia naturale' dell'Istituto delle Scienze," pp. 147, 158–159, 161. For more on Spallanzani's collecting activities, see idem, *La collezione naturalistica di Lazzaro Spallanzani: I modi e i tempi della sua formazione* (Reggio Emilia, 1985).

[28]From the constitution of the University Museum of Turin, in Cuaz, *Intellettuali, potere e circolazioni delle idee,* pp. 88–90.

[29]In fact, Buffon used this term when describing the ostensibly unsystematic nature of Renaissance taxonomy; "Initial Discourse," in Lyon and Sloane, *From Natural History to the History of Nature,* p. 101.

[30]Gian Guiseppe Bianconi, ed., *Alcune lettere inedite del Generale Conte Luigi Ferdinando Marsigli al Canonico Lelio Trionfetti per la fondazione dell'Istituto dell Scienze di Bologna* (Bologna, 1849), pp. 27–28. For the reorganization of Aldrovandi and Cospi's collections, see ASB, *Assunteria dell'Istituto. Diversorum,* Busta 12, n. 13; Busta 13, n. 35; also Giuseppe Gaetano Bolletti, *Dell'origine e de' progressi dell'Instituto delle Scienze di Bologna* (Bologna, 1751), p. 25.

"God willing that they serve some purpose and that the Institute, not here for many years, does not end like the *Studio Aldrovandi.*"[31] The dispersal of many collections shortly after the death of their owners reminded collectors of the temporal nature of their enterprise. In eighteenth-century parlance, only proven utility could counter the fragility of memory and ensure a museum's continuance.

The fortunes of Renaissance collections in the hands of Enlightenment naturalists mirrored the universal rewriting of natural history that occurred throughout Europe during this period. When Giuseppe Monti described the organization of the Institute collection as "near the laws of nature," he reflected the commonly held belief that Renaissance natural history had distorted rather than accurately represented nature, cloaking her in allegory and mystery instead of making her parts apparent.[32] The objects in the Enlightenment museum now "spoke" for themselves rather than being touchstones for the humanist exercise of erudition. Spallanzani also arranged his collection to reflect the new relationship with nature; as he explained in 1778, "The method I used in classifying and ordering the natural products I took from the most illustrious modern Naturalists, or better to say, from nature herself."[33] Caught up in the new taxonomic systems that promised a clearer vision of nature, it was left only to skeptics such as Diderot to suggest that "the order of a cabinet cannot be that of nature; above all nature affects a sublime disorder." Nonetheless, his prescriptions for the organization of a collection also stressed the importance of methodical order, praising museums "distributed in a way most favorable to the study of Natural History."[34] The reorganization of natural history by Marsili, Buffon, Linnaeus, Spallanzani, and their contemporaries was effected as much through the new image they gave collecting as through the new classification systems they proposed. "Indeed, a well-organized Museum is like a universal Natural History," observed the author of the constitution of the University Museum of Turin, "seen at a glance . . . as one great and well accomplished open book."[35] Through new taxonomies and "reasoned experiments," eighteenth-century collectors claimed to have wrenched the book of nature wide open. By comparison, the approach of the Renaissance naturalists seemed tentative, a reflection of their

[31]BUB, *Benedetto XIV,* ms. 4331 (II, 7: 57), c. 116 (Rome, 24 July 1754). For Benedict XIV's donations to the Istituto, see ASB, *Assunteria dell'Istituto. Diversorum,* Busta 12, n. 5; Busta 13, n. 5.

[32]In Spallanzani, "Le 'Camere di storia naturale' dell'Istituto delle Scienze," p. 157.

[33]In idem, "La collezione naturalistica di Lazzaro Spallanzani," p. 595. For example, the biologist cited Linnaeus' *Instructio musei rerum naturalium* (Uppsala, 1751) with approval. It is impossible for us to take such claims of transparency at face value. If anything, the use of objects to prove and disprove different taxonomic systems only made their malleability more evident.

[34]Diderot, "Cabinet d'histoire naturelle," p. 490.

[35]In Cuaz, *Intellettuali, potere e circolazioni delle idee,* p. 87.

ambivalence toward the opening of the book of nature to indiscrete and prying eyes.

By claiming to make the book of nature transparent, Enlightenment collectors satisfied their requirement that a museum be dedicated "to the perfection of the arts and sciences."[36] Thus, the new public museums of eighteenth-century Europe were bound to the Enlightenment ideal of progress. Felice Fontana, the custodian of the science museum in eighteenth-century Florence, promised visitors

> Those [machines] which one sees in the new Museum of Florence are so grand and perfect that they greatly surpass the Machines in other Cabinets and miraculously serve to demonstrate the true and just laws of nature and to discover the truth until now unknown to the Philosopher.

In contrast to the museums of bygone days, "this immense Collection of Scientific Materials" was the inspiration of an enlightened ruler such as "Peter Leopold, who wholly intends to make the Sciences grow in Tuscany, opening his Treasures with acknowledged liberality in order to enlighten the People and make them happy by making them more learned."[37] Pangloss could not have put it better himself.

The language of eighteenth-century collecting dramatically revealed the patent inability of Enlightenment naturalists to understand the premise of the Renaissance and Baroque museums. Despite the fact that collectors such as Aldrovandi and Kircher clearly designated their museums as repositories of all nature, if not all knowledge, Enlightenment collectors refused to see the similarity of purpose that they espoused. With different taxonomies came a different language of nature, and such a language did not offer much room for translation. The perplexity evidenced by eighteenth-century naturalists over the endless analogies woven by their predecessors did not simply constitute a dismissal of their activities but expressed a genuine bewilderment at the premise of such a system. "Although they had no theoretical system, they needed to have one," wrote Albrecht von Haller in his forward to the 1750 German edition of Buffon's *Natural History*.[38] The cosmological significance of the Renaissance museum had no place in the Enlightenment worldview; it was socially and intellectually incommensurable with the new place of natural history in the eighteenth century. Rather than representing an alternative taxonomy of nature, it had ceased to be a system at all.

[36]At least this is what Leibniz advised Peter I to do with his collections in a letter of 1708; Neverov, "'His Majesty's Cabinet' and Peter I's *Kunstkammer*," in *Origins of Museums*, pp. 55–56.

[37]Fontana, *Saggio del real gabinetto*, pp. 1–2.

[38]My argument here has benefited from a reading of Mario Biagioli's "The Anthropology of Incommensurability," *Studies in the History and Philosophy of Science* 21 (1990): 183–209; Albrecht Von Haller, "Forward," in Lyon and Sloane, *From Natural History to the History of Nature*, p. 301. Elsewhere he stated, "these great men had no system" (p. 302).

Eighteenth-century evaluations of the work of Aldrovandi and Kircher support the notion that Enlightenment naturalists genuinely did not comprehend the point of Renaissance natural history. In the preface to his *Natural History* (1749), Buffon offered his opinion on Aldrovandi in great detail:

> Aldrovandi, the most hardworking and knowledgeable of all naturalists, after laboring for sixty years, left behind some immense volumes on Natural History, the majority of which were printed successively after his death. One can reduce them to a tenth of their total size by removing all of the useless and strange things on this subject from them. Despite this prolixity, which I confess is nearly overwhelming, his books must be regarded among the best on the whole of Natural History; the plan of his work is good, his distributions sensible, his divisions well marked and his descriptions rather exact, monotonous but faithful to the truth. [However] the history is less good. Often fables are interspersed, allowing us to see too much of the author's penchant for credulity.[39]

Enlightenment naturalists simultaneously admired the breadth of the Renaissance Aristotelian encyclopedism and chastized it for the inclusion of unnecessary information, from their epistemological perspective. The material that Buffon found so distracting—emblems, adages, morals, and copious literary and poetic citations—was precisely what made Aldrovandi's natural history a humanist text. What Enlightenment philosophes rejected most vehemently was the culture of erudition that made nature an object of humanistic inquiry.

While praising Kircher's condemnation of alchemy and other superstitious arts, seventeenth-century British natural philosophers abhorred his indiscriminant curiosity. "On the other side, he is reputed very credulous, apt to put into print any strange if plausible story, that is brought unto him," wrote Robert Southwell to Robert Boyle in 1661.[40] Within the next half century, natural philosophers further revised their opinion of Kircher to distance themselves from his credulous practices. "He saw the truth through a fog," wrote the Paduan professor of medicine Antonio Vallisneri,

> he knew it but confused it with ancient and ruinous lies, having drunk those prejudices from the doctrines of the old schools in his head. Therefore he is worthy of eternal praise because he distinguished himself from the mob, surpassing many by imagining correctly and laying the groundwork for following generations to think better.[41]

By virtue of his chronological proximity to the Enlightenment, Kircher was portrayed as less of an "other" than Aldrovandi, who—from the distance of

[39]In Cermenati, "Ulisse Aldrovandi e l'America," p. 7, n. 1.
[40]In Murray, *The Museum: Its History and Its Use,* Vol. I, p. 106.
[41]In Paola Lanzara, "Kircher un botanico?" in *Enciclopedismo in Roma barocca,* p. 338.

two centuries—appeared firmly imbedded in the scientific lore that naturalists such as Buffon and Vallisneri zealously dismissed.[42]

Despite the persistence of this largely eighteenth-century perception of the stagnation of the life sciences before 1750, it is obvious that natural history was an active and vital enterprise long before the arrival of Buffon and Linnaeus. In many respects, Enlightenment natural history was the logical culmination of the tradition of critical inquiry initiated by Renaissance naturalists. The emphasis on experience, the concern with method and order, and the ambivalent attitude toward authority were all constituent features of sixteenth- and seventeenth-century natural history. They were products of humanist culture long before they brought about its demise. Notwithstanding the Enlightenment characterization of Renaissance natural history, it was profoundly more "experimental"—tactile, sensory, and empirical—than the rewriting of nature that followed. The eighteenth-century strategy, at least for the discipline of natural history, was to *stop seeing*. Rather than seeing more, Enlightenment science saw less, and in doing so, redefined the purpose of a museum of nature, which became an exclusive rather than an inclusive construct. While this claim initially may seem counterintuitive, given our common perception of modern science as an empirically grounded enterprise, it is borne out by the trajectory of natural history as a form of inquiry. As an encyclopedic enterprise, Renaissance and Baroque natural history strove to accommodate every new "fact" that came to the attention of the early collectors of nature.[43] The concern with "economy," to invoke one of Linnaeus's favorite words, was an inevitable reaction to an excessively material culture whose interpreters never quite seemed to complete the encyclopedias of nature that they started.

If the roots of "modern" attitudes toward studying nature lay in this profoundly premodern culture, then what are we to make of its social configuration? Natural history began as a subject for scholarly debate and became an object of courtly leisure. With the demise of court culture in the course of the eighteenth century, it remained an important patrician activity, worthy of the gentlemen who frequented the academies, salons, and reading clubs. Despite the tendency to view natural history apart from other forms of scientific learning, for example, astronomy and mathematics, it developed within the same social context. Naturalists, anatomists, astronomers, mathematicians, and many other scientific practitioners moved readily between the courts, academies, and universities. Their ability to coexist makes it difficult for us to label one group "ancients" and another "moderns," as

[42]For an interesting treatment of an eighteenth-century natural philosopher between these two worlds, see Cesare Vasoli, "L'Abate Gimma e la 'Nova Encyclopedia' (Cabbalismo, lullismo, magia e 'nuova scienza' in un testo della fine del Seicento)," in *Studi in onore di Antonio Corsano* (Bari, 1970), pp. 787–846.

[43]Daston, "The Factual Sensibility," pp. 452–470.

standard textbook images of this period continue to do. To take the most relevant example, Galileo participated in the *same social system of science* as the naturalists who collected, providing a perfect bridge between the worlds of Aldrovandi and Kircher. He, too, courted Renaissance princes such as the Grand Duke Ferdinando I and depended upon brokers such as Pinelli, Mercuriale, and Vinta to make his way in the world; like Kircher, he also recognized the significance of Barberini patronage as an entry to the papal court.[44] Just as Galileo looked for a "marvelous conjuncture" to solidify the patronage network that supported him, Aldrovandi described the ascension of a new patron, Ferdinando I, in 1587 as a "very opportune conjuncture" (*congiuntura molto opportuna*), thereby utilizing the same social language and rhetorical strategies as his ostensibly more modern counterpart.[45] Rather than removing himself from the world of humanist learning and courtly patronage, as something alien to his more "modern" sensibilities, Galileo placed himself centerstage.

If we can draw any distinction between the social position of collectors such as Aldrovandi and Kircher and a mathematician such as Galileo, we should point to Galileo's decision to enter the court as a means of severing his connection with the academic culture of learning. While Aldrovandi and Kircher entered the courts without leaving their institutional identities behind, Galileo chose to make the court the exclusive locus from which he worked. Secure in his position within the prestigious faculty of medicine at Bologna, Aldrovandi did not have to effect a revolution to gain status and respect; Kircher shared a similar comfort in his association with the Roman College. As a group, naturalists had a more secure place at court than did mathematicians, due to the more apparent aesthetic function of their discipline, as seen in the close associations of naturalists, collectors, and artists. Conversely, as a practitioner of a less aesthetically pleasing and philosophically inferior discipline such as mathematics, Galileo needed a new social identity to legitimize his intellectual practices. His greater success as a philosopher was not scientific but *rhetorical*; and his greater failure as a client was *political.* The numerous intersections of the worlds of Aldrovandi, Galileo, and Kircher further underscore the Florentine mathematician's indebtedness to humanist culture. Humanism was the language of the courts, and regardless of one's philosophical position, one had to speak it to be heard.

The social commensurability of Renaissance and Baroque natural history and Galilean science further reinforces the significance of the Enlightenment reorganization of the sciences as an important moment in the history of science, though not for the reasons it commonly is celebrated. By 1750, the social function of the Renaissance museum no longer fit the image of the naturalist; the naturalist–courtier had been replaced by a more

[44]Biagioli, *Galileo Courtier*; and Redondi, *Galileo Heretic.* The comments that I make in the remainder of the epilogue are based largely on a reading of their work.
[45]BUB, *Aldrovandi*, ms. 21, Vol. II, c. 11, n.d.

"professionally" defined naturalist, tied to the patronage systems of institutions rather than to the personal munificence of individual rulers. Just as the political culture of Renaissance and Baroque Italy provided naturalists with opportunities for greater status and recognition, the political climate of Enlightenment Italy created new positions of authority. Here again it was through carefully constructed images and a strategic use of language that Enlightenment naturalists established their social and epistemological superiority.

Early modern science was a product of varying social and political circumstances, institutional matrices, and the cultural expectations of the urban elite. The appearance of museums of natural history serves as an exemplary case study of the intersections of these different elements. Museums, like the knowledge they produced, were complex entities, and there are surely many other ways to consider their significance for early modern as well as modern cultures. We may never fully understand what motivated Aldrovandi to start one of the first museums of natural history in Bologna nor what led Kircher to feel he was destined to complete the humanist encyclopedia in the heart of Rome. Their reverence for authority and their creative identification with the past produced a remarkably novel institution that became the setting for some of the most innovative scientific and intellectual activities of the sixteenth and seventeenth centuries. They, more than many others, popularized the study of nature—"science" in its broadest sense—for the urban elite through their willingness to make learning a form of display. In their museums, a new scientific culture was formed.

Even if we, like the Enlightenment naturalists, find it difficult to understand the form that this "novelty" took, we cannot deny its presence. Throughout the sixteenth and seventeenth centuries, Europe's leading intellectuals conducted their inquiry into nature in museums. The designation of this setting as a unique space of knowledge made it possible for naturalists throughout Europe to establish common protocols through which to communicate their discoveries and opinions. Linking the etiquette of the republic of letters with the etiquette of the courts, they positioned themselves as the men most able to make scientific knowledge sociable. In broadening the appeal of learning, naturalists also drew upon their understanding of the contemporary political climate. Mastery of nature went hand in hand with the rhetoric of absolutism; museums were an eminently visible reminder of how political might, new forms of knowledge, and power over nature could be combined. None of these transformations fit easily into the standard dichotomies of the old and the new, and we would do well to reexamine the significance of this dialectic. By particularizing our characterizations of early modern science through closer attention to context as well as content, to the connections between social status and intellectual identity as well as the cultural implications of scientific production, we can begin to rethink our ideas about what defines the scientific revolution.

BIBLIOGRAPHY

I. MANUSCRIPT SOURCES

Archivio Isolani, Bologna

Fondo Paleotti. 59 [F. 30]
    29/1–14. Ulisse Aldrovandi, *Lettere.*
    30/1. Ulisse Aldrovandi to the Assunti dello Studio.
    30/2. Camillo Paleotti, *Memoriale del Dottore Aldrovandi.*
    30/3. Camillo Paleotti to the Congregatione sopra l'Indice.
    32. Camillo Paleotti and Ulisse Aldrovandi. *Mazzo di cataloghi di piante di diversi paesi.*

Archivio di Stato, Bologna

Archivio dello Studio Bolognese. Collegi di medicina e d'arti. Nucleo antico.
    197. *Misc. Protomedicatus 1559–1600.*
    217–218. *Libri segreti.*
    248. *Recapiti per il protomedico e per la fabbrica della triaca.*
Assunteria dell'Istituto. Diversorum.
Assunteria di Studio. Diversorum, 10. n. 6. *Carte relative allo Studio Aldrovandi.*
Assunteria di Studio. Requisiti dei lettori.

Biblioteca Comunale dell'Archiginnasio, Bologna

B. 164. Pompeo Viziano, "Del Museo del S[igno]r Dottore Aldrovandi" (1604).
s. XIX. B. 3803. Ulisse Aldrovandi, *Informazione del rotulo del studio di Bologna de Ph[ilosoph]i e Medici all'Ill[ustrissi]mo Card. Paleotti* (1573).
Coll. Autogr., XLVIII, 12705. Ovidio Montalbani. *Lettera.*

Biblioteca Universitaria, Bologna

Mss. Aldrovandi.
    Cod. 384 (408), Busta VI, f. II. *Inventario dei mobili grossi, che si trovano nello Studio Aldrovandi e Museo Cospiano* (12 March 1696).

Cod. 559 (770). *Memorie antiche manuscritte di Bologna raccolte et accresciute sino a'tempi presenti dall'Ab[ate] Ant[onio] Francesco Ghiselli.*

Cod. 595-Y, n. 1. *Catalogo dei libri dello Studio di Ulisse Aldrovandi* (25 May 1742).

Cod. 738 (1071), XXIII, n. 14. *Decreto per la concessione di una sala al Marchese Ferdinando Cospi appresso lo Studio Aldrovandi* (28 June 1660).

Ms. 4312. Jacopo Tosì, *Testacei cioe nicchi chiocciole e conchiglie di piu spezie con piante marine &c. Regalo del Ser[enissi]mo Cosimo III Gran Duca di Toscana al Senator, Marchese, Bali, e Decano Ferdinando Cospi; da questo collocati a publico commodo nel Museo Cospiano fra le altre Curiosità de l'Arte, e della Natura da esso adunate* (1683).

Biblioteca Medicea-Laurenziana, Florence

Codici Ashburnhamiani, 1211. *Ulysses Aldrovandi Opera Varia Inedita.*

Mss. Redi 203, 209, 221, 222, 224.

Biblioteca Nazionale Centrale, Florence

Autografi Palatini, II, 70. Athanasius Kircher to Cardinal Leopoldo de'Medici.

Magl. II, 1, 13. Agostino del Riccio, *Arte della memoria* (1595).

Magl. VIII, 496. Paolo Boccone, *Lettere ad Antonio Magliabecchi* (1677–1699).

Magl. VIII, 505. Filippo Bonanni, *Lettere ad Antonio Magliabecchi* (1688–1714).

Magl. VIII, 1112. Manfredo Settala, *Lettere ad Antonio Magliabecchi* (1662–1678).

Magl. XIV, 1. *Discorso brevissimo sopra le quattro figure et frutto Indiano mandati al Sereniss[i]mo Granduca di Toscana dal Dottor Aldrovandi.*

Magl. XVI, 63(1–4). *Apparato della Fonderia dell'Illustrissimo et Eccellentiss. Sig. D. Antonio Medici. Nel quale si contiene tutta l'arte Spagirica di Teofrasto Paracelso & sue medicine. Et altri segreti bellissime* (1604).

Targioni Tozzetti, 56(1). *Agricoltura teorica del Padre Agostino del Riccio.*

Targioni Tozzetti, 56(2). *Agricoltura sperimentale del P. Agostino del Riccio.*

Biblioteca Riccardiana, Florence

Cod. 2438, Pt. I, lett. 66, 89, 91–93. Ulisse Aldrovandi and Antonio Giganti to Lorenzo Giacomini.

Biblioteca Comunale, Forlì

Autografi Piancastelli. 51. Ulisse Aldrovandi to Camillo Paleotti and Giovan Vincenzo Pinelli.

Autografi Piancastelli. 1214. Two letters of Athanasius Kircher.

British Library, London

ADD. MSS. 10268. Letters to Pietro Vettori.

ADD. MSS. 22804. Kircher to Alessandro Segni (1677–1678).

Sloane. 3322, 4063, and 4064. Filippo Bonanni to James Petiver (1703–1713).

Biblioteca Ambrosiana, Milan

Cod. 387–389 sup. *Del Museo Settala.*

D. 198 inf., ff. 110–118. Ulisse Aldrovandi, *De sepe prudentii.*

D. 332 inf., ff. 68–69. Ulisse Aldrovandi to Ascanio Persio; ff. 165–168 Adriano Spige-
lio to Giovan Vincenzo Pinelli.

G. 140 inf. (37); 141 inf. (15); 144 inf. (49); 186 inf. (118); 188 inf. (233). Ulisse
Aldrovandi and Michele Mercati to Federico Borromeo.

R. 119 sup., f. 133. Ulisse Aldrovandi to Girolamo Mercuriale.

S. 80 sup., f. 260. Letter of Ulisse Aldrovandi.

S. 85 sup., f. 235r. *Il museo di Antonio Giganti.*

Archivio di Stato, Modena

Archivio per le materie. Letterati. Giovan Battista della Porta and Michele Mercati.

Archivio per le materie. Medici. b. 19(95). Leonardo Fioravanti.

Archivio per le materie. Storia naturale.

Biblioteca Estense, Modena

Mss. Campori y.H.1.21–22 (338–339). *Disegni originali che sono descritti nell'Opera scritta
in Latino dal Dott[or] Fis[ico] Collegiato Paolo Maria Terzago, tradotto in Italiano con
aumente dal Dott[or] Fis[ico] Pietro Francesco Scarabelli e stampata in Voghera nel 1666
in un Volume in 4to da Eliseo Viola.*

Mss. Campori y.Y.5.50 (APP. 1694). *Capitoli e statuti del Collegio de Spetiali della Città di
Modena.*

Est. It. 833 (Alpha G I, 15). Ulisse Aldrovandi to the Duke of Modena (1599).

Est. It. 835 (Alpha G I, 17). Giovan Battista della Porta to Cardinale d'Este
(1580–1586).

Biblioteca Nazionale, Naples

Ms. Branc. II F 19, c. 208. *In Ulyssis Aldrovandi museum.*

Ms. Branc. I E 1, cc. 309–315. *Dubitationes aliquot observantq[ue] in Itinerarius Extatico
Doctiss. Patris Athanasii Chircheri S. J.*

Ms. Branc. IV B 13, cc. 91–119. Athanasius Kircher to Domenico Magri.

Archivio di Stato, Parma

Epistolario scelto, b. 1 (Ulisse Aldrovandi).

Biblioteca Palatina, Parma

Ms. Pal. 1010; 1012, f. 1. *Lettere di Ludovico Beccadelli.*

Archivio di Stato, Pisa

Università. 530, 4. *Spese occorse nel viaggio fatto da un simplicista per ritrovare piante e minerali d'ordine di S. A. S.*

Università. 531, 5. *Inventario della galleria e giardino de semplici di S. A. S. in Pisa.*

Università. Versamento II. Sez. G. 77, cc. 363–364 (333–334). *Professori alla cura del giardino dei semplici 1593–1614* and *Semplicisti 1547–1628.*

Archivio Secreto Vaticano, Rome

Fondo Borghese. ser. III, t. 72a. *Legislazione in Polonia del Card. Ippolito Aldrobrandini. Lettere diverse 1588–89,* ff. 379–501 (Michele Mercati).

Archivum Romanum Societatis Iesu, Rome

Fondo Gesuitico. 1069/5, cassetto III, n. 1. *Atto originale antico di consegna al N[ost]ro Museo della Galleria di Alfonso Donnino* (1651).

Rom. 138. *Historia* (1704–1729). XVI. Filippo Bonanni. *Notizie circa la Galleria del Collegio Romano* (10 January 1716).

Biblioteca dell'Accademia Nazionale dei Lincei, Rome

Archivio Linceo. ms. 18. Jan Eck, *Epistolarum medicinalium.*

Archivio Linceo. ms. 31. *Rerum medicarum Novae Hispaniae thesaurus seu plantarum animalium mineralium mexicanorum historia ex Francisci Hernandez* (Rome, 1649).

Archivio Linceo. ms. 32. *Inventario dei beni appartenenti all eredità di Federico Cesi con i relativi prezzi di stima.*

Archivio di S. Maria in Aquiro. 412. *Interessi di Giovanni Faber medico.*

Archivio di S. Maria in Aquiro. 420. *Lettere a Giovanni Faber.*

Biblioteca Angelica, Rome

Ms. 1545. *Il Museo di Michel Mercati compendiato, e riformato in cui si contengono disposte con nuovo ordine, e brevemente spiegato tutte le figure da lui fatte incidere e poi date alle stampe.*

Biblioteca Apostolica Vaticana, Rome

Autograft Ferrajoli. Raccolta prima. vol. 6, ff. 182–188. Athanasius Kircher to Raffaele Maffei.

Barberini Latini. 4252. *Botanologia esotica.*

Barberini Latini. 4265. *Giardinetto secreto del em. sig. card. Barberini.*

Barberini Latini. 6467, ff. 37–39. Athanasius Kircher to Francesco Barberini.

Barberini Latini. 6499, ff. 19–20. Athanasius Kircher to Lucas Holstenius.

Mss. Chigi. F.IV.49. *Beatissimo Patri Alexandro Septimo Pont. Opt. Maximo Athanasius Kircherus infirmus servus felicitatem.*

Mss. Chigi. F.IV.64. *Diatribe arithemetica de priscis numerorum notis earumque origine et fabrica.*

Mss. Chigi. J.VI.225. Athanasius Kircher to Alexander VII.

Vaticani Latini. 6192, vol. 2, ff. 656–657. Ulisse Aldrovandi to Cardinal Guglielmo Sirleto.

Vaticani Latini. 8258, ff. 17–22. Francesco Stelluti, *Breve trattato della natura, e qualita del legno fossile minerale ondato.*

Vaticani Latini. 9064, ff. 83–92. Kircher misc.

Biblioteca Nazionale Vittorio Emmanuele, Rome

Fondo Gesuitico. 893. Daniele Bartoli. *Lettere.*

Fondo Gesuitico. 1334. Filippo Buonanni. *Biblioteca Scriptorum Societatis Jesu.*

Pontifica Università Gregoriana, Rome

Mss. 555–568 (I–XIV). *Carteggio Kircheriano.*

Biblioteca Comunale degli Intronati, Siena

Ms. 116 (A IV, 7).
 cc. 26–42. *Statute & capitoli dell'Arte degli Speziali, del 1560.*
 cc. 45–48r. *Memoria sull'origine della Accademia degli Ardenti* (1602).
Autografi Porri. filza 5a. n. 85 (K XI, 53). Athanasius Kircher to Raffaele Maffei.

Biblioteca Nazionale Marciana, Venice

Archivio Morelliano. 103 (= *Marciana* 12609). *Catalogo delle cose naturali mandate al Serenissimo Francesco de' Medici Gran Duca di Toscana dal Dottore Ulisse Aldrovandi Bolognese.*

Mss. Italiani IV, 133 (= *Marciana* 5103). *Raccolta delle inscrittioni, cossì antiche, come moderne, quadri, e pitture, statue, bronzi, manui, medaglie, gemme, minere, animali, petriti, libri, instrumenti methematici che si trovano in Pustiria nella casa et horti, che sono di me Girolamo de Galdo q. Emilio Dr. che serve anco per Inventario MDCXLIII nel mese di Decembre 27.*

Archivio di Stato, Verona

Arte speziale, n. 25. *Raccolta di parti, ordini, giudici e terminazioni in favor de' speziali da medicinali.*

Arte speziale, n. 26. *Ordini, statuti, capituli formati per regola e governo della magnifica arte specieri* (1568).

Arte speziale, n. 31. *Libro dell'arte degli speziali* (1589–1658).

Biblioteca Civica, Verona

Ms. 151 O. Ognibene Rigotti, *De Ponae familiae nobilitate historicum documentum.*

Ms. 2047. *Produzioni marine cioé cochle, altioni, turbineti, coralloide madrepore, fuchi, e simili. Raccolte, e delineate da me Fra Petronio da Verona capuccino infermiere nel santiss[i]mo Rendetore di Venezia* (1724).

II. SELECTED PRIMARY SOURCES

Aldrovandi, Ulisse. *De animalibus insectis libri septum* (Bologna, 1602).
———. "Avvertimenti del Dottor Aldrovandi." In *Trattati d'arte del Cinquecento,* Paola Barocchi, ed. (Bari: Laterza, 1961), vol. 2, pp. 511–517.
———. *Dendrologia* (Bologna, 1648).
———. *De mollibus, crustaceis, testaceis, et zoophytis* (Bologna, 1606).
———. *Monstrorum historia* (Bologna, 1642).
———. *Musaeum metallicum* (Bologna, 1648).
———. *Ornithologiae hoc est de avibus historiae libri XII* (Bologna, 1599).
———. *De piscibus libri V et de cetis lib[rus] unus* (Bologna, 1613).
Alifer, Joannes Andrea. *Manfredo Septalio Academia funebris publice habita in classe rhetoricae Collegi Braydensis Societatis Jesu* (Milan, 1680).
Anecdota litteraria ex mss. codicibus eruta, vol. 4 (Rome, 1783).
Anguillara, Luigi. *Semplici dell'Eccellente M. Luigi Anguillara, liquali in piu pareri a diversi nobili huomini scritti appaiono* (Venice, 1561).
Bacci, Andrea. *L'Alicorno* (Florence, 1573).

Balfour, Andrew. *Letters Write to a Friend, . . . Containing Excellent Direction and Advices for Travelling thro' France and Italy* (Edinburgh, 1700).

Bartholin, Thomas. *On the Burning of His Library and On Medical Travel,* Charles D. O'Malley, trans. (Lawrence: University of Kansas Press, 1961).

Bartoli, Daniele. *La ricreatione del savio in discorso con la natura e con Dio* (Rome, 1659).

Bellori, Giovan Pietro. *Nota delli musei, librerie, galerie, et ornamenti di statue e pitture ne' palazzi, nelle case, e ne' Giardini di Roma* (Rome, 1664).

Bertioli, Antonio. *Delle considerazioni di Antonio Berthioli Mantovano sopra l'olio di scorpioni dell'eccellentissimo Matthioli* (Mantua, 1585).

Bertioli, Giovan Paolo, and Antonio Bertioli. *Breve avviso del vero balsamo, theriaca et mithridato* (Mantua, 1596).

Bocchi, Zenobio. *Giardino de' semplici in Mantova* (Mantua, 1603).

Boccone, Paolo. *Museo di fisica e di esperienze variato, e decorato di osservazioni naturali* (Venice, 1697).

————. *Museo di piante rare della Sicilia, Malta, Corsica, Italia, Piemonte, e Germania* (Venice, 1697).

————. *Osservazioni naturali* (Bologna, 1684).

Bolletti, Giuseppe Gaetano. *Dell'origine e de' progresso dell'Istituto delle Scienze di Bologna* (Bologna, 1751).

Bonanni, Filippo. *Musaeum Kircherianum* (Rome, 1709).

————. *Rerum naturalium historia,* Giovan Antonio Battarra, ed. (Rome, 1773).

————. *Ricreatione dell'occhio e della mente nell'osservatione delle chiocciole* (Rome, 1681).

———— (pseud. Godefrido Fulberti). *Riflessioni sopra la relatione del ritrovamento dell'uova di chiocciole di A. F. M. in una lettera al sig. Marcello Malpighi* (Rome, 1683).

Borgarucci, Prospero. *La fabrica de gli speziali* (Venice, 1566).

Borromeo, Federico. *Federici Cardinalis Borromei Archiepisc. Mediolani Musaeum (Il museo di Cardinale Federigo Borromeo, Arcivescovo di Milano),* Luigi Grasselli, trans. (Milan, 1909).

Bromley, William. *Remarks made in Travels through France and Italy* (London, 1693).

Buoni, Giacomo Antonio. *Del terremoto* (Modena, 1572).

Burnet, Gilbert. *Some Letters Containing an Account of what seemed most Remarkable in Travelling through Switzerland, Italy, Some Parts of German, &c. In the Years 1685 and 1686* (London, 1689).

Calzolari, Francesco. *Lettera di M. Francesco Calceolari spetiale al segno della campagna d'oro, in Verona. Intorno ad alcune menzogne & calonnie date alla sua theriaca da certo Scalcina perugino* (Cremona, 1566).

————. *Il viaggio di Monte Baldo* (Venice, 1566).

Camillo, Giulio. *L'idea del theatro dell'eccellent. M. Giulio Camillo* (Florence, 1550).

Campanella, Tommaso. *Del senso delle cose e della magia,* Antonio Bruers, ed. (Bari: Laterza, 1925).

Capparoni, Pietro. "Una lettera inedita di Manfredo Settala." *Rivista di storia critica delle scienze mediche e naturali* 5 (1914): 348–350.

Cardano, Girolamo. *The Book of My Life,* Jean Stoner, trans. (New York: Dover, 1962).

Castelli, Pietro. *Discorso della differenza tra gli semplici freschi et i secchi con il modo di seccarli* (Rome, 1629).

Ceñal, Ramon. "Juan Caramuel. Su epistolario con Atanasio Kircher, S. J." *Revista de filosofia* 12 (1953): 101–147.

Cermenati, Mario. "Francesco Calzolari e le sue lettere all'Aldrovandi." *Annali di botanica* 7 (1908): 83–108.

Ceruti, Benedetto, and Andrea Chiocco. *Musaeum Francisci Calceolari Iunioris Veronensis* (Verona, 1622).

Cesalpino, Andrea. *De metallicus libri tres* (Rome, 1596).

―――. *De plantis libri XVI* (Florence, 1583).

―――. *Questions péripatéticiennes*, Maurice Dorolle, trans. (Paris, 1929).

Cesi, Federico. "Del natural desiderio di sapere et instituzione de' Lincei per adempimento di esso." In *Scienziati del Seicento*, Maria Altieri Biagi and Bruno Basile, eds. (Milan: Ricciardi, 1980).

Clemens, Claude. *Musei, sive bibliothecae tam privatae quam publicae extructio, instructio, cura, usus* (Lugduni, 1635).

Cortesi, Fabrizio. "Alcune lettere inedite di Ferrante Imperato." *Annali di botanica* 6 (1907): 121–130.

―――. "Alcune lettere inedite di Giovanni Pona." *Annali di botanica* 6 (1908): 411 425.

―――. "Una lettera inedita di Tobia Aldini a Giovan Battista Faber." *Annali di botanica* 6 (1908): 403–405.

Cortesi, Paolo. *The Renaissance Cardinal's Ideal Palace: A Chapter from Cortesi's De Cardinalatu*, Kathleen Weil-Garris and John F. d'Amico, eds. and trans. (Rome: Edizioni dell'Elefante, 1980).

Costa, Filippo. *Discorsi di M. Filippo Costa sopra le compositioni degli antidoti, & medicamenti, che piu si costumano di dar per bocca* (Mantua, 1586 ed.).

Dati, Carlo. *Delle lodi del commendatore Cassiano dal Pozzo* (Florence, 1664).

Della Casa, Giovanni. *Galateo*, Konrad Eisenbichler and Kenneth R. Bartlett, trans. (Toronto: Center for Renaissance and Reformation Studies, 1990).

Della Porta, Giovan Battista. *Criptologia*, Gabriela Belloni, ed. and trans. (Rome: Centro Internazionale di Studi Umanistici, 1982).

―――. *Della fisonomia dell'huomo di Giovan Battista Della Porta napolitano*, Giovanni di Rosa, trans. (Naples, 1598 ed.).

―――. *Natural Magick*, Derek J. Price, ed. (New York: Dover, 1957).

―――. *Phytognomonica* (Naples, 1588).

De Rosa, Stefano. "Ulisse Aldrovandi e la Toscana. Quattro lettre inedite dello scienziato a Francesco I e Ferdinando I de'Medici e a Belisario Vinta." *Annali dell'Istituto e Museo di Storia della Scienza* 6 (1981): 203–215.

De Sepi, Giorgio. *Romani Collegii Societatus Jesu Musaeum Celeberrimum* (Amsterdam, 1678).

De Toni, G. B. "Il carteggio degli italiani col botanico Carlo Clusio nella Biblioteca Leidense." *Memorie della R. Accademia di scienze, lettere ed arti di Modena*, ser. 3, 10, Pt. II (1912): 113–270.

―――. *Cinque lettere di Luca Ghini ad Ulisse Aldrovandi* (Padua: Tipografia Seminario, 1905).

―――. "Spigolature aldrovandiane I. I placiti inediti di Luca Ghini nei manoscritti aldrovandiani di Bologna." In *Atti del Congresso dei naturalisti italiani* (Milan: Tipografia degli Operai, 1907), pp. 3–5.

―――. "Spigolature aldrovandiane II. Scritti aldrovandiani nella Biblioteca Ambrosiana di Milano." In *Atti del Congresso dei naturalisti italiani* (Milan: Tipografia degli Operai, 1907), pp. 5–7.

_____. "Spigolature aldrovandiane III. Nuovi dati intorno alle relazioni tra Ulisse Aldrovandi e Gherardo Cibo." *Memorie della R. Accademia di scienze, lettere ed arti di Modena*, ser. 3, 7 (1907): 3–12.

_____. "Spigolature aldrovandiane V. Ricordi d'antiche collezioni veronesi nei manoscritti aldrovandiani." *Madonna verona* 1, f. 1 (1907): 18–26.

_____. "Spigolature aldrovandiane VI. Le piante dell'antico orto botanico di Pisa ai tempi di Luca Ghini." *Annali di botanica* 5 (1907): 421–425.

_____. "Spigolature aldrovandiane VII. Notizie intorno ad un erbario perduto del medico Francesco Petrollini e contribuzione alla storia dell'erbario di Ulisse Aldrovandi." *Nuovo giornale botanico italiano*, n.s. 14, 4 (1907): 506–518.

_____. "Spigolature aldrovandiane VIII. Nuovi documenti intorno a Giacomo Raynaud farmacista di Marsiglia ed alle sue relazioni con Ulisse Aldrovandi." *Atti della Reale Istituto Veneto di scienze, lettere ed arti* 68, Pt. II (1908): 117–131.

_____. "Spigolature aldrovandiane IX. Nuovi documenti intorno Francesco Petrollini, prima guida di Ulisse Aldrovandi nello studio delle piante." *Atti del Reale Istituto Veneto di scienze, lettere ed arti* 69, Pt. II (1909–1910): 815–825.

_____. "Spigolature aldrovandiane X. Alcune lettere di Gabriele Falloppia ad Ulisse Aldrovandi." *Atti e memorie della R. Deputazione di storia patria per le provincie modenesi*, ser. 5, 7 (1913): 34–46.

_____. "Spigolature aldrovandiane XI. Intorno alle relazioni del botanico Melchiorre Guillandino con Ulisse Aldrovandi." *Atti della R. Accademia di scienze, lettere ed arti degli Agiati in Rovereto*, ser. 3, 17, f. 2 (1911): 3–25.

_____. "Spigolature aldrovandiane XII. Di Tommaso Bonaretti, medico reggiano, corrispondente di Ulisse Aldrovandi." *Atti e memorie della R. Deputazione di storia patria per le provincie modenesi*, ser. 5, 7 (1913): 82–99.

_____. "Spigolature aldrovandiane XIV. Cinque lettere inedite di Antonio Compagnoni di Macerata ad Ulisse Aldrovandi." *Rivista di storia critica delle scienze mediche e naturali* 6, 3 (1915): 479–486.

_____. "Spigolature aldrovandiane XVI. Intorno alcune lettere di Ulisse Aldrovandi esistenti in Modena." *Atti e memorie della R. Deputazione di storia patria per le provincie modenesi*, ser. 5, 13 (1920): 1–10.

_____. "Spigolature aldrovandiane XVII. Lettere inedite di Francesco Barozzi." *Ateneo veneto* 40 (1917): 133–140.

_____. "Spigolature aldrovandiane XVIII. Lettere di Giovanni Vincenzo Pinelli." *Archivio di storia della scienza* 1 (1919–1920): 297–312.

_____. "Spigolature aldrovandiane XIX. Il botanico padovano Giacomo Antonio Cortusa nelle sue relazioni con Ulisse Aldrovandi e con altri naturalisti." In *Contributo del R. Istituto Veneto di scienze, lettere ed arti alla celebrazione del VII centenario della Università di Padova* (Venice: Carlo Ferrari, 1922), pp. 217–249.

_____. "Spigolature aldrovandiane XX. Gentile dalla Torre veronese e le sue relazioni con Ulisse Aldrovandi." *Atti dell'Accademia d'agricoltura, scienze e lettere di Verona*, ser. 4, 25 (1923): 147–151.

_____. "Spigolature aldrovandiane XXI. Un pugilio di lettere di Giovanni Odorico Melchiori Trentino a Ulisse Aldrovandi." *Atti del Reale Istituto Veneto di scienze, lettere ed arti* 84, Pt. II (1924–1925): 599–624.

Diderot, Denis. "Cabinet d'histoire naturelle." *Encyclopédie ou dictionnaire raisonné des sciences, des arts et des métiers* (Paris, 1751), vol. 2, pp. 489–492.

Documenti inediti per servire alla storia dei musei d'Italia, 4 vols. (Florence: Bencini, 1878–1880).

"Elogio di P. Philippo Buonanni." *Giornale de' letterati d'Italia* 37 (1725): 360–388.

Fantuzzi, Giovanni. *Memorie della vita di Ulisse Aldrovandi* (Bologna, 1774).

_____. "Ulisse Aldrovandi." In his *Notizie degli scrittori bolognesi* (Bologna, 1781), vol. 1, pp. 165–190.

Fedeli, Carlo. "Un nuovo documento sul primo orto botanico pisano." *Rivista di storia delle scienze mediche e naturali*, ser. 3, n. 7–8 (1923): 177–181.

Felici, Costanzo. *Lettere a Ulisse Aldrovandi*, Giorgio Nonni, ed. (Urbino: Quattro Venti, 1982).

Fioravanti, Leonardo. *De' capricci medicinali dell'eccellente medico, & cirugico M. Leonardo Fioravanti Bolognese* (Venice, 1573).

_____. *Del compendio dei secreti rationali* (Turin, 1580).

_____. *Dello specchio di scientia universale* (Venice, 1564).

_____. *Il tesoro dell vita umana* (Venice, 1582).

Florio, Luigi di. "Una lettera inedita di Ulisse Aldrovandi." *Pagine di storia della medicina* 9 (1965): 40–45.

Fontana, Felice. *Saggio del real gabinetto di fisica e di storia naturale di Firenze* (Rome, 1775).

"Frammenti del processo di Ulisse Aldrovandi (Bologna, 5 luglio 1549)." In Camillo Renato, *Opere*, Antonio Rotondò, ed. (Florence: Sansoni, 1968), pp. 224–227.

Frati, Ludovico, ed. "La vita di Ulisse Aldrovandi scritta da lui medesimo." In *Intorno alla vita e alle opere di Ulisse Aldrovandi* (Imola, 1907).

Friedlander, Paul. "Athanasius Kircher und Leibniz. Ein Beitrag zur Geschichte der Polyhistorie im XVII. Jahrhundert." *Rendiconti della Pontificia Accademia Romana di Archeologia* 13, f. 3–4 (1937): 229–247.

Gabrieli, Giuseppe. "Il carteggio linceo della vecchia accademia di Federico Cesi (1603–1630)." *Memorie della R. Accademia Nazionale dei Lincei. Classe di scienze morali, storiche e filologiche*, ser. 6, Vol. VII, f. 1–4 (1938–1942).

Garzoni, Tommaso. *La piazza universale di tutte le professioni del mondo* (Venice, 1651 ed.).

Gassendi, Pierre. *The Mirrour of True Nobility & Gentility. Being a Life of the Renowned Nicolaus Claudo Fabricius Lord of Peiresk, Senator of the Parliament at Aix*, W. Rand, trans. (London, 1657).

Ginori Conti, Piero. *Lettere inedite di Charles de l'Escluse (Carolus Clusius) a Matteo Caccini* (Florence: Olschki, 1939).

Giovio, Paolo. *Libro di mons. Paolo Giovio de' pesci romani*, Carlo Zancaruolo, trans. (Venice, 1560).

Gualtieri, Nicolo. *Index testarum conchyliorum quae adservantur in musaeo Nicolai Gualtieri* (Florence, 1742).

Hall, A. Rupert, and Marie Boas Hall, eds. and trans. *The Correspondence of Henry Oldenburg*, 9 vols. (Madison: University of Wisconsin Press, 1965–1973).

Imperato, Ferrante. *Dell'historia naturale* (Naples, 1599).

Imperato, Francesco. *Discorsi intorno a diverse cose naturali* (Naples, 1628).

_____. *De fossilibus opusculum* (Naples, 1610).

Jaucourt, Chevalier de. "Musée." *Encyclopédie ou dictionnaire raisonné des sciences, des arts et des métiers* (Neufchastel, 1765), vol. 10, pp. 893–894.

Kestler, Johann. *Physiologia Kircheriana experimentalis* (Amsterdam, 1680).

Kircher, Athanasius. *Arca Noë* (Amsterdam, 1675).

———. *Arithmologia sive de abditis numerorum mysteriis* (Rome, 1665).

———. *Ars magna lucis et umbrae* (Amsterdam, 1671).

———. *Ars magna sciendi* (Amsterdam, 1669).

———. *Ars magnesia* (Würzburg, 1631).

———. *China Illustrata*, Charles D. Van Tuyl, trans. (Muskegee, OK: Indian University Press, 1987).

———. *Diatribe de prodigiosis crucibus, quae tam supra vestes hominum, quam res alias, non pridem post ultimum incendium Vesuvii Montis* (Rome, 1661).

———. *Lingua Aegyptiaca restituta* (Rome, 1643).

———. *Magnes sive de arte magnetica libri tres* (Rome, 1641).

———. *Magneticum naturae regnum sive Disceptatio physiologica de triplice in natura rerum magnete* (Amsterdam, 1667).

———. *Mundus subterraneus* (Amsterdam, 1664).

———. *Oedipus Aegyptiacus*, 4 vols. (Rome, 1652–1654).

———. *Prodromus coptus sive Aegyptiacus* (Rome, 1636).

———. *Turris Babel* (Amsterdam, 1679).

———. *Vita admodum P. Athanasii Kircher Societatis Iesu Viri toto orbe celebratissimi* (Augsburg, 1684).

Kirchmayer, Georg Caspar. *Un-Natural History, or Myths of Ancient Science*, Edmund Goldsmid, ed. (Edinburgh, 1886).

Lana Terzi, Francesco. *Prodromo overo saggio di alcune inventioni nuove* (Brescia, 1670).

Lassels, Richard. *The Voyage of Italy, or A Complete Journey through Italy. In Two Parts* (London, 1670).

Legati, Lorenzo. *Breve descrizione del museo dell'Illustriss. Sig. Cav. Commend. dell'Ordine di S. Stefano Ferdinando Cospi* (Bologna, 1667).

———. *Museo Cospiano annesso a quello del famoso Ulisse Aldrovandi e donato alla sua patria dall'illustrissimo Signor Ferdinando Cospi* (Bologna, 1677).

Lumbroso, Giacomo. *Notizie sulla vita di Cassiano dal Pozzo* (Turin, 1874).

Mabillon, Jean, and D. Michael Germain. *Museum Italicum seu Collectio veterum scriptorum ex Bibliotecis Italicus*, 2 vols. (Paris, 1687–1689).

Maffei, Scipione. *Epistolario (1700–1755)*, Celestino Garibotto, ed., 2 vols. (Milan: Giuffrè, 1955).

———. *Verona illustrata*, 4 vols. (Verona, 1731–1732).

Marani, Alberto. "Lettere di Muzio Calini a Ludovico Beccadelli." *Commentari dell'Ateneo di Brescia* 168 (1969): 59–143.

Maranta, Bartolomeo. *Della theriaca et del mithridato libri due* (Venice, 1572).

Mattioli, Pier Andrea. *I discorsi ne i sei libri della materia medicinale di Pedacio Dioscoride Anazarbeo* (Venice, 1557).

Mattirolo, Oreste. "Le lettere di Ulisse Aldrovandi a Francesco I e Ferdinando I Granduchi di Toscana e a Francesco Maria II Duca di Urbino." *Memorie della Reale Accademia delle Scienze di Torino*, ser. II, 54 (1903–1904): 355–401.

Mercati, Angelo. "Lettere di scienziati dall'Archivio Segreto Vaticano." *Commentationes Pontificia Academia Scientiarum* 5, vol. V, n. 2 (1941): 61–209.

Mercati, Michele, *Metallotheca* (Rome, 1717).

Middleton, W. E. Knowles, ed. and trans. *Lorenzo Magolotti at the Court of Charles II. His Relazione d'Inghilterra of 1668* (Waterloo, Ontario, 1980).

Migliorato-Garavini, E. "Appunti di storia della scienza nel Seicento. I. Tre lettere inedite del naturalista Ferrante Imperato ed alcune notizie sul suo erbario." *Rendiconti della R. Accademia Nazionale dei Lincei. Classe di scienze morali, storiche e filologiche,* ser. 8, vol. 7, f. 1–2 (1952): 33–39.

Misson, Maximilian. *A New Voyage to Italy,* English trans., 2 vols. (London, 1695).

Montalbani, Ovidio. *Curae analyticae aliquot naturalium observationum Aldrovandicas circa historias* (Bologna, 1671).

Morhof, Daniel Georg. *Polyhistor literarius, philosophicus et praticus* (Lubeck, 1747, 1688).

Moscardo, Lodovico. *Note overo memorie del museo di Lodovico Moscardo nobile veronese, academico filarmonico, dal medesimo descritte, et in tre libri distinte* (Padua, 1651).

Neickelius, C. F. *Museographia* (Leipzig, 1727).

Olivi, Giovan Battista. *De reconditis et praecipuis collectaneis ab honestissimo, et solertiss.mo Francisco Calceolari Veronensi in musaeo adservatis* (Venice, 1584).

Pastorini, Giovan Battista, S. J. *Orazion funebre per la morte dell'illustriss. sig. can. Manfredo Settala* (Milan, 1680).

Peiresc, Nicolas Claude Fabri de. *Lettres à Cassiano dal Pozzo (1626–1637),* Jean-François Lhote and Danielle Joyal, eds. (Clermont-Ferrand, 1989).

Petrucci, Gioseffo. *Prodromo apologetico alli studi chircheriani* (Amsterdam, 1677).

Pietro, Pericle di. "Epistolario di Gabriele Falloppia." *Quaderni di storia della scienza e della medicina* 10 (1970).

Pliny. *Natural History,* John Bostock and H. T. Riley, trans., 7 vols. (London, 1855).

Pona, Francesco. *L'Amalthea overo della pietra belzoar orientale* (Venice, 1626).

Porro, Girolamo. *L'Horto de i semplici di Padova* (Venice, 1592).

Quiccheberg, Samuel. *Inscriptiones vel tituli theatri amplissimi* (Munich, 1565).

Raimondi, C. "Lettere di P. A. Mattioli ad Ulisse Aldrovandi." *Bullettino senese di storia patria* 13, f. 1–2 (1906): 3–67.

Ray, John. *Travels through the Low-Countries, Germany, Italy, and France. With curious observations* (London, 1738).

Redi, Francesco. *Esperienze intorno a diverse cose naturali, e particolarmente a quelle che ci son portate dall'Indie* (Florence, 1686).

———. *Esperienze intorno alla generazione degl'insetti* (Florence, 1668).

———. *Experiments on the Generation of Insects,* Mab Bigelow, trans. (Chicago, 1909).

———. *Francesco Redi on Vipers,* Peter Knoefel, ed. and trans. (Leiden: Brill, 1988).

———. *Osservazioni intorno alle vipere* (Florence, 1664).

Robinet, André. *G. W. Leibniz Iter Italicum (Mars 1689–Mars 1690). La dynamique de la République des lettres. Nombreux textes inédits,* Studia dell'Accademia Toscana di Scienze e Lettere "La Colombaria," vol. 90 (Florence: Olschki, 1987).

Ronchini, A. "Ulisse Aldrovandi e i Farnese." *Atti e memorie delle RR. Deputazioni di storia patria per le provinicie dell'Emilia,* n.s. 5, Pt. II (1880): 1–14.

Salviani, Ippolito. *Acquatilium animalium historiae* (Rome, 1554).

Scarabelli, Pietro (see Terzago, Paolo).

Schott, Gaspar. *Ioco-seriorum naturae et artis, sive magiae naturalis centuriae tres* (Würzburg, 1666).

———. *Physica curiosa, sive mirabilia naturae et artis libris XII,* 2 vols. (Würzburg, 1697).

———. *Technica curiosa, sive mirabilia artis, libris XII* (Würzburg, 1664).

Scilla, Agostino. *La vana speculazione disingannata dal senso. Lettera risponsiva circa i corpi marini, che petrificati si trovano in varii luoghi terrestri* (Naples, 1670).

Serpetro, Nicolò. *Il mercato delle maraviglie della natura, overo istoria naturale* (Venice, 1659).

Simili, Alessandro. "Alcune lettere inedite di Andrea Bacci a Ulisse Aldrovandi." In *Atti del XXIV Congresso Nazionale di Storia della Medicina (Taranto-Bari, 1969)* (Rome: Società italiana di storia della medicina, 1969), pp. 428–437.

Skippon, Philip. *An Account of a Journey Made thro' Part of the Low-Countries, Germany, Italy and France.* In *A Collection of Voyages and Travels*, A. Churchill and S. Churchill, eds. (London, 1752 ed.), vol. 6.

Spon, Jacob. *Voyage d'Italie, de Dalmatie, de Grece, et du Lévant, fait aux années 1675 & 1676*, 2 vols. (The Hague, 1724 ed.).

Stelluti, Francesco. *Persio tradotto* (Rome, 1630).

———. *Trattato del legno fossile minerale* (Rome, 1637).

Steno, Nicolaus. *The Earliest Geological Treatise (1667)*, Avel Garboe, trans. (London, 1958).

Terzago, Paolo Maria. *Musaeum Septalianum Manfredi Septalae Patritii Mediolanensis* (Dertonae, 1664).

———. *Museo o galleria adunata dal sapere, e dallo studio del sig. Canonico Manfredo Settala nobile milanese*, Pietro Francesco Scarabelli, trans. (Tortona, 1666).

Tesauro, Emanuele. *Il cannocchiale aristotelico* (Turin, 1670 ed.).

———. *La filosofia morale* (Venice, 1792 ed.).

Tosi, Alessandro, ed. *Ulisse Aldrovandi e la Toscana: Carteggio e testimonianze documentarie* (Florence: Olschki, 1989).

Valentini, Michael Bernhard. *Museum museorum*, 2 vols. (Frankfort, 1704–1714).

Vallieri, Werner. "Le 22 lettere di Bartolomeo Maranta all'Aldrovandi." In *Atti del XIX Congresso Nazionale di Storia della Medicina (L'Acquilia, 1963)* (Rome: Società italiana di storia della medicina, 1965), pp. 738–770.

Varchi, Benedetto. *Questione sull'alchimia*, Domenico Morani, ed. (Florence: Stamperia Magheri, 1827).

Visconti, Giovanni Maria. *Exequiae in tempio S. Nazarii Manfredo Septalio Patritio Mediolanensi* (Milan, 1680).

Zanoni, Giacomo. *Istoria botanica* (Bologna, 1675).

III. SELECTED SECONDARY SOURCES

Accordi, Bruno. "Ferrante Imperato (Napoli 1550–1625) e il suo contributo alla storia della geologia." *Geologica Romana* 20 (1981): 43–56.

———. "Illustrators of the Kircher Museum Naturalistic Collections." *Geologica Romana* 15 (1976): 113–126.

———. "Michele Mercati (1541–1593) e la Metallotheca." *Geologica Romana* 19 (1980): 1–50.

———. "The Musaeum Calceolarium of Verona Illustrated in 1622 by Ceruti and Chiocco." *Geologica Romana* 16 (1977): 21–54.

———. "Paolo Boccone (1633–1704)." *Geologica Romana* 14 (1975): 353–359.

Adversi, Aldo. "Ulisse Aldrovandi bibliofilo, bibliografo e bibliologo del Cinquecento." *Annali di scuola speciale per archivisti e bibliotecari dell'Università di Roma* 8, n. 1–2 (1968): 85–181.

Aimi, Antonio, Vincenzo De Michele, and Alessandro Morandotti. *Musaeum Septalianum: una collezione scientifica nella Milano del Seicento* (Florence: Giunti Marzocco, 1984).

Åkerman, Susanna. *Queen Christina of Sweden and Her Circle: The Transformation of a Seventeenth-Century Philosophical Libertine* (Leiden: Brill, 1991).

Andreoli, Aldo. "Un inedito breve di Gregorio XII a Ulisse Aldrovandi." *Atti e memorie dell'Accademia Nazionale di scienze, lettere e arti di Modena*, ser. 6, vol. 4 (1962): 133–149.

———. "Ulisse Aldrovandi e Gregorio XIII (e la teriaca)." *Strenna storica bolognese* 11 (1961): 11–19.

Arnold, Ken. *Cabinets for the Curious: Practicing Science in Early Modern English Museums* (Ph.D. diss., Princeton University, 1991).

Ashworth, William B., Jr. "Catholicism and Early Modern Science." In *God and Nature: Historical Essays on the Encounter Between Christianity and Science*, David C. Lindberg and Ronald L. Numbers, eds. (Berkeley, 1987), pp. 136–166.

———. "Natural History and the Emblematic World View." In *Reappraisals of the Scientific Revolution*, David C. Lindberg and Robert S. Westman, eds. (Cambridge, U.K.: Cambridge University Press, 1990), pp. 303–332.

Atran, Scott. *Cognitive Foundations of Natural History: Towards an Anthropology of Science* (Cambridge, U.K.: Cambridge University Press, 1989).

Azzi Visentini, Margherita. *L'Orto Botanico di Padova e il giardino del Rinascimento* (Milan: Edizioni il Polifilo, 1984).

Baldwin, Martha R. "Alchemy in the Society of Jesus." In *Alchemy Revisited: Proceedings of the International Conference on the History of Alchemy at the University of Groningen 17–19 April 1989*, Z. R. W. M. von Martels, ed. (Leiden: Brill, 1990), pp. 182–187.

———. "Magnetism and the Anti-Copernican Polemic." *Journal of the History of Astronomy* 16 (1985): 155–174.

Balsiger, Barbara. *The 'Kunst- und Wunderkammern': A Catalogue Raisonné of Collecting in Germany, France and England 1565–1750* (Ph.D. diss., University of Pittsburgh, 1970).

Barocchi, Paola, and Giovanna Ragionieri, eds. *Gli Uffizi. Quattro secoli di una galleria*, 2 vols. (Florence: Olschki, 1983).

Basile, Bruno. *L'invenzione del vero. La letteratura scientific da Galileo ad Algarotti* (Rome: Salerno, 1987).

Battisti, Eugenio. *L'Antirinascimento* (Milan: Feltrinelli, 1962).

Bedini, Silvio. "The Evolution of Science Museums." *Technology and Culture* 6 (1965): 1–29.

Benzoni, Gino. *Gli affanni della cultura: Intellettuali e potere nell'Italia della Controriforma e Barocca* (Milan: Feltrinelli, 1978).

Berti, Luciano. *Il Principe dello studiolo: Francesco I dei Medici e la fine del Rinascimento fiorentino* (Florence: EDAM, 1967).

Biagioli, Mario. "Absolutism, the Modern State and the Development of Scientific Manners." *Critical Inquiry* 19 (1993, in press).

———. *Galileo, Courtier.* (Chicago: The University of Chicago Press, 1993).

———. "Galileo's System of Patronage." *History of Science* 28 (1990): 1–62.

———. "Galileo the Emblem Maker." *Isis* 81 (1990): 230–258.

———. "Scientific Revolution, Social Bricolage and Etiquette." In *The Scientific Revolution in National Context*, Roy Porter and Mikulas Teich, eds. (Cambridge, U.K.: Cambridge University Press, 1992), pp. 11–54.

————. "The Social Status of Italian Mathematicians, 1450–1600." *History of Science* 27 (1989): 41–95.

Bianchi, Massimo Luigi. *Signatura rerum. Segni, magia e conoscenza da Paracelso a Leibniz* (Rome: Salerno, 1987).

Blair, Ann. *Restaging Jean Bodin: The "Universae naturae theatrum" (1596) in its Cultural Context* (Ph.D. diss., Princeton University, 1990).

Blumenberg, Hans. *La leggibilità del mondo. Il libro come metafora della natura*, Bruno Argenton, trans. (Bologna: Il Mulino, 1984).

Boehm, Laetitia, and Ezio Raimondi, eds. *Università, accademie e società scientifiche in Italia e in Germania dal Cinquecento al Settecento* (Bologna: Il Mulino, 1981).

Bolzoni, Lina. "L'"invenzione' dello stanzino di Francesco I." In *Le arti del principato mediceo* (Florence: Studio per le edizioni scelte, 1980), pp. 255–299.

————. "Parole e immagini per il ritratto di un nuovo Ulisse: l'"invenzione' dell' Aldrovandi per la sua villa di campagna." In Elisabeth Cropper, ed., *Documentary Culture: Florence and Rome from Grand Duke Ferdinando I to Pope Alexander VII*, Villa Spelman Colloquia, vol. 3 (Bologna: La Nuova Alfa, 1992), pp. 317-348.

————. "Teatro, pittura e fisiognomica nell'arte della memoria di Giovan Battista della Porta." *Intersezioni* 8 (1988): 477–509.

Brizzi, Gian Paolo. *La formazione della classe dirigente nel Sei-Settecento: I seminaria nobilium nell'Italia centro-settentrionale* (Bologna: Il Mulino, 1976).

————, ed. *La 'Ratio studiorum.' Modelli culturali e pratiche educative dei Gesuiti in Italia tra Cinque e Seicento* (Rome: Bulzoni, 1981).

Burke, Peter. *The Historical Anthropology of Early Modern Italy* (Cambridge, U.K.: Cambridge University Press, 1987).

Capparoni, Pietro. *Profili bio-bibliografici di medici e naturalisti celebri italiani dal sec. XVo al sec XVIIIo*, 2 vols. (Rome: Istituto Nazionale Medico Farmacologico, 1925–1928).

Casciato, Maristella, Maria Grazia Ianniello, and Maria Vitale, eds. *Enciclopedismo in Roma barocca: Athanasius Kircher e il Museo del Collegio Romano tra Wunderkammer e museo scientifico* (Venice: Marsili, 1986).

Cascio Pratilli, Giovanni. *L'Università e il Principe: Gli Studi di Siena e di Pisa tra Rinascimento e Controriforma* (Florence: Olschki, 1975).

Cavazza, Marta. *Settecento Inquieto. Alle origini dell'Istituto delle Scienze di Bologna* (Bologna: Il Mulino, 1990).

Céard, Jean. *La nature et les prodiges: l'insolite au XVIe siècle, en France* (Geneva: Droz, 1977).

Céard, Jean, et al. *La curiosité à la Renaissance* (Paris: Société Française des Seizièmistes, 1986).

Cermenati, Mario. "Ulisse Aldrovandi e l'America." *Annali di botanica* 4 (1906): 3–56.

Chiovenda, Emilo. "Francesco Petrollini botanico del secolo XVI." *Annali di botanica* 7 (1909): 339–447.

Cochrane, Eric. *Florence in the Forgotten Centuries 1527–1800* (Chicago: The University of Chicago Press, 1973).

————. *Italy 1530–1630*, Julius Kirshner, ed. (London: Longman, 1988).

————. "The Renaissance Academies in Their Italian and European Setting." In *The Fairest Flower: The Emergence of Linguistic National Consciousness in Renaissance Europe* (Florence: Crusca, 1985), pp. 21–39.

Colie, Rosalie. *Paradoxia Epidemica: The Renaissance Tradition of Paradox* (Princeton, 1966).

Colliva, Paolo. "Bologna dal XIV al XVIII secolo: 'governo misto' o signoria senatoria?" In *Storia della Emilia Romagna*, Aldo Berselli, ed. (Imola: Edizioni Santerno, 1977), pp. 13–34.

Comelli, G. B. "Ferdinando Cospi e le origini del Museo Civico di Bologna." *Atti e memorie della Reale Deputazione di storia patria per le provincie della Romagna*, ser. 3, 7 (1889): 96–127.

Copenhaver, Brian. "Natural Magic, Hermeticism and Occultism in Early Modern Science." In *Reappraisals of the Scientific Revolution*, David L. Lindberg and Robert S. Westman, eds. (Cambridge, U.K.: Cambridge University Press, 1990), pp. 261–301.

––––––. "A Tale of Two Fishes: Magical Objects in Natural History from Antiquity to the Scientific Revolution." *Journal of the History of Ideas* 52 (1991): 373–398.

Cortese, Nino. *Cultura e politica a Napoli dal Cinquecento al Settecento* (Naples: Edizioni scientifiche italiane, 1965).

Covoni, P. F. *Don Antonio de' Medici al Casino di San Marco* (Florence: Tipografia Cooperativa, 1892).

Cuaz, Marco. *Intellettuali, potere e circolazioni delle idee nell'italia moderna 1500–1700* (Turin: Loescher, 1982).

Cultura popolare nell'Emilia Romagna. Medicina, erbe e magia (Milan: Silvana Editoriale, 1981).

Dannenfeldt, Karl H. *Leonhardt Rauwolf: Sixteenth-Century Physician, Botanist and Traveller* (Cambridge, MA: Harvard University Press, 1968).

Daston, Lorraine. "Baconian Facts, Academic Civility and the Prehistory of Objectivity." *Annals of Scholarship* 8 (1991): 337–363.

––––––. "The Factual Sensibility." *Isis* 7 (1988): 452–470.

––––––. "Marvelous Facts and Miraculous Evidence in Early Modern Europe." *Critical Inquiry* 18 (1991): 93–124.

Davis, Natalie Zemon. "Beyond the Market: Books as Gifts in Sixteenth-Century France." *Transactions of the Royal Historical Society*, ser. 5, 33 (1983): 69–88.

Dear, Peter. "Jesuit Mathematical Science and the Reconstitution of Experience in the Early Seventeenth Century." *Studies in the History and Philosophy of Science* 18 (1987): 133–175.

––––––. "Narratives, Anecdotes and Experience: Turning Experience into Science in the Seventeenth Century." In *The Literary Structure of a Scientific Argument*, Peter Dear, ed. (Philadelphia: University of Pennsylvania Press, 1991).

––––––. "*Totius in verba*: Rhetoric and Authority in the Early Royal Society." *Isis* 76 (1985): 145–161.

Defert, Daniel. "The Collection of the World: Accounts of Voyages from the Sixteenth to the Eighteenth Centuries." *Dialectical Anthropology* 7 (1982): 11–20.

De Michele, Vincenzo, Luigi Cagnolaro, Antonio Aimi, and Laura Laurencich. *Il museo di Manfredo Settala nella Milano del XVII secolo* (Milan: Museo Civico di Storia Naturale di Milano, 1983).

De Rosa, Stefano. "Alcuni aspetti della 'commitenza' scientifica medicea prima di Galileo." In *Firenze e la Toscana dei Medici nell'Europa del '500* (Florence: Olschki, 1983), vol. 2, pp. 777–782.

De Toni, G. B. "Notizie bio-bibliografiche intorno Evangelista Quattrami." *Atti del Reale Istituto Veneto di scienze, lettere ed arti* 77, Pt. II (1917–1918): 373–396.

Dionisotti, Carlo. "La galleria degli uomini illustri." In *Cultura e società nel Rinascimento tra riforme e manierismi,* Vittore Branca and Carlo Ossola, eds. (Florence: Olschki, 1984), pp. 449–461.

Dollo, Corrado. *Filosofia e scienze in Sicilia* (Padua: CEDAM, 1979).

————. *Modelli scientific e filosofici nella Sicilia spagnola* (Naples: Guida, 1984).

Dooley, Brendan. "Revisiting the Forgotten Centuries: Recent Work on Early Modern Tuscany." *European History Quarterly* 20 (1990): 519–550.

————. "Social Control and the Italian Universities: From Renaissance to Illuminismo." *Journal of Modern History* 61 (1989): 205–239.

Durling, Richard J. "Conrad Gesner's *Liber amicorum* 1555–1565." *Gesnerus* 22 (1965): 134–159.

Eamon, William. "Arcana Disclosed: The Advent of Printing, the Books of Secrets Tradition and the Development of Experimental Science in the Sixteenth Century." *History of Science* 22 (1984): 111–150.

————. "Books of Secrets in Medieval and Early Modern Europe." *Sudhoffs Archiv* 69 (1985): 26–49.

————. "From the Secrets of Nature to Public Knowledge: The Origins of the Concept of Openness in Science." *Minerva* 23 (1985): 321–347.

————. "Science and Popular Culture in Sixteenth-Century Italy." *Sixteenth-Century Journal* 16 (1985): 471–485.

————. "The *Secreti* of Alexis of Piedmont, 1555." *Res Publica Litterarum* 2 (1979): 43–55.

Eamon, William, and Françoise Paheau. "The Accademia Segreta of Girolamo Ruscelli: A Sixteenth-Century Italian Scientific Society." *Isis* 75 (1984): 327–342.

Elias, Norbert. *The Court Society,* Edmund Jephcott, trans. (New York: Pantheon, 1983).

————. *The History of Manners,* Edmund Jephcott, trans. (New York: Pantheon, 1978).

Evans, R. J. W. *The Making of the Habsburg Monarchy: An Interpretation* (Oxford: Clarendon Press, 1979).

Fanti, Mario. "La villeggiatura di Ulisse Aldrovandi." *Strenna storica bolognese* 8 (1958): 17–43.

Fantoni, Marcello. "Feticci di prestigio: il dono alla corte medicea." In *Rituale, ceremoniale, etichetta,* Sergio Bertelli and Giuliano Crifo, eds. (Milan: Bompiani, 1985), pp. 141–161.

Feldhay, Rivka. "Knowledge and Salvation in Jesuit Culture." *Science in Context* 1 (1987): 195–213.

Ferrari, Giovanna. "Public Anatomy Lessons and the Carnival: The Anatomy Theatre of Bologna." *Past and Present* 117 (1987): 50–106.

Findlen, Paula. "Controlling the Experiment: Rhetoric, Court Patronage and the Experimental Method of Francesco Redi (1626–1697)." *History of Science* 31 (1993): 35–64.

————. "The Economy of Scientific Exchange in Early Modern Italy." In *Patronage and Institutions,* Bruce Moran, ed. (Woodbridge, U.K.: Boydell and Brewer, 1991), pp. 5–24.

————. "Empty Signs? Reading the Book of Nature in Renaissance Science." *Studies in the History and Philosophy of Science* 21 (1990): 511–518.

————. "Jokes of Nature and Jokes of Knowledge: The Playfulness of Scientific Discourse in Early Modern Europe." *Renaissance Quarterly* 43 (1990): 292–331.

———. "The Museum: Its Classical Etymology and Renaissance Genealogy." *Journal of the History of Collections* 1 (1989): 59–78.

Fletcher, John E. "Astronomy in the Life and Correspondence of Athanasius Kircher." *Isis* 61 (1970): 52–67.

———. "Athanasius Kircher and the Distribution of His Books." *The Library*, ser. 5, 23 (1968): 108–117.

———. "A Brief Survey of the Unpublished Correspondence of Athanasius Kircher, S. J. (1602–1680)." *Manuscripta* 13 (1969): 150–160.

———. "Claude Fabri de Peiresc and the Other French Correspondents of Athanasius Kircher (1602–1680)." *Australian Journal of French Studies* 9 (1972): 250–273.

———. "Johann Marcus Marci Writes to Athanasius Kircher." *Janus* 59 (1972): 95–118.

———. "Medical Men and Medicine in the Correspondence of Athanasius Kircher (1602–1680)." *Janus* 56 (1969): 259–277.

———, ed. *Athanasius Kircher und seine Beziehungen zum gelehrten Europa seiner Zeit*, Wolfenbüttler Arbeit zur Barockforschung, 17 (Wiesbaden, 1988).

Fogalari, Gino. "Il Museo Settala contributo per la storia della coltura in Milano nel secolo XVII." *Archivo storico lombardo* XIV, 27 (1900): 58–126.

Foucault, Michel. *The Order of Things*, English trans. (New York: Vintage, 1970).

Fragnito, Gigliola. *In museo e in villa: saggi sul Rinascimento perduto* (Venice: Arsenale, 1988).

Franchini, Dario, et al. *La scienza a corte. Collezionismo eclettico, natura e immagine a Mantova fra Rinascimento e Manierismo* (Rome: Bulzoni, 1979).

Franzoni, Claudio. "'Rimembranze d'infinite cose.' Le collezioni rinascimentali di antichità." In *Memoria dell'antico nell'arte italiana*, Salvatore Settis, ed. (Turin: Einaudi, 1984), vol. 1, pp. 299–360.

Frati, Ludovico. *Catalogo dei manoscritti di Ulisse Aldrovandi* (Bologna: Zanichelli, 1907).

———. "Le edizioni delle opere di Ulisse Aldrovandi." *Rivista delle biblioteche e degli archivi* 9 (1898): 161–164.

Fulco, Giorgio. "Per il 'museo' dei fratelli Della Porta." In *Il Rinascimento meridionale. Raccolta di studi pubblicata in onore di Mario Santoro* (Naples: Società Editrice Napoletana, 1986), pp. 3–73.

Fulmerton, Patricia. *Cultural Aesthetics: Renaissance Literature and the Practice of Social Ornament* (Chicago: The University of Chicago Press, 1991).

Gabrieli, Giuseppe. "L'Archivio di S. Maria in Aquiro o 'degli Orfani' e le carte di Giovanni Faber Linceo." *Archivo della società romana di storia patria* 51 (1929): 61–77.

———. "Il Carteggio Kircheriano." *Atti della Reale Accademia d'Italia*, ser. 7, 2 (1941–1942): 10–17.

———. *Contributi alla storia dell'Accademia dei Lincei*, 2 vols. (Rome, 1990).

———. "L'orizzonte intellettuale e morale di Federico Cesi illustrato da un suo zibaldone inedito." *Rendiconti della R. Accademia Nazionale dei Lincei. Classe di scienze morali, storiche e filologiche*, ser. 6, Vol. XIV, f. 7–12 (1938–1939): 663–725.

———. "Verbali delle adunanze e cronaca della prima Accademia Lincea (1603–1630)." *Atti della Reale Accademia Nazionale dei Lincei. Classe di scienze morali, storiche e filologiche*, ser. 6, vol. 2, f. 6 (1927): 463–510.

Galluzzi, Paolo. "L'Accademia del Cimento: 'Gusti' del Principe, filosofia e ideologia dell'esperimento." *Quaderni storici* 48 (1972): 788–844.

————. "Il mecenatismo mediceo e le scienze." In *Idee, istituzioni, scienze ed arti nella Firenze dei Medici* (Florence: Giunti Martello, 1980), pp. 189–215.

Garin, Eugenio. "La nuova scienza e il simbolo del 'libro.'" In his *La cultura filosofica del Rinascimento italiano* (Florence: Sansoni, 1961), pp. 451–465.

Garrucci, R. "Origine e vicende del Museo Kircheriano dal 1651 al 1773." *Civiltà cattolica* 30, Vol. XI, ser. 10 (1879): 727–739.

Gherardo Cibo alias "Ulisse Servino da Cingoli" (Florence, 1989).

Ginzburg, Carlo. "High and Low: The Theme of Forbidden Knowledge in the Sixteenth and Seventeenth Centuries." *Past and Present* 73 (1976): 28–41.

Gioelli, Felice. "Gaspare Gabrieli. Primo lettore dei semplici nello Studio di Ferrara (1543)." *Atti e memorie. Deputazione provinciale ferrarese di storia patria*, ser. 3, 10 (1970): 5–74.

Godwin, Joselyn. *Athanasius Kircher: A Renaissance Man and the Quest for Lost Knowledge* (London: Thames and Hudson, 1979).

Goldberg, Edward L. *After Vasari: History, Art and Patronage in Late Medici Florence* (Princeton: Princeton University Press, 1988).

————. *Patterns in Late Medici Art Patronage* (Princeton: Princeton University Press, 1983).

Goodman, David C. *Power and Penury: Government, Technology and Science in Philip II's Spain* (Cambridge, U.K.: Cambridge University Press, 1988).

Grafton, Anthony. *Defenders of the Text: The Traditions of Scholarship in an Age of Science 1450–1800* (Cambridge, MA: Harvard University Press, 1991).

————. "The World of the Polyhistors: Humanism and Encyclopedism." *Central European History* 28 (1985): 31–47.

Greene, Edward Lee. *Landmarks of Botanical History*, Frank N. Egerton, ed., 2 vols. (Stanford: Stanford University Press, 1983).

Gröte, Andreas, ed. *Macrocosmos im Microcosmos: Die Welt in der Stube* (in press).

Hampton, Timothy. *Writing from History: The Rhetoric of Exemplarity in Renaissance Literature* (Ithaca, NY: Cornell Press, 1990).

Hanafi, Zakiya. *Matters of Monstrosity in the Seicento* (Ph.D. diss., Stanford University, 1991).

Hannaway, Owen. "Laboratory Design and the Aim of Science: Andreas Libavius versus Tycho Brahe." *Isis* 77 (1986): 585–610.

Harris, Steven J. "Transposing the Merton Thesis: Apostolic Spirituality and the Establishment of the Jesuit Scientific Tradition." *Science in Context* 3 (1989): 29–65.

Haskell, Francis. *Patrons and Painters: Art and Society in Baroque Italy*, rev. ed. (New Haven: Yale University Press, 1980).

Houghton, Walter. "The English Virtuoso in the Seventeenth Century." *Journal of the History of Ideas* 3 (1942): 51–73, 190–219.

Immagine e natura. L'immagine naturalistica nei codici e libri a stampa della Biblioteca Estense e Universitari secoli XV–XVII (Modena: Edizioni Panini, 1984).

Impey, Oliver, and Arthur MacGregor, eds. *The Origins of Museums: The Cabinet of Curiosities in Sixteenth- and Seventeenth-Century Europe* (Oxford: Oxford University Press, 1985).

Incisa della Rocchetta, G. "Il museo di curiosità del card. Flavio I Chigi." *Archivio della società romana di storia patria*, ser. 3, 20 (1966): 141–192.

Intorno alla vita e alle opere di Ulisse Aldrovandi (Imola, 1907).

Iverson, Eric. *The Myth of Egypt and Its Hieroglyphs* (Copenhagen, 1961).

Jed, Stephanie. "Making History Straight: Collecting and Recording in Sixteenth-Century Italy." In Jonathan Crewe, ed., *Reconfiguring the Renaissance: Essays in Critical Materialism,* Bucknell Review, vol. 35, n. 2 (Lewisburg, PA, 1992), pp. 104–120.

Kenseth, Joy, ed., *The Age of the Marvelous* (Hanover, NH: Hood Museum of Art, 1991).

Kent, F. W., and Patricia Simons, eds. *Patronage, Art and Society in the Renaissance* (Oxford: Clarendon Press, 1987).

Kidwell, Clara Sue. *The Accademia dei Lincei and the "Apiarium." A Case Study in the Activities of a Seventeenth-Century Scientific Society* (Ph.D. diss., University of Oklahoma, 1970).

Laissus, Yves. "Les cabinets d'histoire naturelle." In *Enseignements et diffusion des sciences en France au XVIIIe siècle,* René Taton, ed. (Paris, 1964), pp. 659–712.

Laurencich-Minelli, Laura. "L'Indice del Museo Giganti: Interessi etnografici e ordinamento di un museo cinquecentesco." *Museologia scientifica* 1 (1984): 191–242.

Lebrot, Gerard. *Baroni in città. Residenze e comportamenti dell'aristocrazia napoletana 1530–1734* (Naples: Società editrice napoletana, 1979).

Lensi Orlandi, Giulio. *Cosimo e Francesco de' Medici alchemisti* (Florence: Nardini, 1978).

Liebenwein, Wolfgang. *Studiolo. Storia e tipologia di uno spazio culturale,* Claudia Cieri Via, ed. (Modena: Istituto di Studi Rinascimentali, 1989).

Litchfield, R. Burr. *The Emergence of a Bureaucracy: The Florentine Patricians 1530–1790* (Princeton: Princeton University Press, 1986).

Lombardo, Richard. *"With the Eyes of a Lynx": Honor and Prestige in the Accademia dei Lincei* (M.A. thesis, University of Florida, Gainesville, 1990).

Lomonaco, Fabrizio, and Maurizio Torrini, ed., *Galileo e Napoli* (Naples: Guida, 1987).

Lugli, Adalgisa. "Inquiry as Collection." *RES* 12 (1986): 109–124.

———. *Naturalia et mirabilia: Il collezionismo enciclopedico nelle Wunderkammern d'Europa* (Milan: Mazzotta, 1983).

Lytle, Guy Fitch, and Stephen Orgel, eds., *Patronage in the Renaissance* (Princeton: Princeton University Press, 1981).

McCracken, George E. "Athanasius Kircher's Universal Polygraphy." *Isis* 39 (1948): 215–228.

MacGregor, Arthur, ed. *Tradescant's Rarities: Essays on the Foundation of the Ashmolean Museum 1683* (Oxford: Oxford University Press, 1983).

Maragi, Mario. "Corrispondenze mediche di Ulisse Aldrovandi coi Paesi Germanici." *Pagine di storia della medicina* 13 (1969): 102–110.

Maravall, José Antonio. *Culture of the Baroque: Analysis of a Historical Structure,* Terry Cochran, trans. (Minneapolis: University of Minnesota Press, 1986).

Marrara, Danilo. *Lo Studio di Siena nelle riforme del Granduca Ferdinando I (1589 e 1591)* (Milan: Giuffrè, 1970).

———. *L'Università di Pisa come università statale nel Granducato mediceo* (Milan: Giuffrè, 1965).

Martinoni, Renato. *Gian Vincenzo Imperiali, politico, letterato e collezionista genovese del Seicento* (Padua: Antenore, 1983).

I materiali dell'Istituto delle Scienze (Bologna: CLUEB, 1979).

Mauss, Marcel. *The Gift: Forms and Functions of Exchange in Archaic Societies,* Ian Cunnison, trans. (New York: Norton, 1967).

Middleton, W. E. Knowles. *The Experimenters: A Study of the Accademia del Cimento* (Baltimore: Johns Hopkins University Press, 1971).

Moran, Bruce T. "German Prince–Practitioners: Aspects in the Development of Courtly Science, Technology and Proceduers in the Renaissance." *Technology and Culture* 22 (1981): 253–274.

————, ed. *Patronage and Institutions: Science, Technology and Medicine at the European Court, 1500–1750* (Woodbridge, U.K.: Boydell and Brewer, 1991).

Morello, Nicoletta. *La nascita della paleontologia nel Seicento: Colonna, Stenone e Scilla* (Milan: Franco Angeli, 1979).

Muraro, Luisa. *Giambattista della Porta mago e scienziato* (Milan: Feltrinelli, 1978).

Murray, David. *Museums: Their History and Their Use*, 3 vols. (Glasgow: Jackson, Wylie and Co., 1904).

Nauert, Charles G., Jr. "Humanists, Scientists and Pliny: Changing Approaches to a Classical Author." *American Historical Review* 84 (1979): 72–85.

Negri, Lionello, Nicoletta Morello, and Paolo Galluzzi. *Niccolò Stenone e la scienza in Toscana alla fine del '600* (Florence, 1986).

Neviani, Antonio. "Di alcuni minerali ed altre rocce spedite da Michele Mercati ad Ulisse Aldrovandi." *Bullettino della società geologica italiana* 53, f. 2 (1934): 211–214.

————. "Un episodio della lotta fra spontaneisti ed ovulisti. Il padre Filippo Bonanni e l'abate Anton Felice Marsili." *Rivista di storia delle scienze mediche e naturali* 26, f. 7–8 (1935): 211–232.

————. "Ferrante Imperato speziale e naturalista napoletano con documenti inediti." *Atti e memorie dell'Accademia di Storia dell'Arte Sanitaria* 35, f. 2–5 (1936): 3–86.

Nussdorfer, Laurie. *Civic Politics in the Rome of Urban VIII* (Princeton: Princeton University Press, 1992).

Olmi, Giuseppe. "La colonia lincea di Napoli." In *Galileo e Napoli*, F. Lomonaco and M. Torrini, eds. (Naples: Guida, 1987), pp. 23–57.

————. "Farmacopea antica e medicina moderna: La disputa sulla teriaca del Cinquecento." *Physis* 19 (1977): 197–246.

————. "'Molti amici in varii luoghi.' Studio della natura e rapporti epistolari nel secolo XVI." *Nuncius* 6 (1991): 3–31.

————. "Ordine e fama: il museo naturalistico in Italia nei secoli XVI e XVII." *Annali dell'Istituto storico italo-germanico in Trento* 8 (1982): 225–274.

————. "Alle origini della politica cultura dello stato moderno: dal collezionismo privato al *Cabinet du Roy*." *La Cultura* 16 (1978): 471–484.

————. "Osservazione della natura e raffigurazione in Ulisse Aldrovandi (1522–1605)." *Annali dell'Istituto storico italo-germanico in Trento* 3 (1977): 105–181.

————. *Ulisse Aldrovandi. Scienza e natura nel secondo Cinquecento* (Trent: Libera Università degli Studi di Trento, 1976).

Olmi, Giuseppe, and Paolo Prodi. "Gabriele Paleotti, Ulisse Aldrovandi e la cultura a Bologna nel secondo Cinquecento." In *Nell'età di Correggio e dei Caracci: Pittura in Emilia dei secoli XVI e XVII* (Bologna: Nuova Alfa, 1986), pp. 213–235.

Ong, Walter. "System, Space and Intellect in Renaissance Symbolism." In his *The Barbarian Within and Other Fugitive Essays and Studies* (New York: Macmillan, 1962), pp. 68–87.

Ophir, Adi. "A Place of Knowledge Re-Created: The Library of Michel de Montaigne." *Science in Context* 4 (1991): 163–189.

Palmer, Richard. "Medical Botany in Northern Italy in the Renaissance." *Journal of the Royal Society of Medicine* 78 (1985): 149–157.

——. "Medicine at the Papal Court in the Sixteenth Century." In *Medicine at the Courts of Europe 1500–1837*, Vivian Nutton, ed. (London: Routledge, 1990), pp. 49–78.

——. "Pharmacy in the Republic of Venice in the Sixteenth Century." In *The Medical Renaissance of the Sixteenth Century*, A. Wear, R. K. French, and I. M. Lonie, eds. (Cambridge, U.K.: Cambridge University Press, 1985), pp. 100–117.

——. "Physicians and the State in Post-Medieval Italy." In *The Town and State Physician in Europe from the Middle Ages to the Enlightenment*, Andrew W. Russell, ed. (Wolfenbüttel, 1981), pp. 47–61.

Park, Katharine. *Doctors and Medicine in Early Renaissance Florence* (Princeton: Princeton University Press, 1985).

Park, Katharine, and Lorraine J. Daston. "Unnatural Conceptions: The Study of Monsters in Sixteenth- and Seventeenth-Century France and England." *Past and Present* 92 (1981): 20–54.

Pastine, Dino. *La nascita dell'idolatria: l'Oriente religioso di Athanasius Kircher* (Florence: La Nuova Italia, 1978).

Pighetti, Clelia. "Francesco Lana Terzi e la scienza barocca." *Commentari dell'Ateneo di Brescia per il 1985* (Brescia, 1986): 97–117.

——. *L'influsso scientifico di Robert Boyle nel tardo '600 italiano* (Milan: Franco Angeli, 1988).

Pomian, Krzysztof. *Collectors and Curiosities: Paris and Venice, 1500–1800* (London: Polity Press, 1990).

Ponlet, Dominique. "Musée et société dans l'Europe moderne." *Mélanges de l'école française de Rome* 98 (1986): 991–1096.

Prest, John. *The Garden of Eden: The Botanic Garden and the Re-Creation of Paradise* (New Haven: Yale University Press, 1981).

Prinz, Wolfram. *Galleria. Storia e tipologia di uno spazio architettonico*, Claudia Cieri Via, ed. (Modena: Istituto di Studi Rinascimentali, 1988).

Prodi, Paolo. *Il Cardinale Gabriele Paleotti (1522–1597)*, 2 vols. (Rome: Edizione di Storia e Letteratura, 1959–1967).

——. *The Papal Prince, One Body and Two Souls: The Papal Monarchy in Early Modern Europe*, Susan Haskins, trans. (Cambridge, U.K.: Cambridge University Press, 1987).

Quint, Arlene. *Cardinal Federico Borromeo as a Patron and a Critic of the Arts and his Musaeum of 1625* (New York: Garland, 1986).

Redondi, Pietro. *Galileo Heretic*, Raymond Rosenthal, trans. (Princeton: Princeton University Press, 1987).

Reeds, Karen. *Botany in Medieval and Renaissance Universities* (Ph.D. diss., Harvard University, 1975).

——. "Renaissance Humanism and Botany." *Annals of Science* 33 (1976): 519–542.

Reilly, P. Conor, S. J. *Athanasius Kircher S. J. Master of a Hundred Arts 1602–1680* (Wiesbaden: Edizioni del Mondo, 1974).

Res Public Litteraria: Die Institionen der Gelehrsamkeit in der frühen Neuzeit, 2 vols. (Wiesbaden, 1987).

Riddle, John M. *Dioscorides on Pharmacy and Medicine* (Austin: University of Texas Press, 1985).

Rivosecchi, Valerio. *Esotismo in Roma barocca: studi sul Padre Kircher* (Rome: Bulzoni, 1982).

Rizza, Cecilia. *Peiresc e l'Italia* (Turin, 1965).

Rodriguez, Ferdinando. "Il Museo Aldrovandiano nella Biblioteca Universitaria di Bologna." *Archiginnasio* 49 (1954–1955): 207–223.

Rosa, Edoardo. "La teriaca panacea dell'antichità approda all'Archiginnasio." In *L'Archginnasio. Il palazzo, l'università, la biblioteca*, Giancarlo Roversi, ed. (Bologna: Credito Romagnolo, 1987), vol. 1, pp. 320–340.

Rose, Paul Lawrence. "Jacomo Contarini (1536–1595), a Venetian Patron and Collector of Mathematical Instruments and Books." *Physis* 18 (1976): 117–130.

Rossi, Paolo. "The Aristotelians and the 'Moderns': Hypothesis and Nature." *Annali dell'Istituto e Museo di Storia della Scienza di Firenze* 7 (1982): 1–28.

———. *The Dark Abyss of Time: The History of the Earth and the History of the Nations from Hooke to Vico*, Lydia G. Cochrane, trans. (Chicago: The University of Chicago Press, 1984).

Rota Ghibaudi, Silvia. *Ricerche su Lodovico Settala* (Florence: Sansoni, 1959).

Rudwick, Martin J. S. *The Meaning of Fossils: Episodes in the History of Paleontology*, 2d ed. (Chicago: The University of Chicago Press, 1985).

Ryan, Michael T. "Assimilating New Worlds in the Sixteenth and Seventeenth Centuries." *Comparative Studies in Society and History* 23 (1981): 519–538.

Sabbatani, Luigi. "La cattedra dei semplici fondata a Bologna da Luca Ghini." *Studi e memorie per la storia dell'Università di Bologna*, ser. 1, 9 (1926): 13–53.

———. "Il Ghini e l'Anguillara negli orti di Pisa e di Padova." *Rivista di storia delle scienze mediche e naturali*, ser. 3, n. 11–12 (1923): 307–309.

Saccardo, Pier Andrea. *La botanica in Italia* (reprint; Bologna: Forni, n.d.).

Salerno, Luigi. "Arte, scienza e collezioni nel Manierismo." In *Scritti di storia dell'arte in onore di Mario Salmi*, A. Marabottini Marabotti, ed. (Rome, 1963), vol. 3, pp. 193–214.

Scappini, Cristiana, and Maria Pia Torricelli. *Lo Studio Aldrovandi in Palazzo Pubblico (1617-1742)*, Sandra Tugnoli Pattaro, ed. (Bologna: CLUEB, 1993).

Schaefer, Scott Jay. *The Studiolo of Francesco I de' Medici in the Palazzo Vecchio in Florence* (Ph.D. diss., Bryn Mawr College, 1976).

Schlosser, Julius Von. *Raccolte d'arte e di meraviglie del tardo Rinascimento*, Paola di Paolo, trans. (Florence: Sansoni, 1974).

Schmitt, Charles. *The Aristotelian Tradition and Renaissance Universities* (London: Variorum, 1984).

———. *Aristotle and the Renaissance* (Cambridge, MA: Harvard University Press, 1983).

———. *Studies in Renaissance Philosophy and Science* (London: Variorum, 1981).

Schnapper, Antoine. *La géant, la licorne, la tulipe: Collections françaises au XVIIe siècle. I. Histoire et histoire naturelle* (Paris: Flammarion, 1988).

———. "The King of France as Collector in the Seventeenth Century." *Journal of Interdisciplinary History* 17 (1986): 185–202.

Schriff, Ugo. "Il museo di storia naturale e la facoltà di scienze fische e naturale di Firenze." *Archeion* 9 (1928): 88–95, 290–324.

Scienze, credenze occulte, livelli di cultura (Florence: Olscki, 1982).

Shapin, Steven. "The House of Experiment in Seventeenth-Century England." *Isis* 79 (1988): 373–404.

———. "The Invisible Technician." *American Scientist* 77 (1989): 554–563.

_____. "'The Mind Is Its Own Place': Science and Solitude in Seventeenth-Century England." *Science in Context* 4 (1991): 191–218.

_____. " 'A Scholar and a Gentleman': The Problematic Identity of the Scientific Practitioner in Early Modern England." *History of Science* 29 (1991): 279–327.

Shapin, Steven, and Simon Schaffer. *Leviathan and the Air-Pump: Hobbes, Boyle and the Experimental Life* (Princeton: Princeton University Press, 1985).

Shapiro, Barbara, and Ross G. Frank, Jr. *English Scientific Virtuosi in the Sixteenth and Seventeenth Centuries* (Los Angeles: Clark Library, 1979).

Simcock, A. V. *The Ashmolean Museum and Oxford Science 1683–1983* (Oxford, 1984).

Simili, Alessandro. "Spigolature mediche fra gli inediti aldrovandiani." *L'Archiginnasio* 63–65 (1968–1970): 361–488.

Siraisi, Nancy G. *Avicenna in Renaissance Italy: The Canon and Medical Teaching in Italian Universities after 1500* (Princeton: Princeton University Press, 1987).

Slaughter, M. M. *Universal Languages and Scientific Taxonomy in the Seventeenth Century* (Cambridge, U.K.: Cambridge University Press, 1982).

Smith, Pamela H. *The Business of Alchemy: Science and Culture in Baroque Europe* (Princeton: Princeton University Press, in press).

Solinas, Francesco, ed. *Cassiano dal Pozzo. Atti del Seminario di Studi* (Rome: De Luca, 1989).

Spallanzani, Mariafranca. "Le 'camere di storia naturale' dell'Istituto delle Scienze di Bologna nel Settecento." In *Scienza e letteratura nella cultura italiana del Settecento,* Renzo Cremente and Walter Tega, ed. (Bologna: Il Mulino, 1984), pp. 149–183.

_____. "La collezione naturalistica di Lazzaro Spallanzani." In *Lazzaro Spallanzani e la biologia del Settecento,* Giuseppe Montalenti and Paolo Rossi, eds. (Florence: Olschki, 1982), pp. 589–601.

_____. *La collezione naturalistica di Lazzaro Spallanzani: i modi e i tempi della sua formazione* (Reggio Emilia: Comune di Reggio nell'Emilia, 1985).

Stannard, Jerry. "Dioscorides and Renaissance Materia Medica." In M. Florkin, ed., *Materia Medica in the XVIth Century,* Analecta Medico-Historica, vol. 1 (Oxford, 1966), pp. 1–21.

_____. "P. A. Mattioli: Sixteenth Century Commentator on Dioscorides." *Bibliographical Contributions, University of Kansas Libraries* 1 (1969): 59–81.

Stella, Rudolf. "Mecenati a Firenze tra Sei e Settecento." *Arte illustrata* 54 (1973): 213–238.

Stewart, Susan. *On Longing: Narratives of the Miniature, the Gigantic, the Souvenir, the Collection* (Baltimore: Johns Hopkins University Press, 1984).

Tagliabue, Guido Morpurgo. "Aristotelismo e Barocco." In Enrico Castelli, ed., *Retovica e barocca,* Atti del III Congresso Internazionale di Studi Umanistici, vol. 3 (Rome 1955), pp. 119–195.

Tavernari, Carla. "Manfredo Settala, collezionista e scienziato milanese dell '600." *Annali dell'Istituto e Museo di Storia delle Scienze* 1 (1976): 43–61.

_____. "Il Museo Settala. Presupposti e storia." *Museologia scientifica* 7 (1980): 12–46.

Teach Gnudi, Maria, and Jerome Pierce Webster. *The Life and Times of Gaspare Tagliacozzi Surgeon of Bologna 1545–1599* (New York: Herbert Reichner, 1950).

Temkin, Owsei. *Galenism: The Rise and Decline of a Medical Philosophy* (Ithaca: Cornell University Press, 1973).

Tergolina-Gislanzoni-Brasco, Umberto. "Francesco Calzolari speziale veronese." *Bollettino storico italiano dell'arte sanitaria* 33, f. 6 (1934): 3–20.

Thorndike, Lynn. *A History of Magic and Experimental Science*, 8 vols. (New York: Columbia University Press, 1923–1958).

Tongiorgi Tomasi, Lucia. "Gherardo Cibo: Visions of Landscape and the Botanical Sciences in a Sixteenth-Century Artist." *Journal of Garden History* 9 (1989): 199–216.

————. "Il giardino dei semplici dello studio pisano. Collezionismo, scienza e immagine tra Cinque e Seicento." In *Livorno e Pisa: due città e un territorio nella politica dei Medici* (Pisa: Nistri-Lischi e Pacini, 1980), pp. 514–526.

————. "Immagine della natura e collezionismo scientifico nella Pisa medicea." In *Firenze e la Toscana dei Medici nell'Europa del '500* (Florence, 1983), vol. 1, pp. 95–108.

————. "Inventari della galleria e attività iconografica dell'orto dei semplici dello Studio pisano tra Cinque e Seicento." *Annali dell'Istituto e Museo di Storia della Scienza* 4 (1979): 21–27.

————. "L'isola dei semplici." *KOS* 1 (1984): 61–78.

————. "Projects for Botanical and Other Gardens: A Sixteenth-Century Manual." *Journal of Garden History* 3 (1983): 1–34.

Tribby, Jay. "Body/Building: Living the Museum Life in Early Modern Europe." *Rhetorica* 10 (1992): 139–163.

————. "Cooking (with) Clio and Cleo: Eloquence and Experiment in Seventeenth-Century Florence." *Journal of the History of Ideas* 52 (1991): 417–439.

Tugnoli Pattaro, Sandra. *Metodo e sistema delle scienze nel pensiero di Ulisse Aldrovandi* (Bologna: CLUEB, 1981).

Vasoli, Cesare. *L'Enciclopedismo del Seicento* (Naples: Bibliopolis, 1978 ed.).

Villoslada, Riccardo. *Storia del Collegio Romano dal suo inizio (1551) alla soppressione della Compagnia di Gesù (1773)* (Rome: Università Gregoriana, 1954).

Violi, Cesarina. *Antonio Giganti da Fossombrone* (Modena: Ferraguti, 1911).

Viviani, Ugo. "La vita di Andrea Cesalpino." *Atti e memorie della R. Accademia Petrarcha di lettere, arti e scienze*, n.s. XVIII–XIX (1935): 17–84.

Watson, Gilbert. *Theriac and Mithridatium: A Study in Therapeutics* (London: Wellcome Historical Medical Library, 1966).

Westfall, Richard. "Science and Patronage: Galileo and the Telescope." *Isis* 76 (1985): 11–30.

Westman, Robert S. "The Astronomer's Role in the Sixteenth Century." *History of Science* 18 (1980): 105–147.

Whitehead, P. J. P. "Museums in the History of Zoology." *Museum Journal* 70 (1971): 50–57, 159–160.

Yates, Frances. *The Art of Memory* (Chicago: University of Chicago Press, 1966).

————. *Giordano Bruno and the Hermetic Tradition* (Chicago: The University of Chicago Press, 1964).

Zaccagnini, Guido. *Storia dello studio di Bologna durante il Rinascimento* (Geneva: Olschki, 1930).

Zanca, Attilio. "Il 'Giardino de' semplici in Mantova' di Zenobio Bocchi." *Quadrante padano* II, 2 (1981): 32–37.

INDEX

Milton Keynes UK
Ingram Content Group UK Ltd.
UKHW040148141024
449609UK00001B/67